T0074208

Birkhäuser Advanced Texts
Basler Lehrbücher

Edited by
Herbert Amann, Zürich University

Hermann Sohr

The Navier-Stokes Equations

An Elementary Functional Analytic Approach

Springer Basel AG

Author:
Hermann Sohr
Fachbereich Mathematik/Informatik
Universität Paderborn
Warburger Str. 100
D-33098 Paderborn

e-mail: hsohr@uni-paderborn.ch

2000 Mathematics Subject Classification 35Q30, 76D03, 76D05; 76D06, 76D07, 35Q35

A CIP catalogue record for this book is available from the
Library of Congress, Washington D.C., USA

Deutsche Bibliothek Cataloging-in-Publication Data
Sohr, Hermann:
The Navier-Stokes equations : an elementary functional analytic approach / Hermann Sohr. - Boston ; Basel ; Berlin : Birkhäuser, 2001
(Birkhäuser advanced texts)
ISBN 978-3-0348-9493-7 ISBN 978-3-0348-8255-2 (eBook)
DOI 10.1007/978-3-0348-8255-2

Originally published by Birkhäuser Verlag, Basel, Switzerland in 2001
Softcover reprint of the hardcover 1st edition 2001
Printed on acid-free paper produced from chlorine-free pulp. TCF ∞
ISBN 978-3-0348-9493-7

9 8 7 6 5 4 3 2 1

www.birkhasuer-science.com

Contents

IV The Linearized Nonstationary Theory

V The Full Nonlinear Navier-Stokes Equations

Preface

The primary objective of this monograph is to develop an elementary and self-contained approach to the mathematical theory of a viscous incompressible fluid in a domain Ω of the Euclidean space $\mathbb{R}^n$, described by the equations of Navier-Stokes.

The book is mainly directed to students familiar with basic functional analytic tools in Hilbert and Banach spaces. However, for readers' convenience, in the first two chapters we collect without proof some fundamental properties of Sobolev spaces, distributions, operators, etc.

Another important objective is to formulate the theory for a completely general domain Ω. In particular, the theory applies to arbitrary unbounded, non-smooth domains. For this reason, in the nonlinear case, we have to restrict ourselves to space dimensions $n = 2, 3$ that are also most significant from the physical point of view. For mathematical generality, we will develop the linearized theory for all $n \geq 2$.

Although the functional-analytic approach developed here is, in principle, known to specialists, its systematic treatment is not available, and even the diverse aspects available are spread out in the literature. However, the literature is very wide, and I did not even try to include a full list of related papers, also because this could be confusing for the student. In this regard, I would like to apologize for not quoting all the works that, directly or indirectly, have inspired this monograph.

Nevertheless, there are some books, in particular, which I think can be useful for a more complete understanding of the subject. Specifically, for functional analysis I refer the reader to the book of Yosida, and to the books of Nečas and Adams for Sobolev spaces. Concerning the Navier-Stokes equations, the reader is referred to the monographs of Ladyzhenskaya, Temam and Galdi. In the latter in particular one can find more specific information on flow in domains with (smooth) bounded and unbounded boundaries.

I conveyed my research interest to the Navier-Stokes equations more than fifteen years ago, stimulated by my colleague Wolf von Wahl. Since then, we started a fruitful collaboration which lasted for several years.

I owe special thanks to Paolo Galdi and Christian Simader for collaboration, encouragement, motivation and good friendship.

I thank all my co-workers Wolfgang Borchers, Reinhard Farwig, Yoshikazu Giga, Hideo Kozono, Tetsuro Miyakawa, Jan Prüss, Maria Specovius, Gudrun Thäter, Werner Varnhorn, and Michael Wiegner, with whom I had a long-lasting cooperation.

I also wish to thank my colleagues K. Pileckas, R. Rautmann, V. A. Solonnikov for helpful collaboration.

I am deeply indebted to H. Amann for inviting me to write this monograph for Birkhäuser-Verlag, and for constant encouragement during the preparation of this work.

Last but not least, I would like to thank my dear wife Sigrid for keeping away from me all non-mathematical problems, and for giving me a quiet time to elaborate and complete this book.

Altenbeken-Buke, May 1999

Chapter I

Introduction

1 Basic notations

1.1 The equations of Navier-Stokes

Throughout this book, $\Omega \subseteq \mathbb{R}^n$ means a general domain, that is any open nonempty connected subset of the n-dimensional Euclidean space $\mathbb{R}^n$. In the linearized theory we admit that $n \geq 2$, the nonlinear theory is restricted to $n = 2$ and $n = 3$; in the preliminaries, see Chapters I and II, we sometimes admit the case $n = 1$. $\partial\Omega$ always means the boundary of Ω. Ω may be unbounded and $\partial\Omega$ may be also unbounded. In the sections on regularity properties, we suppose certain smoothness conditions on the boundary $\partial\Omega$. The variables $x = (x_1, \ldots, x_n) \in \Omega$ are called space variables. T is always given with $0 < T \leq \infty$ and $[0, T)$ is called the time interval; $t \in [0, T)$ is called the time variable. We admit the case $T = \infty$.

In the cases $n = 2$ and $n = 3$, we assume that the domain Ω is filled up with some fluid like water, air, oil, ...

Let $u(t, x) = (u_1(t, x), \ldots, u_n(t, x))$ be the velocity of the fluid at $(t, x) = (t, x_1, \ldots, x_n)$, $t \in [0, T)$, $x \in \Omega$, and let $p(t, x)$ denote the pressure at (t, x). The given external force is denoted by $f(t, x) = (f_1(t, x), \ldots, f_n(t, x))$.

In our physical model we assume that the motion of the fluid is described by the equations

$$\begin{aligned} u_t - \nu\Delta u + u \cdot \nabla u + \nabla p &= f, \\ \operatorname{div} u &= 0 \end{aligned} \tag{1.1.1}$$

with $t \in [0, T), x \in \Omega$. These equations are called the **Navier-Stokes equations**.

The first condition means the balance of forces according to Newton's law. The condition $\operatorname{div} u = 0$ means that the fluid is homogeneous and incompressible.

The constant $\nu > 0$ is called the **viscosity** of the fluid; it depends on physical properties and is a fixed value throughout this book.

u_t means the derivative in time direction; we also write

$$u_t = u' = \frac{d}{dt}u = \frac{\partial}{\partial t}u\,.$$

The term

$$u_t + u \cdot \nabla u = u_t + \left(u_1 \frac{\partial}{\partial x_1} + \cdots + u_n \frac{\partial}{\partial x_n}\right) u$$

describes the total acceleration of a particle in the fluid. We write

$$D_j = \frac{\partial}{\partial x_j} \;,\quad j = 1, \ldots, n \;,\quad \nabla = (D_1, \ldots, D_n).$$

The term

$$-\nu \Delta u = -\nu(D_1^2 + \ldots + D_n^2)u$$

describes the friction between the particles of the fluid.

$\nabla p = (D_1, \ldots, D_n)p$ is the gradient of the pressure p. We refer to [Tem83] for more details on the physical background. The equations (1.1.1) are a system of $n+1$ partial differential equations with $n+1$ variables $(t, x_1, \ldots, x_n)$ and $n+1$ unknown functions $(p, u_1, \ldots, u_n)$.

To these equations we will add the **boundary condition**

$$u|_{\partial\Omega} = 0 \tag{1.1.2}$$

if $\partial\Omega \neq \emptyset$. This means that $u(t,x) = 0$ holds for all $t \in [0,T)$ and $x \in \partial\Omega$. Further we will add the **initial condition**

$$u(0) = u_0 \tag{1.1.3}$$

with given initial velocity u_0 at $t = 0$. This means that $u(0,x) = u_0(x)$ holds for all $x \in \Omega$. We always use the notation

$$u(t, \cdot) = u(t) \;,\quad t \in [0,T)$$

and therefore (1.1.3) can be written in the form $u(0,\cdot) = u_0(\cdot)$.

If Ω is unbounded we further suppose a condition of the form

$$u(t,x) \longrightarrow 0 \;\text{ as }\; |x| \to \infty. \tag{1.1.4}$$

In the following this condition is always satisfied in a certain weak sense by the fact that u is contained in a special function space. Thus we will omit this condition in the following. The equations (1.1.1) together with the conditions (1.1.2), (1.1.3) are the **Navier-Stokes system** with data f, u_0.

Our aim is to study the solvability theory of this system. In particular we investigate existence, uniqueness and regularity properties of solutions u. Another important problem is the asymptotic behavior of solutions as $t \to \infty$ (decay properties).

The Navier-Stokes system was introduced by Navier [Nav1827]. The first rigorous treatise of this system goes back to Leray [Ler33], [Ler34], see Navier [Nav1827] and Stokes [Sto1845] for the historical background. Concerning the full nonlinear equations we refer in particular to Ladyzhenskaya [Lad69], Temam [Tem77], [Tem83], Solonnikov [Sol77], Heywood [Hey80] and von Wahl [vWa85]. For the stationary case we mainly refer to Galdi [Gal94a], [Gal94b], Girault-Raviart [GiRa86] and Varnhorn [Var94]. See Wiegner [Wie99] concerning recent results and Temam [Tem83] concerning physical explanations.

If $n = 3$, the existence of classical smooth solutions of the equations (1.1.1) is a fundamental open mathematical problem. On the other hand, there are many important applications of these equations, for example in meteorology, in thermo-hydraulics, in plasma physics, in the petroleum industry, etc. Therefore we are very interested to know at least partial solvability results, see the outline in Section 2.

1.2 Further notations

Let $\mathbb{R}$ denote the real numbers, $\mathbb{N} := \{1, 2, \ldots\}$ the natural numbers, and let

$$\mathbb{N}_0 := \{0, 1, 2, \ldots\}.$$

The notation ":=" always means "equal by definition". The Euclidean space

$$\mathbb{R}^n := \{(x_1, \ldots, x_n) \ ; \quad x_j \in \mathbb{R}, \ \ j = 1, \ldots, n\}$$

with norm

$$|x| := (x_1^2 + \cdots + x_n^2)^{\frac{1}{2}}$$

leads to $\mathbb{R}^1 = \mathbb{R}$ if $n = 1$. We write

$$e_1 := (1, 0, \ldots, 0), \ e_2 := (0, 1, 0, \ldots, 0), \ \ldots, \ e_n := (0, \ldots, 0, 1)$$

and $x = (x_1, \ldots, x_n) = x_1 e_1 + \cdots + x_n e_n \in \Omega \subseteq \mathbb{R}^n$.

$$D_j := \frac{\partial}{\partial x_j} \ , \quad j = 1, \ldots, n$$

means the jth partial derivative and

$$\nabla := (D_1, \dots, D_n)$$

the gradient.

Given any multi-index $\alpha = (\alpha_1, \dots, \alpha_n) \in \mathbb{N}_0^n$, we define the operator

$$D^\alpha := D_1^{\alpha_1} D_2^{\alpha_2} \dots D_n^{\alpha_n} = \left(\frac{\partial}{\partial x_1}\right)^{\alpha_1} \left(\frac{\partial}{\partial x_2}\right)^{\alpha_2} \cdots \left(\frac{\partial}{\partial x_n}\right)^{\alpha_n}$$

where $D_j^{\alpha_j} = I$ means the identity if $\alpha_j = 0, \;\; j = 1, \dots, n$.

In each context, the letter I always denotes the identity.

$$\nabla^2 := (D_j D_k)_{j,k=1}^n$$

means the matrix of the second order derivatives.

As usual we write

$$|\alpha| := \alpha_1 + \dots + \alpha_n$$

if $\alpha = (\alpha_1, \dots, \alpha_n) \in \mathbb{N}_0^n$. However, if $x = (x_1, \dots, x_n) \in \mathbb{R}^n$, $y = (y_1, \dots, y_n) \in \mathbb{R}^n$, we always use the notation

$$|x| := (x_1^2 + \dots + x_n^2)^{\frac{1}{2}} \;, \;\; |x - y| := \left(\sum_{j=1}^n (x_j - y_j)^2\right)^{\frac{1}{2}} .$$

for the Euclidean norm.

Further we write

$$x \cdot y := x_1 y_1 + \dots + x_n y_n$$

for the scalar product. Thus we get $|x|^2 = x \cdot x$.

Correspondingly we set

$$A \cdot B := \sum_{j,k=1}^n a_{jk} b_{jk}$$

if $A = (a_{jk})_{j,k=1}^n$ and $B = (b_{jk})_{j,k=1}^n$ are matrices, while AB is written for the usual matrix product.

Let

$$u : \Omega \to \mathbb{R}^n, \quad x \mapsto u(x) = (u_1(x), \dots, u_n(x)), \quad x \in \Omega$$

be a vector field. Then we set

$$\begin{aligned} \text{div } u &:= \nabla \cdot u = D_1 u_1 + \ldots + D_n u_n, \\ \Delta u &:= \text{div } \nabla u = (D_1^2 + \ldots + D_n^2) u = (\Delta u_1, \ldots, \Delta u_n), \\ \nabla u &:= (D_1, \ldots, D_n) u = (D_j u_k)_{j,k=1}^n, \\ \nabla^2 u &:= (D_j D_k)_{j,k=1}^n u = (D_j D_k u_l)_{j,k,l=1}^n, \end{aligned}$$

and

$$\begin{aligned} u \cdot \nabla u &= (u \cdot \nabla) u := (u_1 D_1 + \cdots + u_n D_n) u \\ &= (u_1 D_1 u_k + \cdots + u_n D_n u_k)_{k=1}^n \end{aligned}$$

whenever this is meaningful. Further we set

$$\begin{aligned} \text{div } (u\, u) &= D_1(u_1 u) + \cdots + D_n(u_n u) \\ &= (D_1(u_1 u_k) + \cdots + D_n(u_n u_k))_{k=1}^n \end{aligned}$$

where the matrix $u\, u = u \otimes u = (u_j u_k)_{j,k=1}^n$ means the usual tensor product. We prefer the simple notation $u\, u$.

If

$$p : \Omega \to \mathbb{R} \ , \quad x \mapsto p(x) \ , \quad x \in \Omega$$

is a scalar field, we set

$$\nabla p = (D_1, \ldots, D_n) p = (D_1 p, \ldots, D_n p).$$

If div $u = 0$, we call u **divergence-free** or **solenoidal**. In this case we get

$$\begin{aligned} u \cdot \nabla u &= D_1(u_1 u) + \cdots + D_n(u_n u) - (D_1 u_1 + \cdots + D_n u_n) u \\ &= D_1(u_1 u) + \cdots + D_n(u_n u) \\ &= \text{div } (u\, u) . \end{aligned}$$

More generally, let

$$F : \Omega \to \mathbb{R}^{n^2} \ , \quad F = (F_{jk})_{j,k=1}^n ,$$

be any matrix field. Then we define the vector field

$$\text{div } F = (D_1 F_{1k} + \cdots + D_n F_{nk})_{k=1}^n$$

which means that div applies to the columns of F.

1.3 Linearized equations

Neglecting the nonlinear term $u \cdot \nabla u$ in the Navier-Stokes equations (1.1.1), we get the linearized system

$$u_t - \nu\Delta u + \nabla p = f \quad , \quad \text{div } u = 0, \qquad (1.3.1)$$
$$u|_{\partial\Omega} = 0 \quad , \quad u(0) = u_0$$

which is called the **nonstationary Stokes system.**

If f, u and p are independent of t, we get the **stationary Stokes system**

$$-\nu\Delta u + \nabla p = f \quad , \quad \text{div } u = 0\,, \qquad (1.3.2)$$
$$u|_{\partial\Omega} = 0.$$

Omitting the term u_t in the equations (1.1.1), we obtain the **stationary Navier-Stokes system**

$$-\nu\Delta u + u \cdot \nabla u + \nabla p = f \quad , \quad \text{div } u = 0, \qquad (1.3.3)$$
$$u|_{\partial\Omega} = 0.$$

The mathematical approach to these equations, given in this book, essentially rests on the use of the Stokes operator A defined in the Hilbert space $L^2_\sigma(\Omega)$, see Section 2. The Hilbert space theory needed here requires some functional analytic tools. It enables us to admit completely general domains $\Omega \subseteq \mathbb{R}^n$, where all $n \geq 2$ are admitted in the linearized theory. In the nonlinear theory we need the restriction to $n = 2$ and $n = 3$ because of the structure of the nonlinear term $u \cdot \nabla u$.

The stationary theory is developed in Chapter III, the linear nonstationary theory in Chapter IV, and the full nonlinear theory in Chapter V. In all cases the theory starts with the concept of weak solutions. Such solutions exist in all cases globally in time. In the linearized cases, in the stationary nonlinear case, and if $n = 2$ in the nonstationary nonlinear case, there exists a complete regularity theory of solutions u yielding smoothness properties of u if the given data and the domain Ω are sufficiently smooth.

However, in the three-dimensional nonstationary nonlinear case, we can only prove a local-in-time existence result for regular solutions, see the next section for more explanations.

We will always use positive constants $C, C', C_1, C_2, \ldots$ which are not specified and which may differ from occurence to occurence. If they depend on certain quantities $\alpha, \beta, \ldots$, we write $C = C(\alpha, \beta, \ldots), \ldots$.

Each chapter of this book is divided into sections and each section consists of several subsections. We use the following citations: If we write "Theorem

3.1.2", this theorem is contained in Section 3, Subsection 1 of the **same** chapter. If we write "Theorem 2.3.1, III", this theorem is contained in Section 2, Subsection 3 of Chapter III. Similarly, the formula (2.3.4) is contained in Section 2, Subsection 3 of the same chapter, while the formula (3.1.4), II, is contained in Section 3, Subsection 1 of Chapter II.

The notation "iff" always means "if and only if". All function spaces we consider in this book are real excepting those which are needed for the Fourier transform. Then we use the corresponding complexifications of real spaces.

2 Description of the functional analytic approach

2.1 The role of the Stokes operator A

In this section we will give a short description of the functional analytic approach to the Navier-Stokes system

$$\begin{aligned} u_t - \nu\Delta u + u\cdot\nabla u + \nabla p &= f \ , \quad \text{div } u = 0, \\ u|_{\partial\Omega} &= 0 \ , \quad u(0) = u_0 \end{aligned} \tag{2.1.1}$$

as well as an outline of the whole book. This approach is based on the Stokes operator A. This Section 2 can be omitted during the first reading. Our aim is here to inform the advanced reader on the basic ideas.

In order to give a short explanation we will use in this section all notations and in particular all function spaces introduced later on. For example we use the following spaces:

Let $\Omega \subseteq \mathbb{R}^n$, $n \geq 2$, be any domain and let $0 < T \leq \infty$. Then we set

$$L^2_\sigma(\Omega) := \overline{C^\infty_{0,\sigma}(\Omega)}^{\|v\|_2} \subseteq L^2(\Omega)^n,$$

see (3.5.13), I,

$$W^{1,2}_{0,\sigma}(\Omega) := \overline{C^\infty_{0,\sigma}(\Omega)}^{\|v\|_2+\|\nabla v\|_2} \subseteq W^{1,2}_0(\Omega)^n,$$

see (1.2.1), III, and

$$\widehat{W}^{1,2}_{0,\sigma}(\Omega) := \overline{C^\infty_{0,\sigma}(\Omega)}^{\|\nabla v\|_2},$$

see (1.1.2), III, where $C^\infty_{0,\sigma}(\Omega) := \{v \in C^\infty_0(\Omega)^n;\ \text{div } v = 0\}$, see (3.5.11), I. The latter space possesses the following continuous embeddings:

$$\widehat{W}^{1,2}_{0,\sigma}(\Omega) \subseteq L^q(\Omega)^n \ , \quad q = \tfrac{2n}{n-2} \tag{2.1.2}$$

if $n \geq 3$, see (1.2.4), III, and

$$\widehat{W}^{1,2}_{0,\sigma}(\Omega) \subseteq L^q_{loc}(\overline{\Omega})^2 \ , \quad 1 < q < \infty \tag{2.1.3}$$

if $n = 2$ and $\overline{\Omega} \neq \mathbb{R}^2$, see (1.2.6), III.

We always suppose that $u_0 \in L^2_\sigma(\Omega)$ and that f has the form $f = f_0 + \operatorname{div} F$, $F = (F_{jk})^n_{j,k=1}$, with

$$f_0 \in L^s_{loc}([0,T); L^2(\Omega)^n) \ , \quad F \in L^s_{loc}([0,T); L^2(\Omega)^{n^2}), \tag{2.1.4}$$

$1 \leq s < \infty$, see Section 1.1, V.

Besides the Stokes operator

$$A : D(A) \to L^2_\sigma(\Omega) \ , \quad D(A) \subseteq L^2_\sigma(\Omega),$$

we need its fractional powers

$$A^\alpha : D(A^\alpha) \to L^2_\sigma(\Omega) \ , \quad D(A^\alpha) \subseteq L^2_\sigma(\Omega),$$

$-\frac{1}{2} \leq \alpha \leq \frac{1}{2}$, see Section 2, III, and the semigroup operators

$$S(t) := e^{-tA},$$

$t \geq 0$, see (1.5.7), IV. The complete theory of the system (2.1.1) will be reformulated in terms of these operators. To explain the procedure we start with the basic notion of a weak solution

$$u \in L^\infty_{loc}([0,T);\ L^2_\sigma(\Omega)) \cap L^2_{loc}([0,T);\ W^{1,2}_{0,\sigma}(\Omega)) \tag{2.1.5}$$

of the system (2.1.1), see Definition 1.1.1, V, and assume for the moment that

$$u_t,\ \Delta u,\ u \cdot \nabla u,\ \nabla p,\ f \in L^s_{loc}([0,T); L^2(\Omega)^n). \tag{2.1.6}$$

Then we apply the Helmholtz projection $P : L^2(\Omega)^n \to L^2_\sigma(\Omega)$, see Section 2.5, II, to the first equation of (2.1.1), and with

$$A = -\nu P \Delta \ , \quad P(\nabla p) = 0,$$

the system (2.1.1) can be reformulated as

$$u_t + Au + Pu \cdot \nabla u = Pf \ , \quad u(0) = u_0. \tag{2.1.7}$$

This system can be written, see Section 1.6, IV, in the integral form

$$u(t) = S(t)u_0 + \int_0^t S(t-\tau)(Pf - Pu \cdot \nabla u)d\tau \ , \quad t \in [0,T). \tag{2.1.8}$$

Since (2.1.6) is not true in general for a weak solution u, we give each term of (2.1.7) a precise meaning in a generalized sense. For this purpose we extend the operators A^α, $-\frac{1}{2} \le \alpha \le \frac{1}{2}$, and P in a natural way as follows, see Section 2.5, III:

For any distribution $g \in (C_0^\infty(\Omega)^n)' = C_0^\infty(\Omega)^{n\prime}$,

$$Pg := g|_{C_{0,\sigma}^\infty(\Omega)} \tag{2.1.9}$$

simply means the restriction of the functional g to the test space $C_{0,\sigma}^\infty(\Omega)$. The operator $A^\alpha : D(A^\alpha) \to L_\sigma^2(\Omega)$ will be extended by a closure argument from the original domain $D(A^\alpha) \subseteq L_\sigma^2(\Omega)$ to the larger domain

$$\widehat{D}(A^\alpha) := \overline{D(A^\alpha)}^{\|A^\alpha v\|_2}, \tag{2.1.10}$$

which is the completion of $D(A^\alpha)$ in the norm $\|A^\alpha v\|_2$. This is the so-called homogeneous graph norm of $D(A^\alpha)$. We mainly need the cases $\alpha = \frac{1}{2}$ and $\alpha = -\frac{1}{2}$. Lemma 2.5.1, III, yields a direct characterization of $\widehat{D}(A^{-\frac{1}{2}})$ as a space of functionals. In particular we get

$$\widehat{D}(A^{\frac{1}{2}}) = \widehat{W}_{0,\sigma}^{1,2}(\Omega), \tag{2.1.11}$$

see (2.5.6), III, and this space is needed in the stationary theory, see Chapter III. Lemma 2.5.2, III, contains embedding properties of (2.1.10) in the case $-\frac{1}{2} \le \alpha \le 0$, and Lemma 2.4.2, III, in the case $0 \le \alpha \le \frac{1}{2}$.

Then we extend the operator A^α by the usual closure procedure from $D(A^\alpha)$ to $\widehat{D}(A^\alpha) \supseteq D(A^\alpha)$ keeping the notation A^α. This yields the extended operator

$$A^\alpha : \widehat{D}(A^\alpha) \to L_\sigma^2(\Omega), \tag{2.1.12}$$

see (2.5.13), III.

Using these extensions, we obtain in particular that

$$A^{-\frac{1}{2}} P \operatorname{div} \,: F \mapsto A^{-\frac{1}{2}} P \operatorname{div} F, \tag{2.1.13}$$

$F = (F_{jk})_{j,k=1}^n \in L^2(\Omega)^{n^2}$, is a bounded operator from $L^2(\Omega)^{n^2}$ to $L_\sigma^2(\Omega)$ with operator norm

$$\|A^{-\frac{1}{2}} P \operatorname{div}\| \le \nu^{-\frac{1}{2}}, \tag{2.1.14}$$

see Lemma 2.6.1 and Lemma 2.6.2 in Chapter III.

This enables us to give the integral equation (2.1.8) a precise meaning in the form

$$\begin{aligned} u(t) = S(t)u_0 &+ \int_0^t S(t-\tau)Pf_0(\tau)d\tau \\ &+ A^{\frac{1}{2}} \int_0^t S(t-\tau)A^{-\frac{1}{2}} P \operatorname{div}\,(F(\tau) - u(\tau)u(\tau))d\tau, \end{aligned} \tag{2.1.15}$$

$t \in [0, T)$, for each weak solution u.

This integral equation characterizes completely the weak solutions u, see Theorem 1.3.1, V; it is basic for the nonstationary theory, see Chapter IV and Chapter V. In the linearized theory the term $u(\tau)u(\tau)$ has to be omitted.

The theory of (2.1.15) rests on the investigation of the integral operator $\mathcal{J} : g \mapsto \mathcal{J}g$ defined by

$$(\mathcal{J}g)(t) := \int_0^t S(t-\tau)g(\tau)d\tau \ , \quad t \in [0,T), \tag{2.1.16}$$

see Section 1.6, IV. The basic property which we need is the estimate (1.6.18) in Lemma 1.6.2, IV. This is called the estimate of maximal regularity, see [DoVe87].

In the stationary theory we have to omit u_t in (2.1.7), and we apply the extended operator $A^{-\frac{1}{2}}$. Using

$$u \cdot \nabla u = \text{ div } (u\,u), \tag{2.1.17}$$

see Section 1.2, this leads to the equation

$$A^{\frac{1}{2}}u + A^{-\frac{1}{2}}P \text{ div } (u\,u) = A^{-\frac{1}{2}}Pf \tag{2.1.18}$$

which is basic for the stationary nonlinear theory, see (3.5.9), III. It follows that

$$A^{\frac{1}{2}}u = A^{-\frac{1}{2}}Pf \tag{2.1.19}$$

in the linearized case, see (2.6.9), III. By these equations the stationary theory can be reformulated completely in terms of the Stokes operator.

The next subsections contain more details on this approach.

2.2 The stationary linearized case

Theorem 1.3.1, III, yields the existence of a unique weak solution $u \in \widehat{W}^{1,2}_{0,\sigma}(\Omega)$ of the stationary linearized system

$$-\nu\Delta u + \nabla p = f \ , \quad \text{div } u = 0 \ , \quad u|_{\partial\Omega} = 0 \tag{2.2.1}$$

with given $f = \text{div } F$, $F \in L^2(\Omega)^{n^2}$. The equation (2.1.19) is equivalent to (2.2.1), see Lemma 2.6.3, III.

Theorem 1.5.1, III, yields the regularity properties of the system (2.2.1). It rests on the elementary method of difference quotients.

As a consequence we get the characterization of the Stokes operator A in the form

$$D(A) = W^{1,2}_{0,\sigma}(\Omega) \cap W^{2,2}(\Omega)^n \ , \quad Au = -\nu P\Delta u \ , \quad u \in D(A)$$

if Ω is a uniform C^2-domain or if $\Omega = \mathbb{R}^n, n \geq 2$, see Theorem 2.1.1, III.

The following embedding properties play a basic role in the theory of the Navier-Stokes system. For arbitrary domains $\Omega \subseteq \mathbb{R}^n, n \geq 2$, we will prove the inequalities

$$\|v\|_{L^q(\Omega)^n} \leq C\nu^{-\alpha}\|A^\alpha v\|_{L^2(\Omega)^n} \ , \quad v \in \widehat{D}(A^\alpha), \tag{2.2.2}$$

with $0 \leq \alpha \leq \frac{1}{2}$, $2 \leq q < \infty$, $2\alpha + \frac{n}{q} = \frac{n}{2}$, see (2.4.6), III, and

$$\|A^{-\alpha}Pv\|_{L^2(\Omega)^n} \leq C\nu^{-\alpha}\|v\|_{L^q(\Omega)^n} \ , \quad v \in L^q(\Omega)^n, \tag{2.2.3}$$

with $0 \leq \alpha \leq \frac{1}{2}$, $1 < q \leq 2$, $2\alpha + \frac{n}{2} = \frac{n}{q}$, see (2.5.27), III. If $\Omega \subseteq \mathbb{R}^n$, $n \geq 2$, is a uniform C^2- domain, we get

$$\|v\|_{W^{1,q}(\Omega)^n} \leq C(\nu^{-\alpha}\|A^\alpha v\|_{L^2(\Omega)^n} + \|v\|_{L^2(\Omega)}) \ , \quad v \in D(A^\alpha), \tag{2.2.4}$$

with $\frac{1}{2} \leq \alpha \leq 1$, $2 \leq q < \infty$, $2\alpha + \frac{n}{q} = 1 + \frac{n}{2}$, see (2.4.17), III.

The constant $C > 0$ in (2.2.2) and (2.2.3) depends only on α, n while $C > 0$ in (2.2.4) also depends on Ω.

To prove (2.2.2) we first consider the case $\Omega = \mathbb{R}^n$, where the problem is reduced to the Laplace operator Δ, see Lemma 2.3.3, III. In this case the result rests on the estimate of the Riesz potential (3.3.7), II, for which we refer to [Ste70, Chapter V], see also [Tri78]. To carry over the result from $\mathbb{R}^n$ to Ω we use the Heinz inequality, see Lemma 3.2.3, II, as an appropriate interpolation result.

To prove (2.2.4) we argue similarly. For $\Omega = \mathbb{R}^n$ the result follows by estimating the Bessel potential (3.3.14), II, see Lemma 3.3.3, II. Again the Heinz inequality carries over the result from $\mathbb{R}^n$ to Ω.

Inequality (2.2.3) follows from (2.2.2) by a functional analytic duality argument, see Lemma 2.5.2, III.

2.3 The stationary nonlinear case

Theorem 3.5.1, III, yields the existence of at least one weak solution $u \in \widehat{W}^{1,2}_{0,\sigma}(\Omega)$ of the stationary Navier-Stokes system

$$-\nu\Delta u + u \cdot \nabla u + \nabla p = f \ , \quad \text{div } u = 0 \ , \quad u|_{\partial\Omega} = 0 \tag{2.3.1}$$

with given $f = \text{div } F$, $F \in L^2(\Omega)^{n^2}$. In this case, $\Omega \subseteq \mathbb{R}^n$ is any domain with $n = 2, 3$, and we need the restriction that $\overline{\Omega} \neq \mathbb{R}^2$ if $n = 2$.

Lemma 3.5.2, III, shows that (2.3.1) is equivalent to (2.1.18).

This existence result will be shown first for bounded domains, see Theorem 3.4.1, III, using the Leray-Schauder principle similarly as in [Lad69, Chap. I, Sec. 3]. See [Gal94a], [Gal94b] for another approach.

The regularity properties for (2.3.1) are shown first for a (smooth) bounded domain, see Theorem 3.6.1, III. The first regularity step is the crucial one.

To discuss some details, we assume that $f \in L^2(\Omega)^n$, and that Ω is a C^2-domain. Then we write (2.3.1) in the form (2.1.18),

$$A^{\frac{1}{2}}u = A^{-\frac{1}{2}}Pf - A^{-\frac{1}{2}}P \operatorname{div}(u\,u), \tag{2.3.2}$$

use (2.1.14), the inequalities (2.2.2)–(2.2.4), and the selfadjointness of $A^{\frac{1}{4}}$. This enables us to improve the regularity of u first only "slightly" and to show that

$$A^{\frac{1}{2}}u \in D(A^{\frac{1}{4}}), \quad u \in D(A^{\frac{3}{4}}),$$

see (3.6.4), III. Using this property again on the right side of (2.3.2), we next obtain that $u \in D(A)$. Using (2.1.8), III, we conclude that $u \in W^{2,2}(\Omega)^n$ which yields the first regularity step. Now the regularity properties of higher order easily follow from the linear result, see Theorem 1.5.1, III.

Theorem 3.6.2, III, yields the local regularity properties of (2.3.1) for a smooth unbounded domain Ω. The proof rests on the "cut-off"method which reduces the problem to the case of a bounded domain. This method yields a "local" equation for φu, where $\varphi \in C_0^\infty(\mathbb{R}^n)$ is a so-called cut-off function.

Since div $(\varphi u) \neq 0$ in general, we have to use the method of "removing" the nonzero divergence. For this purpose we need results on the divergence equation

$$\operatorname{div} u = g\,, \quad u|_{\partial\Omega} = 0, \tag{2.3.3}$$

where g is given.The main results on (2.3.3), see Lemma 2.1.1, II, and Lemma 2.3.1, II, are based on functional analytic arguments. We essentially use an estimate of gradients, see Lemma 1.5.4, II, and the functional analytic duality principle, see the proof of Lemma 2.1.1, II. This approach seems to be more elementary than those mainly used in the literature which rest on Bogovski's theory [Bog79], [Bog80], see also [Gal94a]. See von Wahl [vWa88] for another approach to (2.3.3). A further approach to (2.3.3) is implicitly contained in [Sol77, Theorem 2.3], [FaS94a, Corollary 2.3], [GiSo91]. A completely independent approach to (2.3.3) goes back to Pileckas [Pil80]. Lemma 2.3.1, II, yields regularity properties for (2.3.3).

To construct the associated pressure p for (2.2.1) and for (2.3.1), we need results on the gradient ∇, see Section 2.1, II, and Section 2.2, II. Since div and ∇ are connected by a duality principle, the theory for div essentially follows from that for ∇ and conversely.

Lemma 2.2.1, II, contains the main result on gradients. The elementary proof given here does not use de Rham's argument, see [Tem77, Ch. I, Prop. 1.1], [dRh60], and is independent of Bogovski's theory [Bog79], [Bog80].

Another important consequence of this theory is the density property

$$\overline{C_{0,\sigma}^{\infty}(\Omega)}^{\|v\|_{W^{1,q}(\Omega)^n}} = \{v \in W_0^{1,q}(\Omega)^n;\ \operatorname{div} v = 0\},$$

$1 < q < \infty$, where $\Omega \subseteq \mathbb{R}^n, n \geq 2$, is a bounded Lipschitz domain, see Lemma 2.2.3, II.

Theorem 3.7.3, III, yields a uniqueness result for (2.3.1) for domains Ω which have a finite width $d > 0$.

Section 3, III, contains a very short treatise on the stationary Navier-Stokes equations (2.3.1), see [Gal94a], [Gal94b] for further results.

In the literature there are investigated special domains like the aperture domain, see [Gal94a, VI, (0.7)]. In this domain the so-called flux condition plays an important role, see [Gal94a, VI, (0.1)]. Note that the flux is always zero for the solutions $u \in \widehat{W}_{0,\sigma}^{1,2}(\Omega)$ treated here. This easily follows from the definition of this solution space.

2.4 The nonstationary linearized case

The theory of the nonstationary linearized system

$$u_t - \nu\Delta u + \nabla p = f \quad , \quad \operatorname{div} u = 0, \qquad (2.4.1)$$
$$u|_{\partial\Omega} = 0 \quad , \quad u(0) = u_0,$$

see Chapter IV, can be reduced completely to the linear evolution equation

$$u_t + Au = Pf \quad , \quad u(0) = u_0. \qquad (2.4.2)$$

Here the domain $\Omega \subseteq \mathbb{R}^n$ may be arbitrary with $n \geq 2$, f has the general form $f = f_0 + \operatorname{div} F$,

$$f_0 \in L^1_{loc}([0,T);\ L^2(\Omega)^n)\ , \quad F \in L^1_{loc}([0,T);\ L^2(\Omega)^{n^2}), \qquad (2.4.3)$$

and $u_0 \in L^2_\sigma(\Omega)$.

The Definition 2.1.1, IV, of a weak solution u is very general for the linearized system (2.4.1). We only suppose that

$$u \in L^1_{loc}([0,T);\ W_{0,\sigma}^{1,2}(\Omega)) \qquad (2.4.4)$$

which is sufficient that the (usual) integral relation (2.1.4.), IV, is well defined.

If $f_0 \in L^1(0,T;\ L^2(\Omega)^n)$, $F \in L^2(0,T;\ L^2(\Omega)^{n^2})$, then such a weak solution u satisfies the energy equality

$$\frac{1}{2}\|u(t)\|_2^2 + \nu\int_0^t \|\nabla u\|_2^2 d\tau = \frac{1}{2}\|u_0\|_2^2 \qquad (2.4.5)$$
$$+ \int_0^t < f_0, u >_\Omega d\tau - \int_0^t < F, \nabla u >_\Omega d\tau,$$

$0 \le t < T$, and it holds the energy estimate

$$\frac{1}{2}\|u(t)\|_2^2 + \nu \int_0^t \|\nabla u\|_2^2 d\tau \ \le\ 2\|u_0\|_2^2 + 8\|f_0\|_{2,1;T}^2 + 4\nu^{-1}\|F\|_{2,2;T}^2 \,, \qquad (2.4.6)$$

see Theorem 2.3.1, IV; $< \cdot, \cdot >_\Omega$ means the scalar product.

We use the notation

$$\|u\|_{q,s;T} := \left(\int_0^T \|u\|_q^s dt \right)^{\frac{1}{s}} \quad , \quad 1 \le q \le \infty \,, \quad 1 \le s \le \infty$$

which means the norm of the space $L^s(0,T;\, L^q(\Omega)^n)$, see Section 1.2, IV. Thus q is the integration exponent in space and s in time.

To prove (2.4.5) and (2.4.6), we use Yosida's approximation procedure, see Section 3.4, II, defined by

$$u_k := J_k u \quad , \quad J_k := \left(I + \frac{1}{k}\, A^{\frac{1}{2}} \right)^{-1} \quad , \quad k \in \mathbb{N},$$

I denotes the identity. See [Soh83], [Soh84] for the application of Yosida's procedure in this context.

If $f_0 \in L^1(0,T;\, L^2(\Omega)^n)$, $F \in L^s(0,T;\, L^2(\Omega)^{n^2})$, $1 < s < \infty$, each weak solution u of (2.4.1) satisfies the basic integral equation

$$\begin{aligned} u(t) = S(t)u_0 &+ \int_0^t S(t-\tau)Pf_0(\tau)d\tau \\ &+ A^{\frac{1}{2}} \int_0^t S(t-\tau)A^{-\frac{1}{2}}P \ \mathrm{div}\ F(\tau)d\tau \end{aligned} \qquad (2.4.7)$$

for almost all $t \in [0,T)$, and conversely, this formula yields a weak solution u of (2.4.1), see Theorem 2.4.1, IV. In particular this proves the existence and uniqueness results.

The representation formula (2.4.7) enables us to apply the general theory of the evolution equation (2.4.2), see Section 1.5, IV, and Section 1.6, IV. In particular we can apply the maximal regularity estimate (1.6.18), IV, and the estimates in Theorem 1.6.3, IV. To explain this, let $u_0 = 0$ and let $f = f_0$ with $f_0 \in L^s(0,T;\, L^2(\Omega)^n)$, $1 < s < \infty$. Then (1.6.18), IV, leads to the inequality

$$\|Au\|_{2,s;T} \ \le\ C\,\|Pf\|_{2,s;T} \qquad (2.4.8)$$

with $C = C(s) > 0$, see Lemma 1.6.2, IV. In order to apply (2.4.8) with $u_0 = 0$, $f_0 = 0$, $F \in L^s(0,T; L^2(\Omega)^{n^2})$ we write (2.4.7) in the form

$$A^{-\frac{1}{2}}u(t) = \int_0^t S(t-\tau)A^{-\frac{1}{2}}P \ \mathrm{div}\ F(\tau)d\tau,$$

and use the property of the (extended) operator $A^{-\frac{1}{2}}P$ div, see (2.1.13). This yields the estimate

$$\|A^{\frac{1}{2}}u\|_{2,s;T} \leq C\nu^{-\frac{1}{2}}\|F\|_{2,s;T}. \tag{2.4.9}$$

Another proof of (2.4.8) follows from the Dore-Venni theory [DoVe87] in the extended version given in [PrS90] and [GiSo91]. A potential theoretic proof for smooth bounded and exterior domains has been given by Maremonti-Solonnikov [MSol97].

Consider (2.4.1) with $u(0) = 0$, $f_0 = 0$, $f = \operatorname{div} F$. Then from Theorem 2.5.3, IV, we obtain with $C = C(s,\rho,q,\alpha) > 0$ the following basic estimates:

$$\|(A^{-\frac{1}{2}}u)_t\|_{2,s;T} + \|A^{\frac{1}{2}}u\|_{2,s;T} \leq C\nu^{-\frac{1}{2}}\|F\|_{2,s;T}\,, \tag{2.4.10}$$

$$\|A^{\alpha-\frac{1}{2}}u\|_{2,\rho;T} \leq C\nu^{-\frac{1}{2}}\|F\|_{2,s;T}\,, \tag{2.4.11}$$

$$\|u\|_{q,\rho;T} \leq C\nu^{-\alpha}\|F\|_{2,s;T} \tag{2.4.12}$$

if $1 < s \leq \rho < \infty$, $\alpha = 1 + \frac{1}{\rho} - \frac{1}{s}$, $2 \leq q < \infty$, $1 + \frac{n}{q} + \frac{2}{\rho} = \frac{n}{2} + \frac{2}{s}$, $\|F\|_{2,s;T} < \infty$,

$$\frac{1}{2}\|A^{-\frac{1}{2}}u\|^2_{2,\infty;T} + \|u\|^2_{2,2;T} \leq 8\nu^{-1}\|F\|^2_{2,1;T} \tag{2.4.13}$$

if $\|F\|_{2,1;T} < \infty$, and

$$\frac{1}{2}\|u\|^2_{2,\infty;T} + \|A^{\frac{1}{2}}u\|^2_{2,2;T} \leq 4\nu^{-1}\|F\|^2_{2,2;T} \tag{2.4.14}$$

if $\|F\|_{2,2;T} < \infty$.

Some cases of these inequalities, partially improved for bounded and exterior domains, are known and distributed in the literature, see [Sol68], [Sol77], [Mar84], [GiSo89], [GiSo91], [MSol97] and [Soh99]. See also [Hey76], [Hey80], [Tem83], [Miy82], [KOS92], [KoO94], [KoY95].

These inequalities are basic for the nonlinear theory. In particular they will be used to prove decay properties for weak solutions of nonlinear Navier-Stokes equations in Chapter V.

In order to include initial values $u_0 \neq 0$, we need estimates of the term $S(t)u_0$ in (2.4.7), see Theorem 2.5.1, IV. Certain estimates of this type are contained in the book [AsSo94].

The associated pressure p of a weak solution u is investigated in Section 2.6, IV. Caused by the presence of the term u_t in (2.4.1) we can only show that p has the form

$$p = \frac{\partial}{\partial t}\widehat{p}$$

with some $\widehat{p} \in L^s_{loc}([0,T);\, L^2_{loc}(\Omega))$, see Theorem 2.6.1, IV. This means that there is lack of regularity of p compared with ∇u, which is typical for the nonstationary system (2.4.1), see [HeW94] concerning this problem.

Section 2.7, IV, yields the regularity theory for the system (2.4.1). We use the method of differentiating (2.4.2) in the time direction, see Theorem 2.7.2, IV, and Theorem 2.7.3, IV.

2.5 The full nonlinear case

The full nonlinear Navier-Stokes system

$$u_t - \nu\Delta u + u\cdot\nabla u + \nabla p = f \quad , \quad \operatorname{div} u = 0, \qquad u|_{\partial\Omega} = 0 \quad , \quad u(0) = u_0, \tag{2.5.1}$$

is investigated in Chapter V. Here $\Omega \subseteq \mathbb{R}^n$ is a general domain with $n = 2, 3$, and $0 < T \leq \infty$.

We always suppose that $u_0 \in L^2_\sigma(\Omega)$, $f = f_0 + \operatorname{div} F$, and that at least

$$f_0 \in L^1_{loc}([0,T);\, L^2(\Omega)^n) \quad , \quad F \in L^1_{loc}([0,T);\, L^2(\Omega)^{n^2}).$$

By Definition 1.1.1, V, a weak solution

$$u \in L^\infty_{loc}([0,T);\, L^2_\sigma(\Omega)) \cap L^2_{loc}([0,T);\, W^{1,2}_{0,\sigma}(\Omega)) \tag{2.5.2}$$

of the system (2.5.1) satisfies the integral relation (1.1.6), V, for all test functions $v \in C^\infty_0([0,T); C^\infty_{0,\sigma}(\Omega))$, see (3.1.5). Note that u need not satisfy the energy inequality. However, from (2.5.2) it follows that the total energy

$$E_{T'}(u) := \frac{1}{2}\|u\|^2_{2,\infty;T'} + \nu\|\nabla u\|^2_{2,2;T'} < \infty \tag{2.5.3}$$

of the system is finite for each T' with $0 < T' < T$.

A consequence of (2.5.3) is the inequality

$$\|u\|_{q,s;T'} \leq C\nu^{-\frac{1}{s}} E_{T'}(u)^{\frac{1}{2}} < \infty \tag{2.5.4}$$

with $2 \leq s \leq \infty$, $2 \leq q < \infty$, $\frac{n}{q} + \frac{2}{s} = \frac{n}{2}$, $C = C(s,n) > 0$, see Lemma 1.2.1, b), V.

Section 1, V, contains several important properties of weak solutions u without assuming the energy inequality. Most important is the representation formula (2.5.5) below.

Let u_0, f_0 be as above and assume that

$$F \in L^s_{loc}([0,T);\, L^2(\Omega)^{n^2}) \quad , \quad 1 < s \leq \frac{4}{n}.$$

Then each weak solution u is weakly continuous, after possibly a redefinition on a null set of $[0,T)$, and satisfies the integral equation

$$u(t) = S(t)u_0 + \int_0^t S(t-\tau)Pf_0(\tau)d\tau \qquad (2.5.5)$$
$$+ A^{\frac{1}{2}} \int_0^t S(t-\tau)A^{-\frac{1}{2}}P \operatorname{div}(F(\tau) - u(\tau)u(\tau))d\tau$$

even for all $t \in [0,T)$. Conversely, if u with (2.5.2) satisfies (2.5.5) at least for almost all $t \in [0,T)$, then u is a weak solution of (2.5.1), see Theorem 1.3.1, V.

We mention some further properties. If a weak solution u satisfies the condition

$$u\,u \in L^2_{loc}([0,T);\, L^2(\Omega)^{n^2}) \qquad (2.5.6)$$

and if $F \in L^2_{loc}([0,T);\, L^2(\Omega)^{n^2})$, then, after a redefinition as above, u is even strongly continuous and satisfies the energy equality

$$\frac{1}{2}\|u(t)\|_2^2 + \nu \int_0^t \|\nabla u\|_2^2 d\tau \qquad (2.5.7)$$
$$= \frac{1}{2}\|u_0\|_2^2 + \int_0^t < f_0, u >_\Omega d\tau - \int_0^t < F, \nabla u >_\Omega d\tau$$

for all $t \in [0,T)$, see Theorem 1.4.1, V.

If $n = 2$, the condition (2.5.6) is a consequence of (2.5.4) with $q = s = 4$. Indeed, using Hölder's inequality we get

$$\|u\,u\|_{2,2;T'} \le C\|u\|_{4,4;T'}\,\|u\|_{4,4,T'} < \infty$$

with some $C > 0$. Therefore, (2.5.6) always holds in the case $n = 2$, see Theorem 1.4.2, V.

If a weak solution u of (2.5.1) satisfies Serrin's condition

$$u \in L^s_{loc}([0,T);\, L^q(\Omega)^n) \qquad (2.5.8)$$

with $n < q < \infty$, $2 < s < \infty$, $\frac{n}{q} + \frac{2}{s} \le 1$, and if $F \in L^2_{loc}([0,T);\, L^2(\Omega)^{n^2})$, then we will see that u is uniquely determined by the data f, u_0, see Theorem 1.5.1, V. If $n = 2$, Serrin's condition is always satisfied. This follows from (2.5.4) again with $q = s = 4$.

In order to investigate the asymptotic behaviour of a weak solution u as $t \to \infty$, we use an argument which is known in principle. We write

$$u_t - \nu\Delta u + \nabla p = \tilde{f} \qquad (2.5.9)$$

with $\tilde{f} := f - u \cdot \nabla u$, use (2.5.3) and (2.5.4) with $T' = \infty$ to obtain certain properties of $\tilde{f}$, and apply the linear theory. This yields properties of u which can be used again on the right side of (2.5.9). This leads to several integrability properties of weak solutions u, see Theorem 1.6.2, V, for $n = 3$ and Theorem 1.6.3, V, for $n = 2$.

If $T = \infty$, $u_0 \in D(A^{-\frac{1}{2}})$, and $f = \operatorname{div} F$ with

$$F \in L^1(0, \infty; L^2(\Omega)^{n^2}) \cap L^2(0, \infty; L^2(\Omega)^{n^2})$$

then we obtain from these theorems that a weak solution u of (2.5.1) satisfies the following integrability properties:

$$\|(A^{-\frac{1}{2}}u)_t\|_{2,s;\infty} + \|A^{\frac{1}{2}}u\|_{2,s;\infty} < \infty \tag{2.5.10}$$

where $1 < s \leq \frac{4}{n}$,

$$\|A^{-\frac{1}{2}}u\|_{2,\infty;\infty} + \|u\|_{2,2;\infty} < \infty, \tag{2.5.11}$$

$$\|A^{\alpha}u\|_{2,\rho;\infty} < \infty \tag{2.5.12}$$

where $1 < \rho < \infty$, $-\frac{1}{2} + \frac{1}{\rho} < \alpha \leq \frac{1}{\rho}$, $0 \leq \alpha \leq \frac{1}{2}$, and

$$\|u\|_{q,\rho;\infty} < \infty \tag{2.5.13}$$

where $1 < \rho < \infty$, $2 \leq q \leq 6$ if $n = 3$, $2 \leq q < \infty$ if $n = 2$, and

$$\frac{n}{2} \leq \frac{n}{q} + \frac{2}{\rho} < \frac{n}{2} + 1.$$

Some special cases of these ineqalities are known in the literature, see [KoO94], [BMi91], [BMi92] for exterior domains.

Consider a weak solution u as above. Then we see that Serrin's quantity

$$S(q, \rho) := \frac{n}{q} + \frac{2}{\rho}$$

occurs again also in this context. From (2.5.13) we get

$$\|u\|_{q,\rho;\infty} = \left(\int_0^\infty \|u\|_q^\rho dt \right)^{\frac{1}{\rho}} < \infty \tag{2.5.14}$$

with $\frac{n}{2} \leq S(q, \rho) < \frac{n}{2} + 1$. If $q = \rho = 2$, we see from (2.5.11) that (2.5.14) also holds with $S(q, \rho) = S(2, 2) = \frac{n}{2} + 1$. Obviously, the property (2.5.14) measures the asymptotic decay of u as $t \to \infty$ and $|x| \to \infty$ in a certain sense; it becomes better if (2.5.14) holds with smaller values q, ρ and therefore with larger values $S(q, \rho)$.

On the other hand, if (2.5.14) holds with $S(q, \rho) \le 1$, we obtain the local regularity properties of u if the data and Ω are sufficiently smooth, see Section 1.8, V. Therefore, we need (2.5.14) with small values $S(q, \rho) \le 1$ in order to get (local) regularity properties of u, and we are interested in (2.5.14) for large values $S(q, \rho) \ge \frac{n}{2}$ in order to get (global) asymptotic properties of u.

The validity of the energy equality (2.5.7) can also be considered as a certain regularity property of u. We know, see Theorem 1.4.1, V, and (1.4.6), V, that the energy equality is satisfied if (2.5.14) holds with $q = s = 4$, $S(q, \rho) = S(4, 4) = \frac{n}{4} + \frac{1}{2}$. Thus if $n = 3$, this property is obtained even if (2.5.14) holds with $S(4, 4) = 1 + \frac{1}{4}$.

We see that the scale of Serrin's values $S(q, \rho)$ such that (2.5.14) is satisfied measures important properties of u. The region

$$1 < S(q, \rho) < \frac{n}{2}$$

can be considered as the **regularity gap** of u (between that which we know, namely (2.5.14) with $S(q, \rho) = \frac{n}{2}$, and that which we would like to know, namely (2.5.14) with $S(q, \rho) = 1$.) Thus if $n = 2$, this gap is empty and therefore, each given u satisfies the energy equality and is regular if the data and Ω are smooth, see Theorem 1.8.3, V. If $n = 3$, the energy equality holds even in the middle of this gap, namely if (2.5.14) is satisfied with $S(4, 4) = 1 + \frac{1}{4}$. Thus we have a smaller gap of information concerning the energy equality.

As an application of the estimates (2.5.10)–(2.5.13) we prove an algebraic decay property of weak solutions.

Let u be any weak solution of (2.5.1) if $n = 2$, or a special weak solution as in Theorem 3.1.1, V, if $n = 3$. Suppose $T = \infty$, $u_0 \in D(A^{-\frac{1}{2}})$, $f = \operatorname{div} F$, $F \in L^1(0, \infty;\ L^2(\Omega)^{n^2}) \cap L^2(0, \infty;\ L^2(\Omega)^{n^2})$,

$$E_\infty(u) = \frac{1}{2}\|u\|^2_{2,\infty;\infty} + \nu\|\nabla u\|^2_{2,2;\infty} < \infty\,,$$

and let $\phi : [0, \infty) \to \mathbb{R}$ be a continuous function satisfying

$$\dot{\phi} \in L^\infty_{loc}([0, \infty); \mathbb{R})$$

with $\dot{\phi} = \frac{d}{dt}\phi$.

Then holds the weighted energy inequality

$$\phi^2(t)\|u(t)\|_2^2 + \nu \int_0^t \phi^2 \|\nabla u\|_2^2 d\tau \qquad (2.5.15)$$

$$\leq \phi^2(0)\|u_0\|_2^2 + \nu^{-1} \int_0^t \phi^2 \|F\|_2^2 d\tau + C \sup_{0\leq\tau\leq t} |\dot{\phi}(\tau)\phi(\tau)|,$$

$0 \leq t < \infty$, with $C = C(f, u_0) > 0$, see Theorem 3.4.1, V.

An immediate consequence, choose $\phi(t) = (1+t)^{\frac{1}{2}}$, is the decay estimate

$$\|u(t)\|_2 \leq C(1+t)^{-\frac{1}{2}} \quad , \quad t \geq 0, \qquad (2.5.16)$$

with $C = C(f, u_0) > 0$, and the property

$$\int_0^\infty (1+\tau)\|\nabla u\|_2^2 d\tau < \infty$$

under the additional assumption $\int_0^\infty (1+\tau)\|F\|_2^2 d\tau < \infty$.

For special domains like exterior domains or the entire space, there are many results on decay properties, see [Mar84], [Sch86], [Wie87], [BMi91], [BMi92], [KOS92]. The slightly weaker estimate

$$\|u(t)\|_2 \leq C(1+t)^{-\alpha} \quad , \quad t \geq 0$$

with $0 < \alpha < \frac{1}{2}$ is known in the literature, see [BMi92]. For bounded domains we can prove a certain exponential decay, see Theorem 3.5.1, V.

Under the additional assumption of Serrin's condition (2.5.8), the regularity results for a weak solution u of the system (2.5.1) are essentially the same as in the linear case, see Section 1.8, V. There are several regularity results, see [Lad69], [Sol77], [Tem77], [Hey80], [Gig86], [GaM88].

If $n = 2$, Serrin's condition (2.5.8) is always satisfied and therefore we get a complete regularity theory as in the linear case, see Theorem 1.8.3, V.

In the regularity theory under Serrin's condition (2.5.8), the first regularity step is the crucial one, see Theorem 1.8.1, V. The proof given here rests on Yosida's approximation procedure. The higher order regularity properties essentially follow from the linear regularity theory, see Theorem 1.8.2, V. For this purpose we use (2.5.9) and improve the regularity step by step. The result obtained on the left side of (2.5.9) is again used on the right side, and so on.

Section 2, V, contains an approximation method for the Navier-Stokes system (2.5.1). It is based again on Yosida's smoothing procedure. With $J_k = (I + k^{-1}A^{\frac{1}{2}})^{-1}$, $k \in \mathbb{N}$, we define the approximate Navier-Stokes system

$$u_t - \nu\Delta u + (J_k u)\cdot\nabla u + \nabla p = f \ , \quad \operatorname{div} u = 0, \qquad (2.5.17)$$

$$u|_{\partial\Omega} = 0 \ , \quad u(0) = u_0,$$

see Definition 2.1.1, V. The nonlinear term $u \cdot \nabla u$ is now replaced by the "smoothing" term $(J_k u) \cdot \nabla u$.

For each $k \in \mathbb{N}$ we obtain a unique weak solution $u = u_k$ of this system; u is regular if the data and Ω are sufficiently smooth, see Theorem 2.5.1, V, and Theorem 2.3.1, V. The proof rests on Banach's fixed point theorem, applied to the integral equation (2.5.5) with $u(\tau)u(\tau)$ replaced by $(J_k u)(\tau)u(\tau)$, and on the estimates (2.4.10)–(2.4.12) from the linear theory.

In the literature there are several other approximation procedures for the system (2.5.1). Mainly used is the Galerkin method, see [Hop41], [Hop50], [Lad69], [Tem77], [Hey80], [Gal94b]. See [Mas84], [CKN82] for other procedures. The use of the Yosida approximation in this context goes back to [Soh83], [Soh84].

The Yosida approximation has several advantages. We obtain some special properties of the solutions u_k, see Lemma 2.2.1, V, and Lemma 2.6.1, V, which are needed, for example, for the decay theory of weak solutions.

The approximate system (2.5.17) has another important consequence. Setting

$$f_k := f + r_k \ , \quad r_k := \frac{1}{k}(A^{\frac{1}{2}} J_k u) \cdot \nabla u \ , \quad k \in \mathbb{N},$$

we can write the system (2.5.17) in the form

$$\begin{aligned} u_t - \nu \Delta u + u \cdot \nabla u + \nabla p &= f_k \ , \quad \text{div}\, u = 0, \\ u|_{\partial\Omega} &= 0 \ , \quad u(0) = u_0, \end{aligned} \tag{2.5.18}$$

and we see that each $u = u_k$ is a regular solution of the original Navier-Stokes system (2.5.1) with the modified force $f_k = f + r_k$.

Theorem 2.4.1, V, yields an estimate of the "error" term r_k in the interesting case $n = 3$. In particular,

$$\lim_{k \to \infty} \|r_k\|_{q,s;T} = 0 \tag{2.5.19}$$

where $1 < q < 2$, $1 < s < 2$, $4 < \frac{3}{q} + \frac{2}{s} < 5$, see (2.4.8), V.

If $n = 3$, we do not know whether the system (2.5.1) has a unique smooth solution in the whole given interval [0,T). However, (2.5.19) shows that such a solution always exists after a "small" modification of the given exterior force f. In other words, if (2.5.1) has a weak solution u for some smooth f which is not regular – we do not know whether this is possible – then u becomes regular after such a "small" modification of f. Therefore, non-regular solutions, if they exist, must have a certain non-stable character, singularities disappear after small perturbations of f.

Similar problems were investigated first by Fursikov [Fur80], see also [Tem83, 3.4, Theorem 3.3] and [SvW87].

The approximate solutions of (2.5.17) are needed later on to prove the existence of at least one weak solution u of the Navier-Stokes system (2.5.1) which satisfies the energy inequality, see Theorem 3.1.1,V.

Section 4, V, yields an existence result for local-in-time strong solutions in the case $n = 3$. By definition, a weak solution u of (2.5.1) is called a strong solution if Serrin's condition (2.5.8) is satisfied. Thus a strong solution is unique and regular if the data and Ω are sufficiently smooth. The following main result on strong solutions, see Theorem 4.2.2, V, extends a result by Fujita-Kato [FuK64] from smooth bounded domains to arbitrary domains Ω and to more general forces f:

Let $\Omega \subseteq \mathbb{R}^3$ be an arbitrary domain, and let $0 < T \leq \infty$, $u_0 \in D(A^{\frac{1}{4}}), f = f_0 + \operatorname{div} F$ with $f_0 \in L^{\frac{4}{3}}(0,T;\, L^2(\Omega)^3)$, $F \in L^4(0,T;\, L^2(\Omega)^9)$. Then there exist some T', $0 < T' \leq T$, depending on f, u_0, and a strong solution u in the subinterval $[0, T')$.

The existence interval $[0, T')$ of u is determined by condition (4.2.5), V. The constant $K > 0$ in this condition does not depend on the domain Ω. Therefore we can construct a strong solution for each domain $\Omega \subseteq \mathbb{R}^3$ simultaneously on the same existence interval $[0, T')$ for all Ω, see Corollary 4.2.4, V.

The proof of Theorem 4.2.2, V, rests on the estimates (2.4.10)–(2.4.12) from the linear theory. Therefore, the strong solution u above possesses the maximal regularity property within $[0, T')$. This means, the regularity class for u cannot be improved for the given class of data. In this sense the result is sharp. However, it is not known whether it is possible to extend u from $[0, T')$ to a strong solution on the whole interval $[0, T)$.

Depending on the regularity, there are several versions of local-in-time solution results, see [Lad69], [Tem77], [Tem83], [Sol77], [vWa85], [Miy82], [KoO94]. The first result in this context goes back to Kiselev-Ladyzhenskaya [KiL63].

3 Function spaces

3.1 Smooth functions

The purpose of this section is to recall the definition and the elementary properties of some basic function spaces. First we have to introduce the spaces of smooth functions. In this subsection, $\Omega \subseteq \mathbb{R}^n$ means a domain with $n \geq 1$. If $n = 1$, $\Omega = (a, b)$ is an open interval with $-\infty \leq a < b \leq +\infty$. By definition, domains and subdomains are always nonempty.

Let $k \in \mathbb{N}$. Then $C^k(\Omega)$ means the space of all functions

$$u : \Omega \to \mathbb{R} \ , \quad x \mapsto u(x)$$

such that $D^\alpha u$ exists and is continuous in Ω for all $\alpha \in \mathbb{N}_0^n$ with $0 \leq |\alpha| \leq k$. $C^0(\Omega)$ denotes the space of all continuous functions $u : \Omega \to \mathbb{R}$.

$$C^\infty(\Omega) := \bigcap_{k=0}^{\infty} C^k(\Omega)$$

is called the space of smooth functions in Ω.

Let $\overline{M}$ denote the closure of any set $M \subseteq \mathbb{R}^n$. Then

$$\operatorname{supp} u := \overline{\{x \in \Omega;\ u(x) \neq 0\}}$$

is called the support of the function $u : \Omega \to \mathbb{R}$.

If $k \in \mathbb{N}_0$ or $k = \infty$, then we set

$$C_0^k(\Omega) := \{u \in C^k(\Omega);\ \operatorname{supp} u \text{ compact, } \operatorname{supp} u \in \Omega\}.$$

Thus $u \in C_0^k(\Omega)$ means that $u \in C^k(\Omega)$ and that $u = 0$ in Ω outside of some compact subset of Ω.

In particular, $C_0^\infty(\Omega)$ is the space of all smooth functions u which are zero outside of some compact subset depending on u.

Let $u|_M$ denote the restriction of a function u to a subset M. Let $k \in \mathbb{N}_0$ or $k = \infty$. Then $C^k(\overline{\Omega})$ means the space of all restrictions $u|_{\overline{\Omega}}$ to $\overline{\Omega}$ of functions $u \in C^k(\mathbb{R}^n)$ such that

$$\sup_{|\alpha| \leq k,\, x \in \mathbb{R}^n} |D^\alpha u(x)| < \infty.$$

Here "$|\alpha| \leq k$" is replaced by "$|\alpha| < \infty$" if $k = \infty$.

In this case we define the norm

$$\|u\|_{C^k} = \|u\|_{C^k(\overline{\Omega})} := \sup_{|\alpha| \leq k,\, x \in \overline{\Omega}} |D^\alpha u(x)|\,, \tag{3.1.1}$$

where "$|\alpha| \leq k$" is replaced as above if $k = \infty$.

The corresponding loc-space is defined (without norm) by

$$C_{loc}^k(\overline{\Omega}); = \{u|_{\overline{\Omega}}\,;\ u \in C^k(\mathbb{R}^n)\}.$$

Further we define the subspace

$$C_0^k(\overline{\Omega}) := \{u \in C^k(\overline{\Omega});\ \operatorname{supp} u \text{ compact, } \operatorname{supp} u \subseteq \overline{\Omega}\}.$$

Thus $u \in C_0^k(\overline{\Omega})$ means that $u = 0$ outside of some compact set $K \subseteq \overline{\Omega}$ depending on u, but now it is possible that $u \neq 0$ on the boundary $\partial\Omega \subseteq \overline{\Omega}$.

A continuous function $u : \overline{\Omega} \to \mathbb{R}$ is called **Lipschitz continuous** or a **Lipschitz function** iff

$$\|u\|_{C^{0,1}} = \|u\|_{C^{0,1}(\overline{\Omega})} := \sup_{x\in\overline{\Omega}} |u(x)| + \sup_{x,y\in\overline{\Omega},\, x\neq y} \frac{|u(x)-u(y)|}{|x-y|} \tag{3.1.2}$$

is finite. Let $C^{0,1}(\overline{\Omega})$ be the space of all Lipschitz functions defined on $\overline{\Omega}$ with norm $\|u\|_{C^{0,1}(\overline{\Omega})}$.

The corresponding spaces of vector fields $u = (u_1, \dots, u_m)$, $m \in \mathbb{N}$, are now defined in an obvious way. We get the spaces

$$\begin{aligned} C^k(\Omega)^m &:= \{(u_1,\dots,u_m)\,; u_j \in C^k(\Omega)\,, j=1,\dots,m\}, \\ C_0^k(\Omega)^m &:= \{(u_1,\dots,u_m)\,; u_j \in C_0^k(\Omega)\,, j=1,\dots,m\}, \\ C^k(\overline{\Omega})^m &:= \{(u_1,\dots,u_m)\,; u_j \in C^k(\overline{\Omega})\,, j=1,\dots,m\}, \end{aligned}$$

with

$$\|u\|_{C^k} = \|u\|_{C^k(\overline{\Omega})^m} := \sup_{j=1,\dots,m} \|u_j\|_{C^k(\overline{\Omega})}\,,$$

the spaces

$$\begin{aligned} C_0^k(\overline{\Omega})^m &:= \{(u_1,\dots,u_m)\,; u_j \in C_0^k(\overline{\Omega})\,, j=1,\dots,m\}, \\ C^{0,1}(\overline{\Omega})^m &:= \{(u_1,\dots,u_m)\,; u_j \in C^{0,1}(\overline{\Omega})\,, j=1,\dots,m\}, \end{aligned}$$

with

$$\|u\|_{C^{0,1}} = \|u\|_{C^{0,1}(\overline{\Omega})^m} := \sup_{j=1,\dots,m} \|u_j\|_{C^{0,1}(\overline{\Omega})}\,,$$

and the loc-space

$$C_{loc}^k(\overline{\Omega})^m := \{(u_1,\dots,u_m)\,; u_j \in C_{loc}^k(\overline{\Omega}),\ j=1,\dots,m\}.$$

The following spaces (without norm) play a special role as "test" spaces in the theory of weak solutions, "σ" stands for "divergence-free (solenoidal)".

Let $n \geq 2$ and $0 < T \leq \infty$. Then we define the space

$$C_{0,\sigma}^\infty(\Omega) := \{u \in C_0^\infty(\Omega)^n\ ;\ \operatorname{div} u = 0\} \tag{3.1.3}$$

of smooth divergence-free vector fields. In the nonstationary theory we need the test spaces

$$C_0^\infty((0,T);\ C_{0,\sigma}^\infty(\Omega)) := \{u \in C_0^\infty((0,T)\times\Omega)^n\ ;\ \operatorname{div} u = 0\} \tag{3.1.4}$$

where div applies to the variables $x = (x_1, \ldots, x_n) \in \Omega$, and

$$C_0^\infty([0,T);\ C_{0,\sigma}^\infty(\Omega)) \tag{3.1.5}$$
$$:= \{u|_{[0,T)\times\Omega}\ ;\ u \in C_0^\infty((-1,T)\times\Omega)^n\ ,\ \operatorname{div} u = 0\} \tag{3.1.6}$$

where $(0,T)\times\Omega := \{(t,x);\ t\in(0,T),\ x\in\Omega\}$ and $(-1,T)\times\Omega := \{(t,x);t\in(-1,T),\ x\in\Omega\}$. Recall that div only concerns the space variables.

Thus each $u \in C_0^\infty([0,T); C_{0,\sigma}^\infty(\Omega))$ is obtained as the restriction of a smooth solenoidal vector field defined in $(-1,T)\times\Omega$ to the subset $[0,T)\times\Omega$. For each fixed $t\in[0,T)$ we define the vector field

$$u(t) := u(t,\cdot) \in C_{0,\sigma}^\infty(\Omega),$$

and we see that $u(0) = u(0,\cdot)$ is nonzero in general; $u(0)$ is called the initial value of u.

3.2 Smoothness properties of the boundary $\partial\Omega$

In this subsection, $\Omega \subseteq \mathbb{R}^n$ means a domain with $n \geq 2$ and $\partial\Omega \neq \emptyset$.

$$B_r(x_0) := \{x \in \mathbb{R}^n; |x - x_0| < r\}\ ,\quad x_0 \in \mathbb{R}^n\ ,\quad r > 0 \tag{3.2.1}$$

means the open ball with radius r and center x_0. To define smoothness properties of the boundary $\partial\Omega$, we introduce for each $x_0 \in \partial\Omega$ an appropriate new coordinate system with origin in x_0, obtained by a rotation and a translation of the original system. The new coordinates of $x = (x_1, \ldots, x_n)$ are denoted by $y = (y_1, \ldots, y_n)$; they are called local coordinates and the new system is called a local coordinate system in x_0. We set $y = (y', y_n)$ with $y' = (y_1, \ldots, y_{n-1})$, and indentify for simplicity each x with its new coordinates y.

Given $x_0 \in \partial\Omega$, $r > 0$, $\beta > 0$, a local coordinate system in x_0 with coordinates $y = (y', y_n)$, and a real continuous function

$$h : y' \mapsto h(y')\ ,\quad y' = (y_1, \ldots, y_{n-1})\ ,\quad |y'| < r\ , \tag{3.2.2}$$

we define the open set

$$U = U_{r,\beta,h}(x_0) \tag{3.2.3}$$
$$:= \{(y', y_n) \in \mathbb{R}^n;\ h(y') - \beta < y_n < h(y') + \beta,\ |y'| < r\}.$$

Then the domain Ω is called a **Lipschitz domain**, iff for each $x_0 \in \partial\Omega$, there exist a local coordinate system in x_0, constants $r > 0$, $\beta > 0$, and a Lipschitz function $h : y' \mapsto h(y')$, $|y'| \leq r$, with the following properties:

$$U_{r,\beta,h}(x_0) \cap \partial\Omega = \{(y', y_n);\ y_n = h(y'), |y'| < r\} \tag{3.2.4}$$

and

$$U_{r,\beta,h}(x_0) \cap \Omega = \{(y', y_n)\,;\ h(y') - \beta < y_n < h(y'), |y'| < r\}. \tag{3.2.5}$$

Let $k \in \mathbb{N}_0$ or $k = \infty$, and let $B'_r := \{y' \in \mathbb{R}^{n-1}; |y'| < r\}$, $r > 0$. Then, similarly, Ω is called a **$\mathbf{C^k}$-domain**, iff for each $x_0 \in \partial\Omega$, there exist a local coordinate system in x_0, constants $r > 0$, $\beta > 0$, and a function $h \in C^k(\overline{B'}_r)$ with the properties (3.2.4) and (3.2.5). Obviously, each C^k-domain with $k \geq 1$ is also a Lipschitz domain, but a C^0-domain need not be a Lipschitz domain.

The constants $r = r_{x_0}$, $\beta = \beta_{x_0}$, and the function $h = h_{x_0}$ in these definitions may depend on $x_0 \in \partial\Omega$. A Lipschitz domain Ω is called a **uniform Lipschitz domain** iff these constants $r = r_{x_0}$, $\beta = \beta_{x_0}$ can be chosen independently of $x_0 \in \partial\Omega$, and there is some $\gamma = \gamma(\Omega) > 0$ depending only on Ω with

$$\|h_{x_0}\|_{C^{0,1}(\overline{B'}_r)} \leq \gamma \tag{3.2.6}$$

for each $x_0 \in \partial\Omega$.

Correspondingly, a C^k-domain is called a **uniform $\mathbf{C^k}$-domain** iff the constants $r = r_{x_0}$, $\beta = \beta_{x_0}$ in the above definition can be chosen independently of $x_0 \in \partial\Omega$, and there is some $\gamma = \gamma(\Omega, k) > 0$ with

$$\|h_{x_0}\|_{C^k(\overline{B'}_r)} \leq \gamma \tag{3.2.7}$$

for each $x_0 \in \partial\Omega$. We refer to [Ada75, IV, 4.6], [Agm65, Definition 9.2], [Nec67, Chap. 2, 1.1], and [Hey80, p. 645] concerning these definitions.

The neighbourhood $U = U_{r,\beta,h}(x_0)$, $x_0 \in \partial\Omega$, in (3.2.3) depends on r, β, h. If Ω is a bounded Lipschitz domain or a bounded C^k-domain, $k \in \mathbb{N}$, then $\partial\Omega$ is compact, and therefore, we find $x_1, \ldots, x_m \in \partial\Omega$ in such a way that the corresponding neighbourhoods (3.2.3) cover the boundary $\partial\Omega$. Setting

$$h_j := h_{x_j}\ ,\ \ r_j := r_{x_j}\ ,\ \ \beta_j = \beta_{x_j}\ ,\ \ U_j := U_{r_j,\beta_j,h_j}(x_j)\,, \tag{3.2.8}$$

$j = 1, \ldots, m$, we thus get

$$\partial\Omega \subseteq \bigcup_{j=1}^{m} U_j.$$

Thus in this case, Ω is also a uniform Lipschitz domain or a uniform C^k-domain, respectively. The same holds for an unbounded domain Ω which has a compact boundary $\partial\Omega$. Such a domain is called an **exterior domain**.

If Ω is any unbounded uniform Lipschitz domain, we always find countably many vectors $x_j \in \partial\Omega$, $j \in \mathbb{N}$, fixed constants $r, \beta, \gamma > 0$, functions $h_j = h_{x_j} \in$

$C^{0,1}(\overline{B'}_r)$, and neighbourhoods $U_j = U_{r,\beta,h_j}(x_j)$ satisfying (3.2.4), (3.2.5) such that

$$\partial\Omega \subseteq \bigcup_{j=1}^{\infty} U_j \quad , \quad \sup_{j\in\mathbb{N}} \|h_j\|_{C^{0,1}(\overline{B'}_r)} \leq \gamma. \tag{3.2.9}$$

Setting

$$S := \bigcup_{j=1}^{\infty} U_j \tag{3.2.10}$$

we obtain the boundary strip $S \cap \Omega$ inside and the boundary strip $S \cap (\mathbb{R}^n \backslash \Omega)$ outside of Ω. $\mathbb{R}^n \backslash \Omega$ denotes the complement of Ω.

Let Ω be a uniform Lipschitz domain, and let $r, \beta, \gamma, h_j, x_j \in \partial\Omega$, U_j, $j \in \mathbb{N}$, be as above with (3.2.9). Then there are functions $\varphi_j \in C_0^\infty(\mathbb{R}^n)$, $j \in \mathbb{N}$, with the following properties:

$$\operatorname{supp} \varphi_j \subseteq U_j \quad , \quad 0 \leq \varphi_j \leq 1 \, , \; j \in \mathbb{N}, \tag{3.2.11}$$

$$\sum_{j=1}^{\infty} \varphi_j(x) = 1 \quad \text{for all } x \in \partial\Omega. \tag{3.2.12}$$

Additionally, in this case there exist a sequence of open balls $(B_j)_{j=1}^\infty$, $\overline{B}_j \subseteq \Omega$, $j \in \mathbb{N}$, with fixed radius $r' > 0$, and a sequence $(\psi_j)_{j=1}^\infty$, $\psi_j \in C_0^\infty(\mathbb{R}^n)$, $j \in \mathbb{N}$, with the following properties:

$$\operatorname{supp} \psi_j \subseteq B_j \quad , \quad 0 \leq \psi_j \leq 1, \; j \in \mathbb{N}, \tag{3.2.13}$$

$$\sum_{j=1}^{\infty} \varphi_j(x) + \psi_j(x) \;=\; 1 \quad \text{for all } x \in \overline{\Omega}. \tag{3.2.14}$$

The functions φ_j, ψ_j, $j \in \mathbb{N}$, are called a partition of unity for $\overline{\Omega}$.

3.3 L^q-spaces

In this subsection we need some elementary facts from measure theory, see [Apo74, Chap. 10], [Yos80, 0.3], [HiPh57, Sec. 3.7]. For the L^q-theory we refer to [Nec67], [Ada75], [HiPh57], [Agm65], [Miz73].

Let $\Omega \subseteq \mathbb{R}^n, n \geq 1$, be a domain, and let $1 \leq q < \infty$. Then $L^q(\Omega)$ denotes the Banach space of all (equivalence classes of) Lebesgue measurable real functions u defined on Ω which have a finite norm

$$\|u\|_q \;=\; \|u\|_{q,\Omega} \;=\; \|u\|_{L^q(\Omega)} \;=\; \|u\|_{L^q} \;:=\; \left(\int_\Omega |u(x)|^q dx\right)^{\frac{1}{q}}. \tag{3.3.1}$$

For brevity we say "function" instead of "equivalence class of functions", and we say "measurable" instead of "Lebesgue measurable". A null set always means a set which has the Lebesgue measure zero. Sometimes we write "a.a." for "almost all" and "a.e." for "almost everywhere" concerning the Lebesgue measure. We use each of the above notations for this norm provided there is no confusion.

If $q = 2$, $L^q(\Omega) = L^2(\Omega)$ becomes a Hilbert space with scalar product

$$< u, v > = < u, v >_\Omega := \int_\Omega u(x)v(x)dx \tag{3.3.2}$$

for $u, v \in L^2(\Omega)$.

If $q = \infty$, we let $L^q(\Omega) = L^\infty(\Omega)$ be the usual Banach space of all measurable functions u with finite essential supremum

$$\|u\|_\infty \;=\; \|u\|_{\infty,\Omega} \;=\; \|u\|_{L^\infty(\Omega)} \;=\; \|u\|_{L^\infty} \;:=\; \operatorname*{ess\text{-}sup}_{x\in\Omega} |u(x)|\,. \tag{3.3.3}$$

All function spaces we consider here are real except for those in subsections where the Fourier transform is used. There we take the usual complexifications of all these spaces, keeping the same notations.

Let $q' := \frac{q}{q-1}$ be the conjugate (dual) exponent of q, we set $q' = \infty$ if $q = 1$ and $q' = 1$ if $q = \infty$. Setting $\frac{1}{q'} = 0$ if $q' = \infty$, $\frac{1}{q} = 0$ if $q = \infty$, we always obtain

$$\frac{1}{q} + \frac{1}{q'} = 1. \tag{3.3.4}$$

Next we collect some basic facts on L^q-spaces, see [Ada75], [Nec67], [Fri69], [Agm65], and introduce further notations.

If $u \in L^q(\Omega)$, $v \in L^{q'}(\Omega)$, then $u\,v \in L^1(\Omega)$ and **Hölder's inequality** holds,

$$\|u\,v\|_1 \;\le\; \|u\|_q \|v\|_{q'}. \tag{3.3.5}$$

The following more general formulation is an easy consequence of (3.3.5). Let $1 \le \gamma \le \infty$, $\gamma \le q \le \infty$, $\gamma \le r \le \infty$, such that

$$\frac{1}{\gamma} = \frac{1}{q} + \frac{1}{r}$$

and let $u \in L^q(\Omega)$, $v \in L^r(\Omega)$. Then $u\,v \in L^\gamma(\Omega)$ and

$$\|u\,v\|_\gamma \;\le\; \|u\|_q \,\|v\|_r\,. \tag{3.3.6}$$

To deduce (3.3.6) from (3.3.5) we set $\tilde q := \frac{q}{\gamma}$ so that $\tilde q' = \frac{r}{\gamma}$, and apply (3.3.5).

A consequence of (3.3.6) is the following **interpolation inequality**. Let $1 \leq q \leq \gamma \leq r \leq \infty$, $0 \leq \alpha \leq 1$ such that

$$\frac{1}{\gamma} = \frac{\alpha}{q} + \frac{1-\alpha}{r},$$

and let $u \in L^q(\Omega) \cap L^r(\Omega)$. Then $u \in L^\gamma(\Omega)$ and

$$\|u\|_\gamma \leq \|u\|_q^\alpha \|u\|_r^{1-\alpha} \leq \|u\|_q + \|u\|_r. \tag{3.3.7}$$

To deduce (3.3.7) we write $|u| = |u|^\alpha |u|^{1-\alpha}$ and apply (3.3.6). Then we use **Young's inequality**

$$a^\alpha b^{1-\alpha} \leq \alpha a + (1-\alpha)b \leq a + b \tag{3.3.8}$$

with $a, b \geq 0$, see [Yos80, I, 3, (4)].

We consider two types of L^q_{loc}-spaces, $1 \leq q \leq \infty$. We write

$$u \in L^q_{loc}(\Omega) \tag{3.3.9}$$

iff $u \in L^q(B)$ for each open ball $B \subseteq \Omega$ with closure $\overline{B} \subseteq \Omega$. Further, we write

$$u \in L^q_{loc}(\overline{\Omega}) \tag{3.3.10}$$

iff $u \in L^q(B \cap \Omega)$ for each open ball $B \subseteq \mathbb{R}^n$ with $B \cap \Omega \neq \emptyset$. If no confusion is likely, we simply write u instead of $u|_B$ or of $u|_{B\cap\Omega}$.

Thus we get

$$L^q(\Omega) \subseteq L^q_{loc}(\overline{\Omega}) \subseteq L^q_{loc}(\Omega),$$

and if Ω is bounded it follows that

$$L^q(\Omega) = L^q_{loc}(\overline{\Omega}), \quad L^q(\Omega) \subseteq L^q_{loc}(\Omega).$$

Let $(u_j) = (u_j)_{j=1}^\infty$ be a sequence in $L^q(\Omega)$. Then we write

$$u = \lim_{j\to\infty} u_j \quad \text{in } L^q(\Omega)$$

iff $u \in L^q(\Omega)$ and $\lim_{j\to\infty} \|u - u_j\|_q = 0$. Correspondingly, we write

$$u = \lim_{j\to\infty} u_j \quad \text{in } L^q_{loc}(\Omega) \text{ or in } L^q_{loc}(\overline{\Omega})$$

iff

$$\lim_{j\to\infty} \|u - u_j\|_{L^q(B)} = 0 \ \text{ or } \lim_{j\to\infty} \|u - u_j\|_{L^q(B\cap\Omega)} = 0$$

holds for all open balls $B \subseteq \Omega$ with $\overline{B} \subseteq \Omega$, or all open balls $B \subseteq \mathbb{R}^n$ with $B \cap \Omega \neq \emptyset$, respectively.

Let $m \in \mathbb{N}$. Then we define the L^q-spaces for vector fields $u = (u_1, \ldots, u_m)$ as follows:

$$L^q(\Omega)^m := \{(u_1, \ldots, u_m)\,;\; u_j \in L^q(\Omega),\; j = 1, \ldots, m\}$$

is the Banach space with norm

$$\|u\|_q \;=\; \|u\|_{q,\Omega} \;=\; \|u\|_{L^q(\Omega)^m} \;=\; \|u\|_{L^q} \;:=\; \left(\sum_{j=1}^m \|u_j\|_q^q\right)^{\frac{1}{q}} . \tag{3.3.11}$$

Then the space $L^2(\Omega)^m$ is a Hilbert space with scalar product

$$< u, v > = < u, v >_\Omega := \sum_{j=1}^m < u_j, v_j >_\Omega$$

for $u = (u_1, \ldots, u_m)$, $v = (v_1, \ldots, v_m) \in L^2(\Omega)^m$; we also write

$$< u, v >_\Omega = \int_\Omega u \cdot v \, dx$$

with $u \cdot v = u_1 v_1 + \cdots + u_m v_m$.

The inequalities (3.3.5), (3.3.6) and (3.3.7) remain valid in the vector valued case.

Sometimes it is convenient to use equivalent norms in $L^q(\Omega)^m$. For example the norm $\sum_{j=1}^m \|u_j\|_q$ is equivalent to (3.3.11). The L^q_{loc}-spaces

$$L^q_{loc}(\Omega)^m \text{ and } L^q_{loc}(\overline{\Omega})^m$$

are defined in a completely analogous way, which means that each component is contained in the corresponding scalar space.

3.4 The boundary spaces $L^q(\partial\Omega)$

In this subsection, $\Omega \subseteq \mathbb{R}^n$ means a bounded Lipschitz domain with $n \geq 2$ and boundary $\partial\Omega$.

Let $1 \leq q \leq \infty$. In order to define the Banach space $L^q(\partial\Omega)$ on the boundary $\partial\Omega$, we use the local coordinate systems with local coordinates

$$y = (y', y_n) \;,\quad y' = (y_1, \ldots, y_{n-1})$$

defined in Section 3.2. Let h_j, r_j, β_j and U_j, $j = 1,\dots,m$, be chosen as in (3.2.8). Since Ω is bounded we may assume that $r = r_j$ and $\beta = \beta_j$ are independent of $j = 1,\dots,m$. Thus with $B'_r := \{y' \in \mathbb{R}^{n-1};\ |y'| < r\}$ we get $h_j \in C^{0,1}(\overline{B'}_r)$ for $j = 1,\dots,m$.

Further we can choose functions $\varphi_j \in C_0^\infty(\mathbb{R}^n)$, $j = 1,\dots,m$, with the properties

$$\begin{aligned} \operatorname{supp}\varphi_j &\subseteq U_j\,, \quad 0 \le \varphi_j \le 1, \quad j = 1,\dots m, \\ \sum_{j=1}^m \varphi_j(x) &= 1 \quad \text{for all } x \in \partial\Omega \end{aligned} \tag{3.4.1}$$

as in (3.2.11), (3.2.12).

Consider some fixed $j \in \{1,\dots,m\}$, any function

$$u : \partial\Omega \to \mathbb{R}\,,$$

and the local coordinate system in U_j with coordinates $y = (y', y_n)$, $y' = (y_1,\dots,y_{n-1})$. The transformation (rotation and translation) from x to y has the form

$$x = \phi_j(y) = \phi_j(y', y_n)\,, \quad x \in U_j\,, \quad |y'| < r$$

with some (smooth) function ϕ_j. The boundary part $\partial\Omega \cap U_j$ is determined by the equation $y_n = h_j(y')$. Let

$$\nabla' h_j(y') := \left(\frac{\partial}{\partial y_1} h_j(y'), \dots, \frac{\partial}{\partial y_{n-1}} h_j(y')\right) \tag{3.4.2}$$

denote the gradient of h_j in the variables y'.

The surface integral $\int_{\partial\Omega} u\, dS$, for the given function $u : \partial\Omega \to \mathbb{R}$, is defined in the following way, see [Nec67, Chap. 3; 1.1]. Using the representation

$$u(x) = \sum_{j=1}^m u(x)\,\varphi_j(x)\,, \quad x \in \partial\Omega$$

we call u **measurable** or **integrable** on $\partial\Omega$ iff each function

$$y' \mapsto u(\phi_j(y', h_j(y')))\,\varphi_j(\phi_j(y', h_j(y')))\,(1 + |\nabla' h_j(y')|^2)^{\frac{1}{2}}\,, \quad y' \in B'_r$$

is measurable or integrable, respectively, for $j = 1,\dots m$.

$$dS := (1 + |\nabla' h_j(y')|^2)^{\frac{1}{2}}\, dy'$$

is called the "surface element" on $\partial\Omega \cap U_j$. If u is integrable on $\partial\Omega$, then

$$\int_{\partial\Omega} u\, dS := \sum_{j=1}^{m} \int_{\partial\Omega\cap U_j} u\varphi_j\, dS \tag{3.4.3}$$

$$:= \sum_{j=1}^{m} \int_{B'_r} u(\phi_j(y', h_j(y')))\varphi_j(\phi_j(y', h_j(y')))\,(1 + |\nabla' h_j(y')|^2)^{\frac{1}{2}}\, dy',$$

$dy' = dy_1 \dots dy_{n-1}$, is called the **surface integral** of u over $\partial\Omega$. An elementary consideration shows that this integral only depends on (u and) $\partial\Omega$. In particular it does not depend on the local coordinates.

Setting $u = 1$ for all $x \in \partial\Omega$ we obtain the **surface measure** of $\partial\Omega$ denoted by $|\partial\Omega|$.

If $1 \leq q < \infty$, the Banach space $L^q(\partial\Omega)$ is the space of all measurable (classes of) functions u defined on $\partial\Omega$ such that $x \mapsto |u(x)|^q$ is integrable over $\partial\Omega$. The norm of $L^q(\partial\Omega)$ is defined by

$$\|u\|_{L^q(\partial\Omega)} = \|u\|_{q,\partial\Omega} := \left(\int_{\partial\Omega} |u|^q\, dS\right)^{\frac{1}{q}}. \tag{3.4.4}$$

Similarly, the Banach space $L^\infty(\partial\Omega)$ is the space of all measurable (classes of) functions u on $\partial\Omega$ with finite norm

$$\|u\|_{L^\infty(\partial\Omega)} = \|u\|_{\infty,\partial\Omega} := \operatorname*{ess\text{-}sup}_{x\in\partial\Omega} |u(x)|. \tag{3.4.5}$$

The space $L^2(\partial\Omega)$ is a Hilbert space with scalar product

$$< u, v >_{\partial\Omega} := \int_{\partial\Omega} uv\, dS\,, \tag{3.4.6}$$

$u, v \in L^2(\partial\Omega)$. See [Nec67, Chap. 2, 4.1] concerning these spaces.

Let $m \in \mathbb{N}$. Then the corresponding spaces of vector fields $u = (u_1, \dots, u_m)$ are defined by

$$L^q(\partial\Omega)^m := \{(u_1, \dots, u_m);\ u_j \in L^q(\partial\Omega),\ j = 1, \dots, m\}$$

with norm

$$\|u\|_{L^q(\partial\Omega)^m} = \|u\|_{L^q} = \|u\|_{q,\partial\Omega} := \left(\sum_{j=1}^{m} \|u_j\|_{q,\partial\Omega}^q\right)^{\frac{1}{q}},$$

if $1 \le q < \infty$, and

$$\|u\|_{L^\infty(\partial\Omega)^m} = \|u\|_{\infty,\partial\Omega} := \sup_j \|u_j\|_{\infty,\partial\Omega}$$

if $q = \infty$.

$L^2(\partial\Omega)^m$ is a Hilbert space with scalar product

$$< u, v >_{\partial\Omega} := \int_{\partial\Omega} u \cdot v \, dS$$

where $u \cdot v = u_1 v_1 + \cdots + u_m v_m$, $u = (u_1, \ldots, u_m)$, $v = (v_1, \ldots, v_m)$.

Since Ω is a Lipschitz domain we are able to define the **exterior normal unit vector**

$$N(x) = (N_1(x), \ldots, N_n(x)) \tag{3.4.7}$$

for almost all $x \in \partial\Omega$ by using the local coordinates $y' = (y_1, \ldots, y_{n-1})$. For this purpose consider $x_j \in \partial\Omega$, U_j, $h_j \in C^{0,1}(\overline{B'_r})$, $j = 1, \ldots, m$, with $r > 0$ as in (3.2.8). Then each $x \in \partial\Omega$ is contained in some U_j, and there exists a new coordinate system with origin x_j, obtained by a rotation and translation of the original one, such that the new coordinates $y = (y', y_n)$ of $x \in \partial\Omega \cap U_j$ satisfy the equation

$$y_n = h_j(y') \, , \quad |y'| < r.$$

An elementary consideration shows that

$$\begin{aligned} &(1 + |\nabla' h_j(y')|^2)^{-\frac{1}{2}} (-\nabla' h_j(y'), 1) \\ &= (1 + |\nabla' h_j(y')|^2)^{-\frac{1}{2}} (-\frac{\partial}{\partial y_1} h_j(y') \, , \ldots , -\frac{\partial}{\partial y_{n-1}} h_j(y') \, , 1) \end{aligned} \tag{3.4.8}$$

is the exterior normal unit vector at $x \in \partial\Omega \cap U_j$, written in the new coordinates $(y', y_n) = (y', h_j(y'))$. Since h_j is a Lipschitz function, the gradient $\nabla' h_j(y')$ is well defined for almost all y', $|y'| < r$, and

$$\operatorname*{ess-sup}_{|y'|<r} |\nabla' h_j(y')| < \infty \; , \quad j = 1, \ldots, m,$$

see [Nec67, Chap. 1, 2.4] and [Nec67, Chap. 3, 1.1]. Therefore, (3.4.8) is well defined for almost all y', $|y'| < r$. Going back to the original coordinates, we obtain from (3.4.8) the uniquely determined exterior normal unit vector (3.4.7) for almost all $x \in \partial\Omega$.

Using (3.4.8) we see that $N : x \mapsto N(x)$, $x \in \partial\Omega$, is contained in $L^\infty(\partial\Omega)^n$ and that

$$\|N\|_{\infty,\partial\Omega} = 1. \tag{3.4.9}$$

3.5 Distributions

In this subsection, $\Omega \subseteq \mathbb{R}^n$ means any domain with $n \geq 1$. We need some basic functional analytic facts and refer mainly to [Yos80] and [Miz73].

In the theory of distributions, the linear space $C_0^\infty(\Omega)$ of smooth functions on Ω is called a test space and each $\varphi \in C_0^\infty(\Omega)$ is called a test function. $C_0^\infty(\Omega)$ is equipped with a certain topology which we do not need here, see [Yos80]. We only need the notion of continuity of a linear functional $F : \varphi \mapsto F(\varphi)$, $\varphi \in C_0^\infty(\Omega)$, with respect to this topology. This can be described directly as follows: The functional F is continuous iff for each bounded subdomain $G \subseteq \Omega$ with $\overline{G} \subseteq \Omega$ there exist some $k \in \mathbb{N}_0$ and some $C = C(F,G) > 0$ such that

$$|F(\varphi)| \;\leq\; C \, \|\varphi\|_{C^k(\overline{G})}$$

holds for all $\varphi \in C_0^\infty(\Omega)$.

The linear space $C_0^\infty(\Omega)'$ of all linear functionals

$$F : C_0^\infty(\Omega) \to \mathbb{R} \;, \quad \varphi \mapsto F(\varphi) \;, \quad \varphi \in C_0^\infty(\Omega)$$

which are continuous in this sense, is called the space of **distributions** in Ω. We use the notation

$$F(\varphi) \;=\; [F,\varphi] \;=\; [F,\varphi]_\Omega$$

for the value of F at φ, and write sometimes $F = [F, \cdot\,]$. Each function $f \in L^1_{loc}(\Omega)$ yields a distribution by the definition

$$\varphi \mapsto \, < f,\varphi >_\Omega \, = \, < f,\varphi > \; := \int_\Omega f\varphi \, dx. \tag{3.5.1}$$

We write $< f,\cdot > = < f,\cdot >_\Omega$ for this distribution or denote it simply again by f. Thus we identify f with its distribution $< f,\cdot >$ and get the embedding

$$L^1_{loc}(\Omega) \subseteq C_0^\infty(\Omega)'. \tag{3.5.2}$$

Each $f \in L^1_{loc}(\Omega)$ is called a regular distribution.

Consider any differential operator $D^\alpha = D_1^{\alpha_1} \ldots D_n^{\alpha_n}$ with $\alpha = (\alpha_1,\ldots,\alpha_n) \in \mathbb{N}_0^n$. Then for each $F \in C_0^\infty(\Omega)'$, the distribution $D^\alpha F \in C_0^\infty(\Omega)'$ is well defined by setting

$$[D^\alpha F,\varphi] \;:=\; (-1)^{|\alpha|}\,[F,D^\alpha\varphi] \;, \quad \varphi \in C_0^\infty(\Omega). \tag{3.5.3}$$

In particular, for each $f \in L^1_{loc}(\Omega)$, the distribution $D^\alpha f = [D^\alpha f, \cdot\,] \in C_0^\infty(\Omega)'$ is well defined by

$$[D^\alpha f,\varphi] \;:=\; (-1)^{|\alpha|} < f, D^\alpha\varphi > \, = (-1)^{|\alpha|} \int_\Omega f(D^\alpha\varphi)\, dx\,. \tag{3.5.4}$$

If $D^\alpha f$ is again regular, then there exists a function from $L^1_{loc}(\Omega)$, also denoted by $D^\alpha f$, such that

$$[D^\alpha f, \varphi] = < D^\alpha f, \varphi > = \int_\Omega (D^\alpha f) \varphi \, dx$$

for all $\varphi \in C_0^\infty(\Omega)$. The notation

$$D^\alpha f \in L^1_{loc}(\Omega) \tag{3.5.5}$$

always means that $D^\alpha f$ is regular and is, as a function, contained in $L^1_{loc}(\Omega)$.

More generally, let $F \in C_0^\infty(\Omega)'$, and let

$$D := \sum_{|\alpha| \le k} a_\alpha D^\alpha, \tag{3.5.6}$$

$k \in \mathbb{N}_0$, $a_\alpha \in \mathbb{R}$, be any differential operator. Then $DF \in C_0^\infty(\Omega)'$ is defined by the relation

$$[DF, \varphi] = \sum_{|\alpha| \le k} (-1)^{|\alpha|} a_\alpha [F, D^\alpha \varphi] \ , \quad \varphi \in C_0^\infty(\Omega). \tag{3.5.7}$$

In particular, if $f \in L^1_{loc}(\Omega)$ and if Df, defined by (3.5.7), is a regular distribution determined by a function which is again denoted by Df, then we simply write

$$Df \in L^1_{loc}(\Omega). \tag{3.5.8}$$

Then

$$[Df, \varphi] = < Df, \varphi > = \int_\Omega (Df) \varphi \, dx = \sum_{|\alpha| \le k} (-1)^{|\alpha|} a_\alpha < f, D^\alpha \varphi >$$

for all $\varphi \in C_0^\infty(\Omega)$.

Let $f \in L^1_{loc}(\Omega)$ and $\alpha = (\alpha_1, \dots, \alpha_n) \in \mathbb{N}_0^n$. If $D^\alpha f$ is regular, that is if $D^\alpha f \in L^1_{loc}(\Omega)$, we call $D^\alpha f$ the αth **weak** or **generalized derivative** of f. If $1 \le q \le \infty$, the notation

$$D^\alpha f \in L^q(\Omega) \tag{3.5.9}$$

means that $D^\alpha f$ is regular, and is (identified with) a function contained in $L^q(\Omega)$. Instead of (3.5.9) we sometimes simply write

$$\| D^\alpha f \|_q < \infty.$$

Similarly, $Df \in L^q(\Omega)$ with D as in (3.5.6) means that Df is regular and is as a function contained in $L^q(\Omega)$.

If $f \in C^{|\alpha|}(\Omega)$, then $D^\alpha f$ coincides almost everywhere with the classical operation. This motivates the definition of $D^\alpha F$ in (3.5.3).

Further we need the corresponding spaces for vector fields. Let $m \in \mathbb{N}$ and let

$$C_0^\infty(\Omega)^m := \{(\varphi_1, \dots, \varphi_m);\, \varphi_j \in C_0^\infty(\Omega),\ j = 1, \dots, m\}$$

be the space of vector valued test functions $\varphi = (\varphi_1, \dots, \varphi_m)$, equipped with the corresponding topology.

Then for each

$$F = (F_1, \dots, F_m)\ ,\quad F_j \in C_0^\infty(\Omega)'\ ,\quad j = 1, \dots, m$$

we define the functional

$$F : \varphi \mapsto [F, \varphi]\ ,\quad \varphi = (\varphi_1, \dots, \varphi_m) \in C_0^\infty(\Omega)^m\,,$$

by setting

$$[F, \varphi] \;=\; [F, \varphi]_\Omega \;:=\; [F_1, \varphi_1] + \dots + [F_m, \varphi_m].$$

Each linear continuous functional defined on $C_0^\infty(\Omega)^m$ has this form. Thus we call

$$\begin{aligned} C_0^\infty(\Omega)'^m &= C_0^\infty(\Omega)^{m\prime} \\ &= \{(F_1, \dots, F_m)\ ;\ F_j \in C_0^\infty(\Omega)'\ ,\ j = 1, \dots, m\} \end{aligned}$$

the **distribution space** of the test space $C_0^\infty(\Omega)^m$.

Let $f \in L^1_{loc}(\Omega)^m$ and $\alpha = (\alpha_1, \dots, \alpha_n) \in \mathbb{N}_0^n$. Then $f = (f_1, \dots, f_m)$ defines the distribution

$$\varphi \mapsto [f, \varphi] \;=\; < f, \varphi > \;=\; \int_\Omega f \cdot \varphi\, dx \tag{3.5.10}$$

where $f \cdot \varphi = f_1\varphi_1 + \dots + f_m\varphi_m$, $\varphi = (\varphi_1, \dots, \varphi_m) \in C_0^\infty(\Omega)^m$. In the same way as before, we denote this distribution again by f. This leads to the embedding

$$L^1_{loc}(\Omega)^m \subseteq C_0^\infty(\Omega)^{m\prime}.$$

The notations $D^\alpha F$, $D^\alpha f$, DF and Df are defined analogously as above. For example, let $f = (f_1, \dots, f_m) \in L^1_{loc}(\Omega)^m$. Then

$$D^\alpha f \in L^1_{loc}(\Omega)^m$$

means that $D^\alpha f_1, \dots, D^\alpha f_m \in L^1_{loc}(\Omega)$ and that

$$[D^\alpha f, \varphi] = < D^\alpha f, \varphi > = \int_\Omega (D^\alpha f) \cdot \varphi \, dx = (-1)^{|\alpha|} \int_\Omega f \cdot (D^\alpha \varphi) \, dx$$

with $D^\alpha \varphi = (D^\alpha \varphi_1, \dots, D^\alpha \varphi_m)$, $\varphi = (\varphi_1, \dots, \varphi_m) \in C_0^\infty(\Omega)^m$.

Similarly,

$$D^\alpha f \in L^q(\Omega)^m \quad , \quad 1 \le q \le \infty$$

means that $D^\alpha f_1, \dots, D^\alpha f_m \in L^q(\Omega)$, and that

$$[D^\alpha f, \varphi] = < D^\alpha f, \varphi > = \int_\Omega (D^\alpha f) \cdot \varphi \, dx.$$

For the definition of weak solutions of the Navier-Stokes equations we need the subspace

$$C_{0,\sigma}^\infty(\Omega) := \{\varphi \in C_0^\infty(\Omega)^n \; ; \; \text{div } \varphi = 0\} \subseteq C_0^\infty(\Omega)^n \tag{3.5.11}$$

of the solenoidal test functions. The space $C_{0,\sigma}^\infty(\Omega)'$ of all linear continuous functionals defined on $C_{0,\sigma}^\infty(\Omega)$ is also obtained as the space of all restrictions $F|_{C_{0,\sigma}^\infty(\Omega)}$, $F \in C_0^\infty(\Omega)^{n\prime}$. Thus we get

$$C_{0,\sigma}^\infty(\Omega)' = \{F|_{C_{0,\sigma}^\infty(\Omega)} \; ; \; F \in C_0^\infty(\Omega)^{n\prime}\}. \tag{3.5.12}$$

Consider the Hilbert space $L^2(\Omega)^n$ with scalar product

$$< u, v >_\Omega = < u, v > := \int_\Omega u \cdot v \, dx \,,$$

and the subspace

$$L^2_\sigma(\Omega) := \overline{C_{0,\sigma}^\infty(\Omega)}^{\|\cdot\|_2} \subseteq L^2(\Omega)^n \tag{3.5.13}$$

obtained as the closure in the norm $\|\cdot\|_2$. Identifying each $u \in L^2(\Omega)^n$ with the functional $< u, \cdot >: \varphi \mapsto < u, \varphi >$, $\varphi \in C_0^\infty(\Omega)^n$, we get the natural embedding

$$L^2(\Omega)^n \subseteq C_0^\infty(\Omega)^{n\prime}. \tag{3.5.14}$$

Similarly, identifying each $u \in L^2_\sigma(\Omega)$ with $< u, \cdot >: \varphi \mapsto < u, \varphi >$, $\varphi \in C_{0,\sigma}^\infty(\Omega)$, we obtain the natural embedding

$$L^2_\sigma(\Omega) \subseteq C_{0,\sigma}^\infty(\Omega)'. \tag{3.5.15}$$

Later on, see Section 2.5, II, we will use the orthogonal projection

$$P : L^2(\Omega)^n \to L^2_\sigma(\Omega) \tag{3.5.16}$$

from $L^2(\Omega)^n$ onto $L^2_\sigma(\Omega)$ which is called the Helmholtz projection.

The operator P can be extended in a natural way from $L^2(\Omega)^n$ to $C_0^\infty(\Omega)^{n\prime}$ as follows: For each $F \in C_0^\infty(\Omega)^{n\prime}$ we let

$$PF := F|_{C^\infty_{0,\sigma}(\Omega)}$$

be the restriction to $C^\infty_{0,\sigma}(\Omega)$. This yields the extended operator

$$P : C_0^\infty(\Omega)^{n\prime} \to C^\infty_{0,\sigma}(\Omega)' \tag{3.5.17}$$

which coincides with the Helmholtz projection on the subspace $L^2(\Omega)^n$ according to the embeddings above.

3.6 Sobolev spaces

The theory of the Navier-Stokes equations will be formulated using the following Sobolev spaces. We refer to [Nec67], [Ada75], [Agm65] concerning these spaces. Here $\Omega \subseteq \mathbb{R}^n$ means an arbitrary domain with $n \geq 1$.

Let $k \in \mathbb{N}$ and $1 \leq q \leq \infty$. Then the **$\mathbf{L^q}$-Sobolev space** $W^{k,q}(\Omega)$ of order k is defined as the space of all $u \in L^q(\Omega)$ such that

$$D^\alpha u \in L^q(\Omega) \quad \text{for all } |\alpha| \leq k.$$

Recall, this means that $D^\alpha u$ is a regular distribution defined by a function which is again denoted by $D^\alpha u$.

The norm in $W^{k,q}(\Omega)$ is defined by

$$\begin{aligned} \|u\|_{W^{k,q}(\Omega)} = \|u\|_{W^{k,q}} = \|u\|_{k,q} &= \|u\|_{k,q,\Omega} \\ &:= \Big(\sum_{|\alpha| \leq k} \|D^\alpha u\|_q^q \Big)^{\frac{1}{q}} \end{aligned} \tag{3.6.1}$$

if $1 \leq q < \infty$, and by

$$\begin{aligned} \|u\|_{W^{k,\infty}(\Omega)} = \|u\|_{W^{k,\infty}} = \|u\|_{k,\infty} &= \|u\|_{k,\infty,\Omega} \\ &:= \max_{|\alpha| \leq k} \|D^\alpha u\|_\infty \end{aligned} \tag{3.6.2}$$

if $q = \infty$.

We set $W^{0,q}(\Omega) := L^q(\Omega)$ if $k = 0$. In the cases $k = 1, k = 2$ we use the notations

$$\nabla u := (D_j u)_{j=1}^n \quad , \quad \nabla^2 u := (D_j D_l u)_{j,l=1}^n \; ,$$

$$\|\nabla u\|_q := \left(\sum_{j=1}^n \|D_j u\|_q^q \right)^{\frac{1}{q}} \quad , \quad \|\nabla^2 u\|_q := \left(\sum_{j,l=1}^n \|D_j D_l u\|_q^q \right)^{\frac{1}{q}}$$

if $1 \le q < \infty$, and

$$\|\nabla u\|_\infty := \max_{j=1,\dots,n} \|D_j u\|_\infty \quad , \quad \|\nabla^2 u\|_\infty := \max_{j,l=1,\dots,n} \|D_j D_l u\|_\infty .$$

Sometimes it is convenient to use equivalent norms. For example, $\|u\|_q + \|\nabla u\|_q$ is equivalent to $\|u\|_{W^{1,q}(\Omega)}$, and $\|u\|_q + \|\nabla u\|_q + \|\nabla^2 u\|_q$ is equivalent to $\|u\|_{W^{2,q}(\Omega)}$.

The Sobolev space $W^{k,2}(\Omega)$ is a Hilbert space with scalar product

$$\sum_{|\alpha| \le k} < D^\alpha u, D^\alpha v > \quad , \quad u, v \in W^{k,2}(\Omega) . \tag{3.6.3}$$

We mainly use the Hilbert space $W^{1,2}(\Omega)$ with scalar product

$$< u, v > + < \nabla u, \nabla v > := \int_\Omega uv \, dx + \int_\Omega \nabla u \cdot \nabla v \, dx$$

where $\nabla u \cdot \nabla v = (D_1 u)(D_1 v) + \cdots + (D_n u)(D_n v)$.

The subspace

$$W_0^{k,q}(\Omega) := \overline{C_0^\infty(\Omega)}^{\|\cdot\|_{k,q}} \tag{3.6.4}$$

of $W^{k,q}(\Omega)$ is defined as the closure of the smooth functions $C_0^\infty(\Omega)$ in the norm $\|\cdot\|_{k,q}$.

Let $1 < q < \infty$, $k \in \mathbb{N}$, and let $q' := \frac{q}{q-1}$ so that

$$\frac{1}{q} + \frac{1}{q'} = 1.$$

Then the Sobolev space $W^{-k,q}(\Omega)$ of negative order $-k$ is defined as the dual space of $W_0^{k,q'}(\Omega)$, we write

$$W^{-k,q}(\Omega) := W_0^{k,q'}(\Omega)' . \tag{3.6.5}$$

This means, $W^{-k,q}(\Omega)$ is the space of all linear functionals

$$F : \varphi \mapsto [F, \varphi] \ , \quad \varphi \in W_0^{k,q'}(\Omega)$$

which are continuous in the norm $\|\varphi\|_{k,q'}$. F is continuous in $\|\varphi\|_{k,q'}$ iff there is a constant $C = C(F) > 0$ such that

$$|[F, \varphi]| \le C \|\varphi\|_{k,q'} \tag{3.6.6}$$

for all $\varphi \in C_0^\infty(\Omega)$.

The norm in $W^{-k,q}(\Omega)$ is defined as the functional norm

$$\begin{aligned} \|F\|_{W^{-k,q}(\Omega)} &= \|F\|_{W^{-k,q}} = \|F\|_{-k,q} = \|F\|_{-k,q,\Omega} \\ &:= \sup_{0 \ne \varphi \in C_0^\infty(\Omega)} |[F, \varphi]| \,/\, \|\varphi\|_{k,q'} . \end{aligned}$$

If $1 < q < \infty$, $k \in \mathbb{N}$, the spaces $W^{k,q}(\Omega)$ and $W^{-k,q}(\Omega)$ are reflexive spaces. Therefore, $W_0^{1,q'}(\Omega)$ can be treated as the dual space of $W^{-1,q}(\Omega)$. Thus we get

$$W_0^{1,q'}(\Omega) = W^{-1,q}(\Omega)'. \tag{3.6.7}$$

Here we identify each $u \in W_0^{1,q'}(\Omega)$ with the functional

$$[\,\cdot\,, u] : F \mapsto [F, u] \ , \quad F \in W^{-1,q}(\Omega).$$

We do not use here the more general Sobolev spaces $W^{\beta,q}(\Omega)$ with arbitrary order $\beta \in \mathbb{R}$, see [Nec67, Chap. 2]. However, we need some special cases of the boundary spaces $W^{\beta,q}(\partial\Omega)$ which are introduced for bounded Lipschitz domains, see [Nec67, Chap. 2, §4], [Ada75, VII, 7.51].

Let $\Omega \subseteq \mathbb{R}^n$, $n \ge 2$, be a bounded Lipschitz domain, see Section 3.2, with boundary $\partial\Omega$, and let $0 < \beta < 1$, $1 < q < \infty$.

Then the Sobolev space $W^{\beta,q}(\partial\Omega)$ is defined as the Banach space of all $u \in L^q(\partial\Omega)$ which have the finite norm

$$\begin{aligned} \|u\|_{W^{\beta,q}(\partial\Omega)} &= \|u\|_{\beta,q,\partial\Omega} \\ &:= \left(\|u\|_{q,\partial\Omega}^q + \int_{\partial\Omega} \int_{\partial\Omega} \frac{|u(x) - u(y)|^q}{|x - y|^{n-1+\beta q}} \, dS_x \, dS_y \right)^{\frac{1}{q}} . \end{aligned} \tag{3.6.8}$$

Here we use the surface integrals $\int_{\partial\Omega} \ldots dS_x$ in $x \in \partial\Omega$ and $\int_{\partial\Omega} \ldots dS_y$ in $y \in \partial\Omega$, see Section 3.4.

The Sobolev space $W^{-\beta,q}(\partial\Omega)$ of negative order $-\beta$ is defined as the dual space of $W^{\beta,q'}(\partial\Omega)$, $q' = \frac{q}{q-1}$,

$$W^{-\beta,q}(\partial\Omega) := W^{\beta,q'}(\partial\Omega)' . \tag{3.6.9}$$

Let $F \in W^{-\beta,q}(\partial\Omega)$, and let $[F,v]_{\partial\Omega}$ be the value of the functional F at $v \in W^{\beta,q'}(\partial\Omega)$. Then the norm of F is the functional norm

$$\begin{aligned}\|F\|_{W^{-\beta,q}(\partial\Omega)} &= \|F\|_{-\beta,q,\partial\Omega} \\ &:= \sup_{0\neq v\in W^{\beta,q'}(\partial\Omega)} |[F,v]_{\partial\Omega}| \,/\, \|v\|_{\beta,q',\partial\Omega}\,.\end{aligned} \tag{3.6.10}$$

Next we introduce the $W^{k,q}_{loc}$-spaces. Now let $\Omega \subseteq \mathbb{R}^n$, $n \geq 1$, be again a general domain and let $1 \leq q \leq \infty$, $k \in \mathbb{N}_0$.

Then the spaces $W^{k,q}_{loc}(\Omega)$ and $W^{k,q}_{loc}(\overline{\Omega})$ are defined as follows:
We write

$$u \in W^{k,q}_{loc}(\Omega) \tag{3.6.11}$$

iff $D^\alpha u \in L^q_{loc}(\Omega)$ for all $|\alpha| \leq k$, and we write

$$u \in W^{k,q}_{loc}(\overline{\Omega}) \tag{3.6.12}$$

iff $D^\alpha u \in L^q_{loc}(\overline{\Omega})$ for all $|\alpha| \leq k$. See Section 3.3 for the definition of the L^q_{loc}-spaces.

The linear space $W^{-k,q}_{loc}(\Omega)$ is by definition the space of all distributions

$$F : \varphi \mapsto [F,\varphi]\ ,\quad \varphi \in C_0^\infty(\Omega)$$

such that

$$\|F\|_{-k,q,\Omega_0} := \sup_{0\neq\varphi\in C_0^\infty(\Omega_0)} |[F,\varphi]_{\Omega_0}| \,/\, \|\varphi\|_{k,q',\Omega_0}\ ,\quad q' = \frac{q}{q-1}, \tag{3.6.13}$$

is finite for each bounded subdomain $\Omega_0 \subseteq \Omega$ with $\overline{\Omega}_0 \subseteq \Omega$. This means that (the restriction to $C_0^\infty(\Omega_0)$ of) F is contained in each space $W^{-k,q}(\Omega_0)$ for all such Ω_0.

Let $m \in \mathbb{N}$. All spaces in this subsection can also be defined for vector fields $u = (u_1, \ldots, u_m)$.

Let Ω, k, q, β be chosen as in the corresponding cases above. Then we define the vector valued Sobolev space

$$W^{k,q}(\Omega)^m := \{(u_1,\ldots,u_m);\ u_j \in W^{k,q}(\Omega),\ \ j = 1,\ldots,m\}$$

with norm

$$\|u\|_{W^{k,q}(\Omega)^m} = \|u\|_{W^{k,q}(\Omega)} = \|u\|_{k,q,\Omega} = \|u\|_{k,q} := \left(\sum_{j=1}^m \|u_j\|^q_{k,q}\right)^{\frac{1}{q}}.$$

Correspondingly we obtain the vector valued Banach spaces

$$W_0^{k,q}(\Omega)^m\ ,\quad W^{-k,q}(\Omega)^m\ ,\quad W^{\beta,q}(\partial\Omega)^m\ ,\quad W^{-\beta,q}(\partial\Omega)^m .$$

The definitions of the norms are obvious from above. Further we define the linear spaces

$$W^{k,q}_{loc}(\Omega)^m\,,\quad W^{k,q}_{loc}(\overline{\Omega})^m \quad\text{and}\quad W^{-k,q}_{loc}(\Omega)^m.$$

If $F=(F_1,\dots,F_m)\in W^{-\beta,q}(\partial\Omega)^m$ and $v=(v_1,\dots,v_m)\in W^{\beta,q'}(\partial\Omega)^m$, then

$$[F,v]_{\partial\Omega}:=[F_1,v_1]_{\partial\Omega}+\cdots+[F_m,v_m]_{\partial\Omega}$$

means the value of F at v. Correspondingly, if $F=(F_1,\dots,F_m)\in W^{-k,q}(\Omega)^m$, and $v=(v_1,\dots,v_m)\in W^{k,q'}_0(\Omega)^m$, then

$$[F,v]\;=\;[F,v]_\Omega\;:=\;[F_1v_1]_\Omega+\cdots+[F_m,v_m]_\Omega$$

means the value of F at v.

$W^{1,2}(\Omega)^m$ is a Hilbert space with scalar product

$$\begin{aligned}<u,v>+<\nabla u,\nabla v> \;&=\; <u,v>_\Omega+<\nabla u,\nabla v>_\Omega\\ &:=\; \int_\Omega u\cdot v\,dx+\int_\Omega \nabla u\cdot\nabla v\,dx\end{aligned}$$

where $u\cdot v:=u_1v_1+\cdots+u_mv_m\,,\;\;\nabla u\cdot\nabla v:=\sum_{j,l}^n(D_ju_l)(D_jv_l)$,

$$\nabla u=(D_ju_l)_{\substack{j=1,\dots,n\\ l=1,\dots,m}}\quad\text{and}\quad \nabla v=(D_jv_l)_{\substack{j=1,\dots,n\\ l=1,\dots,m}}.$$

Consider for example the above spaces $W^{-k,q}(\Omega)^m$ and $W^{k,q'}_0(\Omega)^m$ such that

$$W^{-k,q}(\Omega)^m\;=\;W^{k,q'}_0(\Omega)^{m\prime}\,. \tag{3.6.14}$$

Then $[F,v]=[F_1,v_1]+\cdots+[F_m,v_m]$ is the value of the functional $F=(F_1,\dots,F_m)\in W^{-k,q}(\Omega)^m$ at $v=(v_1,\dots,v_m)\in W^{k,q'}_0(\Omega)^m$. It holds that

$$|[F,v]|\;\le\;\|F\|_{-k,q}\,\|v\|_{k,q'}, \tag{3.6.15}$$

and

$$\|F\|_{-k,q}\;=\;\sup_{0\neq v\in C_0^\infty(\Omega)^m}\;|[F,v]|/\|v\|_{k,q'}$$

is equal to the infimum of all constants $C=C(F)>0$ such that the estimate

$$|[F,v]|\;\le\;C\|v\|_{k,q'}\;,\quad v\in C_0^\infty(\Omega)^m \tag{3.6.16}$$

is satisfied.

Finally we consider the case $\Omega=\mathbb{R}^n$, $n\ge 1$. Then we know, see [Nec67, Chap. 2, Proposition 2.6], that

$$W^{k,q}(\mathbb{R}^n)\;=\;W^{k,q}_0(\mathbb{R}^n)\;=\;\overline{C_0^\infty(\mathbb{R}^n)}^{\|\cdot\|_{k,q}} \tag{3.6.17}$$

holds for $1<q<\infty$, $k\in\mathbb{N}_0$. The proof rests on the mollification method, see Section 1.7, II.

Chapter II

Preliminary Results

1 Embedding properties and related facts

1.1 Poincaré inequalities

We consider some basic facts on Sobolev spaces without proof. First we collect several inequalities which compare the L^q-norm of a function u with the L^q-norm of its gradient

$$\nabla u = (D_1 u, \dots, D_n u).$$

Such estimates are called Poincaré estimates. For the proofs we refer to [Nec67], [Agm65], [Ada75], and [Fri69].

1.1.1 Lemma *Let $\Omega \subseteq \mathbb{R}^n$, $n \geq 1$, be any bounded domain, let $1 < q < \infty$, and let*

$$d = d(\Omega) := \sup_{x,y \in \Omega} |x - y|$$

denote the diameter of Ω. Then

$$\|u\|_{L^q(\Omega)} \leq C \, \|\nabla u\|_{L^q(\Omega)^n} \tag{1.1.1}$$

for all $u \in W_0^{1,q}(\Omega)$ where $C = C(q,d) > 0$ depends only on q and d.

Proof. See [Ada75, VI, 6.26]. □

From (1.1.1) we conclude that the norms $\|u\|_{W^{1,q}(\Omega)}$ and $\|\nabla u\|_{L^q(\Omega)}$ are equivalent on the subspace $W_0^{1,q}(\Omega) \subseteq W^{1,q}(\Omega)$. To get estimates for general functions $u \in W^{1,q}(\Omega)$, we need that Ω is a bounded Lipschitz domain, see Section 3.2, I.

1.1.2 Lemma *Let $\Omega \subseteq \mathbb{R}^n$ be a bounded Lipschitz domain with $n \geq 2$, let $\Omega_0 \subseteq \Omega$ be any (nonempty) subdomain, and let $1 < q < \infty$. Then*

$$\|u\|_{L^q(\Omega)} \leq C\,(\|\nabla u\|_{L^q(\Omega)^n} + |\int_{\Omega_0} u\,dx|) \tag{1.1.2}$$

for all $u \in W^{1,q}(\Omega)$ where $C = C(q, \Omega, \Omega_0) > 0$ is a constant.

Proof. See [Nec67, Chap. 1, (1.21)]. Inequality (1.1.2) also holds for $n = 1$ where Ω is a bounded open interval. □

From (1.1.2) we conclude that $\|u\|_{W^{1,q}(\Omega)}$ and $\|\nabla u\|_{L^q(\Omega)^n} + |\int_{\Omega_0} u\,dx|$ are equivalent norms on $W^{1,q}(\Omega)$.

The next result yields a bound for $\|u\|_{L^q(\Omega)}$ using the norms $\|\nabla u\|_{W^{-1,q}(\Omega)^n}$ and $\|u\|_{W^{-1,q}(\Omega)}$. We need some preparations.

Let $\Omega \subseteq \mathbb{R}^n$ be a bounded Lipschitz domain with $n \geq 2$ and let $1 < q < \infty$, $q' := \frac{q}{q-1}$.

Consider the spaces $W^{-1,q}(\Omega)^n$ and $W^{-1,q}(\Omega)$, see Section 3.6, I. Then we identify each $u \in L^q(\Omega)$ with the functional

$$< u, \cdot >: \; v \mapsto < u, v > = \int_\Omega uv\,dx \;, \quad v \in W_0^{1,q'}(\Omega),$$

which yields the embedding

$$L^q(\Omega) \subseteq W^{-1,q}(\Omega) \tag{1.1.3}$$

as usual for distributions. We get

$$|< u, v >| \;\leq\; \|u\|_q \|v\|_{q'} \;\leq\; \|u\|_q \|v\|_{1,q'} \;,$$

and this yields

$$\|u\|_{W^{-1,q}(\Omega)} \;\leq\; \|u\|_{L^q(\Omega)} \tag{1.1.4}$$

which shows that the embedding (1.1.3) is continuous.

Further, for each $u \in L^q(\Omega)$ we define the functional $\nabla u = [\nabla u, \cdot]$ by

$$[\nabla u, v] := \; - < u, \text{ div } v > = - \int_\Omega u \text{ div } v\,dx$$

for all $v = (v_1, \ldots, v_n) \in C_0^\infty(\Omega)^n$. Then we see that

$$\nabla u \in W^{-1,q}(\Omega)^n \;,$$

and we get the estimate

$$|[\nabla u, v]| \; = \; |< u, \; \mathrm{div}\, v >| \; \leq \; \|u\|_q \, \|\nabla v\|_{q'} \; \leq \; \|u\|_q \, \|v\|_{1,q'}$$

which shows that

$$\|\nabla u\|_{-1,q} := \sup_{0 \neq v \in C_0^\infty(\Omega)^n} \left(|[\nabla u, v]| \,/\, \|v\|_{1,q'} \right) \; \leq \; \|u\|_q \, . \tag{1.1.5}$$

The inequality in the next lemma is basic for the theory of the operators div and ∇ in the next section.

1.1.3 Lemma *Let $\Omega \subseteq \mathbb{R}^n$, $n \geq 2$, be a bounded Lipschitz domain and let $1 < q < \infty$. Then*

$$\|u\|_{L^q(\Omega)} \; \leq \; C \left(\|\nabla u\|_{W^{-1,q}(\Omega)^n} + \|u\|_{W^{-1,q}(\Omega)} \right) \tag{1.1.6}$$

for all $u \in L^q(\Omega)$ where $C = C(q, \Omega) > 0$ is a constant.

Proof. See [Nec67, Chap. 3, Lemma 7.1] for $q = 2$ and [Nec67b] for general q. The proof for $q = 2$ can be extended to all $1 < q < \infty$ if we replace the argument based on the Fourier transform by a potential theoretic fact. Here we use this lemma only for $q = 2$. □

Using (1.1.4) and (1.1.5) we see that

$$\|\nabla u\|_{W^{-1,q}(\Omega)^n} + \|u\|_{W^{-1,q}(\Omega)} \; \leq \; 2\|u\|_{L^q(\Omega)}. \tag{1.1.7}$$

Therefore, under the assumptions of Lemma 1.1.3 we conclude that

$$\|u\|_{L^q(\Omega)} \quad \text{and} \quad \|\nabla u\|_{W^{-1,q}(\Omega)^n} + \|u\|_{W^{-1,q}(\Omega)}$$

are equivalent norms in $L^q(\Omega)$.

Inequality (1.1.6) can be extended as follows:

Let $k \in \mathbb{N}$ and consider the spaces

$$W^{-k,q}(\Omega) \;\; , \quad W^{-k-1,q}(\Omega)^n \;\; , \quad W^{-k-1,q}(\Omega)$$

which are the dual spaces of

$$W_0^{k,q'}(\Omega) \;\; , \quad W_0^{k+1,q'}(\Omega)^n \;\; , \quad W_0^{k+1,q'}(\Omega),$$

respectively. Let $u : v \mapsto [u, v]$ be any functional from $W^{-k,q}(\Omega)$. Then the inequality

$$|[u, v]| \; \leq \; \|u\|_{-k,q} \|v\|_{k,q'} \; \leq \; \|u\|_{-k,q} \|v\|_{k+1,q'}$$

shows that

$$\|u\|_{W^{-k-1,q}(\Omega)} \le \|u\|_{W^{-k,q}(\Omega)} \,.$$

The gradient ∇u is treated as a functional $[\nabla u, \cdot\,] : v \mapsto [\nabla u, v]$ defined by

$$[\nabla u, v] := -[u, \text{ div } v] \;, \quad v \in C_0^\infty(\Omega)^n \,,$$

and using

$$\begin{aligned} |[\nabla u, v]| &= |[u, \text{ div } v]| \le \|u\|_{-k,q} \,\|\text{div } v\|_{k,q'} \\ &\le C\|u\|_{-k,q} \,\|v\|_{k+1,q'} \;, \end{aligned}$$

we get $\nabla u \in W^{-k-1,q}(\Omega)^n$ and

$$\|\nabla u\|_{W^{-k-1,q}(\Omega)^n} \le C\|u\|_{W^{-k,q}(\Omega)}$$

with some $C = C(n) > 0$.

1.1.4 Lemma *Let $\Omega \subseteq \mathbb{R}^n$, $n \ge 2$, be a bounded Lipschitz domain and let $1 < q < \infty$, $k \in \mathbb{N}$. Then*

$$\|u\|_{W^{-k,q}(\Omega)} \le C\left(\|\nabla u\|_{W^{-k-1,q}(\Omega)^n} + \|u\|_{W^{-k-1,q}(\Omega)}\right) \qquad (1.1.8)$$

for all $u \in W^{-k,q}(\Omega)$ where $C = C(q, k, \Omega) > 0$ is a constant.

Proof. See [Nec67, Chap. 3, Lemma 7.1]. Using the estimates above we see that the both sides of (1.1.8) define equivalent norms. Lemma 1.1.3 is obtained by setting $k = 0$. □

The next lemma shows that $u \in L^q_{loc}(\Omega)$, $\nabla u \in L^q(\Omega)^n$ even implies $u \in W^{1,q}(\Omega)$ if Ω is a bounded Lipschitz domain.

1.1.5 Lemma *Let $\Omega \subseteq \mathbb{R}^n$, $n \ge 2$, be any Lipschitz domain and let $1 < q < \infty$. Then we have:*

a) *If $u \in L^q_{loc}(\Omega)$ and $\nabla u \in L^q(\Omega)^n$, then*

$$u \in L^q_{loc}(\overline{\Omega}) \text{ and therefore } u \in W^{1,q}_{loc}(\overline{\Omega}). \qquad (1.1.9)$$

b) *If Ω is a bounded Lipschitz domain and $u \in L^q_{loc}(\Omega)$, $\nabla u \in L^q(\Omega)^n$, then*

$$u \in L^q(\Omega) \text{ and therefore } u \in W^{1,q}(\Omega). \qquad (1.1.10)$$

Proof. This result follows by applying [Nec67, Chap. 2, Theorem 7.6] to bounded Lipschitz subdomains of Ω. However, we can argue directly: Indeed, b) is a consequence of a), and a) can be derived using b). It its sufficient to prove the result in a neighbourhood of any $x_0 \in \partial\Omega$. Use a local coordinate system in x_0, see Section 3.2, I, define a translation in the exterior normal direction and apply the estimate of Lemma 1.1.2. This yields the result. □

1.2 Traces and Green's formula

Let $\Omega \subseteq \mathbb{R}^n$, $n \geq 2$, be a bounded Lipschitz domain with boundary $\partial\Omega$, and let $1 < q < \infty$, $q' = \frac{q}{q-1}$.

Our purpose is to introduce a bounded linear operator

$$\Gamma : u \mapsto \Gamma u \tag{1.2.1}$$

from $W^{1,q}(\Omega)$ onto $W^{1-\frac{1}{q},q}(\partial\Omega)$ so that

$$\Gamma u = u|_{\partial\Omega} \tag{1.2.2}$$

holds for all $u \in C^\infty(\overline{\Omega})$. This means, Γu coincides with the restriction of u to the boundary $\partial\Omega$ if u is smooth. In other words, Γ extends the restriction operator $u \mapsto u|_{\partial\Omega}$ from the smooth function space $C^\infty(\overline{\Omega})$ to the larger space $W^{1,q}(\Omega)$. $W^{1-\frac{1}{q},q}(\partial\Omega)$ will be the right space such that this operator is bounded and even surjective.

Γ is called the **trace operator** of Ω. The existence, boundedness, and surjectivity of such an operator

$$\Gamma : W^{1,q}(\Omega) \to W^{1-\frac{1}{q},q}(\partial\Omega)\,,$$

satisfying (1.2.2) for all $u \in C^\infty(\overline{\Omega})$, follows by combining [Nec67, Chap. 2, Theorem 5.5] with [Nec67, Chap. 2, Theorem 5.7]. See also [Ada75, VII, 7.53].

We use the notation (1.2.2) not only for $u \in C^\infty(\overline{\Omega})$ but for all $u \in W^{1,q}(\Omega)$, and call $\Gamma u = u|_{\partial\Omega}$ the **trace** of $u \in W^{1,q}(\Omega)$. We consider the trace of u as the restriction of u to $\partial\Omega$ in the generalized sense.

The construction of Γ rests on the use of the local coordinate systems, see Section 3.2, I. If the boundedness of Γ is shown on the subspace $C^\infty(\overline{\Omega}) \subseteq W^{1,q}(\Omega)$, the density property

$$\overline{C^\infty(\overline{\Omega})}^{\|\cdot\|_{W^{1,q}(\Omega)}} = W^{1,q}(\Omega)\,, \tag{1.2.3}$$

see [Nec67, Chap. 2, Theorem 3.1], then yields boundedness on $W^{1,q}(\Omega)$.

The boundedness of Γ means that there is a constant $C = C(q,\Omega) > 0$ so that the estimate

$$\|\Gamma u\|_{W^{1-\frac{1}{q},q}(\partial\Omega)} \leq C\|u\|_{W^{1,q}(\Omega)} \tag{1.2.4}$$

holds for all $u \in W^{1,q}(\Omega)$. We will simply write

$$\|\Gamma u\|_{W^{1-\frac{1}{q},q}(\partial\Omega)} = \|u\|_{W^{1-\frac{1}{q},q}(\partial\Omega)} = \|u\|_{1-\frac{1}{q},q,\partial\Omega}$$

if there is no confusion. See Section 3.4, I, for the definition of this norm.

Using the trace $\Gamma u = u|_{\partial\Omega}$, we get a direct characterization of the space $W_0^{1,q}(\Omega) = \overline{C_0^\infty(\Omega)}^{\|\cdot\|_{W^{1,q}}}$. It holds that

$$W_0^{1,q}(\Omega) = \{u \in W^{1,q}(\Omega);\ \ u|_{\partial\Omega} = 0\} \tag{1.2.5}$$

for our bounded Lipschitz domain Ω, see [Nec67, Chap. 2, Theorem 4.10] or [Ada75, VII, 7.55].

Since Γ is a surjective operator, each given element $g \in W^{1-\frac{1}{q},q}(\partial\Omega)$ is the trace $g = u|_{\partial\Omega}$ of at least one $u \in W^{1,q}(\Omega)$. Moreover, it is even possible to select some $u \in W^{1,q}(\Omega)$ for each $g \in W^{1-\frac{1}{q},q}(\partial\Omega)$ in such a way that the mapping

$$g \mapsto u \quad \text{with} \quad g = u|_{\partial\Omega}$$

is a bounded linear operator from $W^{1-\frac{1}{q},q}(\partial\Omega)$ into $W^{1,q}(\Omega)$.

Thus there exists a bounded linear operator

$$\Gamma_e :\ W^{1-\frac{1}{q},q}(\partial\Omega) \ \to\ W^{1,q}(\Omega) \tag{1.2.6}$$

with the property

$$\Gamma\Gamma_e g = g \tag{1.2.7}$$

for all $g \in W^{1-\frac{1}{q},q}(\partial\Omega)$. We call $u = \Gamma_e g$ an **extension** of g from $\partial\Omega$ to Ω.

Γ_e is called an **extension operator** from $W^{1-\frac{1}{q},q}(\partial\Omega)$ into $W^{1,q}(\Omega)$, see [Nec67, Chap. 2, Theorem 5.7]. The boundedness of Γ_e means that there is a constant $C = C(q,\Omega) > 0$ such that

$$\|\Gamma_e g\|_{W^{1,q}(\Omega)} \ \leq\ C\|g\|_{W^{1-\frac{1}{q},q}(\partial\Omega)} \tag{1.2.8}$$

holds for all $g \in W^{1-\frac{1}{q},q}(\partial\Omega)$.

Green's formula is well known in elementary classical analysis for smooth functions, see [Miz73, Chap. 3, (3.54)] or [Nec67, Chap. 1, (2.9)]. It extends the elementary rule of partial integration from intervals in $\mathbb{R}$ to higher dimensions $n \geq 2$. The following general formulation can be derived from the classical one by using density and closure arguments, see [Nec67, Chap. 3, 1.2].

Let $u \in C^\infty(\overline{\Omega})$, $v \in C^\infty(\overline{\Omega})^n$, and let $\int_{\partial\Omega} \cdots\, dS$ denote the surface integral, see Section 3.4, I. Then we get

$$\text{div}\ (uv) = (\nabla u) \cdot v + u \ \text{div}\ v$$

by an elementary calculation, and Green's formula reads

$$\int_\Omega u \ \text{div}\ v\, dx = \int_{\partial\Omega} uN \cdot v\, dS - \int_\Omega (\nabla u) \cdot v\, dx\ , \tag{1.2.9}$$

where $N : x \mapsto N(x) = (N_1(x), \dots, N_n(x))$ means the **exterior normal vector field** at the boundary $\partial\Omega$, see (3.4.7), I. We can write this formula in the form

$$< u, \operatorname{div} v >_\Omega \; = \; < u, N \cdot v >_{\partial\Omega} \; - \; < \nabla u, v >_\Omega \, , \qquad (1.2.10)$$

see (3.4.6), I, for this notation.

Using the density property (1.2.3) and the trace operator Γ above, we can extend Green's formula to all $u \in W^{1,q}(\Omega)$ and $v \in W^{1,q'}(\Omega)^n$. Then $< u, N \cdot v >_{\partial\Omega}$ remains well defined as a surface integral, see (3.4.3), I, with the traces

$$u|_{\partial\Omega} \in W^{1-\frac{1}{q},q}(\partial\Omega) \quad \text{and} \quad N \cdot v|_{\partial\Omega} \in W^{1-\frac{1}{q'},q'}(\partial\Omega) \, ; \qquad (1.2.11)$$

we see that $uN \cdot v|_{\partial\Omega} \in L^1(\partial\Omega)$. Note that $|N| \in L^\infty(\partial\Omega)$, see (3.4.9), I. This leads to the following result.

1.2.1 Lemma *Let $\Omega \subseteq \mathbb{R}^n$, $n \geq 2$, be a bounded Lipschitz domain with boundary $\partial\Omega$, and let $1 < q < \infty$, $q' := \frac{q}{q-1}$. Then for all $u \in W^{1,q}(\Omega)$ and $v \in W^{1,q'}(\Omega)^n$, (1.2.11) holds in the trace sense and we get the formula*

$$< u, \operatorname{div} v >_\Omega = < u, N \cdot v >_{\partial\Omega} \; - \; < \nabla u, v >_\Omega \, , \qquad (1.2.12)$$

where N means the exterior normal field at $\partial\Omega$.

Proof. See [Nec67, Chap. 3, Theorem 1.1]. □

Lemma 1.2.3 will give a further extension of Green's formula (1.2.12) to more general functions v. For this purpose we use the more general trace operator Γ_N, see the next lemma, for which we need some preparation.

Inserting $u = \Gamma_e g \in W^{1,q}(\Omega)$ with $u|_{\partial\Omega} = g \in W^{1-\frac{1}{q},q}(\partial\Omega)$ and $v \in W^{1,q'}(\Omega)^n$ in (1.2.12), we get

$$< \Gamma_e g, \; \operatorname{div} v >_\Omega = < g, N \cdot v >_{\partial\Omega} - < \nabla\Gamma_e g, v >_\Omega \, ,$$

and using (1.2.8) yields the estimate

$$\begin{aligned} |< g, N \cdot v >_{\partial\Omega}| \;&\leq\; |< \nabla\Gamma_e g, v >_\Omega| \; + \; |< \Gamma_e g, \operatorname{div} v >_\Omega| \\ &\leq\; C \, \|g\|_{W^{1-\frac{1}{q},q}(\partial\Omega)} (\|v\|_{q'} + \| \operatorname{div} v\|_{q'}), \end{aligned} \qquad (1.2.13)$$

with some constant $C = C(q, \Omega) > 0$. This shows that the functional

$$< \cdot, N \cdot v >_{\partial\Omega} : \;\; g \mapsto < g, N \cdot v >_{\partial\Omega} \;\; , \quad g \in W^{1-\frac{1}{q},q}(\partial\Omega) \qquad (1.2.14)$$

is continuous in the norm $\|g\|_{W^{1-\frac{1}{q},q}(\partial\Omega)}$, for each fixed $v \in W^{1,q'}(\Omega)^n$. Therefore, $< \cdot, N \cdot v >_{\partial\Omega}$ belongs to the dual space of $W^{1-\frac{1}{q},q}(\partial\Omega)$, which is the space

$$W^{1-\frac{1}{q},q}(\partial\Omega)' = W^{-(1-\frac{1}{q}),q'}(\partial\Omega) = W^{-\frac{1}{q'},q'}(\partial\Omega),$$

see (3.6.9), I. Thus we get

$$< \cdot, N \cdot v >_{\partial\Omega} \in W^{-\frac{1}{q'},q'}(\partial\Omega) \quad \text{for all } v \in W^{1,q'}(\Omega)$$

and we may treat the well defined functional (1.2.14) as the trace $N \cdot v|_{\partial\Omega}$ of the normal component of v at $\partial\Omega$ in the generalized sense. Further we get from (1.2.13) that

$$\| < \cdot, N \cdot v >_{\partial\Omega} \|_{W^{-\frac{1}{q'},q'}(\partial\Omega)} \leq C\,(\|v\|_{q'}^{q'} + \|\text{div } v\|_{q'}^{q'})^{\frac{1}{q'}} \tag{1.2.15}$$

holds with some constant $C = C(q, \Omega) > 0$.

Let $E_{q'}(\Omega)$ be the Banach space of all $v \in L^{q'}(\Omega)^n$ with div $v \in L^{q'}(\Omega)$ (in the sense of distributions) and norm $\|v\|_{E_{q'}(\Omega)} := (\|v\|_{q'}^{q'} + \|\text{div } v\|_{q'}^{q'})^{\frac{1}{q'}}$. The same density argument as in (1.2.3) yields that

$$\overline{C^\infty(\overline{\Omega})^n}^{\|\cdot\|_{E_{q'}(\Omega)}} = E_{q'}(\Omega)\,, \tag{1.2.16}$$

and therefore that

$$\overline{W^{1,q'}(\Omega)^n}^{\|\cdot\|_{E_{q'}(\Omega)}} = E_{q'}(\Omega)\,. \tag{1.2.17}$$

Estimate (1.2.15) means that the operator

$$v \mapsto < \cdot, N \cdot v >_{\partial\Omega}\,, \quad v \in W^{1,q'}(\Omega), \tag{1.2.18}$$

from $W^{1,q'}(\Omega)$ to $W^{-\frac{1}{q'},q'}(\partial\Omega)$ is continuous in the norm of $E_{q'}(\Omega)$. Therefore, using (1.2.17) we see that the operator (1.2.18) extends by closure to a bounded linear operator

$$v \mapsto < \cdot, N \cdot v >_{\partial\Omega}\,, \quad v \in E_{q'}(\Omega), \tag{1.2.19}$$

from $E_{q'}(\Omega)$ to $W^{-\frac{1}{q'},q'}(\partial\Omega)$. The functional $< \cdot, N \cdot v >_{\partial\Omega}$ is therefore well defined as an element of $W^{-\frac{1}{q'},q'}(\partial\Omega)$ for each $v \in E_{q'}(\Omega)$.

Replacing q' by q, we thus obtain the following general trace lemma.

1.2.2 Lemma *Let $\Omega \subseteq \mathbb{R}^n$, $n \geq 2$, be a bounded Lipschitz domain with boundary $\partial\Omega$, let $1 < q < \infty$, $q' = \frac{q}{q-1}$, and let*

$$E_q(\Omega) := \{v \in L^q(\Omega)^n\,;\ \text{div } v \in L^q(\Omega)\} \tag{1.2.20}$$

be the Banach space with norm

$$\|v\|_{E_q(\Omega)} := (\|v\|_q^q + \|\text{div } v\|_q^q)^{\frac{1}{q}} . \tag{1.2.21}$$

Then there exists a bounded linear operator

$$\Gamma_N : v \mapsto \Gamma_N v \, , \quad v \in E_q(\Omega), \tag{1.2.22}$$

from $E_q(\Omega)$ into $W^{-\frac{1}{q},q}(\partial\Omega)$ such that $\Gamma_N\, v$ coincides with the functional

$$g \mapsto < g, N \cdot v >_{\partial\Omega} = \int_{\partial\Omega} g(x) N(x) \cdot v(x)\, dS \, , \quad g \in W^{\frac{1}{q},q'}(\partial\Omega) \tag{1.2.23}$$

if $v \in C^\infty(\overline{\Omega})^n$.

Proof. See [SiSo92, Theorem 5.3] or [Tem77, Chap. I, Theorem 1.2]. □

The operator $\Gamma_N : v \mapsto \Gamma_N\, v$ from $E_q(\Omega)$ to $W^{-\frac{1}{q},q}(\partial\Omega)$ is called the **generalized trace operator** for the normal component. For each $v \in E_q(\Omega)$, the functional $\Gamma_N v \in W^{-\frac{1}{q},q}(\partial\Omega)$ is called the **generalized trace** of the normal component $N \cdot v$ at $\partial\Omega$. We use the notation

$$\Gamma_N v \;=\; < \cdot, N \cdot v >_{\partial\Omega} = \; N \cdot v|_{\partial\Omega} \tag{1.2.24}$$

for all $v \in E_q(\Omega)$, although $N \cdot v|_{\partial\Omega}$ need not exist in the sense of usual traces (unless $v \in W^{1,q}(\Omega)^n$). Note that v itself need not have a well defined trace at $\partial\Omega$ in any sense. We refer to [Tem77, Chap. I, 1.2] and to [SiSo92, (5.1)] concerning the space $E_q(\Omega)$.

The next lemma yields the most general formulation of Green's formula.

1.2.3 Lemma *Let $\Omega \subseteq \mathbb{R}^n$, $n \geq 2$, be a bounded Lipschitz domain with boundary $\partial\Omega$, and let $1 < q < \infty$, $q' = \frac{q}{q-1}$.*

Then for all $u \in W^{1,q}(\Omega)$ and $v \in E_{q'}(\Omega)$,

$$< u, \text{div } v >_\Omega \;=\; < u, N \cdot v >_{\partial\Omega} - < \nabla u, v >_\Omega \tag{1.2.25}$$

where $< u, N \cdot v >_{\partial\Omega}$ is well defined in the sense of the generalized trace with

$$N \cdot v|_{\partial\Omega} \in W^{-\frac{1}{q'},q'}(\partial\Omega) \;\; , \quad u|_{\partial\Omega} \in W^{1-\frac{1}{q},q}(\partial\Omega).$$

Proof. Using (1.2.17) we find a sequence $(v_j)_{j=1}^\infty$ in $W^{1,q'}(\Omega)^n$ with $v = \lim_{j\to\infty} v_j$ in the norm of $E_{q'}(\Omega)$. Then we insert v_j for v in formula (1.2.12) and let $j \to \infty$. The estimate (1.2.15), used with v replaced by $v - v_j$, shows that

$$< u, N \cdot v >_{\partial\Omega} = \lim_{j\to\infty} < u, N \cdot v_j >_{\partial\Omega} .$$

This leads to (1.2.25). □

1.3 Embedding properties

The embedding properties below will be used frequently, for example in order to estimate the nonlinear term $u \cdot \nabla u$ of the Navier-Stokes equations. The first lemma contains a special case of Sobolev's embedding theorem. For the proofs we refer to [Nir59], [Fri69], [Nec67], [Ada75].

1.3.1 Lemma *Let $n \in \mathbb{N}$. Then we get:*

a) *If $1 < r \leq n$, $1 < q < \infty$, $1 < \gamma < \infty$, $0 \leq \beta \leq 1$ such that*

$$\beta(\frac{1}{r} - \frac{1}{n}) + (1-\beta)\frac{1}{\gamma} = \frac{1}{q}, \tag{1.3.1}$$

then

$$\begin{aligned} \|u\|_{L^q(\mathbb{R}^n)} &\leq C\|\nabla u\|^{\beta}_{L^r(\mathbb{R}^n)^n}\|u\|^{1-\beta}_{L^\gamma(\mathbb{R}^n)} \\ &\leq C\left(\|\nabla u\|_{L^r(\mathbb{R}^n)^n} + \|u\|_{L^\gamma(\mathbb{R}^n)}\right) \end{aligned} \tag{1.3.2}$$

for all $u \in C_0^\infty(\mathbb{R}^n)$ where $C = C(n,r,q,\gamma) > 0$ is a constant.

b) *If $r > n$, then*

$$\sup_{x,y\in\mathbb{R}^n, x\neq y} \frac{|u(x)-u(y)|}{|x-y|^{1-\frac{n}{r}}} \leq C\|\nabla u\|_{L^r(\mathbb{R}^n)^n} \tag{1.3.3}$$

for all $u \in C_0^\infty(\mathbb{R}^n)$ where $C = C(n,r) > 0$ is a constant.

Proof. See [Nir59], [Fri69, Part 1, Theorem 9.3]. □

Remarks

a) In the special case $r = n$ we get $(1-\beta)\frac{1}{\gamma} = \frac{1}{q}$, $0 \leq \beta < 1$ ($q = \infty$ is excluded), $1 < \gamma \leq q < \infty$, $\beta = 1 - \frac{\gamma}{q}$, and this leads to

$$\|u\|_{L^q(\mathbb{R}^n)} \leq C\,\|\nabla u\|^{1-\frac{\gamma}{q}}_{L^n(\mathbb{R}^n)^n}\,\|u\|^{\frac{\gamma}{q}}_{L^\gamma(\mathbb{R}^n)} \tag{1.3.4}$$

for all $u \in C_0^\infty(\mathbb{R}^n)$. Note that an inequality of the form $\|u\|_\infty \leq C\|\nabla u\|_n$ is excluded.

b) The second inequality in (1.3.2) follows from the first one by Young's inequality (3.3.8), I.

c) Inequality (1.3.2) leads in the case $1 < r < n$, $\beta = 1$, $r < q$, $n \geq 2$, $\frac{1}{n} + \frac{1}{q} = \frac{1}{r}$ to the estimate

$$\|u\|_{L^q(\mathbb{R}^n)} \leq C\,\|\nabla u\|_{L^r(\mathbb{R}^n)^n} \tag{1.3.5}$$

for all $u \in C_0^\infty(\mathbb{R}^n)$ with $C = C(n,q) > 0$.

The following lemma yields a restricted result but includes the important case $q = \infty$. It is a consequence of (1.3.3) and the Poincaré inequality (1.1.2).

1.3.2 Lemma *Let $\Omega \subseteq \mathbb{R}^n$, $n \geq 1$, be an arbitrary domain with $\overline{\Omega} \neq \mathbb{R}^n$, and let $B \subseteq \mathbb{R}^n$ be any open ball with $B \cap \Omega \neq \emptyset$. Then we have:*

a) *If $1 < q < \infty$, then*

$$\|u\|_{L^q(B\cap\Omega)} \leq C\,\|\nabla u\|_{L^q(\Omega)^n} \tag{1.3.6}$$

for all $u \in C_0^\infty(\Omega)$ with $C = C(q, \Omega, B) > 0$.

b) *If $q > n$, then*

$$\|u\|_{L^\infty(B\cap\Omega)} \leq C\,\|\nabla u\|_{L^q(\Omega)^n} \tag{1.3.7}$$

for all $u \in C_0^\infty(\Omega)$ with $C = C(q, \Omega, B) > 0$.

Proof. Since $\overline{\Omega} \neq \mathbb{R}^n$ we can choose some open ball $B_0 \subseteq \mathbb{R}^n$ with $\overline{B_0} \cap \overline{\Omega} = \emptyset$. To prove a) we use Poincaré's inequality in Lemma 1.1.2 with Ω_0, Ω replaced by $B_0, \widetilde{\Omega}$; $\widetilde{\Omega}$ means any bounded Lipschitz domain containing B_0 and $B \cap \Omega$. Extending each $u \in C_0^\infty(\Omega)$ by zero we get $u \in C_0^\infty(\mathbb{R}^n)$, and since $u = 0$ in B_0 we obtain from (1.1.2) that

$$\|u\|_{L^q(B\cap\Omega)} \leq \|u\|_{L^q(\widetilde{\Omega})} \leq C\,\|\nabla u\|_{L^q(\widetilde{\Omega})^n} \leq C\,\|\nabla u\|_{L^q(\Omega)^n}$$

for all $u \in C_0^\infty(\Omega)$ with some $C = C(q, \Omega, B) > 0$. Indeed, C depends only on q, B_0 and B.

To prove b) we apply the above estimate (1.3.3) to $u \in C_0^\infty(\Omega)$ with r replaced by q. Let y_0 be the center of B_0. Then we get, extending u by zero as above, that

$$\begin{aligned}
\|u\|_{L^\infty(B\cap\Omega)} &= \sup_{x\in B\cap\Omega} |u(x)| = \sup_{x\in B\cap\Omega} |u(x) - u(y_0)| \\
&\leq \big(\sup_{x\in B\cap\Omega} |x - y_0|^{1-\frac{n}{q}}\big) \sup_{x\in B\cap\Omega} \frac{|u(x) - u(y_0)|}{|x - y_0|^{1-\frac{n}{q}}} \\
&\leq C\big(\sup_{x\in B\cap\Omega} |x - y_0|^{1-\frac{n}{q}}\big)\|\nabla u\|_{L^q(\Omega)^n}
\end{aligned}$$

with $C = C(n, q) > 0$. This proves the lemma. □

The next two lemmas are special cases of Sobolev's embedding theorem for bounded domains.

1.3.3 Lemma *Let $\Omega \subseteq \mathbb{R}^n$, $n \geq 2$, be a bounded C^1-domain, and let $1 < r \leq n$, $1 < q < \infty$, $1 < \gamma < \infty$, $0 \leq \beta \leq 1$ so that*

$$\beta\Big(\frac{1}{r} - \frac{1}{n}\Big) + (1 - \beta)\frac{1}{\gamma} = \frac{1}{q}. \tag{1.3.8}$$

Then

$$\begin{aligned} \|u\|_{L^q(\Omega)} &\leq C\,\|u\|^{\beta}_{W^{1,r}(\Omega)}\|u\|^{1-\beta}_{L^\gamma(\Omega)} \\ &\leq C\left(\|u\|_{W^{1,r}(\Omega)} + \|u\|_{L^\gamma(\Omega)}\right) \end{aligned} \tag{1.3.9}$$

for all $u \in W^{1,r}(\Omega) \cap L^\gamma(\Omega)$ where $C = C(\Omega, q, r, \gamma) > 0$ is a constant.

Proof. See [Fri69, Part 1, Theorem 10.1]. Note that the case $n = r$ is not excluded. In this case we have $0 \leq \beta < 1$. □

The next lemma concerns the embedding of continuous functions in certain $W^{m,q}$-spaces for bounded domains.

1.3.4 Lemma *Let $k \in \mathbb{N}_0$, $m \in \mathbb{N}$, $1 < q < \infty$ with $m - \frac{n}{q} > k$, $n \geq 2$, and let $\Omega \subseteq \mathbb{R}^n$ be a bounded C^m-domain. Then, after redefinition on a subset of Ω of measure zero, each $u \in W^{m,q}(\Omega)$ is contained in $C^k(\overline{\Omega})$ and*

$$\|u\|_{C^k(\overline{\Omega})} \leq C\,\|u\|_{W^{m,q}(\Omega)} \tag{1.3.10}$$

where $C = C(\Omega, m, q) > 0$ is a constant.

Proof. See [Fri69, Part 1, Theorem 11.1]. □

Finally we mention a special embedding result for the two-dimensional case.

1.3.5 Lemma *Let $\Omega \subseteq \mathbb{R}^2$ be any two-dimensional domain with $\overline{\Omega} \neq \mathbb{R}^2$, let $B_0, B \subseteq \mathbb{R}^2$ be open balls with $\overline{B_0} \cap \overline{\Omega} = \emptyset$, $B \cap \Omega \neq \emptyset$, and let $1 < q < \infty$. Then*

$$\|u\|_{L^q(B\cap\Omega)} \leq C\,\|\nabla u\|_{L^2(\Omega)^2} \tag{1.3.11}$$

for all $u \in C_0^\infty(\Omega)$ where $C = C(B_0, B, q) > 0$ is a constant.

Proof. Let x_0 be the center of B_0, $R > 0$ the radius, and let $u \in C_0^\infty(\Omega)$. Then we use the inequality

$$\left(\int_\Omega \left(\frac{|u(x)|}{|x - x_0| \ln |x - x_0|/R}\right)^2 dx\right)^{\frac{1}{2}} \leq C\,\|\nabla u\|_{L^2(\Omega)^2} \tag{1.3.12}$$

where $C = C(B_0) > 0$ is a constant. An elementary proof of this inequality can be found in [Lad69, Chap. 1, (14)].

Next we use the above inequality (1.3.9) for B with $n = 2$, $2 < q < \infty$, $r = \gamma = 2$, $\beta = 1 - \frac{2}{q}$, and get

$$\|u\|_{L^q(B\cap\Omega)} \leq \|u\|_{L^q(B)} \leq C\left(\|\nabla u\|_{L^2(B)^2} + \|u\|_{L^2(B)}\right) \tag{1.3.13}$$

with some $C = C(B, q) > 0$. On the right side, B can be replaced by $B \cap \Omega$.

If $1 < q \le 2$ we get using (1.3.12) that

$$\begin{aligned}
\|u\|_{L^q(B\cap\Omega)} &\le C_1 \|u\|_{L^2(B\cap\Omega)} \\
&\le C_1 \left(\sup_{x\in B\cap\Omega} (|x-x_0| \ln|x-x_0|/R) \right) \left(\int_{B\cap\Omega} \left(\frac{|u(x)|}{|x-x_0|\ln|x-x_0|/R} \right)^2 dx \right)^{\frac{1}{2}} \\
&\le C_2 \|\nabla u\|_{L^2(\Omega)^2}
\end{aligned}$$

with constants $C_1 = C_1(B,q) > 0$, $C_2 = C_2(B_0, B, q) > 0$. This yields the result for $1 < q \le 2$. If $q > 2$ we deduce from (1.3.13) and the last inequality for $q = 2$ that

$$\begin{aligned}
\|u\|_{L^q(B\cap\Omega)} &\le C \, \|\nabla u\|_{L^2(\Omega)^2} + \|u\|_{L^2(B\cap\Omega)} \\
&\le C \, (\|\nabla u\|_{L^2(\Omega)^2} + C_2 \|\nabla u\|_{L^2(B\cap\Omega)^2}) \\
&\le C \, (1 + C_2) \|\nabla u\|_{L^2(\Omega)^2} .
\end{aligned}$$

This proves the lemma. □

1.4 Decomposition of domains

The decomposition property below will be used later on for technical reasons in order to "approximate" an arbitrary unbounded domain Ω by a sequence of bounded Lipschitz subdomains.

We need it, for example, for the existence proof of weak solutions, see the proof of Theorem 3.5.1, III. A similar result as that in the following lemma is contained in [Gal94a, III, proof of Lemma 1.1].

Recall the definition

$$\mathrm{dist}(A, B) := \inf_{x\in A,\, y\in B} |x - y|$$

for arbitrary subsets $A, B \subseteq \mathbb{R}^n$.

1.4.1 Lemma *Let $\Omega \subseteq \mathbb{R}^n$ be an arbitrary domain with $n \ge 2$. Then there exists a sequence $(\Omega_j)_{j=1}^\infty$ of bounded Lipschitz subdomains of Ω and a sequence $(\varepsilon_j)_{j=1}^\infty$ of positive numbers with the following properties:*

a) $\overline{\Omega}_j \subseteq \Omega_{j+1}$, $j \in \mathbb{N}$,

b) $\mathrm{dist}\,(\partial\Omega_{j+1}, \Omega_j) \ge \varepsilon_{j+1}$, $j \in \mathbb{N}$,

c) $\lim_{j\to\infty} \varepsilon_j = 0$,

d) $\Omega = \bigcup_{j=1}^\infty \Omega_j$.

Proof. The proof rests on the following elementary considerations. Let

$$B_r(x) := \{y \in \mathbb{R}^n;\ |y - x| < r\}$$

be the open ball with center $x \in \mathbb{R}^n$ and radius $r > 0$.

We fix some $x_0 \in \Omega$. Let $\tilde{\Omega}$ be the largest domain concerning inclusions such that

$$\widetilde{\Omega} \subseteq \Omega \cap B_1(x_0) \ , \quad x_0 \in \widetilde{\Omega}.$$

The boundary $\partial\widetilde{\Omega}$ of $\widetilde{\Omega}$ is compact and therefore, for a given $\varepsilon > 0$, we can choose finitely many balls $B_\varepsilon(x_j)$ with $x_j \in \partial\widetilde{\Omega}$, $j = 1, \dots, m$, and

$$\partial\widetilde{\Omega} \subseteq \bigcup_{j=1}^{m} B_\varepsilon(x_j).$$

Let $\hat{\Omega} := \widetilde{\Omega} \backslash \bigcup_{j=1}^m \overline{B}_\varepsilon(x_j)$. We can choose ε with $0 < \varepsilon < 1$ in such a way that $x_0 \in \hat{\Omega}$. Obviously, $\hat{\Omega}$ is a bounded Lipschitz domain, its boundary consists of parts of the boundaries of balls. We set $\Omega_1 := \hat{\Omega}$ and $\varepsilon_1 := \varepsilon$.

Next we choose $\widetilde{\Omega}$ as the largest domain with

$$\widetilde{\Omega} \subseteq \Omega \cap B_2(x_0) \ , \quad x_0 \in \widetilde{\Omega}.$$

Then the domain $\hat{\Omega}$ is constructed in the same way as before with $0 < \varepsilon < \frac{1}{2}$ and $\varepsilon < \frac{1}{2}$ dist $(\partial\widetilde{\Omega}, \Omega_1)$. Now we set $\Omega_2 := \hat{\Omega}$, $\varepsilon_2 := \varepsilon$ and obtain $\overline{\Omega}_1 \subseteq \Omega_2$, dist $(\partial\Omega_2, \Omega_1) > \varepsilon_2$.

Repeating this procedure, we find by induction a sequence $(\Omega_j)_{j=1}^\infty$ of Lipschitz subdomains of Ω and a sequence $(\varepsilon_j)_{j=1}^\infty$ with $0 < \varepsilon_j < \frac{1}{j}$, $j \in \mathbb{N}$. The properties a), b) and c) are satisfied. In order to prove d) we consider any $x \in \Omega$. Since Ω is a domain, we can choose some $j_0 \in \mathbb{N}$ and some subdomain $\Omega_0 \subseteq \Omega$ such that

$$x \in \Omega_0 \subseteq \Omega \cap B_{j_0}(x_0) \ , \quad x_0 \in \Omega_0.$$

Let $d :=$ dist $(\partial\Omega_0, x)$ and choose $j_1 > j_0$ with $\varepsilon_{j_1} < d$. Then the above construction shows that $x \in \Omega_{j_1}$. This proves the lemma. □

1.4.2 Remark The construction above yields the following additional property: To each bounded subdomain $\Omega' \subseteq \Omega$ with $\overline{\Omega'} \subseteq \Omega$ there exists some $j \in \mathbb{N}$ such that $\Omega' \subseteq \Omega_j$.

1.5 Compact embeddings

Such embedding properties are needed later on in the proofs for technical reasons.

Consider a bounded domain $\Omega \subseteq \mathbb{R}^n$ with $n \geq 1$, and let $1 < q < \infty$. Then the natural embedding

$$u \mapsto u \ , \quad u \in W_0^{1,q}(\Omega) \tag{1.5.1}$$

defines a bounded linear operator from $W_0^{1,q}(\Omega)$ into $L^q(\Omega)$ since

$$\|u\|_{L^q(\Omega)} \le \|u\|_{W^{1,q}(\Omega)} \ , \quad u \in W_0^{1,q}(\Omega). \tag{1.5.2}$$

Hence the embedding $W_0^{1,q}(\Omega) \subseteq L^q(\Omega)$ is continuous. The following lemma shows that the embedding operator (1.5.1) is even a compact operator. This means that each sequence $(u_j)_{j=1}^\infty$ in $W_0^{1,q}(\Omega)$, which is bounded in the norm of $W^{1,q}(\Omega)$, contains a subsequence which converges in the norm of $L^q(\Omega)$ to some element $u \in L^q(\Omega)$. Since $\sup_{j\in\mathbb{N}} \|u_j\|_{1,q} < \infty$, it even holds that $u \in W_0^{1,q}(\Omega)$.

1.5.1 Lemma *Let $\Omega \subseteq \mathbb{R}^n$, $n \ge 1$, be any bounded domain, and let $1 < q < \infty$. Then the embedding operator $u \mapsto u$ from $W_0^{1,q}(\Omega)$ into $L^q(\Omega)$ is compact. Therefore, each bounded sequence in $W_0^{1,q}(\Omega)$ contains a subsequence which converges in the norm of $L^q(\Omega)$ to some element of $W_0^{1,q}(\Omega)$.*

Proof. This is a special case of Rellich's theorem [Ada75, VI, Theorem 6.2, Part IV]. See also [Agm65, Sec. 8, Theorem 8.3] or [Tem77, Chap. II, Theorem 1.1]. □

Next we consider the dual space $L^q(\Omega)'$ of $L^q(\Omega), 1 < q < \infty$, consisting of all linear functionals defined on $L^q(\Omega)$ which are continuous in the norm $\|\cdot\|_q$. We know, see [Nec67, Chap. 2, Proposition 2.5], each such functional has the form

$$u \mapsto \langle f, u \rangle = \int_\Omega f u \, dx \ , \quad u \in L^q(\Omega) \tag{1.5.3}$$

with some $f \in L^{q'}(\Omega)$, $q' = \frac{q}{q-1}$. Thus we get

$$L^{q'}(\Omega) = L^q(\Omega)' \tag{1.5.4}$$

if we identify each $f \in L^{q'}(\Omega)$ with the functional

$$\langle f, \cdot \rangle : u \mapsto \langle f, u \rangle \ , \quad u \in L^q(\Omega).$$

Since $1 < q' < \infty$ we get in the same way that

$$L^q(\Omega)'' = L^{q'}(\Omega)' = L^q(\Omega). \tag{1.5.5}$$

Here $u \in L^q(\Omega)$ is identified with the functional

$$\langle \cdot, u \rangle : f \mapsto \langle f, u \rangle \ , \quad f \in L^{q'}(\Omega).$$

Thus $L^q(\Omega)$ is a reflexive Banach space for $1 < q < \infty$. See Section 3.1 for some explanations.

If $u \in W_0^{1,q}(\Omega)$ we can use the Poincaré inequality (1.1.1) and see that

$$|\langle f, u \rangle| \le \|f\|_{q'} \|u\|_q \le C \|f\|_{q'} \|\nabla u\|_q \tag{1.5.6}$$

for all $f \in L^{q'}(\Omega)$ with $C = C(q, \Omega) > 0$.

Consider now the dual space $W^{-1,q'}(\Omega) = W_0^{1,q}(\Omega)'$ of $W_0^{1,q}(\Omega)$, see (3.6.5), I. By (1.5.6) we know that each $f \in L^{q'}(\Omega)$ defines the continuous functional

$$< f, \cdot >: u \mapsto < f, u > \ , \quad u \in W_0^{1,q}(\Omega).$$

Thus, identifying each f with $< f, \cdot >$ we obtain the natural continuous embedding

$$L^{q'}(\Omega) \subseteq W^{-1,q'}(\Omega). \tag{1.5.7}$$

The embedding operator from $L^{q'}(\Omega)$ into $W^{-1,q'}(\Omega)$ can be understood as the dual operator of the embedding operator from $W_0^{1,q}(\Omega)$ into $L^q(\Omega)$. See [Yos80, VII, 1] concerning dual operators. We know, see Schauder's theorem [Yos80, X, 4], that the dual operator of a compact linear operator is again compact. Therefore, (1.5.7) is a compact embedding. Replacing q' by q we thus obtain the following result.

1.5.2 Lemma *Let $\Omega \subseteq \mathbb{R}^n$, $n \geq 1$, be any bounded domain, and let $1 < q < \infty$. Then the embedding*

$$L^q(\Omega) \subseteq W^{-1,q}(\Omega) \tag{1.5.8}$$

is compact. Therefore, each bounded sequence in $L^q(\Omega)$ contains a subsequence which converges in the norm of $W^{-1,q}(\Omega)$ to some element of $L^q(\Omega)$.

Proof. Use Lemma 1.5.1 and apply [Yos80, X, 4]. □

If Ω is a bounded Lipschitz domain, a similar compactness result also holds for the embedding $W^{1,q}(\Omega) \subseteq L^q(\Omega)$.

1.5.3 Lemma *Let $\Omega \subseteq \mathbb{R}^n$, $n \geq 1$, be a bounded Lipschitz domain, and let $1 < q < \infty$. Then the embedding*

$$W^{1,q}(\Omega) \subseteq L^q(\Omega) \tag{1.5.9}$$

is compact. Therefore, each bounded sequence in $W^{1,q}(\Omega)$ contains a subsequence which converges in the norm of $L^q(\Omega)$ to some element of $W^{1,q}(\Omega)$.

Proof. See [Nec67, Chap. 2, Theorem 6.3] □

The compactness of the embedding (1.5.7) can be used to improve the estimate (1.1.6) in Lemma 1.1.3. We can "remove" the second term on the right side of (1.1.6) under an additional condition on u. This leads to the following result.

1.5.4 Lemma *Let $\Omega \subseteq \mathbb{R}^n$, $n \geq 2$, be a bounded Lipschitz domain, let $\Omega_0 \subseteq \Omega$, $\Omega_0 \neq \emptyset$, be any subdomain, and let $1 < q < \infty$. Then*

$$\|u\|_{L^q(\Omega)} \leq C_1 \|\nabla u\|_{W^{-1,q}(\Omega)^n} \leq C_1 C_2 \|u\|_{L^q(\Omega)} \tag{1.5.10}$$

for all $u \in L^q(\Omega)$ *satisfying*

$$\int_{\Omega_0} u\, dx = 0\,; \tag{1.5.11}$$

$C_1 = C_1(q, \Omega, \Omega_0) > 0$ *and* $C_2 = C_2(n) > 0$ *are constants.*

Proof. Recall that $\nabla u \in W^{-1,q}(\Omega)^n$ with $u \in L^q(\Omega)$ means the functional

$$[\nabla u, \cdot\,] : v \mapsto [\nabla u, v] \;=\; -< u, \mathrm{div}\ v >= -\int_\Omega u \ \mathrm{div}\ v\, dx,$$

$v \in W_0^{1,q'}(\Omega)^n$, $q' = \frac{q}{q-1}$, see the proof of Lemma 1.1.3.

The estimate

$$\begin{aligned} |[\nabla u, v]| \;&=\; |< u,\ \mathrm{div}\ v >| \;\le\; \|u\|_q \,\|\mathrm{div}\ v\|_{q'} \\ &\le\; C\,\|u\|_q\,\|v\|_{W^{1,q'}(\Omega)^n} \end{aligned} \tag{1.5.12}$$

for all $v \in W_0^{1,q'}(\Omega)^n$, with $C = C(n) > 0$, proves the second inequality in (1.5.10).

Thus it remains to prove the first inequality in (1.5.10). To prove it we use a contradiction argument. Assume there does not exist a constant $C > 0$ such that

$$\|u\|_q \;\le\; C\,\|\nabla u\|_{-1,q}$$

holds for all $u \in L^q(\Omega)$ with $\int_{\Omega_0} u\, dx = 0$. Then for each $j \in \mathbb{N}$ there is some $u_j \in L^q(\Omega)$ with $\|u_j\|_q > j\|\nabla u_j\|_{-1,q}$, $\int_{\Omega_0} u_j\, dx = 0$. Setting

$$\tilde{u}_j := \|u_j\|_q^{-1} u_j \;\;,\;\;\; j \in \mathbb{N}$$

we obtain a sequence $(\tilde{u}_j)_{j=1}^\infty$ in $L^q(\Omega)$ satisfying

$$\|\tilde{u}_j\|_q = 1 \;\;,\;\;\; \int_{\Omega_0} \tilde{u}_j\, dx = 0 \;\;,\;\;\; \|\nabla \tilde{u}_j\|_{-1,q} < \frac{1}{j}$$

for all $j \in \mathbb{N}$.

Since $L^q(\Omega)$ is reflexive and the sequence $(\tilde{u}_j)_{j=1}^\infty$ is bounded, there exists a subsequence which converges weakly in $L^q(\Omega)$ to some element $u \in L^q(\Omega)$, see Section 3.1. For simplicity we may assume that $(\tilde{u}_j)_{j=1}^\infty$ itself has this property. This means that

$$< u, v > \;=\; \lim_{j\to\infty} < \tilde{u}_j, v >$$

for all $v \in L^{q'}(\Omega)$. In particular, it follows that $\int_{\Omega_0} u\,dx = 0$. Using

$$\lim_{j\to\infty} \|\nabla\tilde{u}_j\|_{-1,q} = 0 \quad ,$$

$$|[\nabla\tilde{u}_j, v]| = |<\tilde{u}_j, \operatorname{div} v>| \leq \|\nabla\tilde{u}_j\|_{-1,q}\,\|v\|_{1,q'} \,,$$

and

$$\begin{aligned}
|[\nabla u, v]| &= |<u,\ \operatorname{div} v>| = |\lim_{j\to\infty} <\tilde{u}_j,\ \operatorname{div} v>| \\
&= \lim_{j\to\infty} |<\tilde{u}_j,\ \operatorname{div} v>| = \liminf_{j\to\infty} |<\tilde{u}_j,\ \operatorname{div} v>| \\
&\leq \liminf_{j\to\infty} (\|\nabla\tilde{u}_j\|_{-1,q}\,\|v\|_{1,q'}) \\
&= (\liminf_{j\to\infty} \|\nabla\tilde{u}_j\|_{-1,q})\,\|v\|_{1,q'} \\
&= (\lim_{j\to\infty} \|\nabla\tilde{u}_j\|_{-1,q})\,\|v\|_{1,q'} = 0,
\end{aligned}$$

$v \in W_0^{1,q'}(\Omega)$, we see that $\|\nabla u\|_{-1,q} = 0$. Therefore, it holds that $\nabla u = 0$ in the sense of distributions and therefore, u is a constant. The mollification method in Section 1.7 will give a proof of this property, see (1.7.18). Since $\int_{\Omega_0} u\,dx = 0$ we conclude that $u = 0$.

On the other hand, applying inequality (1.1.6) to $\tilde{u}_j$ yields

$$\|\tilde{u}_j\|_q = 1 \leq C(\|\nabla\tilde{u}_j\|_{-1,q} + \|\tilde{u}_j\|_{-1,q}) \tag{1.5.13}$$

for all $j \in \mathbb{N}$, where $C > 0$ is the constant in (1.1.6). Since $(\tilde{u}_j)_{j=1}^{\infty}$ is bounded in $L^q(\Omega)$ and since the embedding $L^q(\Omega) \subseteq W^{-1,q}(\Omega)$ is compact, see Lemma 1.5.2, there is a subsequence of $(\tilde{u}_j)_{j=1}^{\infty}$ which converges in $W^{-1,q}(\Omega)$ to some $\tilde{u} \in L^q(\Omega)$. It also converges weakly to $\tilde{u} \in L^q(\Omega)$, and therefore we get $\tilde{u} = u = 0$. We may assume that the sequence $(\tilde{u}_j)_{j=1}^{\infty}$ itself converges in $W^{-1,q}(\Omega)$ to $u = 0$. Therefore,

$$\lim_{j\to\infty} \|\tilde{u}_j\|_{-1,q} = 0.$$

However, from (1.5.13) we get that

$$1 \leq \lim_{j\to\infty} C\,(\|\nabla\tilde{u}_j\|_{-1,q} + \|\tilde{u}_j\|_{-1,q}) = 0.$$

This is a contradiction and the lemma is proved. The argument used here is well known, see Peetre's lemma [LiMa72, Chap. 2, Lemma 5.1]. □

1.6 Representation of functionals

In the theory of the Navier-Stokes equations we are interested in the case that the external force $f = (f_1, \dots, f_n)$ has the special form

$$f = \operatorname{div} F \tag{1.6.1}$$

in the sense of distributions. Here $F = (F_{jl})_{j,l=1}^n$ means a matrix and (1.6.1) means by definition that

$$f_l = \operatorname{div}(F_{1l}, \dots, F_{nl}) = \sum_{j=1}^n D_j F_{jl},$$

$l = 1, \dots, n$. Thus the operation div applies to the columns of the matrix F.

Below we consider some conditions which are sufficient for the representation (1.6.1). If Ω is bounded, we may use the Poincaré inequality and get the following easy fact.

1.6.1 Lemma *Let $\Omega \subseteq \mathbb{R}^n$ be any bounded domain with $n \geq 2$, and let $f \in W^{-1,2}(\Omega)^n$.*

Then there exists at least one matrix $F \in L^2(\Omega)^{n^2}$ satisfying

$$f = \operatorname{div} F$$

in the sense of distributions, and

$$\|f\|_{W^{-1,2}(\Omega)^n} \leq \|F\|_{L^2(\Omega)^{n^2}} \leq C\|f\|_{W^{-1,2}(\Omega)^n} \tag{1.6.2}$$

with $C = C(\Omega) > 0$.

Proof. Consider the closed subspace

$$D := \{\nabla v \in L^2(\Omega)^{n^2}; v \in W_0^{1,2}(\Omega)^n\} \subseteq L^2(\Omega)^{n^2} \tag{1.6.3}$$

of all gradients $\nabla v = (D_j v_l)_{j,l=1}^n$ of functions $v = (v_1, \dots, v_n) \in W_0^{1,2}(\Omega)^n$. Let the functional

$$\tilde{f} : \nabla v \mapsto [\tilde{f}, \nabla v], \quad \nabla v \in D$$

be defined by $[\tilde{f}, \nabla v] := [f, v]$ for all $v \in W_0^{1,2}(\Omega)^n$. Then the Poincaré inequality (1.1.1) yields some $C = C(\Omega) > 0$ such that

$$|[\tilde{f}, \nabla v]| = |[f, v]| \leq \|f\|_{-1,2} \|v\|_{1,2} \leq C\|f\|_{-1,2} \|\nabla v\|_2$$

for all $\nabla v \in D$. Therefore, $\tilde{f}$ is a continuous functional defined on the subspace $D \subseteq L^2(\Omega)^{n^2}$.

The Hahn-Banach theorem, see [Yos80, IV, 1], yields a linear extension of $\tilde{f}$ from D to $L^2(\Omega)^{n^2}$ with the same functional norm. Then we may use the Riesz representation theorem, see [Yos80, III, 6], and obtain a matrix $F \in L^2(\Omega)^{n^2}$ satisfying

$$< F, \nabla v > = \sum_{j,l=1}^{n} \int_\Omega F_{jl}(D_j v_l)\, dx = \int_\Omega F \cdot \nabla v\, dx \;=\; [\tilde{f}, \nabla v] \;=\; [f, v]\,,$$

$v = (v_1, \dots, v_n) \in W_0^{1,2}(\Omega)^n$, and

$$\|F\|_{L^2(\Omega)^{n^2}} \;\le\; C\, \|f\|_{-1,2}\,.$$

Further we get

$$|[f,v]| = |< F, \nabla v >| \;\le\; \|F\|_2\, \|\nabla v\|_2 \;\le\; \|F\|_2 (\|v\|_2^2 + \|\nabla v\|_2^2)^{\frac{1}{2}}$$

for all $v \in W_0^{1,2}(\Omega)^n$ which shows that

$$\|f\|_{W^{-1,2}(\Omega)} \;\le\; \|F\|_2.$$

If $v \in C_0^\infty(\Omega)^n$ we see that

$$\begin{aligned} < F, \nabla v > \;&=\; \sum_{j,l=1}^{n} < F_{jl}, D_j v_l > \;=\; -\sum_{j,l=1}^{n} < D_j F_{jl}, v_l > \\ &=\; -[\, \mathrm{div}\; F, v] \;=\; [f, v] \end{aligned}$$

holds in the sense of distributions. This yields the representation $\mathrm{div}(-F) = f$ and (1.6.1) holds with F replaced by $-F$. This proves the lemma. □

Consider the bounded domain Ω as in Lemma 1.6.1 and let $f \in L^2(\Omega)^n$. Then we identify f with the functional $< f, \cdot >$ and get

$$f \in W^{-1,2}(\Omega)^n \;\;,\;\; \|f\|_{-1,2} \le C\, \|f\|_2, \tag{1.6.4}$$

with C from (1.1.1). This yields the continuous embedding

$$L^2(\Omega)^n \subseteq W^{-1,2}(\Omega)^n. \tag{1.6.5}$$

Using the above lemma we see that for each $f \in L^2(\Omega)^n$ there exists some $F \in L^2(\Omega)^{n^2}$ satisfying

$$f \;=\; \mathrm{div}\; F \tag{1.6.6}$$

in the sense of distributions, and

$$\|F\|_{L^2(\Omega)^{n^2}} \leq C\,\|f\|_{L^2(\Omega)^n} \tag{1.6.7}$$

where $C = C(\Omega) > 0$ is a constant.

If Ω is not bounded, then, in general, $\|\nabla v\|_{L^2(\Omega)^{n^2}}$ and $\|v\|_{W^{1,2}(\Omega)^n}$ are not equivalent norms in $W_0^{1,2}(\Omega)^n$. Therefore, we cannot expect that each $f \in W^{-1,2}(\Omega)^n$ has a representation $f = \operatorname{div} F$ with $F \in L^2(\Omega)^{n^2}$. The following lemma yields a criterion for this property. We have to distinguish the cases $n \geq 3$ and $n = 2$. If $n = 2$ we need an open ball $B_R(x_0)$ with center x_0 and radius R.

1.6.2 Lemma

a) *Let $\Omega \subseteq \mathbb{R}^n$ be any unbounded domain with $n \geq 3$ and let $f \in L^q(\Omega)^n$ with $q = \frac{2n}{n+2}$. Then there exists a matrix function $F \in L^2(\Omega)^{n^2}$ satisfying*

$$f = \operatorname{div} F \tag{1.6.8}$$

in the sense of distributions, and

$$\|f\|_{W^{-1,2}(\Omega)^n} \leq \|F\|_{L^2(\Omega)^{n^2}} \leq C\|f\|_{L^q(\Omega)^n} \tag{1.6.9}$$

with some constant $C = C(n) > 0$.

b) *Let $\Omega \subseteq \mathbb{R}^2$ be any unbounded domain with $\overline{\Omega} \neq \mathbb{R}^2$, let $x_0 \notin \overline{\Omega}$, $R > 0$, $\overline{B}_R(x_0) \cap \overline{\Omega} = \emptyset$, $f \in L^2_{loc}(\overline{\Omega})^2$, and suppose that*

$$\|f\|_\wedge^2 := \int_\Omega |f(x)|^2\,|x - x_0|^2\,(\ln|x - x_0|/R)^2\,dx < \infty. \tag{1.6.10}$$

Then there exists a matrix function $F \in L^2(\Omega)^4$ satisfying

$$f = \operatorname{div} F \tag{1.6.11}$$

in the sense of distributions, and

$$\|f\|_{W^{-1,2}(\Omega)^2} \leq \|F\|_{L^2(\Omega)^{n^4}} \leq C\,\|f\|_\wedge \tag{1.6.12}$$

with some constant $C = C(\Omega) > 0$.

Proof. To prove a) we use Sobolev's inequality (1.3.5) with $q' = \frac{q}{q-1} = \frac{2n}{n-2}$, $\frac{1}{n} + \frac{1}{q'} = \frac{1}{2}$. This yields

$$\|v\|_{q'} \leq C\|\nabla v\|_2\,, \quad v \in C_0^\infty(\Omega)^n, \tag{1.6.13}$$

with $C = C(n, q) > 0$.

Since $\frac{1}{q} + \frac{1}{q'} = \frac{n+2}{2n} + \frac{n-2}{2n} = 1$, we get the estimate

$$|< f, v >| \;\leq\; \|f\|_q \, \|v\|_{q'} \;\leq\; C\, \|f\|_q \, \|\nabla v\|_2. \tag{1.6.14}$$

This shows that the functional defined by $\nabla v \mapsto < f, v >$ is continuous on the subspace $D \subseteq L^2(\Omega)^{n^2}$, see (1.6.3), and the same argument as in the proof of Lemma 1.6.1 yields some F satisfying (1.6.8) and (1.6.9).

To prove b) we may assume that $R = 1$. Then we use the embedding inequality (1.3.12) and obtain

$$\left(\int_\Omega \left(\frac{|v(x)|}{|x - x_0| \ln |x - x_0|} \right)^2 dx \right)^{\frac{1}{2}} \leq C \, \|\nabla v\|_{L^2(\Omega)^4}$$

for all $v \in C_0^\infty(\Omega)^n$ with $C = C(\Omega) > 0$. This leads to

$$\begin{aligned} &|< f, v >| \\ &= \; \left| \int_\Omega (f(x)|x - x_0| \ln|x - x_0|) \cdot (v(x)|x - x_0|^{-1} (\ln|x - x_0|)^{-1}) \right| dx \\ &\leq \; C \|f\|_\wedge \|\nabla v\|_2, \end{aligned}$$

and the assertion in b) follows in the same way as before. □

1.7 Mollification method

This method enables us to approximate L^q-functions by C^∞-functions. It will be used later on in the proofs. See [Ada75, II, 2.17], [Nec67, Chap. 2, 1.3], [Yos80, I, Prop. 8], [Fri69, Part 1, (6.3)], [Miz73, Chap. 1, end of 7, and Chap. 2, Prop. 2.4, (3)], [Agm65, Sec. 1, Def. 1.7].

Let $\Omega \subseteq \mathbb{R}^n$ be a domain with $n \geq 1$ and let $\Omega_0 \subseteq \Omega$, $\Omega_0 \neq \emptyset$, be a bounded subdomain with $\overline{\Omega}_0 \subseteq \Omega$. Let

$$B_r(x) := \{y \in \mathbb{R}^n; \; |x - y| < r\} \tag{1.7.1}$$

be the open ball with center x and radius $r > 0$, and let the function $\mathcal{F} \in C_0^\infty(\mathbb{R})^n$ satisfy the following properties:

$$\begin{aligned} \operatorname{supp} \mathcal{F} \;&\subseteq\; B_1(0) \;, \quad 0 \leq \mathcal{F} \leq 1 \;, \quad \int_{B_1(0)} \mathcal{F} dx = 1, \\ \mathcal{F}(x) \;&=\; \mathcal{F}(-x) \quad \text{for all } \; x \in \mathbb{R}^n. \end{aligned} \tag{1.7.2}$$

Let $\mathcal{F}_\varepsilon \in C_0^\infty(\mathbb{R}^n)$, $\varepsilon > 0$, be defined by

$$\mathcal{F}_\varepsilon(x) := \varepsilon^{-n} \mathcal{F}(\varepsilon^{-1} x) \;, \quad x \in \mathbb{R}^n. \tag{1.7.3}$$

Then supp $\mathcal{F}_\varepsilon \subseteq B_\varepsilon(0)$ and the transformation formula for integrals, see [Apo74, Theorem 15.11], yields

$$\int_{\mathbb{R}^n} \mathcal{F}_\varepsilon(x)\,dx = \int_{\mathbb{R}^n} \mathcal{F}(y)\,dy = 1 \tag{1.7.4}$$

with $y = \frac{1}{\varepsilon}x$, $dy = \varepsilon^{-n}\,dx$.

Consider any function $u \in L^1_{loc}(\overline{\Omega})$ and set $u(x) := 0$ for all $x \notin \Omega$. Then we get $u \in L^1_{loc}(\mathbb{R}^n)$. Let $u^\varepsilon = \mathcal{F}_\varepsilon \star u$ be defined by

$$u^\varepsilon(x) = (\mathcal{F}_\varepsilon \star u)(x) := \int_{\mathbb{R}^n} \mathcal{F}_\varepsilon(x-y)u(y)\,dy \ , \quad x \in \mathbb{R}^n. \tag{1.7.5}$$

Using again the transformation formula for integrals we see that

$$u^\varepsilon(x) = \int_{\mathbb{R}^n} \mathcal{F}_\varepsilon(x-y)u(y)\,dy = \int_{\mathbb{R}^n} \mathcal{F}_\varepsilon(z)u(x-z)\,dz \tag{1.7.6}$$

with $x - y = z$, $dy = dz$, and that

$$u^\varepsilon(x) = \int_{\mathbb{R}^n} \mathcal{F}_\varepsilon(x-y)u(y)\,dy = \int_{\mathbb{R}^n} \mathcal{F}(z)u(x-\varepsilon z)\,dz \tag{1.7.7}$$

with $\varepsilon^{-1}(x-y) = z$, $y = x - \varepsilon z$, $dy = \varepsilon^n dz$.

If u is continuous in Ω, then

$$\lim_{\varepsilon\to 0} u^\varepsilon(x) = u(x) \quad \text{uniformly for all } x \in \Omega_0. \tag{1.7.8}$$

The proof of this fact rests on the representation

$$u^\varepsilon(x) - u(x) = \int_{\mathbb{R}^n} \mathcal{F}(z)(u(x-\varepsilon z) - u(x))\,dz \ , \quad x \in \Omega_0.$$

Let $u \in L^q(\Omega)$, $1 < q < \infty$, and $q' = \frac{q}{q-1}$. Then by Hölder's inequality and Fubini's theorem, see [Apo74], we get

$$\begin{aligned}
\|\mathcal{F}_\varepsilon \star u\|_{L^q(\Omega)} &= \left(\int_\Omega \left| \int_{|z|\le 1} \mathcal{F}(z)^{\frac{1}{q'}} \mathcal{F}(z)^{\frac{1}{q}} u(x-\varepsilon z)\,dz \right|^q dx\right)^{\frac{1}{q}} \\
&\le \left(\int_{|z|\le 1} \mathcal{F}(z)\,dz\right)^{\frac{1}{q'}} \left(\int_{|z|\le 1} \mathcal{F}(z) \left(\int_{\mathbb{R}^n} |u(x-\varepsilon z)|^q\,dx\right) dz\right)^{\frac{1}{q}} \\
&\le \left(\int_{|z|\le 1} \mathcal{F}\,dz\right)^{\frac{1}{q'}} \left(\int_{|z|\le 1} \mathcal{F}\,dz\right)^{\frac{1}{q}} \|u\|_{L^q(\Omega)} \\
&= \|u\|_{L^q(\Omega)}.
\end{aligned}$$

This estimate

$$\|\mathcal{F}_\varepsilon \star u\|_{L^q(\Omega)} \leq \|u\|_{L^q(\Omega)}$$

also holds if $q = 1$.

This shows,

$$\mathcal{F}_\varepsilon\star : u \mapsto \mathcal{F}_\varepsilon \star u \ , \quad u \in L^q(\Omega) \tag{1.7.9}$$

is a bounded linear operator from $L^q(\Omega)$ to $L^q(\Omega)$ with operator norm

$$\|\mathcal{F}_\varepsilon \star\| \leq 1 \ , \quad \varepsilon > 0. \tag{1.7.10}$$

Next we use the density

$$\overline{C_0^\infty(\Omega)}^{\|\cdot\|_{L^q(\Omega)}} = L^q(\Omega) \ , \quad 1 \leq q < \infty, \tag{1.7.11}$$

the property (1.7.8), which holds for each $u \in C_0^\infty(\Omega)$, and the uniform boundedness (1.7.10). This leads by an elementary calculation to

$$\lim_{\varepsilon\to 0} \|(\mathcal{F}_\varepsilon \star u) - u\|_{L^q(\Omega)} = 0$$

for all $u \in L^q(\Omega)$, $1 \leq q < \infty$.

Collecting these facts yields the following result:

1.7.1 Lemma *Let $\Omega \subseteq \mathbb{R}^n, n \geq 1$, be any domain, and let $1 \leq q < \infty$, $\varepsilon > 0$. Then for all $u \in L^q(\Omega)$ we get*

$$\|(\mathcal{F}_\varepsilon \star u)\|_{L^q(\Omega)} \leq \|u\|_{L^q(\Omega)} \tag{1.7.12}$$

and

$$\lim_{\varepsilon\to 0}(\mathcal{F}_\varepsilon \star u) = u \tag{1.7.13}$$

with respect to the norm $\|\cdot\|_{L^q(\Omega)}$.

Proof. See [Ada75, II, Lemma 2.18]. □

We mention some further properties of the operator $\mathcal{F}_\varepsilon\star$, see [Ada75, II, 2.17–2.19]. Let Ω and $\Omega_0 \subseteq \Omega$ be as above. Let $x \in \Omega_0$ and

$$0 < \varepsilon < \text{ dist } (\partial\Omega, \Omega_0) := \inf_{x\in\partial\Omega, y\in\Omega_0} |x - y| \tag{1.7.14}$$

with $0 < \varepsilon < \infty$ if $\partial\Omega = \emptyset$.

Consider any distribution $u \in C_0^\infty(\Omega)'$ in Ω, for example $u \in L^1_{loc}(\Omega)$. Then for each fixed $x \in \Omega_0$, we let $\mathcal{F}_\varepsilon(x - \cdot)$ be the test function

$$\mathcal{F}_\varepsilon(x - \cdot) : y \mapsto \mathcal{F}_\varepsilon(x - y) \ , \quad y \in \Omega,$$

and we see,

$$u^\varepsilon(x) = (\mathcal{F}_\varepsilon \star u)(x) = \int_\Omega \mathcal{F}_\varepsilon(x-y)u(y)\,dy \;:=\; [u, \mathcal{F}_\varepsilon(x - \cdot)] \tag{1.7.15}$$

is well defined in the sense of distributions. In this case, the "integral" $\int_\Omega \cdots dy$ is only used formally as a notation. An easy calculation yields the properties

$$u^\varepsilon \;=\; \mathcal{F}_\varepsilon \star u \;\in C^\infty(\overline{\Omega}_0) \tag{1.7.16}$$

and

$$(D^\alpha u^\varepsilon)(x) \;=\; (\mathcal{F}_\varepsilon \star (D^\alpha u))(x) \;=\; ((D^\alpha \mathcal{F}_\varepsilon) \star u)(x) \tag{1.7.17}$$

for all $x \in \Omega_0$, where $D^\alpha = D_1^{\alpha_1} \dots D_n^{\alpha_n}$, $\alpha = (\alpha_1, \dots, \alpha_n) \in \mathbb{N}_0^n$. Thus if $x \in \Omega_0$, and ε satisfies (1.7.14), D^α commutes with the operator $\mathcal{F}_\varepsilon \star$.

As an application of this method we prove the following property:

$$\left.\begin{array}{l}\text{If } u \in L^1_{loc}(\Omega) \text{ and } \nabla u = 0 \text{ in the sense of}\\ \text{distributions, then } u \text{ is a constant.}\end{array}\right\} \tag{1.7.18}$$

Indeed, we see that $\nabla u^\varepsilon(x) = (\nabla u)^\varepsilon(x) = 0$ for all $x \in \Omega_0$ and all ε as in (1.7.14). Since u^ε is smooth, see (1.7.16), an elementary argument shows that $u^\varepsilon = C_\varepsilon$ holds in Ω_0 with a constant C_ε depending on ε. Letting $\varepsilon \to 0$ and using (1.7.13) we see that C_ε converges to some constant C. Replacing Ω_0 by the subdomains Ω_j, $j \in \mathbb{N}$, in Lemma 1.4.1, we conclude that u is constant on the whole domain Ω.

The results of this subsection can also be used if u is replaced by a vector field $u = (u_1, \dots, u_m)$, $m \in \mathbb{N}$. If $n = 1$, $\Omega \subseteq \mathbb{R}$ means any open interval.

2 The operators ∇ and div

2.1 Solvability of div $v = g$ and $\nabla p = f$

The investigation of these operators is the first important step in the theory of the Navier-Stokes system. The construction of the pressure p rests on properties of ∇ and div. Both operators div and ∇ are connected by a duality principle, see the proof of the lemma below. Therefore, it is sufficient to know the basic properties of one of these operators. The approach which we use here is based on the estimates of gradients in Lemma 1.5.4. There are several other approaches to these operators, see [Bog79], [Bog80], [Gal94a, III, Lemma 3.1], [vWa88], and [Pil80].

2.1.1 Lemma *Let $\Omega \subseteq \mathbb{R}^n, n \geq 2$, be a bounded Lipschitz domain, let $\Omega_0 \subseteq \Omega$, $\Omega_0 \neq \emptyset$, be any subdomain, and let $1 < q < \infty$, $q' = \frac{q}{q-1}$. Then we have:*

a) *For each $g \in L^q(\Omega)$ with $\int_\Omega g dx = 0$, there exists at least one $v \in W_0^{1,q}(\Omega)^n$ satisfying*

$$\text{div } v = g \ , \quad \|\nabla v\|_{L^q(\Omega)^n} \leq C\|g\|_{L^q(\Omega)} \, , \tag{2.1.1}$$

where $C = C(q, \Omega) > 0$ is a constant.

b) *For each $f \in W^{-1,q}(\Omega)^n$ so that*

$$[f, v] = 0 \quad \textit{for all} \quad v \in W_0^{1,q'}(\Omega)^n \text{ with } \text{div } v = 0,$$

there exists a unique $p \in L^q(\Omega)$ satisfying

$$\nabla p = f \ , \quad \int_{\Omega_0} p\, dx = 0 \ , \quad \|p\|_{L^q(\Omega)} \leq C \, \|f\|_{W^{-1,q}(\Omega)^n}, \tag{2.1.2}$$

where $C = C(q, \Omega, \Omega_0) > 0$ is a constant.

Proof. First let $\Omega_0 = \Omega$. We set

$$< p, g > := \int_\Omega pg \, dx \ , \quad p \in L^q(\Omega) \, , \ \ g \in L^{q'}(\Omega) \, .$$

Since $1 < q < \infty$, $L^q(\Omega)$ and $L^{q'}(\Omega)$ are reflexive Banach spaces. Therefore, $L^q(\Omega)$ is the dual space of $L^{q'}(\Omega)$ if we identify each $p \in L^q(\Omega)$ with the functional $< p, \cdot >$, and $L^{q'}(\Omega)$ is the dual space of $L^q(\Omega)$ if we identify each $g \in L^{q'}(\Omega)$ with the functional $< \cdot, g >$. See [Yos80, IV, 9, (3)] for these notions.

Consider now the closed subspaces

$$\begin{aligned} L_0^q(\Omega) \ &:= \ \{p \in L^q(\Omega); \int_\Omega p\, dx = 0\} \subseteq L^q(\Omega), \\ L_0^{q'}(\Omega) \ &:= \ \{g \in L^{q'}(\Omega); \int_\Omega g\, dx = 0\} \subseteq L^{q'}(\Omega) \, . \end{aligned}$$

As before we set

$$< p, g > = \int_\Omega pg \, dx \ , \quad p \in L_0^q(\Omega) \, , \ \ g \in L_0^{q'}(\Omega).$$

Each continuous linear functional defined on $L_0^{q'}(\Omega)$ has a continuous linear extension to $L^{q'}(\Omega)$, see the Hahn-Banach theorem [Yos80, IV, 1], see also Section 3.1. Therefore, each such functional has the form

$$g \mapsto < p, g > \ , \quad g \in L_0^{q'}(\Omega),$$

with some $p \in L^q(\Omega)$. We choose $p_0 \in \mathbb{R}$ in such a way that $\int_\Omega (p - p_0)\,dx = 0$. Then $< p, g > = < p - p_0, g >$ for $g \in L_0^{q'}(\Omega)$ and it holds that $p - p_0 \in L_0^q(\Omega)$.

This shows that $L_0^q(\Omega)$ is the dual space of $L_0^{q'}(\Omega)$ if each $p \in L_0^q(\Omega)$ is identified with the functional $< p, \cdot >$. Correspondingly, $L_0^{q'}(\Omega)$ is the dual space of $L_0^q(\Omega)$. Thus we get

$$L_0^q(\Omega) = L_0^{q'}(\Omega)' \ , \quad L_0^{q'}(\Omega) = L_0^q(\Omega)' \, . \tag{2.1.3}$$

Next we consider the space $W_0^{1,q'}(\Omega)^n$ and its dual space

$$W^{-1,q}(\Omega)^n \ = \ W_0^{1,q'}(\Omega)^{n\prime} \, ,$$

see (3.6.5), I. Let $[f, v]$ denote the value of $f \in W^{-1,q}(\Omega)^n$ at $v \in W_0^{1,q'}(\Omega)^n$. Then $W_0^{1,q'}(\Omega)^n$ is the dual space of $W^{-1,q}(\Omega)^n$ if each $v \in W_0^{1,q'}(\Omega)^n$ is identified with the functional $[\,\cdot\,, v] : f \mapsto [f, v]$.

Let $v \in W_0^{1,q'}(\Omega)^n$. Then from (1.2.5) we see that $v|_{\partial\Omega} = 0$ holds in the sense of traces, and Green's formula (1.2.12), applied with $u = 1$ in Ω, shows that

$$\int_\Omega \text{div } v\,dx = 0 \ , \quad \text{div } v \in L_0^{q'}(\Omega) \, .$$

The linear operator

$$\text{div } : v \mapsto \text{div } v \ , \quad v \in W_0^{1,q'}(\Omega)^n \tag{2.1.4}$$

from $W_0^{1,q'}(\Omega)^n$ to $L_0^{q'}(\Omega)$ is bounded since

$$\|\text{div } v\|_{q'} \ \leq \ C_1 \, \|v\|_{W^{1,q'}(\Omega)^n} \tag{2.1.5}$$

with $C_1 = C_1(n) > 0$. Let

$$R(\text{div}) := \{\text{div } v \in L_0^{q'}(\Omega) \ ; \ v \in W_0^{1,q'}(\Omega)^n\}$$

denote the range space and let

$$N(\text{div}) := \{v \in W_0^{1,q'}(\Omega)^n \ ; \ \text{div } v = 0\}$$

be the null space of div.

Further we consider the operator

$$\nabla : p \mapsto \nabla p \ , \quad p \in L_0^q(\Omega) \tag{2.1.6}$$

from $L^q_0(\Omega)$ to $W^{-1,q}(\Omega)^n$, defined by the relation

$$[\nabla p, v] := - < p, \text{div } v > \ , \quad v \in W_0^{1,q'}(\Omega)^n \ , \ \ p \in L^q_0(\Omega), \tag{2.1.7}$$

with range

$$R(\nabla) := \{\nabla p \in W^{-1,q}(\Omega)^n \ ; \ p \in L^q_0(\Omega)\}. \tag{2.1.8}$$

If $\nabla p = 0$ we see that p is a constant, see (1.7.18), and therefore $p = 0$ since $\int_\Omega p \, dx = 0$. Thus

$$N(\nabla) := \{p \in L^q_0(\Omega) \ ; \ \nabla p = 0\} = \{0\}. \tag{2.1.9}$$

From the estimate

$$\begin{aligned} |[\nabla p, v]| \ &= \ |< p, \text{div } v >| \ \leq \ \|p\|_q \, \|\text{div } v\|_{q'} \\ &\leq \ C_1 \, \|p\|_q \, \|v\|_{1,q'} \ , \end{aligned}$$

with C_1 as in (2.1.5), we see that ∇ is a bounded operator from $L^q_0(\Omega)$ to $W^{-1,q}(\Omega)^n$. It holds that

$$\|\nabla p\|_{-1,q} \ \leq \ C_1 \|p\|_q \ , \quad p \in L^q_0(\Omega). \tag{2.1.10}$$

Next we use a functional analytic argument. The relation (2.1.7) means that $-\nabla$ is the dual operator of div, we write

$$-\nabla = \ \text{div}' \ , \tag{2.1.11}$$

see [Yos80, VII, 1] for this notion.

From Lemma 1.5.4, see (1.5.10), we obtain the estimate

$$\|p\|_q \ \leq \ C_2 \|\nabla p\|_{-1,q} \ , \quad p \in L^q_0(\Omega) \tag{2.1.12}$$

with some constant $C_2 = C_2(q, \Omega) > 0$. This shows that the range $R(-\nabla) = R(\nabla)$ of $-\nabla$ is a closed subspace of $W^{-1,q}(\Omega)^n$. Therefore we conclude that the inverse

$$\nabla^{-1} : \nabla p \mapsto p \ , \quad \nabla p \in R(\nabla)$$

from $R(\nabla)$ onto $L^q_0(\Omega)$ is a bounded operator, see [Yos80, II, 6, Theorem 1].

The closed range theorem, see [Yos80, VII, 5], yields now the following result:

$R(\text{div})$ is a closed subspace of $L^{q'}_0(\Omega)$, we have

$$R(\text{div}) \ = \ \{g \in L^{q'}_0(\Omega); \ < p, g > = 0 \quad \text{for all } p \in N(\nabla)\}, \tag{2.1.13}$$

and

$$R(\nabla) = \{f \in W^{-1,q}(\Omega)^n;\ [f,v] = 0 \quad \text{for all } v \in N(\text{div})\}. \tag{2.1.14}$$

Since $N(\nabla) = \{0\}$ we conclude that

$$R(\text{div}) = L_0^{q'}(\Omega). \tag{2.1.15}$$

Let

$$W_0^{1,q'}(\Omega)^n / N(\text{div}) := \{[v];\ v \in W_0^{1,q'}(\Omega)^n\} \tag{2.1.16}$$

denote the quotient space (see [Yos80, I, 11]) of all classes $[v] := v + N(\text{div})$, $v \in W_0^{1,q'}(\Omega)^n$, equipped with the norm

$$\|[v]\|_{W_0^{1,q'}(\Omega)^n / N(\text{div})} := \inf_{w \in [v]} \|\nabla(v + w)\|_{q'}. \tag{2.1.17}$$

Recall that $\|\nabla v\|_{q'}$ is an equivalent norm of $W_0^{1,q'}(\Omega)^n$ since Ω is bounded, see (1.1.1).

We see that there exists the well defined inverse operator

$$\text{div}^{-1}:\ \text{div } v \mapsto [v] \tag{2.1.18}$$

from $R(\text{div}) = L_0^{q'}(\Omega)$ onto $W_0^{1,q'}(\Omega)^n / N(\text{div})$. The operator div in (2.1.4) is bounded and therefore closed, which means its graph is closed. From the closed graph theorem, see [Yos80, II, 6, Theorem 1], we can now conclude that the operator div^{-1} in (2.1.18) is bounded. This means that

$$\|\,[v]\,\|_{W_0^{1,q'}(\Omega)/N(\text{div})} \leq C_3 \,\|\text{div } v\|_{q'} \tag{2.1.19}$$

for all $v \in W_0^{1,q'}(\Omega)^n$ with some constant $C_3 = C_3(q, \Omega) > 0$.

Therefore, for each $g \in L_0^{q'}(\Omega)$ we can select a representative $v \in W_0^{1,q'}(\Omega)^n$ such that $\text{div } v = g$ and

$$\|\nabla v\|_{q'} \leq C_3 \,\|g\|_{q'}.$$

Note that this mapping $g \mapsto v$ need not be linear. This proves assertion a) with q replaced by q'.

To prove b) we use (2.1.14). If $f \in W^{-1,q}(\Omega)^n$ satisfies $[f, v] = 0$ for all $v \in N(\text{div})$, then from (2.1.14) we see that $f \in R(\nabla)$, and therefore there exists some $p \in L_0^q(\Omega)$ with $f = \nabla p$; p is unique since $N(\nabla) = 0$, and the estimate in (2.1.2) follows from (2.1.12) with $C := C_2$.

This proves b) in the case $\Omega_0 = \Omega$. If $\Omega_0 \subseteq \Omega$ is any subdomain, then for given $f \in R(\nabla)$ we first choose $p \in L^q_0(\Omega)$ as above, and then we set $\tilde{p} := p - p_0$ so that

$$p_0 := |\Omega_0|^{-1} \int_{\Omega_0} p\,dx, \tag{2.1.20}$$

where $|\Omega_0|$ means the Lebesgue measure of Ω_0. Then $\int_{\Omega_0} \tilde{p}\,dx = 0$, and using Hölder's inequality we get

$$\begin{aligned}
\|\tilde{p}\|_q &\leq \|p\|_q + \|p_0\|_q \\
&\leq \|p\|_q + |\Omega_0|^{-1} \Big| \int_{\Omega_0} p\,dx \Big| \, |\Omega|^{\frac{1}{q}} \\
&\leq \|p\|_q (1 + |\Omega_0|^{-\frac{1}{q}} \, |\Omega|^{\frac{1}{q}}) \\
&\leq C \, \|f\|_{W^{-1,q}(\Omega)^n}
\end{aligned}$$

with $C = C(q, \Omega, \Omega_0) > 0$. The proof is complete. □

2.2 A criterion for gradients

Lemma 2.1.1 contains in particular a criterion for the property that

$$f \in W^{-1,q}(\Omega)^n$$

is a gradient of the form $f = \nabla p$ with $p \in L^q(\Omega)$. A sufficient condition is that

$$[f, v] = 0 \quad \text{for all} \quad v \in N(\text{div}) := \{v \in W_0^{1,q'}(\Omega)^n; \ \text{div}\, v = 0\}$$

where $[f, v]$ means the value of the functional f at v.

Our aim is to improve this criterion and to show that it is sufficient to require $[f, v] = 0$ only for all

$$v \in C^\infty_{0,\sigma}(\Omega) = \{v \in C_0^\infty(\Omega)^n; \ \text{div}\, v = 0\}.$$

This is important since $C^\infty_{0,\sigma}(\Omega)$ is the appropriate space of test functions in the theory of Navier-Stokes equations.

There are several approaches to such criterions. They are based on de Rham's theory [dRh60], see [Tem77, Chap. I, Prop. 1.1], on Bogovski's theory, see [Bog80], or on an elementary argument in [SiSo92]. Here we essentially follow [SiSo92], see also [Gal94a, III, proof of Lemma 1.1].

Further we will admit a general domain $\Omega \subseteq \mathbb{R}^n, n \geq 2$, in the next result. Recall that by definition, see (3.6.13), I, the following holds:

$$f \in W^{-1,q}_{loc}(\Omega)^n \quad \text{iff} \quad f \in W^{-1,q}(\Omega_0)^n$$

for all bounded subdomains $\Omega_0 \subseteq \Omega$ with $\overline{\Omega}_0 \subseteq \Omega$.

2.2.1 Lemma *Let $\Omega \subseteq \mathbb{R}^n$, $n \geq 2$, be an arbitrary domain, let $\Omega_0 \subseteq \Omega$ be a bounded subdomain with $\overline{\Omega}_0 \subseteq \Omega$, $\Omega_0 \neq \emptyset$, and let $1 < q < \infty$. Suppose $f \in W_{loc}^{-1,q}(\Omega)^n$ satisfies*

$$[f, v] = 0 \quad \textit{for all} \quad v \in C_{0,\sigma}^{\infty}(\Omega). \tag{2.2.1}$$

Then there exists a unique $p \in L_{loc}^q(\Omega)$ satisfying $\nabla p = f$ in the sense of distributions and

$$\int_{\Omega_0} p\, dx = 0. \tag{2.2.2}$$

Proof. The lemma is proved if we show the following property:
For any bounded Lipschitz subdomain $\Omega_1 \subseteq \Omega$ with $\overline{\Omega}_0 \subseteq \Omega_1$, $\overline{\Omega}_1 \subseteq \Omega$, there exists a unique $p \in L^q(\Omega_1)$ with $\nabla p = f$ in the sense of distributions in Ω_1, and with $\int_{\Omega_0} p\, dx = 0$.

Indeed, using a representation of Ω as a union of bounded Lipschitz domains, see Lemma 1.4.1, and the uniqueness of p in Ω_1, we will see that p can be extended to a well defined function defined on Ω with the desired properties.

Let Ω_1 be such a subdomain. Then we choose, using a similar construction as in the proof of Lemma 1.4.1, another bounded Lipschitz subdomain $\Omega_2 \subseteq \Omega$ satisfying

$$\overline{\Omega}_1 \subseteq \Omega_2 \;, \quad \overline{\Omega}_2 \subseteq \Omega.$$

From $f \in W_{loc}^{-1,q}(\Omega)^n$ we see that $f \in W^{-1,q}(\Omega_2)^n$, and since Ω_2 is bounded we get by Lemma 1.6.1 a representation of the form

$$f = \text{ div } F \quad \text{with} \quad F = (F_{jl})_{j,l=1}^n \in L^q(\Omega_2)^{n^2} .$$

This was shown in Lemma 1.6.1 only for $q = 2$, however the same proof holds for $1 < q < \infty$.

Next we use the mollification method, see Section 1.7, and set $F^{\varepsilon} := \mathcal{F}_{\varepsilon} \star F = (\mathcal{F}_{\varepsilon} \star F_{jl})_{j,l=1}^n$ with $0 < \varepsilon <$ dist $(\partial\Omega_2, \Omega_1)$. This yields $F^{\varepsilon} \in C^{\infty}(\overline{\Omega}_1)^{n^2}$.

Our purpose is to prove the representation

$$\text{div } F^{\varepsilon} \;=\; \nabla U_{\varepsilon} \tag{2.2.3}$$

with some function $U_{\varepsilon} \in C^{\infty}(\overline{\Omega}_1)$. To prove this we use the following elementary procedure from [SiSo92].

Let $w : \tau \mapsto w(\tau)$, $0 \leq \tau \leq 1$, be a continuous mapping from $[0, 1]$ to $\overline{\Omega}_1$. We assume that the derivative w' exists and is piecewise continuous on $[0, 1]$. Such a function w is called a curve in $\overline{\Omega}_1$; w is called a closed curve if $w(0) = w(1)$.

Further we consider a vector field $g = (g_1, \dots, g_n) \in C^\infty(\overline{\Omega}_1)^n$, and define the curve integral

$$\int_0^1 g(w(\tau)) \cdot w'(\tau)\, d\tau := \int_0^1 \sum_{j=1}^n g_j(w(\tau)) w_j'(\tau)\, d\tau$$

with $w(\tau) = (w_1(\tau), \dots, w_n(\tau))$, $w'(\tau) = (w_1'(\tau), \dots, w_n'(\tau))$.

An elementary classical argument shows that if this integral is zero for each closed curve in $\overline{\Omega}_1$, then g has the form $g = \nabla U$ with $U \in C^\infty(\overline{\Omega}_1)$.

To apply this argument for the proof of (2.2.3), we have to show that

$$\int_0^1 (\text{ div } F^\varepsilon)(w(\tau)) \cdot w'(\tau)\, d\tau = 0 \tag{2.2.4}$$

for each closed curve w in $\overline{\Omega}_1$. To prove this we set

$$V_{w,\varepsilon}(x) := \int_0^1 \mathcal{F}_\varepsilon(x - w(\tau)) w'(\tau)\, d\tau \ , \quad x \in \Omega_2,$$

and get $V_{w,\varepsilon} \in C_0^\infty(\Omega_2)^n$,

$$\begin{aligned} \text{div } V_{w,\varepsilon}(x) &= \int_0^1 \sum_{j=1}^n (D_j \mathcal{F}_\varepsilon)(x - w(\tau)) w_j'(\tau)\, d\tau \\ &= -\int_0^1 \frac{d}{d\tau} \mathcal{F}_\varepsilon(x - w(\tau))\, d\tau \\ &= \mathcal{F}_\varepsilon(x - w(0)) - \mathcal{F}_\varepsilon(x - w(1)) = 0 \end{aligned}$$

if w is a closed curve in $\overline{\Omega}_1$. This leads to $V_{w,\varepsilon} \in C_{0,\sigma}^\infty(\Omega_2)^n$, and using the assumption (2.2.1) and Fubini's theorem we obtain

$$\begin{aligned} 0 &= [f, V_{w,\varepsilon}] = [\text{div } F, V_{w,\varepsilon}] \\ &= \sum_{j,l=1}^n \int_{\Omega_2} D_j\, F_{jl}(x) \left(\int_0^1 \mathcal{F}_\varepsilon(x - w(\tau))\, w_l'(\tau)\, d\tau \right) dx \\ &= \int_0^1 \left(\sum_{j,l=1}^n \int_{\Omega_2} \mathcal{F}_\varepsilon(w(\tau) - x) D_j\, F_{jl}(x) dx \right) w_l'(\tau)\, d\tau \\ &= \int_0^1 \left(\sum_{j,l=1}^n \int_{\Omega_2} (D_j\, \mathcal{F}_\varepsilon)(w(\tau) - x)\, F_{jl}(x)) dx \right) w_l'(\tau)\, d\tau \\ &= \int_0^1 (\text{ div } F^\varepsilon(w(\tau)) \cdot w'(\tau)\, d\tau. \end{aligned}$$

This proves (2.2.4).

Thus we get the representation (2.2.3) with some $U_\varepsilon \in C^\infty(\overline{\Omega}_1)$ which is determined up to a constant. Choosing this constant in an appropriate way we can conclude that $\int_{\Omega_0} U_\varepsilon dx = 0$. Using Lemma 1.5.4, (1.5.10), we obtain

$$\begin{aligned} \|U_\varepsilon\|_{L^q(\Omega_1)} &\leq C\,\|\nabla U_\varepsilon\|_{W^{-1,q}(\Omega_1)} \\ &= C \sup_{0\neq v\in C_0^\infty(\Omega_1)^n} \left(|[\nabla U_\varepsilon, v]| \,/\, \|\nabla v\|_{q'}\right) \\ &= C \sup_{0\neq v\in C_0^\infty(\Omega_1)^n} \left(| < F^\varepsilon, \nabla v > | \,/\, \|\nabla v\|_{q'}\right) \\ &\leq C\|F^\varepsilon\|_{L^q(\Omega_1)} \end{aligned}$$

with $C = C(q, \Omega_0, \Omega_1) > 0$ independent of ε.

Since $\|F - F^\varepsilon\|_{L^q(\Omega_1)} \to 0$ as $\varepsilon \to 0$, see Lemma 1.7.1, we obtain, letting $\varepsilon \to 0$, some $U \in L^q(\Omega_1)$ satisfying

$$\int_{\Omega_0} U\,dx = 0 \;\;,\quad \lim_{\varepsilon\to 0} \|U - U_\varepsilon\|_{L^q(\Omega_1)} = 0 \;\;,\quad f = \operatorname{div} F = \nabla U$$

in Ω_1. To prove this, we choose $0 < \eta < \varepsilon$ and replace F^ε by $F^\varepsilon - F^\eta$, U_ε by $U_\varepsilon - U_\eta$ in the last estimate. U is uniquely determined.

Consider now all possible Lipschitz subdomains Ω_1 as defined above with $\overline{\Omega}_0 \subseteq \Omega_1$. Each bounded subdomain $\Omega' \subseteq \Omega$ with $\overline{\Omega'} \subseteq \Omega$ is contained in such a domain Ω_1, see Remark 1.4.2.

Defining p by U constructed above in each such Ω_1, the uniqueness of U because of $\int_{\Omega_0} U\,dx = 0$ yields in this way a uniquely determined function $p \in L^q_{loc}(\Omega)$ with $f = \nabla p$ in the whole domain Ω. This proves the lemma. □

If in particular Ω is a bounded Lipschitz domain, we can improve the above result, see the next lemma, and show that even $p \in L^q(\Omega)$. Moreover p satisfies the important estimate (1.5.10). For the proof we use the **scaling argument**, see, e.g., the proof of [Tem77, Chap. I, Theorem 1.1].

2.2.2 Lemma *Let $\Omega \subseteq \mathbb{R}^n, n \geq 2$, be a bounded Lipschitz domain, let $\Omega_0 \subseteq \Omega$, $\Omega_0 \neq \emptyset$, be any subdomain, and let $1 < q < \infty$. Suppose $f \in W^{-1,q}(\Omega)^n$ satisfies*

$$[f, v] = 0 \quad \textit{for all} \quad v \in C^\infty_{0,\sigma}(\Omega). \tag{2.2.5}$$

Then there exists a unique $p \in L^q(\Omega)$ satisfying

$$\int_{\Omega_0} p\,dx = 0 \;\;,\quad f = \nabla p$$

in the sense of distributions. The estimate

$$\|p\|_{L^q(\Omega)} \;\leq\; C_1\,\|f\|_{W^{-1,q}(\Omega)^n} \;\leq\; C_1 C_2\,\|p\|_{L^q(\Omega)} \tag{2.2.6}$$

holds with constants $C_1 = C_1(q, \Omega_0, \Omega) > 0$ and $C_2 = C_2(n) > 0$.

Proof. First we assume additionally that Ω is **starlike** with respect to some $x_0 \in \Omega$. This means that the line $\{x_0 + te;\ t \in \mathbb{R}\}$ intersects the boundary $\partial\Omega$ in exactly two points for each vector $e \in \mathbb{R}^n$. We may assume, for simplicity, that $x_0 = 0$. This property enables us to apply the following scaling argument.

Let $0 < \varepsilon < 1$,

$$\Omega_\varepsilon := \{x \in \mathbb{R}^n;\ \ \varepsilon x \in \Omega\}$$

and let the functional $f_\varepsilon \in W^{-1,q}(\Omega_\varepsilon)^n$ be defined by $[f_\varepsilon, v] := [f, v_\varepsilon]$, $v \in W_0^{1,q'}(\Omega_\varepsilon)^n$, where $v_\varepsilon \in W_0^{1,q'}(\Omega)^n$ is defined by $v_\varepsilon(x) := v(\varepsilon^{-1}x)$, $x \in \Omega$.

Let $v \in C_{0,\sigma}^\infty(\Omega_\varepsilon)$. Then $v_\varepsilon \in C_{0,\sigma}^\infty(\Omega)$, and from (2.2.5) we get that $[f_\varepsilon, v] = 0$ for all $v \in C_{0,\sigma}^\infty(\Omega_\varepsilon)$. Applying Lemma 2.2.1 yields a unique $p_\varepsilon \in L^q_{loc}(\Omega_\varepsilon)$ satisfying $\int_{\Omega_0} p_\varepsilon dx = 0$ and $f_\varepsilon = \nabla p_\varepsilon$ in Ω_ε. Note that $\overline{\Omega} \subseteq \Omega_\varepsilon$ and therefore $\overline{\Omega}_0 \subseteq \Omega_\varepsilon$

Since $\overline{\Omega} \subseteq \Omega_\varepsilon$ we get $p_\varepsilon \in L^q(\Omega)$, $0 < \varepsilon < 1$. Therefore we may apply Lemma 1.5.4 and estimate (1.5.10). This yields

$$\|p_\varepsilon\|_{L^q(\Omega)} \ \leq\ C\,\|\nabla p_\varepsilon\|_{W^{-1,q}(\Omega)^n} = C\,\|f_\varepsilon\|_{W^{-1,q}(\Omega)^n}$$

with $C = C(q, \Omega) > 0$ not depending on ε.

Let now $\frac{1}{2} \leq \varepsilon < 1$ and $v \in C_{0,\sigma}^\infty(\Omega)$. Extending v by zero we get $v \in C_{0,\sigma}^\infty(\Omega_\varepsilon)$. Then a calculation shows that

$$\|\nabla v_\varepsilon\|_{L^{q'}(\Omega)^{n^2}} \ \leq\ 2\,\|\nabla v\|_{L^{q'}(\Omega)^{n^2}}\ \ ,\ \ q' = \frac{q}{q-1},$$

and

$$\begin{aligned} |[f_\varepsilon, v]| \ = \ |[f, v_\varepsilon]| \ &\leq\ \|f\|_{W^{-1,q}(\Omega)^n}\,\|\nabla v_\varepsilon\|_{L^{q'}(\Omega)^{n^2}} \\ &\leq\ 2\,\|f\|_{W^{-1,q}(\Omega)^n}\,\|\nabla v\|_{L^{q'}(\Omega)^{n^2}}. \end{aligned}$$

This yields

$$\|p_\varepsilon\|_{L^q(\Omega)} \ \leq\ C\,\|f_\varepsilon\|_{W^{-1,q}(\Omega)^n} \ \leq\ 2C\,\|f\|_{W^{-1,q}(\Omega)^n} \tag{2.2.7}$$

for $\frac{1}{2} \leq \varepsilon < 1$.

Since C does not depend on ε, we are able to let $\varepsilon \to 1$. Choose $\frac{1}{2} \leq \varepsilon_j < 1$, $j \in \mathbb{N}$, with $\lim_{j\to\infty} \varepsilon_j = 1$, and set $p_j := p_{\varepsilon_j}$, $j \in \mathbb{N}$. The uniform boundedness in (2.2.7) shows the existence of a subsequence of $(p_j)_{j=1}^\infty$ which converges weakly in $L^q(\Omega)$ to some $p \in L^q(\Omega)$. We may assume that the sequence itself has this property. With $f_j := f_{\varepsilon_j}$ we get

$$\begin{aligned} < p,\ \text{div}\ v >_\Omega \ &= \ \lim_{j\to\infty} < p_j, \text{div}\ v >_\Omega \ = \ \lim_{j\to\infty}\,[-\nabla p_j, v]_\Omega \\ &= \ \lim_{j\to\infty}\,[-f_j, v]_\Omega = \lim_{j\to\infty}\,[-f, v_j]_\Omega \\ &= \ [-f, v]_\Omega \end{aligned}$$

for all $v \in C_0^\infty(\Omega)^n$, where $v_j := v_{\varepsilon_j}$ is defined as above by $v_{\varepsilon_j}(x) := v(\varepsilon_j^{-1}x)$, $x \in \Omega$. This shows that $f = \nabla p$ in the sense of distributions. The weak convergence of p_j to p yields that $\int_{\Omega_0} p\,dx = 0$. This proves the uniqueness property of p.

The weak convergence property shows, see Section 3.1 or the proof of Lemma 1.5.4, that

$$\|p\|_{L^q(\Omega)} \leq \liminf_{j\to\infty} \|p_j\|_{L^q(\Omega)} \leq 2C\|f\|_{W^{-1,q}(\Omega)^n}.$$

This proves the lemma for starlike domains.

The case of a general bounded Lipschitz domain Ω can be reduced to the case above by the following localization argument. Using the definition of a Lipschitz domain, we easily find bounded starlike subdomains $\Omega_1, \ldots, \Omega_m \subseteq \Omega$ such that

$$\Omega = \Omega_1 \cup \cdots \cup \Omega_m.$$

For $j = 1, \ldots, m$ let $f_j \in W^{-1,q}(\Omega_j)^n$ be the restriction of f to $W_0^{1,q'}(\Omega_j)^n$.

Consider first the case that $\overline{\Omega}_0 \subseteq \Omega$. Then from Lemma 2.2.1 we obtain a unique $p \in L^q_{loc}(\Omega)$ satisfying $f = \nabla p$, $\int_{\Omega_0} p\,dx = 0$. Since $\Omega_j \subseteq \Omega$ we get in particular that $\nabla p = f_j$, $j = 1, \ldots, m$, in the sense of distributions in Ω_j. On the other hand, the result above yields some $p_j \in L^q(\Omega_j)$ with $\nabla p_j = f_j$, $j \in \mathbb{N}$, which is uniquely determined up to a constant. Therefore we get $p + C_j = p_j$, $j = 1, \ldots, m$, where C_j is a constant. This proves that $p \in L^q(\Omega)$. If $\Omega_0 \subseteq \Omega$ is any subdomain, we choose a subdomain $\Omega_0' \subseteq \Omega$ with $\overline{\Omega'}_0 \subseteq \Omega$. This yields as above some $\tilde{p} \in L^q(\Omega)$ with $\nabla \tilde{p} = f$ and $\int_{\Omega_0'} \tilde{p}\,dx = 0$. Subtracting a constant from $\tilde{p}$ yields the desired $p \in L^q(\Omega)$ with $\nabla p = f$ and $\int_{\Omega_0} p\,dx = 0$. Since $p \in L^q(\Omega)$, the estimate (2.2.6) now follows from Lemma 1.5.4, (1.5.10). This completes the proof. □

The following density property is an important consequence of Lemma 2.2.2. Note that this property need not hold in unbounded domains, see [Hey76] for counter examples.

2.2.3 Lemma *Let $\Omega \subseteq \mathbb{R}^n$, $n \geq 2$, be a bounded Lipschitz domain, and let $1 < q < \infty$. Then $C_{0,\sigma}^\infty(\Omega) = \{v \in C_0^\infty(\Omega)^n; \operatorname{div} v = 0\}$ is dense in the space $N(\operatorname{div}) = \{v \in W_0^{1,q}(\Omega)^n; \operatorname{div} v = 0\}$ with respect to the norm $\|\cdot\|_{W^{1,q}(\Omega)^n} = \|\cdot\|_{1,q}$. Thus*

$$\overline{C_{0,\sigma}^\infty(\Omega)}^{\|\cdot\|_{1,q}} = N(\operatorname{div}). \tag{2.2.8}$$

Proof. We use a functional analytic argument. To prove (2.2.8), it suffices to show that each functional $f \in W^{-1,q'}(\Omega)^n$, $q' = \frac{q}{q-1}$, from the dual space

$W^{-1,q'}(\Omega)^n$ of $W_0^{1,q}(\Omega)^n$ which vanishes on $C_{0,\sigma}^\infty(\Omega)$ even vanishes on $N(\text{div})$. Then (2.2.8) must be valid, otherwise we would find by the Hahn-Banach theorem some $f \in W^{-1,q'}(\Omega)^n$ with $[f, v] = 0$ for all $v \in C_{0,\sigma}^\infty(\Omega)$ and $[f, v_0] \neq 0$ for some $v_0 \in N(\text{div})$.

Thus let $f \in W^{-1,q'}(\Omega)^n$ be given with $[f, v] = 0$, $v \in C_{0,\sigma}^\infty(\Omega)$. From Lemma 2.2.2 we see that $f = \nabla p$ with some $p \in L^{q'}(\Omega)$. It follows that

$$[f, v] = [\nabla p, v] = - < p, \text{ div } v > \tag{2.2.9}$$

for all $v \in C_0^\infty(\Omega)^n$. Since f is continuous in $\|\nabla v\|_q$, and since $p \in L^{q'}(\Omega)$, we conclude that (2.2.9) even holds for all $v \in W_0^{1,q}(\Omega)^n$. It follows that

$$[f, v] = - < p, \text{div } v > = 0 \ , \quad v \in N(\text{div}).$$

This proves the lemma. □

2.3 Regularity results on $\text{div}\, v = g$

Lemma 2.1.1 yields a solution $v \in W_0^{1,q}(\Omega)^n$ of the system

$$\text{div } v = g \ , \quad v|_{\partial\Omega} = 0 \tag{2.3.1}$$

for each given $g \in L^q(\Omega)$ with $\int_\Omega g\, dx = 0$. In the regularity theory of the Navier-Stokes equations we need solutions v of (2.3.1) with higher regularity properties if g is sufficiently smooth. The next lemma yields such a result. See [Bog80] or [Gal94a, III, 3] for a different approach to the regularity theory of (2.3.1).

2.3.1 Lemma *Let $\Omega \subseteq \mathbb{R}^n$, $n \geq 2$, be a bounded Lipschitz domain, and let $1 < q < \infty$, $k \in \mathbb{N}$. Then for each $g \in W_0^{k,q}(\Omega)$ with $\int_\Omega g\, dx = 0$, there exists at least one $v \in W_0^{k+1,q}(\Omega)^n$ satisfying*

$$\text{div } v = g \ , \quad \|v\|_{W^{k+1,q}(\Omega)^n} \leq C \|g\|_{W^{k,q}(\Omega)} \tag{2.3.2}$$

with some constant $C = C(q, k, \Omega) > 0$.

Proof. See [Gal94a, III, Theorem 3.2] for another proof. The result also holds for $k = 0$ and is contained in this case in Lemma 2.1.1, a). We use the same argument as for $k = 0$, now for $k \geq 1$. For $k = 0$ the proof rests on inequality (1.5.10) which follows from (1.1.6) by a compactness argument, see the proof of Lemma 1.5.4. The same argument can be used in the case $k \geq 1$. Instead of (1.1.6) we now use the corresponding inequality (1.1.8) for $k \geq 1$. The analogous compactness argument as in the proof of Lemma 1.5.4 yields instead of (1.5.10) the inequality

$$\|u\|_{W^{-k,q}(\Omega)/N(\nabla)} \leq C_1 \|\nabla u\|_{W^{-k-1,q}(\Omega)^n} \leq C_1 C_2 \|u\|_{W^{-k,q}(\Omega)} \tag{2.3.3}$$

for all $u \in W^{-k,q}(\Omega)$ with constants $C_1 = C_1(q,k,\Omega) > 0$, $C_2 = C_2(n,k) > 0$. $W^{-k,q}(\Omega)/N(\nabla)$ means the quotient space modulo the null space $N(\nabla)$, which consists of the constants. If $k = 0$, $W^{-k,q}(\Omega)/N(\nabla) = L^q(\Omega)/N(\nabla)$ can be identified with $L^q_0(\Omega) = \{u \in L^q(\Omega);\ \int_\Omega u\, dx = 0\}$.

The proof of Lemma 2.3.1 follows from (2.3.3) with q replaced by $q' = \frac{q}{q-1}$ by the same duality principle as in the proof of Lemma 2.1.1. It follows that the bounded linear operator

$$\operatorname{div}\ : v \mapsto \operatorname{div}\, v$$

from $W_0^{k+1,q}(\Omega)^n$ to $W_0^{k,q}(\Omega)$ has the closed range $W_0^{k,q}(\Omega) \cap L^q_0(\Omega)$. Therefore, the inverse operator div^{-1} from $W_0^{k,q}(\Omega) \cap L^q_0(\Omega)$ onto the quotient space $W_0^{k+1,q}(\Omega)^n/N(\operatorname{div})$, $N(\operatorname{div}) := \{v \in W_0^{k+1,q}(\Omega)^n;\ \operatorname{div}\, v = 0\}$, is bounded. This proves the existence of some $v \in W_0^{k+1,q}(\Omega)^n$ satisfying (2.3.2). The proof is complete. □

2.4 Further results on the equation $\operatorname{div} v = g$

Modifying the duality argument in the proof of Lemma 2.1.1 we find some other solution classes of this equation. Here we need the traces, see Section 1.2, II, and the exterior normal vector field N at the boundary $\partial\Omega$, see (3.4.7), I.

2.4.1 Lemma *Let $\Omega \subseteq \mathbb{R}^n$, $n \geq 2$, be a bounded Lipschitz domain with boundary $\partial\Omega$, and let $1 < q < \infty$. Then we have:*

a) *For each $g \in W^{-1,q}(\Omega)$ there exists at least one $v \in L^q(\Omega)^n$ satisfying*

$$\operatorname{div}\, v = g$$

in the sense of distributions, and

$$\|v\|_{L^q(\Omega)^n} \leq C\, \|g\|_{W^{-1,q}(\Omega)} \tag{2.4.1}$$

with some constant $C = C(q,\Omega) > 0$.

b) *For each $g \in L^q(\Omega)$ with $\int_\Omega g\, dx = 0$, there exists at least one $v \in L^q(\Omega)^n$ satisfying*

$$\operatorname{div}\, v\ =\ g$$

in the sense of distributions, $N \cdot v|_{\partial\Omega} = 0$ in the sense of generalized traces (1.2.24), and

$$\|v\|_{L^q(\Omega)^n} \leq C\, \|g\|_{L^q(\Omega)} \tag{2.4.2}$$

with some constant $C = C(q,\Omega) > 0$.

Proof. To prove a) we consider the operator

$$\text{div } : v \mapsto \text{ div } v$$

from $L^q(\Omega)^n$ to $W^{-1,q}(\Omega)$, and its dual operator $\text{div}' = -\nabla$,

$$-\nabla : p \mapsto \nabla p,$$

from $W_0^{1,q'}(\Omega)$ to $L^{q'}(\Omega)^n$, $q' = \frac{q}{q-1}$. We get

$$[p, \text{ div } v] \;= < -\nabla p, v >$$

for all $p \in W_0^{1,q'}(\Omega)$ and $v \in L^q(\Omega)^n$. From Poincaré's inequality (1.1.1) we see that $-\nabla$ has a closed range. Therefore, div has also a closed range which is the whole space $W^{-1,q}(\Omega)$, since $\{0\}$ is the null space of $-\nabla$; see the closed range theorem [Yos80].

The inverse operator div^{-1} from $W^{-1,q}(\Omega)$ to the quotient space $L^q(\Omega)^n/ N(\text{div})$, $N(\text{div}) := \{v \in L^q(\Omega)^n;\ \text{div } v = 0\}$, is therefore bounded. This yields a).

To prove b) we define the operator

$$\text{div } : v \mapsto \text{ div } v$$

with domain

$$D(\text{div}) := \{v \in L^q(\Omega)^n;\ \text{div } v \in L^q(\Omega),\ N \cdot v|_{\partial\Omega} = 0\} \;\subseteq\; L^q(\Omega)^n$$

and range $R(\text{div}) \subseteq L^q(\Omega)$. From Green's formula (1.2.25) we conclude that $\int_\Omega \text{div } v \, dx = 0$ for $v \in D(\text{div})$. To see this we set $u \equiv 1$ in (1.2.25). This yields $R(\text{div}) \subseteq L_0^q(\Omega) = \{g \in L^q(\Omega); \int_\Omega g \, dx = 0\}$. The trace $N \cdot v|_{\partial\Omega}$ is well defined since $D(\text{div}) \subseteq E_q(\Omega)$, see Lemma 1.2.2.

$D(\text{div})$ is dense in $L^q(\Omega)^n$ since $C_0^\infty(\Omega)^n \subseteq D(\text{div})$. We consider div as an operator from $D(\text{div})$ to $R(\text{div}) \subseteq L_0^q(\Omega)$.

$L_0^{q'}(\Omega)$ is the dual space of $L_0^q(\Omega)$, see (2.1.3). Next we define the operator

$$\nabla : \; p \mapsto \nabla p$$

with domain $D(\nabla) := \{p \in L_0^{q'}(\Omega);\ \nabla p \in L^{q'}(\Omega)^n\} \subseteq W^{1,q'}(\Omega)$ and range $R(\nabla) \subseteq L^{q'}(\Omega)^n$. It holds that $N(\nabla) = \{p \in L_0^{q'}(\Omega);\ \nabla p = 0\} = \{0\}$ since

$$\nabla p = 0 \ , \quad \int_\Omega p \, dx = 0$$

implies $p = 0$, see (1.7.18). Green's formula (1.2.25) yields

$$< p, \text{div } v > = \; - < \nabla p, v >$$

for all $p \in D(\nabla)$ and $v \in D(\text{div})$. This means, $-\nabla$ is the dual operator of div. Poincaré's inequality (1.1.2) implies that $R(-\nabla)$ is closed in $L^{q'}(\Omega)^n$. Therefore, $R(\text{div}) \subseteq L^q_0(\Omega)$ is closed too, and since $N(-\nabla) = \{0\}$, we conclude that $R(\text{div}) = L^q_0(\Omega)$ and that

$$\inf_{v_o \in N(div)} \|v + v_0\|_q \leq C\,\|\text{div}\; v\|_q$$

with $N(\text{div}) := \{v \in D(\text{div});\ \text{div}\; v = 0\}$, $C = C(q, \Omega) > 0$. Thus we may choose v in such a way that (2.4.2) is satisfied. This proves b). □

2.5 Helmholtz decomposition in L^2-spaces

In this subsection $\Omega \subseteq \mathbb{R}^n$ is an arbitrary domain with $n \geq 2$. We consider the Hilbert space $L^2(\Omega)^n$ with scalar product

$$< f, g >_\Omega = < f, g > = \int_\Omega f \cdot g\, dx,$$

the subspace

$$L^2_\sigma(\Omega) := \overline{C^\infty_{0,\sigma}(\Omega)}^{\|\cdot\|_2}\ ,\quad C^\infty_{0,\sigma}(\Omega) := \{f \in C^\infty_0(\Omega)^n;\ \text{div}\; f = 0\}, \tag{2.5.1}$$

and the space

$$G(\Omega) := \{f \in L^2(\Omega)^n;\ \exists\, p \in L^2_{loc}(\Omega) : f = \nabla p\}. \tag{2.5.2}$$

In other words, $L^2_\sigma(\Omega)$ is the closure of $C^\infty_{0,\sigma}(\Omega)$ in the norm $\|\cdot\|_2 = \|\cdot\|_{L^2(\Omega)^n}$, and $G(\Omega)$ is the space of those $f \in L^2(\Omega)^n$ for which there is some $p \in L^2_{loc}(\Omega)$ satisfying $f = \nabla p$ in the sense of distributions. "$\exists$" means "there exists".

The next lemma shows that $G(\Omega)$ is orthogonal to $L^2_\sigma(\Omega)$, we write

$$G(\Omega) = L^2_\sigma(\Omega)^\perp$$

for this property. This leads to the unique decomposition (2.5.4) of each $f \in L^2(\Omega)^n$ which is called the **Helmholtz decomposition** of f. In particular we see that $G(\Omega)$ is a closed subspace of $L^2(\Omega)^n$. See [Gal94a, III, 1], [FuM77], [SiZ98] concerning the Helmholtz decomposition in L^q-spaces with $1 < q < \infty$.

2.5.1 Lemma *Let $\Omega \subseteq \mathbb{R}^n, n \geq 2$, be any domain. Then*

$$G(\Omega) = \{f \in L^2(\Omega)^n;\ < f, v > = 0 \quad \textit{for all} \quad v \in L^2_\sigma(\Omega)\}, \tag{2.5.3}$$

and each $f \in L^2(\Omega)^n$ has a unique decomposition

$$f \;=\; f_0 + \nabla p \tag{2.5.4}$$

with $f_0 \in L^2_\sigma(\Omega)$, $\nabla p \in G(\Omega)$, $< f_0, \nabla p > = 0$,

$$\|f\|_2^2 = \|f_0\|_2^2 + \|\nabla p\|_2^2. \tag{2.5.5}$$

Remark As a consequence of this lemma we obtain a bounded linear operator $P : f \mapsto Pf$ from $L^2(\Omega)^n$ onto $L^2_\sigma(\Omega)$ defined by $Pf := f_0$ with f_0 as in (2.5.4). P is called the **Helmholtz projection** of $L^2(\Omega)^n$ onto $L^2_\sigma(\Omega)$.

2.5.2 Lemma *Let* $\Omega \subseteq \mathbb{R}^n$, $n \geq 2$, *be any domain, and let* $f = f_0 + \nabla p$ *be the Helmholtz decomposition of* $f \in L^2(\Omega)^n$. *Then*

$$P : L^2(\Omega)^n \to L^2_\sigma(\Omega)\ , \tag{2.5.6}$$

defined by $Pf := f_0$ *for all* $f \in L^2(\Omega)^n$, *is a bounded linear operator with operator norm* $\|P\| \leq 1$. *Thus*

$$\|Pf\|_2 \leq \|f\|_2\ ,\quad f \in L^2(\Omega)^n. \tag{2.5.7}$$

P has the following properties:

$$P(\nabla p) = 0\ ,\qquad (I - P)f = \nabla p\ ,\qquad P^2 f = Pf\ ,$$
$$(I-P)^2 f = (I-P)f,\quad < Pf, g > = < f, Pg >,\quad \|f\|_2^2 = \|Pf\|_2^2 + \|(I-P)f\|_2^2$$

for all $f, g \in L^2(\Omega)^n$.

From these properties we easily conclude that P is a selfadjoint operator, and that $P' = P$, where P' means the dual operator of P, see Section 3.2 for this notion.

Proof of Lemma 2.5.1. First we prove the characterization (2.5.3) of the subspace $G(\Omega)$ in (2.5.2). The space on the right side of (2.5.3) is by definition the orthogonal subspace of $L^2_\sigma(\Omega)$. Thus we have to show that

$$G(\Omega)\ =\ L^2_\sigma(\Omega)^\perp\ . \tag{2.5.8}$$

To prove (2.5.8) let $f \in L^2_\sigma(\Omega)^\perp$. Then for any bounded subdomain $\Omega_0 \subseteq \Omega$ with $\overline{\Omega}_0 \subseteq \Omega$ we get, using Poincaré's inequality (1.1.1), that

$$|< f, v >|\ \leq\ \|f\|_2\,\|v\|_{L^2(\Omega_0)^n}\ \leq\ C\,\|f\|_2\,\|\nabla v\|_{L^2(\Omega_0)^{n^2}}$$

for all $v \in C_0^\infty(\Omega_0)^n$ with $C = C(\Omega_0) > 0$. This shows that

$$f\ \in\ W^{-1,2}_{loc}(\Omega)^n.$$

Next we observe that $[f, v] = < f, v > = 0$ for all $v \in C^\infty_{0,\sigma}(\Omega)$. Lemma 2.2.1 yields some $p \in L^2_{loc}(\Omega)$, uniquely determined up to a constant, which satisfies $f = \nabla p$ in the sense of distributions. This shows that $f \in G(\Omega)$.

Conversely, let $f \in G(\Omega)$ with $f = \nabla p$, $p \in L^2_{loc}(\Omega)$. Then $< \nabla p, v > = - < p, \text{ div } v > = 0$ for all $v \in C^\infty_{0,\sigma}(\Omega)$, and since $\nabla p \in L^2(\Omega)^n$, this even holds for all $v \in L^2_\sigma(\Omega)$. This proves (2.5.8).

Using some elementary Hilbert space properties, see Section 3.2, we get the unique orthogonal decomposition $f = f_0 + \nabla p$ for each $f \in L^2(\Omega)^n$ with $f \in L^2_\sigma(\Omega)$, $\nabla p \in L^2_\sigma(\Omega)^\perp = G(\Omega)$; (2.5.5) is obvious. This proves the Lemma. □

Proof of Lemma 2.5.2. The Hilbert space theory yields a uniquely determined projection operator P from $L^2(\Omega)^n$ onto the subspace $L^2_\sigma(\Omega)$; the properties of P are obvious. This yields the lemma. □

For special domains we can improve the properties of the Helmholtz decomposition $f = f_0 + \nabla p$. In particular we are interested in bounded Lipschitz domains and in the case $\Omega = \mathbb{R}^n$. In these cases we can give special important characterizations of $L^2_\sigma(\Omega)$ and $G(\Omega)$.

In the following lemma, $N \cdot f|_{\partial\Omega}$ means the generalized trace, see (1.2.24), and N the exterior normal field at $\partial\Omega$, see (3.4.7), I. Note that the trace $N \cdot f|_{\partial\Omega}$ in (2.5.9) is well defined since $f \in E_2(\Omega)$, see (1.2.20).

2.5.3 Lemma *Let $\Omega \subseteq \mathbb{R}^n, n \geq 2$, be a bounded Lipschitz domain with boundary $\partial\Omega$. Then*

$$L^2_\sigma(\Omega) = \{f \in L^2(\Omega)^n;\ \text{div } f = 0,\ N \cdot f|_{\partial\Omega} = 0\} \tag{2.5.9}$$

and

$$G(\Omega) := \{f \in L^2(\Omega)^n;\ \exists\, p \in L^2(\Omega) : f = \nabla p\}. \tag{2.5.10}$$

Proof. In other words, $G(\Omega)$ is the space of all $f \in L^2(\Omega)^n$ for which there is some $p \in L^2(\Omega)$ with $f = \nabla p$ in the sense of distributions.

To prove (2.5.10), it suffices to show the following property:

$$p \in L^2_{loc}(\Omega),\ \nabla p \in L^2(\Omega)^n \text{ implies } p \in L^2(\Omega).$$

This is a consequence of Lemma 1.1.5, b). Thus we obtain (2.5.10).

To prove (2.5.9), let L be the space on the right side of (2.5.9). From $G(\Omega) = L^2_\sigma(\Omega)^\perp$ we get by an elementary Hilbert space argument that $G(\Omega)^\perp = L^2_\sigma(\Omega)^{\perp\perp} = L^2_\sigma(\Omega)$. Thus it remains to show that $L = G(\Omega)^\perp$.

To prove this let $f \in G(\Omega)^\perp$. By definition

$$G(\Omega)^\perp := \{f \in L^2(\Omega)^n;\ < f, \nabla p > = 0 \quad \text{for all } \nabla p \in G(\Omega)\},$$

and therefore we obtain in particular $< f, \nabla p > = 0$ for all $p \in C^\infty_0(\Omega)$. This means that div $f = 0$ in the sense of distributions. It follows that $f \in E_2(\Omega)$, see Lemma 1.2.2. Using (2.5.10) we get $< f, \nabla p > = 0$ for all $p \in W^{1,2}(\Omega)$. Green's formula (1.2.25) now yields that

$$0 = < p, \text{ div } f >_\Omega = < p, N \cdot f >_{\partial\Omega} - < \nabla p, f >_\Omega = < p, N \cdot f >_{\partial\Omega}$$

for all $p \in W^{1,2}(\Omega)$. This shows that $N \cdot f|_{\partial\Omega} = 0$ and therefore that $f \in L$. Thus we have $G(\Omega)^\perp \subseteq L$.

Conversely let $f \in L$. Then $f \in E_2(\Omega)$ and Green's formula (1.2.25) yields $< f, \nabla p >_\Omega = < \text{div} f, p >_\Omega = 0$ for all $\nabla p \in G(\Omega)$. This shows that $f \in G(\Omega)^\perp$. Therefore we get $L = G(\Omega)^\perp$ and (2.5.9) holds. The proof is complete. □

In the case $\Omega = \mathbb{R}^n$ we can prove the following characterization of the spaces $L^2_\sigma(\Omega)$ and $G(\Omega)$.

2.5.4 Lemma *Let $n \in \mathbb{N}$, $n \geq 2$. Then*

$$L^2_\sigma(\mathbb{R}^n) = \{f \in L^2(\mathbb{R}^n)^n;\ \text{div } f = 0\}, \tag{2.5.11}$$

and $G(\mathbb{R}^n)$ is the closure of the space

$$\nabla C_0^\infty(\mathbb{R}^n) := \{\nabla p\ ;\ p \in C_0^\infty(\mathbb{R}^n)\} \tag{2.5.12}$$

with respect to the norm $\|\cdot\|_{L^2(\mathbb{R}^n)^n}$. Thus

$$G(\mathbb{R}^n) = \overline{\nabla C_0^\infty(\mathbb{R}^n)}^{\|\cdot\|_2}. \tag{2.5.13}$$

Proof. First we prove (2.5.13). For this purpose we use the scaling method and the mollification method, see Section 1.7.

To prepare the scaling argument we consider a function $\varphi \in C_0^\infty(\mathbb{R}^n)$ with the properties

$$0 \leq \varphi \leq 1\ ,\quad \varphi(x) = 1 \text{ if } |x| \leq 1\ ,\quad \varphi(x) = 0 \text{ if } |x| \geq 2, \tag{2.5.14}$$

and define the functions

$$\varphi_j \in C_0^\infty(\mathbb{R}^n)\ ,\quad \varphi_j(x) := \varphi(j^{-1}x)\ ,\quad x \in \mathbb{R}^n,\ j \in \mathbb{N}. \tag{2.5.15}$$

It follows that $\lim_{j\to\infty} \varphi_j(x) = 1$ for all $x \in \mathbb{R}^n$, and setting

$$B_j := \{x \in \mathbb{R}^n;\ |x| < j\}\ ,\quad G_j := B_{2j} \backslash \overline{B}_j\ , \tag{2.5.16}$$

we get supp $\nabla\varphi_j \subseteq \overline{G}_j$, supp $\varphi_j \subseteq \overline{B}_{2j}$, $j \in \mathbb{N}$. See [SiSo96] for the method concerning φ.

To show (2.5.13) we consider any $\nabla p \in G(\mathbb{R}^n) = \{\nabla p \in L^2(\mathbb{R}^n)^n;\ p \in L^2_{loc}(\mathbb{R}^n)\}$ and choose constants K_j, $j \in \mathbb{N}$, such that

$$\int_{G_j} (p - K_j)\,dx = 0\ ,\quad j \in \mathbb{N}.$$

Applying Poincaré's inequality (1.1.2) to G_1, we get

$$\|p - K_1\|_{L^2(G_1)} \le C\,\|\nabla p\|_{L^2(G_1)^n} \tag{2.5.17}$$

with some constant $C > 0$. Using the transformation formula for integrals with $x = jy$, $dx = j^n dy$, we obtain

$$\begin{aligned}
\|p - K_j\|_{L^2(G_j)} &= \Big(\int_{G_j} |p(x) - K_j|^2 dx\Big)^{\frac{1}{2}} = \Big(\int_{G_1} |p(jy) - K_j|^2 dy\Big)^{\frac{1}{2}} j^{\frac{n}{2}} \\
&\le C j^{\frac{n}{2}} \Big(\int_{G_1} |\nabla_y p(jy)|^2 dy\Big)^{\frac{1}{2}} \\
&= C j^{\frac{n}{2}}\, j^{-\frac{n}{2}}\, j \Big(\int_{G_j} |\nabla p(x)|^2 dx\Big)^{\frac{1}{2}} \\
&= C j\,\|\nabla p\|_{L^2(G_j)^n}
\end{aligned}$$

with C as in (2.5.17) since

$$\int_{G_1} (p(jy) - K_j)\,dy = j^{-n} \int_{G_j} (p(x) - K_j)\,dx = 0.$$

Thus we get

$$\|p - K_j\|_{L^2(G_j)} \le jC\,\|\nabla p\|_{L^2(G_j)^n}\ ,\quad j \in \mathbb{N}. \tag{2.5.18}$$

Setting $p_j := \varphi_j(p - K_j)$ and using $\nabla p_j = (\nabla\varphi_j)(p - K_j) + \varphi_j \nabla(p - K_j) = (\nabla\varphi_j)(p - K_j) + \varphi_j \nabla p$, we obtain

$$\begin{aligned}
\|\nabla p - \nabla p_j\|_{L^2(\mathbb{R}^n)^n} &\le \|\nabla p - \varphi_j \nabla p\|_{L^2(\mathbb{R}^n)^n} + \|(\nabla\varphi_j)(p - K_j)\|_{L^2(\mathbb{R}^n)^n} \\
&\le \|\nabla p - \varphi_j \nabla p\|_{L^2(\mathbb{R}^n)^n} + \frac{C'}{j}\|p - K_j\|_{L^2(G_j)^n}
\end{aligned}$$

with $\nabla\varphi_j(x) = \nabla\varphi(j^{-1}x) = j^{-1}(\nabla\varphi)(j^{-1}x)$ and $C' := \sup_x |\nabla\varphi(x)|$.

Lebesgue's dominated convergence lemma, see [Apo74], yields

$$\begin{aligned}
&\lim_{j\to\infty} \|\nabla p - \varphi_j \nabla p\|_{L^2(\mathbb{R}^n)^n} \\
&\quad = \Big(\int_{\mathbb{R}^n} \big(\lim_{j\to\infty} |1 - \varphi_j(x)|^2\big)\,|\nabla p(x)|^2\,dx\Big)^{\frac{1}{2}} = 0,
\end{aligned} \tag{2.5.19}$$

since $|1-\varphi_j(x)| = |1-\varphi(j^{-1}x)| \leq 2$ and $\lim_{j\to\infty}|1-\varphi(j^{-1}x)| = 0$ for each $x \in \mathbb{R}^n$. Using (2.5.18) we get

$$\|\nabla p - \nabla p_j\|_{L^2(\mathbb{R}^n)^n} \leq \|\nabla p - \varphi_j \nabla p\|_{L^2(\mathbb{R}^n)^n} + C'C\|\nabla p\|_{L^2(G_j)^n}.$$

Together with

$$\lim_{j\to\infty} \|\nabla p\|_{L^2(G_j)^n} = \lim_{j\to\infty} (\int_{G_j} |\nabla p(x)|_2^2 \, dx)^{\frac{1}{2}} = 0$$

and (2.5.19) we conclude that

$$\lim_{j\to\infty} \|\nabla p - \nabla p_j\|_{L^2(\mathbb{R}^n)^n} = 0. \tag{2.5.20}$$

Next we use the mollification method, see Lemma 1.7.1. Since supp $p_j \subseteq \overline{B}_{2j}$ we can approximate each p_j by C_0^∞-functions in the gradient norm. Using the operator $\mathcal{F}_\varepsilon\star$, $\varepsilon > 0$, see (1.7.5), we find for each $j \in \mathbb{N}$ some $\varepsilon_j > 0$ such that

$$\|\nabla p_j - \mathcal{F}_{\varepsilon_j} \star \nabla p_j\|_{L^2(\mathbb{R}^n)^n} \leq \frac{1}{j}.$$

With $\nabla(\mathcal{F}_{\varepsilon_j} \star p_j) = \mathcal{F}_{\varepsilon_j} \star (\nabla p_j)$, see (1.7.17), we get

$$\|\nabla p_j - \nabla(\mathcal{F}_{\varepsilon_j} \star p_j)\|_{L^2(\mathbb{R}^n)^n} \leq \frac{1}{j} \tag{2.5.21}$$

for all $j \in \mathbb{N}$.

Setting $\tilde{p}_j := \mathcal{F}_{\varepsilon_j} \star p_j$ we see that $\tilde{p}_j \in C_0^\infty(\mathbb{R}^n)$, $j \in \mathbb{N}$, and combining (2.5.20) with (2.5.21) leads to

$$\lim_{j\to\infty} \|\nabla p - \nabla \tilde{p}_j\|_{L^2(\mathbb{R}^n)^n} = 0.$$

This proves (2.5.13).

To prove (2.5.11), let L be the space on the right side of (2.5.11). Recall, div $f = 0$ is understood in the sense of distributions. Since

$$L^2_\sigma(\mathbb{R}^n) = \overline{C^\infty_{0,\sigma}(\mathbb{R}^n)}^{\|\cdot\|_2} \subseteq L,$$

we only have to show that $L \subseteq L^2_\sigma(\mathbb{R}^n)$. For this purpose let $f \in L$. Then

$$< f, \nabla p > = -[\text{ div } f, p] = -< \text{ div } f, p > = 0 \tag{2.5.22}$$

for all $p \in C_0^\infty(\mathbb{R}^n)$. Since $f \in L^2(\mathbb{R}^n)^n$ and since the space of all ∇p with $p \in C_0^\infty(\mathbb{R}^n)$ is dense in $G(\mathbb{R}^n)$ in the norm $\|\cdot\|_2$, see (2.5.13), we see that

$< f, \nabla p > = 0$ holds as well for all $\nabla p \in G(\mathbb{R}^n)$. This means that $f \in G(\mathbb{R}^n)^\perp$, and we see that

$$f \in G(\mathbb{R}^n)^\perp = L^2_\sigma(\mathbb{R}^n)^{\perp\perp} = L^2_\sigma(\mathbb{R}^n).$$

Thus we get $f \in L^2_\sigma(\mathbb{R}^n)$ and $L \subseteq L^2_\sigma(\mathbb{R}^n)$ which proves (2.5.11). The proof of the lemma is complete. □

Finally we mention an important density property which follows by the same approximation argument as above.

2.5.5 Lemma *Let $n \in \mathbb{N}$, $n \geq 2$. Then*

$$\overline{C^\infty_{0,\sigma}(\mathbb{R}^n)}^{\|\cdot\|_{W^{1,2}(\mathbb{R}^n)^n}} = \{v \in W^{1,2}(\mathbb{R}^n)^n;\ \text{div } v = 0\}, \tag{2.5.23}$$

Thus $C^\infty_{0,\sigma}(\mathbb{R}^n) = \{v \in C^\infty_0(\mathbb{R}^n)^n;\ \text{div } v = 0\}$ is dense in the space on the right side of (2.5.23) with respect to the norm of $W^{1,2}(\mathbb{R}^n)^n$.

Proof. Recall that

$$W^{1,2}(\mathbb{R}^n)^n = W^{1,2}_0(\mathbb{R}^n)^n = \overline{C^\infty_0(\mathbb{R}^n)^n}^{\|\cdot\|_{W^{1,2}(\mathbb{R}^n)^n}}, \tag{2.5.24}$$

see (3.6.17), I.

To prove (2.5.23), let $v \in W^{1,2}_0(\mathbb{R}^n)^n = W^{1,2}(\mathbb{R}^n)^n$ with $\text{div } v = 0$. Then we have to construct some $v_j \in C^\infty_{0,\sigma}(\mathbb{R}^n)$, $j \in \mathbb{N}$, such that

$$\lim_{i\to\infty} \|v - v_j\|_{W^{1,2}(\mathbb{R}^n)^n} = 0. \tag{2.5.25}$$

For this purpose we use the same approximation method as in the last proof, and consider $\varphi_j, B_j, G_j, j \in \mathbb{N}$, as in (2.5.15), (2.5.16), $\mathcal{F}_{\varepsilon_j}$ as in (2.5.21). Then we construct some $w_j \in W^{1,2}_0(G_j)^n$, $j \in \mathbb{N}$, such that

$$\text{div } w_j = \text{div } (\varphi_j v) = (\nabla \varphi_j) \cdot v \tag{2.5.26}$$

and

$$\lim_{j\to\infty} \|w_j\|_{W^{1,2}(G_j)} = 0. \tag{2.5.27}$$

Assume for a moment that we already have such a sequence $(w_j)^\infty_{j=1}$. Then a similar argument as in (2.5.19) shows that

$$\lim_{j\to\infty} \|v - \varphi_j v\|_{W^{1,2}(\mathbb{R}^n)} = 0,$$

and setting $\tilde{v}_j := \varphi_j v - w_j$, $j \in \mathbb{N}$, we get $\text{div } \tilde{v}_j = 0$ and

$$\lim_{j\to\infty} \|v - \tilde{v}_j\|_{W^{1,2}(\mathbb{R}^n)} = 0.$$

A similar argument as in (2.5.21) leads to

$$\lim_{j\to\infty} \|\tilde{v}_j - \mathcal{F}_{\varepsilon_j} \star \tilde{v}_j\|_{W^{1,2}(\mathbb{R}^n)} = 0.$$

Then we set $v_j := \mathcal{F}_{\varepsilon_j} \star \tilde{v}_j$ and obtain

$$v_j \in C_0^\infty(\mathbb{R}^n)^n \ , \quad \text{div } v_j = \mathcal{F}_{\varepsilon_j} \star \ \text{div } \tilde{v}_j = 0 \ , \quad j \in \mathbb{N},$$

see (1.7.17), and (2.5.25) follows.

Thus it remains to construct the above sequence $(w_j)_{j=1}^\infty$. For this purpose we use Lemma 2.1.1, a). First we observe that

$$\int_{B_{2j}} \text{div } (\varphi_j v)\, dx \ = \ \int_{G_j} (\nabla\varphi_j) \cdot v \, dx = 0. \tag{2.5.28}$$

This follows from Green's formula (1.2.12) with $u \equiv 1$. Then we use the transformation $x = jy$, $x \in G_j$, $y \in G_1$, and setting $\tilde{w}_j(y) = w_j(jy) = w_j(x)$, we get from (2.5.26) the transformed equations

$$\text{div } \tilde{w}_j(y) \ = \ j(\nabla\varphi_j)(jy) \cdot v(jy) \ , \tag{2.5.29}$$

now in G_1 for all $j \in \mathbb{N}$. Using (2.5.28) we see that

$$\begin{aligned}\int_{G_1} (\text{div } \tilde{w}_j)(y)\, dy \ &= \ j \int_{G_1} (\nabla\varphi_j)(jy) \cdot v(jy)\, dy \\ &= \ jj^{-n} \int_{G_j} (\nabla\varphi_j)(x) \cdot v(x)\, dx \ = \ 0 \,,\end{aligned}$$

and Lemma 2.1.1, a), yields a solution $\tilde{w}_j \in W_0^{1,2}(G_1)^n$ satisfying

$$\|\nabla\tilde{w}_j\|_{L^2(G_1)} \ \le \ C \ (\int_{G_1} |j(\nabla\varphi_j)(jy) \cdot v(jy)|^2\, dy)^{\frac{1}{2}}$$

for all $j \in \mathbb{N}$ with some fixed $C = C(G_1) > 0$. Then $w_j \in W_0^{1,2}(G_j)^n$ defined by $w_j(x) = \tilde{w}_j(y)$, $x = jy$, is a solution of (2.5.26), and we get

$$\begin{aligned}\|\nabla w_j\|_{L^2(G_j)} \ &= \ \left(\int_{G_j} |\nabla w_j(x)|^2\, dx\right)^{\frac{1}{2}} = j^{-1} j^{\frac{n}{2}} \left(\int_{G_1} |(\nabla\tilde{w}_j)(y)|^2\, dy\right)^{\frac{1}{2}} \\ &= \ j^{-1} j^{\frac{n}{2}} \|\nabla\tilde{w}_j\|_{L^2(G_1)} \le C\, j^{\frac{n}{2}} \left(\int_{G_1} |(\nabla\varphi_j)(jy) \cdot v(jy)|^2\, dy\right)^{\frac{1}{2}} \\ &= \ C \left(\int_{G_j} |(\nabla\varphi_j)(x) \cdot v(x)|^2\, dx\right)^{\frac{1}{2}}\end{aligned}$$

$$
\begin{aligned}
&= j^{-1} C \left(\int_{G_j} |(\nabla \varphi)(j^{-1}x) \cdot v(x)|^2 \, dx \right)^{\frac{1}{2}} \\
&\le j^{-1} C_1 \|v\|_{L^2(G_j)}
\end{aligned}
$$

for all $j \in \mathbb{N}$, with some $C_1 = C_1(G_1) > 0$ not depending on j.

Then with Poincaré's inequality for G_1 we obtain

$$
\begin{aligned}
\|w_j\|_{L^2(G_j)} &= \left(\int_{G_j} |w_j(x)|^2 \, dx \right)^{\frac{1}{2}} = j^{\frac{n}{2}} \left(\int_{G_1} |\tilde{w}_j(y)|^2 \, dy \right)^{\frac{1}{2}} \\
&\le C_2 \, j^{\frac{n}{2}} \left(\int_{G_1} |\nabla_y \tilde{w}_j(y)|^2 \, dy \right)^{\frac{1}{2}} \\
&= C_2 \, j \left(\int_{G_j} |\nabla w_j(x)|^2 \, dx \right)^{\frac{1}{2}} \\
&= C_2 \, j \, \|\nabla w_j\|_{L^2(G_j)} \\
&\le C_2 \, C_1 \, \|v\|_{L^2(G_j)}
\end{aligned}
$$

with some $C_2 = C_2(G_1) > 0$ and C_1 as above.

Since obviously

$$
\lim_{j \to \infty} \|v\|_{L^2(G_j)^n} = 0 \,,
$$

we conclude from these estimates that (2.5.27) is satisfied. This completes the proof. □

3 Elementary functional analytic properties

3.1 Basic facts on Banach spaces

For the convenience of the reader, and in order to fix notations, we collect some elementary facts on Banach spaces and in particular on Hilbert spaces. We mainly refer to [Yos80], [HiPh57], [Heu75].

Let X be a (real) **Banach space** with norm $\|v\|_X = \|v\|$, $v \in X$. By definition, the **dual space** X' of X is the Banach space of all linear continuous functionals

$$
f : v \mapsto [f, v] \ , \quad v \in X
$$

with norm

$$
\|f\|_{X'} := \sup_{0 \ne v \in X} (|[f, v]| / \|v\|_X).
$$

Sometimes we write $f = [f, \cdot]$; $[f, v]$ always means the value of the functional f at v.

A linear functional $f : v \mapsto [f, v]$, $v \in X$, is continuous iff there is a constant $C = C(f) > 0$ such that

$$|[f, v]| \leq C\|v\|_X \quad \text{for all } v \in X. \tag{3.1.1}$$

It holds that $\|f\|_{X'} = \inf C(f)$, which is the infimum over all such constants $C(f)$ for fixed f. Therefore, if (3.1.1) holds with any $C = C(f)$, then

$$\|f\|_{X'} \leq C. \tag{3.1.2}$$

A sequence $(v_j)_{j=1}^{\infty}$ in X converges **strongly** to some $v \in X$ iff

$$\lim_{j \to \infty} \|v - v_j\| = 0;$$

we write $v = s - \lim_{j\to\infty} v_j$ in this case. The sequence $(v_j)_{j=1}^{\infty}$ in X converges **weakly** to $v \in X$ iff

$$\lim_{j \to \infty} [f, v_j] = [f, v]$$

for all $f \in X'$; we write $v = w - \lim_{j\to\infty} v_j$ in this case.

X is **reflexive** iff each linear continuous functional on X' has the form $f \mapsto [f, v]$, $f \in X'$, with some fixed $v \in X$. We write $[\cdot, v]$ for this functional. Usually we identify each $v \in X$ with the functional $[\cdot, v]$. Then X can be identified with $(X')' = X''$ and we write $X'' = X$ if X is reflexive.

If X is reflexive, each bounded sequence $(v_j)_{j=1}^{\infty}$ in X contains a subsequence which converges weakly to some $v \in X$. For simplicity we will always assume that the sequence itself has this property. In this case

$$\|v\| \leq \liminf_{j \to \infty} \|v_j\| \leq \sup_j \|v_j\|. \tag{3.1.3}$$

Let $D \subseteq X$ be any subspace of X and let $\overline{D}$ denote the closure of D in the norm $\|\cdot\|$. D is called **dense** in X iff $\overline{D} = X$. We also write $\overline{D}^{\|\cdot\|} = X$ in this case.

Consider two Banach spaces X and $\mathcal{Y}$ with norms $\|\cdot\|_X$, $\|\cdot\|_{\mathcal{Y}}$, respectively. Let

$$B : v \mapsto Bv \ , \quad v \in D(B)$$

be any linear operator with **domain** $D(B) \subseteq X$ and **range** $R(B) := \{Bv;\ v \in D(B)\} \subseteq \mathcal{Y}$. $N(B) := \{v \in D(B);\ Bv = 0\}$ means the **null space** of B, and

$G(B) := \{(v, Bv);\ v \in D(B)\} \subseteq X \times \mathcal{Y}$ means the **graph** of B. If $\overline{D(B)} = X$, B is called densely defined. The norm

$$\|v\|_{D(B)} := \|v\|_X + \|Bv\|_{\mathcal{Y}} \ , \quad v \in D(B) \tag{3.1.4}$$

is called the **graph norm** of $D(B)$. B is called **closed** if the graph $G(B)$ is closed in $X \times \mathcal{Y}$ with respect to the norm $\|v\|_X + \|w\|_{\mathcal{Y}}$, $(v, w) \in X \times \mathcal{Y}$. If B is closed, $D(B)$ is a Banach space in the graph norm $\|\cdot\|_{D(B)}$.

Let $N(B) = \{0\}$. Then B is injective and

$$\|v\|_{\widehat{D}(B)} := \|Bv\|_{\mathcal{Y}} \ , \quad v \in D(B) \tag{3.1.5}$$

is called the **homogeneous graph norm** of $D(B)$. Even if B is closed, $D(B)$ need not be a Banach space in this norm. The **completion** $\widehat{D}(B)$ of $D(B)$ consists of all (classes of) Cauchy sequences $(v_j)_{j=1}^\infty$ in $D(B)$ with respect to this norm.

Let $v = (v_j)_{j=1}^\infty$ be any element of $\widehat{D}(B)$. Then, by definition, $(Bv)_{j=1}^\infty$ is a Cauchy sequence in $R(B) \subseteq \mathcal{Y}$. Setting

$$Bv := s - \lim_{j\to\infty} Bv_j \ , \quad v \in \widehat{D}(B) \tag{3.1.6}$$

we get a (well defined) linear operator from $\widehat{D}(B)$ to $\mathcal{Y}$ which is an extension of the given operator $v \mapsto Bv$, $v \in D(B)$. This extension is called the **closure extension** of B from $D(B)$ to $\widehat{D}(B)$, we simply use the same notation B for this extension. Note that $\widehat{D}(B) \supseteq D(B) \subseteq X$, but $\widehat{D}(B)$ need not be a subspace of X.

Let $B : D(B) \to \mathcal{Y}$, $D(B) \subseteq X$, be a densely defined closed operator. Then the **dual operator** $B' : f \mapsto B'f$ with domain $D(B') \subseteq \mathcal{Y}'$ and range $R(B') \subseteq X'$ is well defined by the following property:

It holds that $[f, Bv] = [B'f, v]$ for all $f \in D(B')$, $v \in D(B)$, and B' is maximal with this property (that is, $D(B')$ is the totality of all $f \in \mathcal{Y}'$ such that $v \mapsto [f, Bv]$, $v \in D(B)$, is continuous in $\|v\|_X$).

If one of the spaces $R(B)$, $R(B')$ is closed, then both are closed and $R(B) = \{w \in \mathcal{Y};\ [f, w] = 0 \text{ for all } f \in N(B')\}$, $R(B') = \{g \in X';\ [g, v] = 0 \text{ for all } v \in N(B)\}$; see the closed range theorem [Yos80, VII, 5]. If $R(B)$ is closed, then there is a constant $C > 0$ with

$$\|Bv\|_{\mathcal{Y}} \geq C \|\,[v]\,\|_{X/N(B)} \tag{3.1.7}$$

for all $v \in D(B)$, where

$$\|\,[v]\,\|_{X/N(B)} := \inf_{v_0 \in N(B)} \|v + v_0\|_X$$

means the quotient norm of $[v] = v + N(B)$; see [Yos80, I, 11] and the closed graph theorem [Yos80, II, 6, Theorem 1].

Let X and $\mathcal{Y}$ be reflexive Banach spaces and let $B : v \mapsto Bv,\ v \in D(B)$, be a closed linear operator with dense domain $D(B) \subseteq X$ and range $R(B) \subseteq \mathcal{Y}$. Suppose $(v_j)_{j=1}^{\infty}$ is a sequence in $D(B)$ with the following property:

$$\begin{aligned} &(v_j)_{j=1}^{\infty} \text{ converges weakly in } X \text{ to some } v \in X, \\ &\text{and } \sup_j \|Bv_j\|_{\mathcal{Y}} < \infty. \end{aligned} \tag{3.1.8}$$

Then $v \in D(B)$ and we get the estimate

$$\|Bv\|_{\mathcal{Y}} \le \lim_{j\to\infty} \inf \|Bv_j\|_{\mathcal{Y}} \le \sup_j \|Bv_j\|_{\mathcal{Y}}. \tag{3.1.9}$$

The proof of (3.1.9) rests on the following facts, see [Yos80, V, 1]. The pairs $(v_j, Bv_j),\ j \in \mathbb{N}$, yield a bounded sequence with respect to the graph norm (3.1.4), and the graph $G(B)$ is a reflexive Banach space with this norm. Therefore we get a subsequence which converges weakly in $G(B)$ to some element $(\tilde{v}, B\tilde{v}) \in G(B)$, and we may assume that the sequence itself has this property. Since $(v_j)_{j=1}^{\infty}$ converges to $v \in X$ weakly, we get $\tilde{v} = v,\ B\tilde{v} = Bv$ and $v \in D(B)$; (3.1.9) now follows from (3.1.3).

Let $B : v \mapsto Bv$ be any closed linear operator with dense domain $D(B) \subseteq X$ and range $R(B) \subseteq \mathcal{Y}$, and suppose that $N(B) = \{0\}$. This means that B is injective. Then the inverse operator $B^{-1} : D(B^{-1}) \to X$ with domain $D(B^{-1}) = R(B) \subseteq \mathcal{Y}$ and range $R(B^{-1}) = D(B) \subseteq X$, is well defined by $B^{-1}Bv = v$ for all $v \in D(B)$. B^{-1} is a closed operator.

Suppose $B : v \mapsto Bv$ is a bounded linear operator from X to $\mathcal{Y}$. Thus $D(B) = X$, and

$$\|B\| := \sup_{0\neq v\in X} (\|Bv\|_{\mathcal{Y}}/\|v\|_X) < \infty.$$

Then $\|B\|$ is called the norm of B. B is called **compact** iff for each bounded sequence $(v_j)_{j=1}^{\infty}$ in X, the sequence $(Bv_j)_{j=1}^{\infty}$ contains a subsequence which converges strongly in $\mathcal{Y}$ to some element of $\mathcal{Y}$.

Finally we consider an operator $B : X \to X$ which is only a mapping and need not be linear. B is called **completely continuous** iff

$$\left.\begin{array}{l} B \text{ is continuous and for each bounded sequence } (v_j)_{j=1}^{\infty} \text{ in } X, \\ \text{the sequence } (Bv_j)_{j=1}^{\infty} \text{ contains a subsequence which} \\ \text{converges strongly to some element of } X. \end{array}\right\} \tag{3.1.10}$$

We need the following result.

3.1.1 Lemma (Leray-Schauder principle) *Let X be a Banach space and let $B : X \to X$ be a completely continuous operator. Assume there exists some $r > 0$ with the following property:*

$$\text{If } \ v \in X \ , \ 0 \le \lambda \le 1 \ , \ v = \lambda B v, \ \text{ then } \ \|v\|_X \le r. \tag{3.1.11}$$

Then there exists at least one $v \in X$ with $v = Bv$, $\|v\|_X \le r$.

Proof. See [LeSch34], [Lad69, Chap. 1, Sec. 3], [Zei76, 6.5, Theorem 6.1]. □

3.2 Basic facts on Hilbert spaces

Here we mainly refer to [Yos80], [Kat66], [ReSi75], [Heu75] and [Wei76]. Let H be a (real) **Hilbert space** with scalar product $< u, v >_H = < u, v >$ and norm $\|u\|_H = \|u\| = < u, u >^{\frac{1}{2}}$, $u, v \in H$. Then H' denotes the dual space of all continuous linear functionals defined on H.

The Riesz representation theorem, see [Yos80, III, 6], shows that each element of H' has the form

$$v \mapsto < u, v > \ , \quad v \in H$$

with some fixed $u \in H$. As usual, this functional $< u, \cdot >$ will be identified with u, and we therefore obtain that $H' = H$.

Let $B : v \mapsto Bv$ be a closed linear operator with dense domain $D(B) \subseteq H$ and range $R(B) \subseteq H$. Then the **dual (adjoint)** operator B' with (dense) domain $D(B') \subseteq H$ and range $R(B') \subseteq H$ is determined by the property

$$< u, Bv > = < B'u, v > \quad \text{for all } v \in D(B) \ , \quad u \in D(B'), \tag{3.2.1}$$

and $D(B')$ is the totality of all $u \in H$ such that the functional $v \mapsto < u, Bv >$, $v \in D(B)$, is continuous in $\|v\|_H$.

If $B = B'$, that is if $D(B) = D(B')$ and $Bv = B'v$ for all $v \in D(B)$, B is called a **selfadjoint** operator. A selfadjoint operator B is called **positive** if $< v, Bv > \ \ge 0$ for all $v \in D(B)$.

If $N(B) = \{v \in D(B);\ Bv = 0\} = \{0\}$, B is injective and we define the inverse operator $B^{-1} : D(B^{-1}) \to H$ by $D(B^{-1}) = R(B)$, $R(B^{-1}) = D(B)$, $B^{-1}Bv = v$ for all $v \in D(B)$. If B is positive selfadjoint, B^{-1} is also positive selfadjoint. See [Yos80, VII, 3] concerning these facts.

B is bounded iff $D(B) = H$ and there exists some $C = C(B) > 0$ such that

$$\|Bv\| \le C \, \|v\| \quad \text{for all } \ v \in H. \tag{3.2.2}$$

The operator norm $\|B\|$ is the infimum of all $C(B)$ with (3.2.2). Thus

$$\|B\| \ \le \ C \tag{3.2.3}$$

for all $C = C(B) > 0$ with (3.2.2).

Let $D \subseteq H$ be any closed subspace of H. Then

$$D^{\perp} := \{u \in H;\ < u, v > = 0 \ \text{ for all } v \in D\} \tag{3.2.4}$$

is called the **orthogonal subspace** of D. Each $u \in H$ has a unique decomposition $u = u_1 + u_2$ with $u_1 \in D,\ u_2 \in D^{\perp}$.

The operator $P : u \mapsto Pu$, defined by $Pu := u_1$ for all $u \in H$, is called the **projection** of H onto D. P is a positive selfadjoint operator with $P^2 = P$ and operator norm $\|P\| \leq 1$.

Let I denote the identity. If P is the projection of H onto D, then $I - P$ is the projection onto $D^{\perp}$, and

$$\|u\|^2 = \|Pu\|^2 + \|(I-P)u\|^2 \quad \text{for all } \ u \in H. \tag{3.2.5}$$

Let $D \subseteq H$ be a dense subspace, and let $S(u,v) \in \mathbb{R}$ be defined for all $u, v \in D$ with the following properties:

$$\begin{aligned} &v \mapsto S(u,v),\ v \in D, \text{ is a linear functional for each } u \in D \\ &S(u,v) = S(v,u) \ \text{ and } \ S(u,u) \geq 0 \text{ for all } u, v \in D. \end{aligned}$$

Then $S : (u,v) \mapsto S(u,v)$ is called a **positive symmetric bilinear form** with dense domain $D = D(S) \subseteq H$.

By

$$< u, v > + S(u,v) \ , \quad u, v \in D, \tag{3.2.6}$$

we obtain a scalar product and by

$$(\|u\|^2 + S(u,u))^{\frac{1}{2}} \ , \quad u \in D, \tag{3.2.7}$$

we get the corresponding norm in D. S is called **closed** if D is complete with respect to this norm. This means that D is a Hilbert space with the scalar product (3.2.6). We need the following result:

3.2.1 Lemma *Let H be a Hilbert space with scalar product $< \cdot, \cdot >$ and norm $\| \cdot \|$, and let $S : (u,v) \mapsto S(u,v)$ be a closed positive symmetric bilinear form with dense domain $D = D(S) \subseteq H$.*

Then there exists a uniquely determined positive selfadjoint operator $B : D(B) \to H$ with dense domain $D(B) \subseteq D$, satisfying:

$$\left.\begin{aligned} &D(B) \text{ is the totality of all } \ u \in D \text{ such that the} \\ &\text{functional } v \mapsto S(u,v),\ v \in D, \text{ is continuous in } \|v\|, \\ &\text{and } S(u,v) = < Bu, v > \ \text{ for all } u \in D(B),\ v \in D. \end{aligned}\right\} \tag{3.2.8}$$

Proof. See [Kat84, VI, Theorem 2.6] or [Wei76, Satz 5.37]. The proof rests on the Riesz representation theorem, applied to the scalar product (3.2.6). □

We need this lemma in order to define the Stokes operator A for arbitrary domains $\Omega \subseteq \mathbb{R}^n$, $n \geq 2$.

Next we mention some facts on the spectral representation of selfadjoint operators, see [Yos80, XI, 5-7 and 12], [Kat84, Chap. V], [Wei76, 7.2]. Here we only need the special case of positive selfadjoint operators.

For each $\lambda \in [0, \infty)$, let E_λ be a projection operator which projects H onto a subspace $D_\lambda \subseteq H$. We call $\{E_\lambda;\ \lambda \geq 0\}$ a family of projections. Let $0 \leq \lambda_0 \leq \infty$. Then we write

$$E_{\lambda_0} = s - \lim_{\lambda \to \lambda_0} E_\lambda \tag{3.2.9}$$

iff $E_{\lambda_0} v = s - \lim_{\lambda \to \lambda_0} E_\lambda v$ holds for all $v \in H$ (strong convergence of operators).

Suppose $\{E_\lambda;\ \lambda \geq 0\}$ has the following properties:

a) $E_\lambda E_\mu = E_\mu E_\lambda = E_\lambda \ ,\quad 0 \leq \lambda \leq \mu < \infty$

b) $E_\lambda = s - \lim_{\mu \to \lambda} E_\mu \ ,\quad 0 < \mu < \lambda < \infty$

c) $E_0 = 0 \ ,\quad s - \lim_{\lambda \to \infty} E_\lambda = I$.

Then $\{E_\lambda; \lambda \geq 0\}$ is called a **resolution of the identity** I on $[0, \infty)$. Condition a) means that E_λ and E_μ commute and that $D_\lambda \subseteq D_\mu$ for $\lambda \leq \mu$. It follows that $E_\mu - E_\lambda$, $\lambda \leq \mu$, is again a projection operator, and that $\lambda \mapsto \|E_\lambda v\|^2$ is monotonously increasing for each $v \in H$. Condition b) means that $\lambda \mapsto E_\lambda$ is left continuous in the interval $(0, \infty)$ with respect to the strong convergence of operators. $E_0 = 0$ means zero as an operator, and the last condition means that $\lim_{\lambda \to \infty} \|v - E_\lambda v\| = 0$ for all $v \in H$.

For each continuous function $g : \lambda \mapsto g(\lambda)$, $\lambda \geq 0$, we can define the usual Stieltjes integral

$$\int_0^b g(\lambda)\, d\|E_\lambda v\|^2 \ ,\quad v \in H \ ,\quad 0 < b < \infty$$

as a limit of Riemann-Stieltjes sums of the form

$$\sum_{j=1}^m g(\lambda_j)\,(\|E_{\lambda_j} v\|^2 - \|E_{\lambda_{j-1}} v\|^2) = \sum_{j=1}^m g(\lambda_j)\, \|(E_{\lambda_j} - E_{\lambda_{j-1}}) v\|^2$$

where $0 = \lambda_0 < \lambda_1 < \cdots < \lambda_m = b$, $\max |\lambda_j - \lambda_{j-1}| \to 0$, see [Apo74, 7.3].

If $g(\lambda) \geq 0$ for all $\lambda \geq 0$, and if

$$\int_0^\infty g(\lambda)\, d\|E_\lambda v\|^2 = \lim_{b \to \infty} \int_0^b g(\lambda)\, d\|E_\lambda v\|^2$$

exists for some $v \in H$, we simply write $\int_0^\infty g(\lambda)\, d\|E_\lambda v\|^2 < \infty$.

Let $g : \lambda \mapsto g(\lambda),\ \lambda \geq 0$, be a continuous real function. Then the integral

$$\int_0^b g(\lambda)\, dE_\lambda v \ \in H \quad , \quad 0 < b < \infty \ , \ v \in H$$

is well defined as the strong limit of the usual Riemann sums of the form $\sum_{j=1}^m g(\lambda_j)\,(E_{\lambda_j} - E_{\lambda_{j-1}})v$, $0 = \lambda_0 < \lambda_1 < \cdots < \lambda_m = b$, and

$$\|\int_0^b g(\lambda)\, dE_\lambda v\|^2 = \int_0^b g^2(\lambda)\, d\|E_\lambda v\|^2.$$

If $\int_0^\infty g^2(\lambda)\, d\|E_\lambda v\|^2 < \infty$ for some $v \in H$, then the integral

$$\int_0^\infty g(\lambda)\, dE_\lambda v := s - \lim_{b\to\infty} \int_0^b g(\lambda)\, dE_\lambda v$$

exists. We thus obtain a well defined operator

$$\int_0^\infty g(\lambda)\, dE_\lambda \ : \ v \mapsto \int_0^\infty g(\lambda)\, dE_\lambda v \tag{3.2.10}$$

which is selfadjoint and has the dense domain

$$D\left(\int_0^\infty g(\lambda)\, dE_\lambda\right) := \{v \in H;\ \int_0^\infty g^2(\lambda) d\|E_\lambda v\|^2 < \infty\}. \tag{3.2.11}$$

We see that

$$\|\int_0^\infty g(\lambda)\, dE_\lambda v\|^2 = \int_0^\infty g^2(\lambda) d\|E_\lambda v\|^2 \tag{3.2.12}$$

and that

$$< \left(\int_0^\infty g(\lambda)\, dE_\lambda\right) v, v > = \int_0^\infty g(\lambda) d\|E_\lambda v\|^2 \tag{3.2.13}$$

for all $v \in D(\int_0^\infty g(\lambda)\, dE_\lambda)$. In particular for all $v \in H$ we get

$$v = \int_0^\infty dE_\lambda v \ , \quad \|v\|^2 = \int_0^\infty d\|E_\lambda v\|^2. \tag{3.2.14}$$

If $g(\lambda) \geq 0$ for all $\lambda \geq 0$, then with (3.2.13) we see that $\int_0^\infty g(\lambda)\, dE_\lambda$ is positive selfadjoint, and if

$$\sup_{\lambda \geq 0} |g(\lambda)| < \infty\,,$$

we conclude from (3.2.11) and (3.2.12), that $\int_0^\infty g(\lambda) dE_\lambda$ is a bounded operator with $D(\int_0^\infty g(\lambda)\, dE_\lambda) = H$ and operator norm

$$\| \int_0^\infty g(\lambda)\, dE_\lambda \| \le \sup_{\lambda \ge 0} |g(\lambda)|. \tag{3.2.15}$$

In particular,

$$\int_0^\infty \lambda\, dE_\lambda \quad \text{with } D\left(\int_0^\infty \lambda\, dE_\lambda\right) = \{v \in H;\ \int_0^\infty \lambda^2\, d\|E_\lambda v\|^2 < \infty\} \tag{3.2.16}$$

is a positive selfadjoint operator.

Let now $B : D(B) \to H$ be any positive selfadjoint operator with (dense) domain $D(B) \subseteq H$. Then there exists a uniquely determined resolution

$$\{E_\lambda;\ \lambda \ge 0\}$$

of identity such that

$$B = \int_0^\infty \lambda\, dE_\lambda\ ,\quad D(B) = \{v \in H;\ \int_0^\infty \lambda^2\, d\|E_\lambda v\|^2 < \infty\}. \tag{3.2.17}$$

This is called the **spectral representation** of B; see [Yos80, XI, 5], [Kat66, VI, 5.1].

For each continuous real function $g : [0, \infty) \to \mathbb{R}$, we define as above the selfadjoint operator

$$g(B) := \int_0^\infty g(\lambda)\, dE_\lambda \tag{3.2.18}$$

with domain

$$D(g(B)) = \{v \in H;\ \int_0^\infty g^2(\lambda)\, d\|E_\lambda v\|^2 < \infty\}.$$

If $\sup_{\lambda \ge 0} |g(\lambda)| < \infty$, $g(B)$ is bounded with $D(g(B)) = H$, and we see that

$$v \in D(B) \quad \text{implies} \quad g(B)v \in D(B) \quad \text{and} \quad Bg(B)v = g(B)Bv. \tag{3.2.19}$$

This property means that $g(B)$ **commutes** with B; see [Yos80, XI, 12]. Then

$$Bg(B)v = \int_0^\infty \lambda g(\lambda)\, dE_\lambda v \quad \text{for all } v \in D(B). \tag{3.2.20}$$

In particular we define the **fractional powers**

$$B^\alpha := \int_0^\infty \lambda^\alpha\, dE_\lambda\ ,\quad D(B^\alpha) := \{v \in H;\ \int_0^\infty \lambda^{2\alpha} d\|E_\lambda v\|^2 < \infty\} \tag{3.2.21}$$

for all $\alpha \ge 0$. It holds that $B^\alpha = I$ for $\alpha = 0$.

For all $\mu > 0$, we consider the **resolvent**

$$(\mu I + B)^{-1} = \int_0^\infty (\mu + \lambda)^{-1}\, dE_\lambda, \tag{3.2.22}$$

which is the inverse of $\mu I + B$. This operator is bounded with norm

$$\|(\mu I + B)^{-1}\| \le \sup_{\lambda \ge 0} (\mu + \lambda)^{-1} \le \mu^{-1}. \tag{3.2.23}$$

If there is a $\delta > 0$ with $E_\lambda = 0$ for $0 \le \lambda \le \delta$, then B is obviously invertible and has the bounded inverse operator

$$B^{-1} = \int_\delta^\infty \lambda^{-1}\, dE_\lambda \tag{3.2.24}$$

with $\|B^{-1}\| \le \sup_{\lambda \ge \delta} \lambda^{-1}$.

Let $N(B) = \{v \in D(B); Bv = 0\}$ be the null space of B and let P_0 be the projection operator from H onto $N(B)$. Then we conclude that

$$P_0 = s - \lim_{\lambda \to 0} E_\lambda \ , \quad \lambda > 0, \tag{3.2.25}$$

holds in the strong sense. This means that $N(B) = \bigcap_{\lambda > 0} D_\lambda$.
Therefore, the jump of $\lambda \mapsto E_\lambda$ at $\lambda = 0$ determines the null space $N(B)$ of B. B is injective, i.e., $N(B) = \{0\}$, iff $\lambda \mapsto E_\lambda$ is right continuous at $\lambda = 0$ with respect to the strong convergence.

Let now $N(B) = \{0\}$. Then for each $v \in H$ the function $\lambda \mapsto \|E_\lambda v\|^2, \lambda \ge 0$, is right continuous at $\lambda = 0$. This enables us to obtain an integral representation of the inverse operator

$$B^{-1} : D(B^{-1}) \to H \ , \quad D(B^{-1}) = R(B),$$

although $\lambda \mapsto \lambda^{-1}$ is not a continuous function defined on the whole interval $[0, \infty)$ as in (3.2.18). We obtain (with $\delta > 0$) the representation

$$B^{-1} v = \int_0^\infty \lambda^{-1}\, dE_\lambda v = s - \lim_{\delta \to 0} \int_\delta^\infty \lambda^{-1}\, dE_\lambda v \ , \quad v \in D(B^{-1}), \tag{3.2.26}$$

B^{-1} is positive selfadjoint, and

$$D(B^{-1}) = \{v \in H;\ \|B^{-1} v\|^2 = \int_0^\infty \lambda^{-2}\, d\|E_\lambda v\|^2 < \infty\}. \tag{3.2.27}$$

More generally, in the case $N(B) = \{0\}$ we can define the operator $B^{-\alpha} : D(B^{-\alpha}) \to H$ for $\alpha \ge 0$ by

$$B^{-\alpha} v = \int_0^\infty \lambda^{-\alpha}\, dE_\lambda v := s - \lim_{\delta \to 0} \int_\delta^\infty \lambda^{-\alpha}\, dE_\lambda v \ , \quad v \in D(B^{-\alpha}) \tag{3.2.28}$$

with domain

$$D(B^{-\alpha}) = \{v \in H;\ \|B^{-\alpha}v\|^2 = \int_0^\infty \lambda^{-2\alpha} d\|E_\lambda v\|^2 < \infty\}. \tag{3.2.29}$$

Then $N(B) = \{0\}$ implies $N(B^\alpha) = \{0\}$, $D(B^{-\alpha}) \subseteq H$ is dense, $B^{-\alpha}$ is positive selfadjoint, and

$$B^{-\alpha} = (B^{-1})^\alpha = (B^\alpha)^{-1}.$$

Thus $B^{-\alpha}$ is the inverse operator of B^α, and therefore we get $D(B^\alpha) = R(B^{-\alpha})$ and $D(B^{-\alpha}) = R(B^\alpha)$. If $0 \le \alpha \le 1$ we obtain

$$D(B) \subseteq D(B^\alpha)\ ,\quad D(B^{-1}) \subseteq D(B^{-\alpha}). \tag{3.2.30}$$

These properties follow from the integral representations above.

Next we assume that the given positive selfadjoint operator B is defined by the form S with domain $D(S)$ as in Lemma 3.2.1. In this case we get

$$\begin{aligned} S(u,u) &= \ < Bu, u > \ = \ < B^{\frac{1}{2}}u, B^{\frac{1}{2}}u > \ = \ \|B^{\frac{1}{2}}u\|^2 \\ &= \int_0^\infty \lambda\, d\|E_\lambda u\|^2 \end{aligned}$$

for all $u \in D(B)$. Then a closure argument shows that

$$D(B^{\frac{1}{2}}) = D(S)\ ,\quad S(u,u) = \|B^{\frac{1}{2}}u\|^2 \ \text{ for all } \ u \in D(S). \tag{3.2.31}$$

We conclude from the spectral representation $B = \int_0^\infty \lambda\, dE_\lambda$ that $Bu = 0$ holds for $u \in D(B)$ iff $S(u,u) = 0$. Therefore,

$$N(B) = \{0\} \ \text{ iff } \ \{u \in D(S);\ S(u,u) = 0\} = \{0\}. \tag{3.2.32}$$

This means that B is injective iff $S(u,u) = 0$ implies that $u = 0$.

The next lemma yields the **interpolation inequality** for fractional powers.

3.2.2 Lemma *Let $B: D(B) \to H$, $D(B) \subseteq H$, be a positive selfadjoint operator in the Hilbert space H, and let $0 \le \alpha \le 1$. Then*

$$\|B^\alpha v\| \ \le\ \|Bv\|^\alpha \|v\|^{1-\alpha} \ \le\ \alpha\|Bv\| + (1-\alpha)\|v\| \tag{3.2.33}$$

for all $v \in D(B)$.

Proof. Using the spectral representation and Hölder's inequality, see [Yos80, I, 3, (5)], we obtain

$$\begin{aligned}\|B^\alpha v\|^2 &= \int_0^\infty \lambda^{2\alpha}\, d\|E_\lambda v\|^2 \\ &\le \Big(\int_0^\infty \lambda^2\, d\|E_\lambda v\|^2\Big)^\alpha \Big(\int_0^\infty d\|E_\lambda v\|^2\Big)^{1-\alpha} \\ &= \|Bv\|^{2\alpha}\, \|v\|^{2(1-\alpha)},\end{aligned}$$

and apply Young's inequality (3.3.8), I. This proves the lemma. □

Finally we need a special result on fractional powers which is due to Heinz [Hei51].

3.2.3 Lemma (Heinz) *Let H_1, H_2 be two Hilbert spaces with norms $\|\cdot\|_1, \|\cdot\|_2$, respectively. Let $B : H_1 \to H_2$ be a bounded linear operator from H_1 into H_2 with operator norm $\|B\|$, and let*

$$A_1 : D(A_1) \to H_1 \ , \quad A_2 : D(A_2) \to H_2$$

be positive selfadjoint injective operators with domains $D(A_1) \subseteq H_1$, $D(A_2) \subseteq H_2$. Suppose B maps $D(A_1)$ into $D(A_2)$ and

$$\|A_2 Bv\|_2 \le C\, \|A_1 v\|_1 \quad \textit{for all} \ \ v \in D(A_1) \tag{3.2.34}$$

with some constant $C > 0$.

Then for $0 \le \alpha \le 1$, B maps $D(A_1^\alpha)$ into $D(A_2^\alpha)$, and the inequality

$$\|A_2^\alpha Bv\|_2 \le C^\alpha\, \|B\|^{1-\alpha} \|A_1^\alpha v\|_1 \tag{3.2.35}$$

holds for all $v \in D(A_1^\alpha)$.

Proof. See [Hei51] or [Tan79, Theorem 2.3.3], [Kre71, Chap. I, Theorem 7.1]. Inequality (3.2.35) is called the **Heinz inequality**. □

3.3 The Laplace operator Δ

After discussing the operators div and ∇, see Section 2, the Laplacian

$$\Delta = \operatorname{div} \nabla = D_1^2 + \cdots + D_n^2$$

is the next important operator which occurs in the Navier-Stokes equations (1.1.1), I. The purpose of this subsection is to consider some basic facts on Δ mainly for the whole space $\mathbb{R}^n$, $n \ge 1$. These are potential theoretic properties.

We need the Riesz potential and the Bessel potential. For the proofs we refer to [Ste70], [Tri78], [Ada75], [SiSo96].

First let $\Omega \subseteq \mathbb{R}^n$, $n \geq 1$, be an arbitrary domain. We consider the Hilbert space $L^2(\Omega)$ with scalar product

$$< u, v > = < u, v >_\Omega := \int_\Omega uv\, dx\,,$$

norm $\|u\|_{L^2(\Omega)} = \|u\|_2 = \|u\|_{2,\Omega} =< u, u >^{\frac{1}{2}}$, and define the bilinear form S with domain $D(S) \subseteq L^2(\Omega)$ by setting

$$D(S) := W_0^{1,2}(\Omega)\ ,\quad S(u,v) := < \nabla u, \nabla v > := \int_\Omega (\nabla u)\cdot(\nabla v)\, dx \qquad (3.3.1)$$

for $u, v \in D(S)$. Recall that $< \nabla u, \nabla v >= \sum_{j=1}^n \int_\Omega (D_j u)(D_j v)\, dx$. Since $W_0^{1,2}(\Omega)$ is complete with respect to the norm

$$(\|u\|_2^2 + S(u,u))^{\frac{1}{2}} = (\|u\|_2^2 + \|\nabla u\|_2^2)^{\frac{1}{2}}\ , \qquad (3.3.2)$$

the form S is closed. S is obviously symmetric and positive. Therefore, by Lemma 3.2.1 we obtain a positive selfadjoint operator $B : D(B) \to L^2(\Omega)$ with dense domain $D(B) \subseteq W_0^{1,2}(\Omega)$ satisfying the relation

$$< \nabla u, \nabla v > = < Bu, v > \quad \text{for all } u \in D(B)\,,\ v \in W_0^{1,2}(\Omega).$$

Setting $v \in C_0^\infty(\Omega)$, we see that

$$Bu = -\Delta u = -\text{ div } \nabla u$$

holds in the sense of distributions. Therefore we set $B = -\Delta$. Thus the operator

$$-\Delta : D(-\Delta) \to L^2(\Omega)$$

is defined by

$$D(-\Delta) = \{u \in W_0^{1,2}(\Omega);\ v \mapsto < \nabla u, \nabla v > \ \text{ is continuous in } \|v\|_2\} \qquad (3.3.3)$$

and by

$$< (-\Delta)u, v > = < \nabla u, \nabla v > \quad \text{for } u \in D(-\Delta)\,,\quad v \in W_0^{1,2}(\Omega). \qquad (3.3.4)$$

Obviously $\nabla u = 0$ implies $u = 0$ for all $u \in W_0^{1,2}(\Omega)$. Therefore, see (3.2.21) and (3.2.28), the fractional powers

$$(-\Delta)^{\frac{\alpha}{2}} = \int_0^\infty \lambda^{\frac{\alpha}{2}}\, dE_\lambda\,, \qquad (3.3.5)$$

with domain

$$D((-\Delta)^{\frac{\alpha}{2}}) = \{v \in L^2(\Omega);\ \int_0^\infty \lambda^\alpha \, d\|E_\lambda v\|_2^2 < \infty\},$$

are well defined for all $\alpha \in \mathbb{R}$. Here $\{E_\lambda; \lambda \geq 0\}$ denotes the resolution of identity for $-\Delta$, see Section 3.2.

An equivalent characterization is

$$D(-\Delta) = D(\Delta) = \{u \in W_0^{1,2}(\Omega);\ \Delta u \in L^2(\Omega)\} \tag{3.3.6}$$

with $\Delta u \in L^2(\Omega)$ in the sense of distributions.

Consider now the case $\Omega = \mathbb{R}^n$, $n \geq 1$. Then we have $W_0^{1,2}(\mathbb{R}^n) = W^{1,2}(\mathbb{R}^n)$, see (3.6.17), I. In this case there exists an explicit characterization of the spectral representation (3.3.5) which is obtained by using the **Fourier transform** $\mathcal{F}$. $\mathcal{F}$ is defined by

$$(\mathcal{F}u)(y) := \int_{\mathbb{R}^n} e^{-2\pi i x \cdot y} \, u(x) \, dx \quad , \quad y \in \mathbb{R}^n,$$

in the sense of distributions, see [Yos80, VI, 1], [Ste70, III, 1.2], [Tri78, 2.2.1]. For this purpose we have to work for the moment in the corresponding complex function spaces. This requires us to use complexifications of the real function spaces.

Then a calculation shows, see [Ste70, Chap. V, 1.1, (4)], that u and $(-\Delta)^{\frac{\alpha}{2}} u$ satisfy the integral equation

$$u(x) = \frac{1}{\gamma(\alpha, n)} \int_{\mathbb{R}^n} |x - y|^{-n+\alpha} (-\Delta)^{\frac{\alpha}{2}} u(y) \, dy \quad , \quad x \in \mathbb{R}^n \tag{3.3.7}$$

for $0 < \alpha < n$, where $\gamma(\alpha, n) := \pi^{\frac{n}{2}} 2^\alpha \Gamma(\frac{\alpha}{2}) / \Gamma(\frac{n}{2} - \frac{\alpha}{2})$. Γ means the Gamma function. The expression (3.3.7) is called the **Riesz potential**; it can be directly estimated by the Hardy-Littlewood theorem, see [Tri78, 1.18.8, Theorem 3]. The result is the following lemma.

3.3.1 Lemma *Let $n \in \mathbb{N}$, $0 < \alpha < n$, $2 \leq q < \infty$,*

$$\alpha + \frac{n}{q} = \frac{n}{2} \,, \tag{3.3.8}$$

and suppose that $u \in D((-\Delta)^{\frac{\alpha}{2}})$. Then $u \in L^q(\mathbb{R}^n)$ and

$$\|u\|_{L^q(\mathbb{R}^n)} \leq C \, \|(-\Delta)^{\frac{\alpha}{2}} u\|_{L^2(\mathbb{R}^n)} \tag{3.3.9}$$

with some constant $C = C(\alpha, n) > 0$.

Proof. See [Ste70, Chap. V, 1.2, Theorem 1]. It is shown that in this case the integral (3.3.7) converges absolutely for almost all $x \in \mathbb{R}^n$, the Hardy-Littlewood theorem, see also [Tri78, 1.18.8], yields the result. □

The following lemma concerns the special case $n = 1$. In this case, we write $(-\Delta)^{\frac{\alpha}{2}} u = f$, $u = (-\Delta)^{-\frac{\alpha}{2}} f$, and we are mainly interested in the estimate (3.3.9). Now we admit that $f \in L^r(\mathbb{R})$ with $1 < r < \infty$. The following result rests again on the Hardy-Littlewood theorem.

3.3.2 Lemma *Let* $0 < \alpha < 1$, $1 < r < q < \infty$ *with*

$$\alpha + \frac{1}{q} = \frac{1}{r}, \tag{3.3.10}$$

and suppose $f \in L^r(\mathbb{R})$. *Then the integral*

$$u(t) := \int_{\mathbb{R}} |t - \tau|^{\alpha - 1} f(\tau)\, d\tau$$

converges absolutely for almost all $t \in \mathbb{R}$, *and*

$$\|u\|_{L^q(\mathbb{R})} \ \le\ C\, \|f\|_{L^r(\mathbb{R})} \tag{3.3.11}$$

with some constant $C = C(\alpha, q) > 0$.

Proof. See [Ste70, Chap. V, 1.2] or [Tri78, 1.18.9, Theorem 3]. □

Next we consider the positive selfadjoint operator $I - \Delta$ with domain $D(I - \Delta) = D(\Delta)$. We can define $I - \Delta$ also directly by using the form

$$< u, v > \ + \ < \nabla u, \nabla v > \tag{3.3.12}$$

instead of (3.3.1), see Lemma 3.2.1.

In this case u and $(I - \Delta)^{\frac{\alpha}{2}} u$ satisfy for $\alpha \ge 0$ the integral equation

$$u(x) = \int_{\mathbb{R}^n} G_\alpha(x - y)((I - \Delta)^{\frac{\alpha}{2}} u)(y)\, dy \ , \quad x \in \mathbb{R}^n, \tag{3.3.13}$$

where G_α is defined by

$$G_\alpha(z) := (4\pi)^{-\frac{\alpha}{2}}\, \Gamma(\alpha/2)^{-1} \int_0^\infty e^{-\pi |z|^2/t}\, e^{-t/4\pi}\, t^{-1 + (-n+\alpha)/2}\, dt, \tag{3.3.14}$$

$z \in \mathbb{R}^n$, see [Ste70, Chap. V, 3, (26)]. The expression (3.3.13) is called the **Bessel potential**. There are similar estimates as for the Riesz potential (3.3.7). We only need the following special case.

3.3.3 Lemma *Let* $n \in \mathbb{N}$, $1 \le \alpha \le 2$, $2 \le q < \infty$, *with*

$$\alpha + \frac{n}{q} = 1 + \frac{n}{2}, \tag{3.3.15}$$

and suppose that $u \in D((I - \Delta)^{\frac{\alpha}{2}})$. *Then* $u \in W^{1,q}(\mathbb{R}^n)$ *and*

$$\|u\|_{W^{1,q}(\mathbb{R}^n)} \ \le\ C\, \|(I - \Delta)^{\frac{\alpha}{2}} u\|_{L^2(\mathbb{R}^n)} \tag{3.3.16}$$

with some constant $C = C(\alpha, n) > 0$.

Proof. A direct proof follows using [Ste70, Chap. V, (29), (30)] and the Hardy-Littlewood estimate [Tri78, 1.18.8, Theorem 3] in the same way as before. It is based on the estimate of the potential (3.3.13). Another proof rests on the following argument. First we use [Ste70, V, 3, Theorem 3] or [Tri78, 2.3.3, (2)], [Ada75, Theorem 7.63, (f)] in order to show that the norms

$$\|u\|_{W^{1,q}(\mathbb{R}^n)} \quad \text{and} \quad \|(I-\Delta)^{\frac{\alpha}{2}} u\|_{L^q(\mathbb{R}^n)} \tag{3.3.17}$$

are equivalent. Then we use the embedding inequality

$$\|(I-\Delta)^{\frac{1}{2}} u\|_{L^q(\mathbb{R}^n)} \le C\,\|(I-\Delta)^{\frac{\alpha}{2}} u\|_{L^2(\mathbb{R}^n)} \tag{3.3.18}$$

with q, α as in (3.3.15); this follows from [Ada75, Theorem 7.63, (d)] or [Tri78, 2.8.1, Remark 2]. See also [Tri78, 2.8.1, (15)]. This yields the result. □

3.4 Resolvent and Yosida approximation

In the theory of the Navier-Stokes equations the Yosida approximation is used for technical reasons as a "smoothing" procedure which approximates L^2- functions by more regular functions. See [Ama95, II.6.1] concerning general properties, and see [Soh83], [Soh84], [MiSo88] concerning applications to the Navier-Stokes equations.

Let H be a Hilbert space and let $B : D(B) \to H$ be a positive selfadjoint operator as in (3.2.17). Then we consider the resolvent

$$(\mu I + B)^{-1} = \int_0^\infty (\mu+\lambda)^{-1}\, dE_\lambda \;, \quad \mu > 0 \tag{3.4.1}$$

as defined in (3.2.22). The relation

$$\begin{aligned}(\mu I + B)^{-1}(\mu I + B)v &= (\mu I + B)(\mu I + B)^{-1}v \\ &= \int_0^\infty (\mu+\lambda)(\mu+\lambda)^{-1}\, dE_\lambda v \\ &= \int_0^\infty dE_\lambda v = v\end{aligned}$$

holds for all $v \in D(B)$. For each $k \in \mathbb{N}$ we define the operator

$$J_k = J_{k,B} := (I + k^{-1}B)^{-1} = k(kI+B)^{-1} = \int_0^\infty (1 + k^{-1}\lambda)^{-1} dE_\lambda \,. \tag{3.4.2}$$

This representation shows that

$$J_k v \in D(B) \quad \text{for all} \;\; v \in H \;,\;\; k \in \mathbb{N}, \tag{3.4.3}$$

and that

$$BJ_k = \int_0^\infty \lambda(1+k^{-1}\lambda)^{-1}\,dE_\lambda \tag{3.4.4}$$

is a bounded operator with operator norm

$$\|BJ_k\| \le \sup_{\lambda\ge 0} |\lambda(1+k^{-1}\lambda)^{-1}| \le k, \tag{3.4.5}$$

see (3.2.15). In the same way we get

$$\|J_k\| \le \sup_{\lambda\ge 0} |(1+k^{-1}\lambda)^{-1}| \le 1. \tag{3.4.6}$$

The operators J_k, $k \in \mathbb{N}$, are called the **Yosida approximation** of the identity I. We have the following result; see [Yos80, IX, 9 and 12] or (in a slightly modified formulation) the proof of [Fri69, Part 2, Theorem 1.2] for more details.

3.4.1 Lemma *Let H be a Hilbert space and let $B : D(B) \to H$ be a positive selfadjoint operator with (dense) domain $D(B) \subseteq H$. Let J_k, $k \in \mathbb{N}$, be defined by* (3.4.2).

Then we have:

$$\left.\begin{array}{l} J_k v \in D(B) \textit{ for all } v \in H,\ BJ_k \textit{ is bounded with } (3.4.5), \\ BJ_k v = J_k B v \textit{ for all } v \in D(B),\ J_k \textit{ is bounded with } (3.4.6), \end{array}\right\} \tag{3.4.7}$$

and

$$v = s - \lim_{k\to\infty} J_k v \quad \textit{for all } v \in H, \tag{3.4.8}$$

$$Bv = s - \lim_{k\to\infty} BJ_k v \quad \textit{for all } v \in D(B). \tag{3.4.9}$$

Proof. The properties (3.4.7) immediately follow from the spectral representation (3.4.1), see Section 3.2.

The property (3.4.8) means that $\lim_{k\to\infty}\|v - J_k v\| = 0$. To prove this we use (3.2.12), get

$$\begin{aligned} \|v - J_k v\|^2 &= \|(I - J_k)v\|^2 = \|\int_0^\infty (1-(1+k^{-1}\lambda)^{-1})\,dE_\lambda v\|^2 \\ &= \int_0^\infty (1-(1+k^{-1}\lambda)^{-1})^2\,d\|E_\lambda v\|^2, \end{aligned}$$

$(1-(1+k^{-1}\lambda)^{-1})^2 \le 1$, and obtain

$$\lim_{k\to\infty}(1-(1+k^{-1}\lambda)^{-1})^2 = \lim_{k\to\infty}(\frac{\lambda}{k+\lambda})^2 = 0$$

for all $\lambda \geq 0$. Then we use Lebesgue's dominated convergence theorem [Apo74], and see that

$$\lim_{k\to\infty} \|v - J_k v\|^2 = \int_0^\infty \lim_{k\to\infty} \left(\frac{\lambda}{k+\lambda}\right)^2 d\|E_\lambda v\|^2 = 0.$$

Let $v \in D(B)$. Then $BJ_k v = J_k Bv$, and from above we get

$$\lim_{k\to\infty} \|Bv - BJ_k v\|^2 = \lim_{k\to\infty} \|(I - J_k)Bv\|^2 = 0.$$

This proves the lemma. □

Chapter III

The Stationary Navier-Stokes Equations

1 Weak solutions of the Stokes equations

1.1 The notion of weak solutions

Let $\Omega \subseteq \mathbb{R}^n$, $n \geq 2$, be any domain with boundary $\partial\Omega$. Our purpose is to investigate the **Stokes system**

$$
\begin{aligned}
-\nu\Delta u + \nabla p &= f \quad , \ \mathrm{div}\ u = 0\,, \\
u|_{\partial\Omega} &= 0.
\end{aligned}
\tag{1.1.1}
$$

Recall that $f = (f_1, \ldots, f_n)$ means the given exterior force, $u = (u_1, \ldots, u_n)$ the unknown velocity field, and p the unknown pressure; $\nu > 0$ is the given viscosity constant. We have to treat existence, uniqueness and regularity of solutions u, p of this system. Here we mainly refer to [Gal94a], [Gal94b], [Hey76], [Hey80], [Sol77]. For more information, see, e.g., [Catt61], [Fin65], [Kom67], [Lad69], [SoS73], [Tem77], [Gig86], [GiRa86], [SoV88], [GiSo89], [GSi90], [GiSo91], [GiSe91], [FSS93], [GSS94], [Var94], [Gal98], [SSp98], [NST99].

First we introduce the concept of weak solutions. The idea is the following: It seems to be rather difficult to prove directly the existence of classical regular solutions. Therefore we argue indirectly. In the first step we get rid of the pressure p and construct a so-called weak solution using a Hilbert space argument. In the second step we construct the pressure p and prove regularity properties of u, p under smoothness assumptions on f and Ω.

The appropriate solution space for weak solutions u is the completion

$$
\widehat{W}^{1,2}_{0,\sigma}(\Omega) := \overline{C^\infty_{0,\sigma}(\Omega)}^{\|\nabla v\|_2}
\tag{1.1.2}
$$

of $C^\infty_{0,\sigma}(\Omega) = \{v \in C^\infty_0(\Omega)^n;\ \mathrm{div}\ v = 0\}$ with respect to the norm $\|\nabla v\|_2$, where

$$
\|\nabla v\|_2 = \Big(\int_\Omega |\nabla v|^2\, dx\Big)^{\frac{1}{2}} \ , \quad \nabla v = (D_j v_l)^n_{j,l=1} \ , \quad v = (v_l)^n_{l=1}.
$$

By definition, $\widehat{W}_{0,\sigma}^{1,2}(\Omega)$ is the space of all (classes of) Cauchy sequences $(u_j)_{j=1}^{\infty}$ in $C_{0,\sigma}^{\infty}(\Omega)$ with respect to the norm $\|\nabla v\|_2$. For each such sequence, $(\nabla u_j)_{j=1}^{\infty}$ is a Cauchy sequence with respect to $\|\cdot\|_{L^2(\Omega)}$ and determines uniquely the gradient

$$\nabla u = s - \lim_{j\to\infty} \nabla u_j \in L^2(\Omega)^{n^2}.$$

We will show that $(u_j)_{j=1}^{\infty}$ itself converges to a uniquely determined $u \in L^q(\Omega)^n$ if $n \geq 3$, $q = \frac{2n}{n-2}$, and to a unique $u \in L^2_{loc}(\overline{\Omega})^2$ if $n = 2$, $\overline{\Omega} \neq \mathbb{R}^2$; see Lemma 1.2.1 below. Thus in these cases we can identify each (abstract) element of the space $\widehat{W}_{0,\sigma}^{1,2}(\Omega)$ with such a well defined function u, and we can write $u \in \widehat{W}_{0,\sigma}^{1,2}(\Omega)$. This yields the well defined continuous embeddings

$$\widehat{W}_{0,\sigma}^{1,2}(\Omega) \subseteq L^q(\Omega) \text{ if } n \geq 3, \quad \widehat{W}_{0,\sigma}^{1,2}(\Omega) \subseteq L^2_{loc}(\Omega)^2 \text{ if } n = 2, \quad \overline{\Omega} \neq \mathbb{R}^2,$$

see the next subsection. In these cases, each $u \in \widehat{W}_{0,\sigma}^{1,2}(\Omega)$ yields in particular a well defined distribution. In the remaining cases, the gradient $\nabla u \in L^2(\Omega)^{n^2}$ is always well defined but $u \in L^2_{loc}(\overline{\Omega})^n$ itself may be determined only up to a constant and need not yield a well defined distribution. An example is the case $\Omega = \mathbb{R}^2$, see [Gal94a, II.5], [DLi55] and [HLi56] for a general discussion of this problem.

We omit here a further discussion of the critical case $\overline{\Omega} = \mathbb{R}^2$. The definition below is meaningful in all cases, since ∇u, Δu and div u are well defined distributions in all cases.

In all cases, $\widehat{W}_{0,\sigma}^{1,2}(\Omega)$ becomes a Hilbert space with scalar product

$$< \nabla u, \nabla v >_{\Omega} = < \nabla u, \nabla v > = \int_{\Omega} (\nabla u) \cdot (\nabla v)\, dx$$

and norm $\|\nabla u\|_2$.

The most general exterior force $f = (f_1, \dots, f_n)$ we consider here will be a distribution of the form

$$f = f_0 + \text{div } F \quad \text{with } f_0 \in L^2_{loc}(\Omega)^n \text{ and } F \in L^2(\Omega)^{n^2}$$

defined by

$$\begin{aligned} [f, v]_{\Omega} = [f, v] &= [f_0, v] + [\text{div} F, v] \\ &= < f_0, v > - < F, \nabla v > \\ &= \int_{\Omega} f_0 \cdot v\, dx - \int_{\Omega} F \cdot \nabla v\, dx, \end{aligned}$$

for all $v \in C_0^{\infty}(\Omega)^n$. Recall, $F = (F_{jl})_{j,l=1}^{n}$ is a matrix field, and div $F :=$ $(D_1 F_{1l} + \cdots + D_n F_{nl})_{l=1}^{n}$. This means, div applies to the columns of F and yields the vector field div F. Further recall that $f_0 = (f_{01}, \dots, f_{0n})$, $v = (v_1, \dots, v_n)$, $f_0 \cdot v = f_{01}v_1 + \cdots + f_{0n}v_n$, and $F \cdot \nabla v = \sum_{j,l=1}^n F_{jl} D_j v_l$.

The existence of a weak solution $u \in \widehat{W}_{0,\sigma}^{1,2}(\Omega)$ will be shown only in the case $f =$ div F, $f_0 = 0$. Section 1.6, II, yields sufficient conditions for the property $f =$ div F. The existence proof simply rests on a Hilbert space argument in the space $\widehat{W}_{0,\sigma}^{1,2}(\Omega)$. For each such $f =$ div F we will get a unique $u \in \widehat{W}_{0,\sigma}^{1,2}(\Omega)$ satisfying

$$\nu < \nabla u, \nabla v > = [f, v] = - < F, \nabla v >$$

for all $v \in C_{0,\sigma}^\infty(\Omega)$.

The pressure part ∇p in the Stokes system (1.1.1) will be constructed in a second step. Starting with such $u \in \widehat{W}_{0,\sigma}^{1,2}(\Omega)$, we will consider the functional

$$v \mapsto [f, v] - \nu < \nabla u, \nabla v > \ , \quad v \in C_0^\infty(\Omega)^n \tag{1.1.3}$$

which is an element of the space $W_{loc}^{-1,2}(\Omega)^n$. Now we can apply Lemma 2.2.1, II, and obtain some $p \in L_{loc}^2(\Omega)$, uniquely determined up to a constant, which satisfies (1.1.1) together with u in the sense of distributions. Conversely, if $u \in \widehat{W}_{0,\sigma}^{1,2}(\Omega)$ and $p \in L_{loc}^2(\Omega)$ satisfy (1.1.1) in the sense of distributions, then u is a weak solution.

The condition $u \in \widehat{W}_{0,\sigma}^{1,2}(\Omega)$ also contains some information concerning the boundary condition $u|_{\partial\Omega} = 0$ in a very weak sense. Consider for example a bounded Lipschitz domain $\Omega \subseteq \mathbb{R}^n$. In this case we can use Poincaré's inequality and obtain the characterization

$$\widehat{W}_{0,\sigma}^{1,2}(\Omega) = \{u \in W^{1,2}(\Omega)^n; \text{ div } u = 0, \ u|_{\partial\Omega} = 0\},$$

see Lemma 1.2.1 below. This shows, $u \in \widehat{W}_{0,\sigma}^{1,2}(\Omega)$ implies $u|_{\partial\Omega} = 0$ in the sense of traces. See (1.2.1), II, concerning traces.

1.1.1 Definition *Let $\Omega \subseteq \mathbb{R}^n$, $n \geq 2$, be any domain, and let $f = f_0 +$ div F with*

$$f_0 \in L_{loc}^2(\Omega)^n \ , \quad F \in L^2(\Omega)^{n^2}. \tag{1.1.4}$$

Then $u \in \widehat{W}_{0,\sigma}^{1,2}(\Omega)$ is called a **weak solution** *of the Stokes system* (1.1.1) *with force f iff*

$$\nu < \nabla u, \nabla v > \ = \ [f, v] \tag{1.1.5}$$

holds for all $v \in C_{0,\sigma}^\infty(\Omega)$. If u is such a weak solution and if $p \in L_{loc}^2(\Omega)$ is given such that

$$-\nu\Delta u + \nabla p \ = \ f \tag{1.1.6}$$

holds in the sense of distributions, then (u, p) is called a **weak solution pair** *of* (1.1.1) *with force f, and p is called an* **associated pressure** *of u.*

We see that $f \in W^{-1,2}_{loc}(\Omega)^n$, $p \in L^2_{loc}(\Omega)$, $\nabla u \in L^2(\Omega)^{n^2}$, and $\Delta u = \operatorname{div} \nabla u \in W^{-1,2}_{loc}(\Omega)^n$. Therefore (1.1.6) is a well defined equation in the space $W^{-1,2}_{loc}(\Omega)^n$.

The construction of an associated pressure p is given by the following lemma.

1.1.2 Lemma *Let $\Omega \subseteq \mathbb{R}^n$, $n \geq 2$, be any domain, let $\Omega_0 \subseteq \Omega$, $\Omega_0 \neq \emptyset$, be a bounded subdomain with $\overline{\Omega}_0 \subseteq \Omega$, and let $f = f_0 + \operatorname{div} F$ with (1.1.4).*

Suppose $u \in \widehat{W}^{1,2}_{0,\sigma}(\Omega)$ is a weak solution of the Stokes system (1.1.1) with force f. Then there exists a uniquely determined $p \in L^2_{loc}(\Omega)$ satisfying $\int_{\Omega_0} p\,dx = 0$ and (1.1.6) in the sense of distributions.

A pair $(u,p) \in \widehat{W}^{1,2}_{0,\sigma}(\Omega) \times L^2_{loc}(\Omega)$ is a weak solution pair of (1.1.1) with force f iff (1.1.6) holds in the sense of distributions; p is uniquely determined by u under the additional condition $\int_{\Omega_0} p\,dx = 0$.

Proof. Let $u \in \widehat{W}^{1,2}_{0,\sigma}(\Omega)$ be a weak solution and consider the functional $G : v \mapsto [G,v]$, $v \in C_0^\infty(\Omega)^n$, defined by

$$\begin{aligned}[G,v] \;&:=\; [f + \nu\Delta u, v] \;=\; [f,v] + \nu\,[\Delta u, v] \\ &=\; < f_0, v > \;-\; < F, \nabla v > \;-\nu < \nabla u, \nabla v > .\end{aligned}$$

Then for each bounded subdomain $\Omega' \subseteq \Omega$ with $\overline{\Omega'} \subseteq \Omega$ we obtain

$$|[G,v]| \;\leq\; C\,\|f_0\|_{L^2(\Omega')}\|\nabla v\|_2 \;+\; \|F\|_2\|\nabla v\|_2 \;+\; \nu\|\nabla u\|_2\|\nabla v\|_2$$

for all $v \in C_0^\infty(\Omega')^n$. C means the constant in Poincaré's inequality for Ω'. This shows that $G \in W^{-1,2}_{loc}(\Omega)^n$. Then Lemma 2.2.1, II, yields a unique $p \in L^2_{loc}(\Omega)$ with $G = \nabla p$ in the sense of distributions and with $\int_{\Omega_0} p\,dx = 0$. The first assertion of the lemma follows; the second assertion is obvious. □

1.2 Embedding properties of $\widehat{W}^{1,2}_{0,\sigma}(\Omega)$

The space $\widehat{W}^{1,2}_{0,\sigma}(\Omega)$ of weak solutions is defined in an abstract way as the completion of $C^\infty_{0,\sigma}(\Omega)$ with respect to the norm $\|\nabla v\|_2$. It is continuously embedded in several function spaces, see [Gal94a, II.5] for similar embedding properties. The proof below shows that each "abstract" element of $\widehat{W}^{1,2}_{0,\sigma}(\Omega)$ is identified with a "concrete" function u such that the following embeddings are well defined.

1.2.1 Lemma *Let $\Omega \subseteq \mathbb{R}^n$, $n \geq 2$, be any domain, let $\widehat{W}^{1,2}_{0,\sigma}(\Omega)$ be the space (1.1.2) with norm*

$$\|u\|_{\widehat{W}^{1,2}_{0,\sigma}(\Omega)} = \|\nabla u\|_2\,,$$

and let

$$W^{1,2}_{0,\sigma}(\Omega) := \overline{C^\infty_{0,\sigma}(\Omega)}^{\|u\|_{W^{1,2}(\Omega)^n}} \tag{1.2.1}$$

be the closure of $C^\infty_{0,\sigma}(\Omega)$ *with respect to the norm*

$$\|u\|_{W^{1,2}_{0,\sigma}(\Omega)} := \|u\|_{W^{1,2}(\Omega)^n} = (\|u\|_2^2 + \|\nabla u\|_2^2)^{\frac{1}{2}}.$$

Then we have:

a) *It holds that*

$$W^{1,2}_{0,\sigma}(\Omega) \subseteq \widehat{W}^{1,2}_{0,\sigma}(\Omega) \tag{1.2.2}$$

and

$$\|u\|_{\widehat{W}^{1,2}_{0,\sigma}(\Omega)} \leq \|u\|_{W^{1,2}_{0,\sigma}(\Omega)} \;, \quad u \in W^{1,2}_{0,\sigma}(\Omega).$$

b) *The space*

$$GR(\Omega) := \{\nabla u \in L^2(\Omega)^{n^2};\ u \in \widehat{W}^{1,2}_{0,\sigma}(\Omega)\} \subseteq L^2(\Omega)^{n^2} \tag{1.2.3}$$

is isometric to $\widehat{W}^{1,2}_{0,\sigma}(\Omega)$, *and* $u \mapsto \nabla u$ *defines an isometric mapping from* $\widehat{W}^{1,2}_{0,\sigma}(\Omega)$ *onto* $GR(\Omega)$.

c) *If* $n \geq 3$, $q = \frac{2n}{n-2}$, *then*

$$\widehat{W}^{1,2}_{0,\sigma}(\Omega) \subseteq L^q(\Omega)^n, \tag{1.2.4}$$

and

$$\|u\|_{L^q(\Omega)^n} \leq C\,\|\nabla u\|_{L^2(\Omega)^{n^2}} \;, \quad u \in \widehat{W}^{1,2}_{0,\sigma}(\Omega), \tag{1.2.5}$$

with $C = C(n) > 0$.

d) *If* $n = 2$, $1 < q < \infty$, $\overline{\Omega} \neq \mathbb{R}^2$, *then*

$$\widehat{W}^{1,2}_{0,\sigma}(\Omega) \subseteq L^q_{loc}(\overline{\Omega})^2, \tag{1.2.6}$$

and

$$\|u\|_{L^q(B\cap\Omega)^2} \leq C\,\|\nabla u\|_{L^2(\Omega)^4} \;, \quad u \in \widehat{W}^{1,2}_{0,\sigma}(\Omega), \tag{1.2.7}$$

with $C = C(q, B_0, B) > 0$, *where* $B_0, B \subseteq \mathbb{R}^2$ *are open balls with* $\overline{B}_0 \cap \overline{\Omega} = \emptyset$, $B \cap \Omega \neq \emptyset$.

e) *If* Ω *is bounded, then*

$$\widehat{W}^{1,2}_{0,\sigma}(\Omega) = W^{1,2}_{0,\sigma}(\Omega) \tag{1.2.8}$$

with equivalent norms.

f) *If* Ω *is a bounded Lipschitz domain, then*

$$\begin{aligned} \widehat{W}^{1,2}_{0,\sigma}(\Omega) &= \{u \in W^{1,2}_0(\Omega)^n;\ \text{div}\, u = 0\} \\ &= \{u \in W^{1,2}(\Omega)^n;\ u|_{\partial\Omega} = 0,\ \text{div}\, u = 0\}. \end{aligned} \tag{1.2.9}$$

Proof. Let $u \in \widehat{W}^{1,2}_{0,\sigma}(\Omega)$. By definition, u is represented by a sequence $(u_j)_{j=1}^{\infty}$ in $C^{\infty}_{0,\sigma}(\Omega)$ in such a way that $(\nabla u_j)_{j=1}^{\infty}$ is a Cauchy sequence in $L^2(\Omega)^{n^2}$. In all cases above, we will prove a certain convergence property of $(u_j)_{j=1}^{\infty}$ which yields a uniquely determined limit function at least contained in $L^2_{loc}(\overline{\Omega})^n$. This function will be identified with u and yields the corresponding embedding.

If $u \in W^{1,2}_{0,\sigma}(\Omega)$, $(u_j)_{j=1}^{\infty}$ may be chosen as a Cauchy sequence with respect to $\|\cdot\|_{W^{1,2}}$. Thus $(u_j)_{j=1}^{\infty}$ converges in $L^2(\Omega)^n$. Setting $u := s - \lim_{j\to\infty} u_j$ we get

$$\|\nabla u\|_2 \leq (\|u\|_2^2 + \|\nabla u\|_2^2)^{\frac{1}{2}} ,$$

which proves the continuous embedding (1.2.2). This proves a).

The space $GR(\Omega)$ is closed by the definition of $\widehat{W}^{1,2}_{0,\sigma}(\Omega)$. If $(\nabla u_j)_{j=1}^{\infty}$ is a Cauchy sequence in $L^2(\Omega)^{n^2}$, the element

$$\nabla u = s - \lim_{j\to\infty} \nabla u_j \quad \in GR(\Omega)$$

is uniquely determined. The assertion in b) is obvious.

To prove c) we use Sobolev's embedding inequality (1.3.5), II, with $r = 2$, $q = \frac{2n}{n-2}$, and see that $(u_j)_{j=1}^{\infty}$ is a Cauchy sequence in $L^q(\Omega)^n$ which converges to some $u \in L^q(\Omega)^n$ satisfying (1.2.5).

To prove d) we use the embedding property (1.3.11), II, and get a unique $u \in L^q_{loc}(\overline{\Omega})^n$ satisfying (1.2.7).

To prove e) we use Poincaré's inequality (1.1.1), II, and see that the norms $\|\nabla u\|_2$ and $(\|u\|_2^2 + \|\nabla u\|_2^2)^{\frac{1}{2}}$ are equivalent.

If Ω is a bounded Lipschitz domain, the characterization (1.2.9) follows using e) and Lemma 2.2.3, II, together with (1.2.5), II. □

We do not discuss possible embeddings in the exceptional case $n = 2$, $\overline{\Omega} = \mathbb{R}^2$. In this case $\nabla u \in L^2(\Omega)^{n^2}$ is always well defined for all $u \in \widehat{W}^{1,2}_{0,\sigma}(\Omega)$ but $u \in L^2_{loc}(\overline{\Omega})$ itself may be determined only up to a constant. See [DLi55], [HLi56] concerning this case.

1.3 Existence of weak solutions

The following main theorem yields the existence of weak solutions in the case $f = \text{div } F$. We have no existence result in the general case $f = f_0 + \text{div } F$ with (1.1.4). However, we know several sufficient conditions for a function f to have this special form; see Lemma 1.6.1, II, and Lemma 1.6.2, II. We refer to [Gal94a], [Hey80] for further results.

1.3.1 Theorem *Let $\Omega \subseteq \mathbb{R}^n$, $n \geq 2$, be any domain, let $\Omega_0 \subseteq \Omega$, $\Omega_0 \neq \emptyset$, be a bounded subdomain with $\overline{\Omega}_0 \subseteq \Omega$, and let $f = \text{div } F$ with $F \in L^2(\Omega)^{n^2}$. Then there exists a unique pair*

$$(u,p) \in \widehat{W}^{1,2}_{0,\sigma}(\Omega) \times L^2_{loc}(\Omega)$$

satisfying $\int_{\Omega_0} p\,dx = 0$ and

$$-\nu\Delta u + \nabla p = f \quad , \quad \text{div } u = 0 \tag{1.3.1}$$

in the sense of distributions; u is a weak solution and p an associated pressure of the system (1.1.1) with force f.

Moreover, u satisfies the inequality

$$\|\nabla u\|_{L^2(\Omega)^{n^2}} \leq \nu^{-1}\|F\|_{L^2(\Omega)^{n^2}}. \tag{1.3.2}$$

If Ω is a bounded Lipschitz domain, then $p \in L^2(\Omega)$ and

$$\|p\|_{L^2(\Omega)} \leq C\,\|F\|_{L^2(\Omega)^{n^2}} \tag{1.3.3}$$

with $C = C(\Omega_0, \Omega) > 0$.

Proof. To prove the theorem we use in $\widehat{W}^{1,2}_{0,\sigma}(\Omega)$ the scalar product

$$\nu < \nabla u, \nabla v > \; = \; \nu \int_\Omega (\nabla u) \cdot (\nabla v)\, dx.$$

The functional $f : v \mapsto [f,v] = - < F, \nabla v >$, $v \in C^\infty_{0,\sigma}(\Omega)$, is continuous in $\|\nabla v\|_2$ since

$$|[f,v]| \; = \; |< F, \nabla v >| \; \leq \; \|F\|_2 \, \|\nabla v\|_2.$$

The Riesz representation theorem, see Section 3.2, II, yields a uniquely determined $u \in \widehat{W}^{1,2}_{0,\sigma}(\Omega)$ satisfying

$$\nu < \nabla u, \nabla v > = \; - < F, \nabla v > \;, \quad v \in \widehat{W}^{1,2}_{0,\sigma}(\Omega).$$

Setting $u = v$ yields

$$\nu\|\nabla u\|_2^2 \; \leq \; \|F\|_2 \, \|\nabla u\|_2$$

and therefore $\nu\|\nabla u\|_2 \leq \|F\|_2$ which proves (1.3.2). With p from Lemma 1.1.2, and with $\text{div } u = 0$ we see that (1.3.1) is satisfied.

To prove the uniqueness of (u,p), let $(\tilde{u}, \tilde{p}) \in \widehat{W}^{1,2}_{0,\sigma}(\Omega) \times L^2_{loc}(\Omega)$ be another pair solving (1.3.1). Then we see that

$$\nu \; < \nabla(u - \tilde{u}), \nabla v > = 0 \quad \text{for all } \; v \in C^\infty_{0,\sigma}(\Omega).$$

This holds as well for all $v \in \widehat{W}^{1,2}_{0,\sigma}(\Omega)$. Setting $v = u - \tilde{u}$ we see that $\|\nabla(u-\tilde{u})\|_2 = 0$. Therefore, $u - \tilde{u} = 0$.

If Ω is a bounded Lipschitz domain, we use Lemma 2.2.2, II, and get the inequalities

$$\begin{aligned}
\|p\|_{L^2(\Omega)} &\leq C\,\|\nabla p\|_{W^{-1,2}(\Omega)^n} = C\,\|\nu\Delta u + f\|_{W^{-1,2}(\Omega)^n} \\
&= C \sup_{0\neq v\in C_0^\infty(\Omega)^n} \big(|[\nu\Delta u + f, v]| \,/\, \|\nabla v\|_2\big) \\
&= C \sup_{0\neq v\in C_0^\infty(\Omega)^n} \big(|-\nu <\nabla u, \nabla v> - <F, \nabla v>| \,/\, \|\nabla v\|_2\big) \\
&\leq C\,(\nu\|\nabla u\|_2 + \|F\|_2) \leq 2C\,\|F\|_2
\end{aligned}$$

with $C = C(\Omega_0, \Omega) > 0$. This proves the theorem. □

1.4 The nonhomogeneous case $\operatorname{div} u = g$

We will solve the more general system

$$\begin{aligned}
-\nu\Delta u + \nabla p &= f \;, \quad \operatorname{div} u = g \;, \\
u|_{\partial\Omega} &= 0
\end{aligned} \tag{1.4.1}$$

in a corresponding weak sense. For this purpose we use some properties of the equation

$$\operatorname{div} v = g \;, \quad v|_{\partial\Omega} = 0, \tag{1.4.2}$$

see Lemma 2.1.1, a), II. This enables us to "remove" the divergence g. Since this lemma is restricted to bounded Lipschitz domains, we need that the support supp g is contained in a bounded Lipschitz subdomain $\Omega_1 \subseteq \Omega$. Solving (1.4.2) in Ω_1 by this lemma, we obtain some $v \in W_0^{1,2}(\Omega_1)^n$ satisfying (1.4.2) and

$$\|\nabla v\|_2 \leq C\,\|g\|_2 \tag{1.4.3}$$

with $C = C(\Omega_1) > 0$. Subtracting v leads to the equations

$$\begin{aligned}
-\nu\Delta(u-v) + \nabla p &= f + \nu\Delta v \;, \quad \operatorname{div}\,(u-v) = 0\,, \\
u - v|_{\partial\Omega} &= 0
\end{aligned} \tag{1.4.4}$$

which can be solved by Theorem 1.3.1. This method is known as the method of **subtracting the divergence**. In a similar way we can also solve the corresponding problem with nonhomogeneous boundary condition $u|_{\partial\Omega} \neq 0$. This is omitted here, see [Gal94a], [Gal94b] for this case. The next theorem yields the result on (1.4.1).

1.4.1 Theorem *Let $\Omega \subseteq \mathbb{R}^n, n \geq 2$, be any domain, let $\Omega_0 \subseteq \Omega$, $\Omega_0 \neq \emptyset$, be a bounded subdomain with $\overline{\Omega}_0 \subseteq \Omega$, and let $\Omega_1 \subseteq \Omega$, $\Omega_1 \neq \emptyset$, be a bounded Lipschitz subdomain. Suppose $f = \operatorname{div} F$, $F \in L^2(\Omega)^{n^2}$, and $g \in L^2(\Omega)$ such that*

$$supp\; g \subseteq \Omega_1 \;, \quad \int_{\Omega_1} g\, dx = 0.$$

Then there exists a unique pair (u,p) with the following properties:

u has a decomposition

$$u = u_0 + u_1 \quad with \;\; u_0 \in \widehat{W}^{1,2}_{0,\sigma}(\Omega) \;, \quad u_1 \in W^{1,2}_0(\Omega_1)^n,$$

it holds that

$$p \in L^2_{loc}(\Omega) \;, \quad \int_{\Omega_0} p\, dx = 0,$$

and

$$-\nu\Delta u + \nabla p = f \;, \quad \operatorname{div}\, u = g \tag{1.4.5}$$

in the sense of distributions.

Moreover

$$\|\nabla u\|_{L^2(\Omega)^{n^2}} \;\leq\; C\,(\nu^{-1}\|F\|_{L^2(\Omega)^{n^2}} + \|g\|_{L^2(\Omega)}) \tag{1.4.6}$$

with $C = C(\Omega_1) > 0$.

If Ω is a bounded Lipschitz domain, then $p \in L^2(\Omega)$ and

$$\|p\|_{L^2(\Omega)} \;\leq\; C\,(\|F\|_{L^2(\Omega)^{n^2}} + \nu\|g\|_{L^2(\Omega)}) \tag{1.4.7}$$

with $C = C(\Omega, \Omega_1, \Omega_0) > 0$.

1.4.2 Remark The condition $\int_{\Omega_0} p\, dx = 0$ is only needed to get a unique $p \in L^2_{loc}(\Omega)$. Otherwise p is only determined up to a constant. Therefore, if Ω is a bounded Lipschitz domain, we may choose $\Omega_0 = \Omega$ in the lemma above so that p satisfies the condition

$$\int_{\Omega} p\, dx = 0.$$

Proof of Theorem 1.4.1. We choose some $v \in W^{1,2}_0(\Omega_1)^n$ satisfying

$$\operatorname{div}\, v \;=\; g$$

and inequality (1.4.3). Then using Theorem 1.3.1 we find a unique pair $(\widehat{u}, p) \in \widehat{W}_{0,\sigma}^{1,2}(\Omega) \times L^2_{loc}(\Omega)$ satisfying $\int_{\Omega_0} p\, dx = 0$ and

$$-\nu\Delta\widehat{u} + \nabla p \;=\; f + \nu\Delta v \;=\; \text{div } (F + \nu\nabla v) \tag{1.4.8}$$

in the sense of distributions, see (1.4.4). Setting

$$u := \widehat{u} + v$$

we see that (1.4.5) is satisfied in the sense of distributions.

Using (1.3.2) with F replaced by $F + \nu\nabla v$ we obtain

$$\|\nabla u\|_2 - \|\nabla v\|_2 \;\le\; \|\nabla\widehat{u}\|_2 \;\le\; \nu^{-1}\|F + \nu\nabla v\|_2 \;\le\; \nu^{-1}\|F\|_2 + \|\nabla v\|_2,$$

and together with (1.4.3), (1.4.6) follows. The above decomposition follows with $u_0 = \widehat{u}$ and $u_1 = v$.

To prove the uniqueness property, let $(\tilde{u}, \tilde{p})$ be another pair satisfying with some decomposition $\tilde{u} = \tilde{u}_0 + \tilde{u}_1$ the same properties as (u, p). Then $u_1 - \tilde{u}_1 \in W_0^{1,2}(\Omega_1)^n$, $\int_{\Omega_0} \tilde{p}\, dx = 0$ and div $(u_1 - \tilde{u}_1) = 0$. Since Ω_1 is a bounded Lipschitz domain, we get by Lemma 1.2.1, f), that $u_1 - \tilde{u}_1 \in \widehat{W}_{0,\sigma}^{1,2}(\Omega_1) \subseteq \widehat{W}_{0,\sigma}^{1,2}(\Omega)$. Therefore, $(u - \tilde{u}, p - \tilde{p})$ is a solution pair with div $(u - \tilde{u}) = 0$ as in Theorem 1.3.1, with $f = 0$, $\int_{\Omega_0}(p - \tilde{p})\, dx = 0$. The uniqueness assertion in this theorem now yields $u = \tilde{u}$, $p = \tilde{p}$.

If Ω is a bounded Lipschitz domain, we apply (1.3.3) with F replaced by $F + \nu\nabla v$, and we use (1.4.3). This yields (1.4.7). The proof is complete. □

1.5 Regularity properties of weak solutions

Consider the Stokes system in the general form

$$\begin{aligned} -\nu\Delta u + \nabla p \;&=\; f \;, \quad \text{div } u = g, \\ u|_{\partial\Omega} \;&=\; 0 \end{aligned} \tag{1.5.1}$$

as in Theorem 1.4.1. Our purpose is to prove local regularity properties

$$u \in W_{loc}^{k+2}(\overline{\Omega})^n \;\text{ and }\; p \in W_{loc}^{k+1}(\overline{\Omega}) \;\;, \quad k \in \mathbb{N}$$

under some conditions on f, g, and Ω.

If k is sufficiently large, then the embedding properties of Lemma 1.3.4, II, imply the classical differentiability properties of u and p of arbitrary order. In principle we follow here the theory developed by Solonnikov-Scadilov [SoS73], Heywood [Hey80], and Galdi-Maremonti [GaM88], [Gal94a, IV.4, IV.5]. It rests on the method of differentiating the equations (1.5.1) "along the boundary". This does not destroy the boundary condition $u|_{\partial\Omega} = 0$. Then we may apply the estimates of Theorem 1.4.1. To carry out this method precisely, we use local coordinates in sufficiently small parts of the boundary. First we exclude the case $\Omega = \mathbb{R}^n$.

1.5.1 Theorem *Let $k \in \mathbb{N}_0$, let $\Omega \subseteq \mathbb{R}^n$, $\Omega \neq \mathbb{R}^n$, $n \geq 2$, be any C^{k+2}-domain, and let $\Omega_0 \subseteq \Omega$, $\Omega_0 \neq \emptyset$, be a bounded subdomain with $\overline{\Omega}_0 \subseteq \Omega$. Suppose*

$$f \in W^{k,2}_{loc}(\overline{\Omega})^n\,, \quad g \in W^{k+1,2}_{loc}(\overline{\Omega})\,, \quad u_0 \in \widehat{W}^{1,2}_{0,\sigma}(\Omega)\,,$$
$$u_1 \in W^{1,2}_0(\Omega)^n\,, \qquad p \in L^2_{loc}(\Omega)\,, \qquad \int_{\Omega_0} p\, dx = 0,$$

and assume that $u := u_0 + u_1$ and p solve the system

$$-\nu\Delta u + \nabla p = f \ , \quad \text{div } u = g \tag{1.5.2}$$

in the sense of distributions.

Then we have

$$u \in W^{k+2,2}_{loc}(\overline{\Omega})^n \ , \quad p \in W^{k+1,2}_{loc}(\overline{\Omega}), \tag{1.5.3}$$

and for each bounded subdomain $G_0 \subseteq \Omega$, there exists another bounded subdomain $G_1 \subseteq \Omega$ with $G_0 \subseteq G_1$ such that

$$\begin{aligned} &\|u\|_{W^{k+2,2}(G_0)} + \|p\|_{W^{k+1,2}(G_0)} \\ &\quad \leq C\left(\|f\|_{W^{k,2}(G_1)} + \|g\|_{W^{k+1,2}(G_1)} + \|u\|_{L^2(G_1)} + \|\nabla u\|_{L^2(G_1)}\right) \end{aligned} \tag{1.5.4}$$

with some constant $C = C(\nu, k, \Omega, \Omega_0, G_0, G_1) > 0$.

1.5.2 Remark We cannot prove the existence of the solution u in Theorem 1.5.1. This is possible under additional assumptions on f and g, see Theorem 1.4.1.

Proof. First we consider the case $k = 0$. The general case follows by induction on k. We need several steps where we treat special cases. In a) we consider a part of a half space and use the method of difference quotients. In b) we consider a part of a "bended" half space. This case will be reduced to a) by a transformation of coordinates. The general case for $k = 0$ will be reduced to b) using the localization method.

Here we use the following notations:
$x = (x', x_n)$, $x' = (x_1, \ldots, x_{n-1})$, and correspondingly $u = (u', u_n)$, $u' = (u_1, \ldots, u_{n-1})$, $f = (f', f_n)$, $f' = (f_1, \ldots, f_{n-1})$, $\nabla = (\nabla', D_n)$, $\nabla' = (D_1, \ldots, D_{n-1})$, $\Delta = \Delta' + D_n^2$, $\Delta' = D_1^2 + \cdots + D_{n-1}^2$, $\text{div}' u' = D_1 u_1 + \cdots + D_{n-1} u_{n-1}$.

Set

$$\begin{aligned} Q_\alpha &:= \{(x', x_n) \in \mathbb{R}^n;\ -\alpha \leq x_n \leq 0,\ |x'| \leq \alpha\}, \\ D_\alpha &:= \{x' \in \mathbb{R}^{n-1};\ |x'| \leq \alpha\} \end{aligned}$$

with $\alpha > 0$, and

$$Q_{\alpha,h} := \{(x', x_n) \in \mathbb{R}^n;\ h(x') - \alpha \le x_n \le h(x'),\ |x'| \le \alpha\}$$

where $h : x' \mapsto h(x')$, $|x'| \le \alpha$, is a function contained in $C^2(D_\alpha)$.

a) Suppose Ω is bounded, $u = u_1 \in W_0^{1,2}(\Omega)^n$, and there are constants $0 < \alpha < \beta$ with the properties

$$\text{supp}\, u,\ \text{supp}\, f \subseteq Q_\alpha \ , \quad Q_\beta \subseteq \overline{\Omega}.$$

Then we get

$$u_{|\partial Q_\alpha} = 0 \ , \quad \int_\Omega g\, dx = \int_\Omega \text{div}\ u\, dx = 0\,,$$

and f can be written in the form $f = \text{div}\ F$, $F \in L^2(\Omega)^{n^2}$, see Lemma 1.6.1, II. Therefore, Theorem 1.4.1 is applicable and in particular we obtain that $p \in L^2(\Omega)$. We may set $\Omega_0 = \Omega$, see Remark 1.4.2.

Next we consider the difference quotients

$$(D_j^\delta u)\,(x', x_n) := \delta^{-1}\big(u(x' + \delta e_j, x_n) - u(x', x_n)\big) \tag{1.5.5}$$

in x_j-direction with $0 < \delta < \beta - \alpha$, $j = 1, \dots, n-1$. In the same way we define D_j^δ for other functions.

Using the special assumptions above we see that $D_j^\delta u$ and $D_j^\delta p$ are solutions of the system

$$\begin{aligned} -\nu\Delta(D_j^\delta u) + \nabla(D_j^\delta p) &= D_j^\delta f \ , \quad \text{div}\ (D_j^\delta u) = D_j^\delta g \\ D_j^\delta u|_{\partial\Omega} &= 0\,, \end{aligned} \tag{1.5.6}$$

and we can apply Theorem 1.4.1. To explain this we observe that $D_j^\delta f$ can be written in the form $D_j^\delta f = D_j F_{\delta,j}$ with some $F_{\delta,j} \in L^2(\Omega)^{n^2}$ such that $\|F_{\delta,j}\|_2 \le C\,\|f\|_2$ with $C > 0$ not depending on δ, j. See (2.3.22) concerning this property. We refer to [Agm66, Sec. 3, Th. 3.13- 3.16] and [Fr69, Part I, 15] for details concerning difference quotients.

We may apply (1.4.6), (1.4.7) and obtain the estimate

$$\|\nabla(D_j^\delta u)\|_2 + \|D_j^\delta p\|_2 \le C\,(\|f\|_2 + \|D_j^\delta g\|_2) \tag{1.5.7}$$

with $C = C(\nu, \Omega) > 0$ not depending on δ, j. Letting $\delta \to 0$ we conclude that $D_j \nabla u \in L^2(\Omega)^{n^2}$, $D_j p \in L^2(\Omega)$ for $j = 1, \dots, n-1$. We simply write $\|\nabla' \nabla u\|_2 < \infty$, $\|\nabla' p\|_2 < \infty$, and obtain the estimate

$$\|\nabla' \nabla u\|_2 + \|\nabla' p\|_2 \le C(\|f\|_2 + \|\nabla g\|_2).$$

From the equations $-\nu\Delta u + \nabla p = f$, $\operatorname{div} u = g$ we obtain

$$\begin{aligned} -\nu\Delta' u' - \nu D_n^2 u' + \nabla' p &= f' \ , \quad -\nu\Delta' u_n - \nu D_n^2 u_n + D_n p = f_n \ , \\ \operatorname{div}'(D_n u') + D_n^2 u_n &= D_n g. \end{aligned}$$

This yields $\|D_n^2 u'\|_2 < \infty$, $\|D_n^2 u_n\|_2 < \infty$, $\|D_n p\|_2 < \infty$, and it follows that

$$\|\nabla^2 u\|_{L^2(\Omega)} + \|\nabla p\|_{L^2(\Omega)} \leq C\,(\|f\|_{L^2(\Omega)} + \|\nabla g\|_{L^2(\Omega)}) \tag{1.5.8}$$

with $C = C(\nu, \Omega) > 0$. This proves (1.5.4) with $k = 0$, $G_0 = G_1 = \Omega$.

b) Suppose Ω is bounded, $u = u_1 \in W_0^{1,2}(\Omega)^n$, and there are constants $0 < \alpha < \alpha' < \beta$, and a function $h \in C^2(D_\beta)$ with $h(x') = 0$ for $\alpha' \leq |x'| < \beta$, such that

$$\operatorname{supp} u, \ \operatorname{supp} f \subseteq Q_{\alpha,h} \ , \quad Q_{\beta,h} \subseteq \overline{\Omega}.$$

Further we suppose that

$$\|h\|_{C^1(D_\alpha)} + \|h\|^2_{C^1(D_\alpha)} \leq K \tag{1.5.9}$$

where $K > 0$ is a constant which will be determined later on (smallness assumption).

Similarly as in a) we get

$$u_{|\partial Q_{\alpha,h}} = 0 \ , \quad \int_\Omega g\,dx = 0 \ , \quad p \in L^2(\Omega),$$

and we may apply Theorem 1.4.1 with $\Omega_0 = \Omega$.

We reduce this problem to the case a). For this purpose we use the following transformation to new coordinates $y = (y', y_n)$, $y' = (y_1, \ldots, y_{n-1})$:

$$y' := x' \ , \quad y_n := x_n - h(x') \ , \quad x \in Q_{\beta,h}. \tag{1.5.10}$$

We define $\widehat{u}$, $\widehat{p}$, $\widehat{f}$, $\widehat{g}$ by $\widehat{u}(y) = u(x)$, $\widehat{p}(y) = p(x)$, $\widehat{f}(y) = f(x)$, $\widehat{g}(y) = g(x)$ and an elementary calculation leads from

$$-\nu\Delta u + \nabla p = f \ , \quad \operatorname{div} u = g \tag{1.5.11}$$

to the new equations

$$-\nu\Delta\widehat{u} + \nabla\widehat{p} = \widehat{f} + S_1 \ , \quad \operatorname{div}\widehat{u} = \widehat{g} + S_2 \tag{1.5.12}$$

where S_1, S_2 are given by

$$\begin{aligned} S_1 &:= (h_1, \ldots, h_{n-1}, 0) D_n \widehat{p} - \nu \sum_{j=1}^{n-1} (2h_j D_j D_n - h_j^2 D_n^2 + h_{jj} D_n)\widehat{u}, \\ S_2 &:= \sum_{j=1}^{n-1} h_j D_n \widehat{u}_j \ , \quad h_j = D_j h \ , \quad h_{jj} = D_j^2 h \ , \quad j = 1, \ldots, n-1. \end{aligned}$$

Using (1.5.10) we obtain a bounded C^2-domain $\widehat{\Omega} \subseteq \mathbb{R}^n$ in such a way that $x \in \Omega$ iff $y \in \widehat{\Omega}$, and that

$$\text{supp}\ \widehat{u},\ \text{supp}\ \widehat{f} \subseteq Q_\alpha\ ,\quad Q_\beta \subseteq \overline{\widehat{\Omega}}$$

Note that the calculation for (1.5.12) essentially rests on the relations

$$\begin{aligned}(D_j u)(x) &= (D_j\widehat{u})(y) - h_j(y')(D_n\widehat{u})(y)\ ,\quad j = 1,\dots,n-1,\\ (D_n u)(x) &= (D_n\widehat{u})(y),\end{aligned}$$

and correspondingly for p, f, g.

To explain the next argument we assume for the moment that

$$\|\nabla^2\widehat{u}\|_2 < \infty\ ,\quad \|\nabla\widehat{p}\|_2 < \infty. \tag{1.5.13}$$

Then we may use inequality (1.5.8) from step a) and obtain the estimate

$$\begin{aligned}&\|\nabla^2\widehat{u}\|_2 + \|\nabla\widehat{p}\|_2 \\ &\leq C\Big(\|\widehat{f}\|_2 + \|\nabla\widehat{g}\|_2 + \|h\|_{C^2(D_\alpha)}\|\nabla\widehat{u}\|_2 \\ &\qquad + (\|h\|_{C^1(D_\alpha)} + \|h\|^2_{C^1(D_\alpha)})(\|\nabla^2\widehat{u}\|_2 + \|\nabla\widehat{p}\|_2)\Big)\end{aligned} \tag{1.5.14}$$

with $C = C(\nu,\Omega) > 0$. Now we fix K in (1.5.9) and set $K := \frac{1}{2}C^{-1}$. Then we conclude (**absorption principle**) that with some $C' = C'(\nu,\Omega) > 0$ the inequality

$$\|\nabla^2\widehat{u}\|_2 + \|\nabla\widehat{p}\|_2 \leq C'(\|\widehat{f}\|_2 + \|\nabla\widehat{g}\|_2 + \|\nabla\widehat{u}\|_2) \tag{1.5.15}$$

is satisfied. Going back to the original coordinates x, and using the transformation formula for integrals, we obtain the inequality

$$\begin{aligned}&\|\nabla^2 u\|_{L^2(\Omega)} + \|\nabla p\|_{L^2(\Omega)} \\ &\leq C(\|f\|_{L^2(\Omega)} + \|g\|_{L^2(\Omega)} + \|\nabla g\|_{L^2(\Omega)} + \|u\|_{L^2(\Omega)} + \|\nabla u\|_{L^2(\Omega)})\end{aligned} \tag{1.5.16}$$

with $C = C(\nu,\Omega) > 0$. This yields (1.5.4) with $\Omega = G_0 = G_1$.

Since we do not yet know whether (1.5.13) is valid, we have to use the difference quotients in y_j-direction, $j = 1,\dots,n-1$, similarly as in step a). This yields the estimate (1.5.14) with $\nabla^2\widehat{u}$, $\nabla\widehat{p}$ replaced by $D_j^\delta\nabla\widehat{u}$, $D_j^\delta\widehat{p}$, $j = 1,\dots,n-1$, and letting $\delta \to 0$ we get (1.5.15) first only for $\|\nabla'\nabla\widehat{u}\|_2$, $\|\nabla'\widehat{p}\|_2$. The estimates for $\|D_n\nabla\widehat{u}\|_2$, $\|D_n\widehat{p}\|_2$ follow from the equations in a similar way as in step a). This proves (1.5.16), and the proof of step b) is complete.

c) Consider now the general case for $k = 0$. We will apply the result in b) locally for "small" portions of Ω. For this purpose we consider open balls B_0, $B_1 \subseteq \mathbb{R}^n$

with $\overline{B}_0 \subseteq B_1$, $B_0 \cap \Omega \neq \emptyset$, and choose a "cut-off" function $\varphi \in C_0^\infty(\mathbb{R}^n)$ satisfying

$$0 \le \varphi \le 1 \ , \quad \operatorname{supp} \varphi \subseteq B_1 \ , \quad \varphi(x) = 1 \ \text{in } B_0.$$

Choosing these balls sufficiently small we may assume that $G := B_1 \cap \Omega$ is a domain. Since Ω is a C^2-domain, G is a Lipschitz domain.

First we prove an estimate for p in G. We use (1.5.1) and consider the functional

$$\begin{aligned} \nabla p \,:\, v \mapsto [\nabla p, v] \;&=\; [f + \nu \nabla u, v] \\ &=\; < f, v > \, - \, \nu < \nabla u, \nabla v > \end{aligned} \tag{1.5.17}$$

with $v \in C_0^\infty(G)^n$. Using the Poincaré inequality for G, we obtain

$$\begin{aligned} |[\nabla p, v]| \;&\le\; \|f\|_2 \, \|v\|_2 + \nu \|\nabla u\|_2 \, \|\nabla v\|_2 \\ &\le\; (C \, \|f\|_2 + \nu \|\nabla u\|_2) \, \|\nabla v\|_2 \end{aligned}$$

with $C = C(G) > 0$. This shows $\nabla p \in W^{-1,2}(G)^n$,

$$\|\nabla p\|_{W^{-1,2}(G)} \;\le\; C \, \|f\|_{L^2(G)} + \nu \|\nabla u\|_{L^2(G)} \tag{1.5.18}$$

and Lemma 2.2.2, II, yields $p \in L^2(G)$. In this lemma p is determined up to a constant. In our case we have $\int_{\Omega_0} p \, dx = 0$, and we can choose some $p_0 = p_0(G)$ in such a way that

$$\|p - p_0\|_{L^2(G)} \;\le\; C \left(\|f\|_{L^2(G)} + \|\nabla u\|_{L^2(G)} \right) \tag{1.5.19}$$

with some $C = C(\nu, \Omega_0, G) > 0$.

From (1.5.1) we get

$$-\nu \Delta u + \nabla (p - p_0) = f \ , \quad \operatorname{div} u = 0,$$

and multiplication with φ yields

$$\begin{aligned} -\nu \Delta (\varphi u) + \nabla (\varphi (p - p_0)) \;&=\; \tilde{f} \ , \quad \operatorname{div} (\varphi u) = \tilde{g} \ , \\ \varphi u_{|\partial G} \;&=\; 0, \end{aligned} \tag{1.5.20}$$

where

$$\tilde{f} \;:=\; \varphi f - 2\nu (\nabla \varphi)(\nabla u) - \nu (\Delta \varphi) u + (\nabla \varphi)(p - p_0), \tag{1.5.21}$$

$(\nabla \varphi)(\nabla u) = ((\nabla \varphi) \cdot (\nabla u_1), \ldots, (\nabla \varphi) \cdot (\nabla u_n))$, and

$$\tilde{g} \;:=\; \varphi g + (\nabla \varphi) \cdot u. \tag{1.5.22}$$

Using Green's formula (1.2.12), II, we see that $\int_G \tilde{g} \, dx = 0$.

Using (1.5.19) we obtain the estimates

$$\|\tilde{f}\|_{L^2(G)} \leq C(\|f\|_{L^2(G)} + \|\nabla u\|_{L^2(G)} + \|u\|_{L^2(G)}) \tag{1.5.23}$$

and

$$\begin{aligned} &\|\tilde{g}\|_{L^2(G)} + \|\nabla\tilde{g}\|_{L^2(G)} \\ &\leq C(\|g\|_{L^2(G)} + \|\nabla g\|_{L^2(G)} + \|u\|_{L^2(G)} + \|\nabla u\|_{L^2(G)}) \end{aligned} \tag{1.5.24}$$

with $C = C(\nu, \Omega_0, G) > 0$. We call (1.5.20) the **localized system.**

Consider first the case $\overline{B}_1 \subseteq \Omega$, $G = \Omega \cap B_1 = B_1$. Using a translation and a rotation of the coordinates, we see that the estimate (1.5.8) is applicable. This yields an estimate of the expression

$$\|\nabla^2(\varphi u)\|_{L^2(G)} + \|\nabla\varphi(p - p_0)\|_{L^2(G)} \tag{1.5.25}$$

by the terms of the right sides of (1.5.23), (1.5.24).

Using the relations $D_i\, D_j(\varphi u) = \varphi(D_i\, D_j u) + (D_i D_j \varphi)u + (D_j\varphi)\,(D_i u) + (D_i\varphi)\,(D_j u)$ and $D_j(\varphi(p - p_0)) = \varphi\, D_j p + (D_j\varphi)\,(p - p_0)$ with $i, j = 1, \ldots, n$, we see that

$$\|\varphi\nabla^2 u\|_{L^2(G)} + \|\varphi\nabla p\|_{L^2(G)}$$

can be estimated in the same way as (1.5.25). This yields an estimate which we can write in the form

$$\begin{aligned} &\|\varphi\nabla^2 u\|^2_{L^2(G)} + \|\varphi\nabla p\|^2_{L^2(G)} \\ &\leq C(\|f\|^2_{L^2(G)} + \|g\|^2_{L^2(G)} + \|\nabla g\|^2_{L^2(G)} + \|u\|^2_{L^2(G)} + \|\nabla u\|^2_{L^2(G)}) \end{aligned} \tag{1.5.26}$$

with $C = C(\nu, \Omega_0, G) > 0$.

Next we consider the case $B_1 \cap \partial\Omega \neq \emptyset$. We may assume that the center of B_1 is contained in $\partial\Omega$. Now we use the definition of a C^2-domain in Section 3.2, I. Using this definition, we can carry out a rotation and a translation of the coordinate system in such a way that in some appropriate C^2-subdomain $G' \subseteq G$ with supp $(\varphi u) \subseteq \overline{G'}$ the assumptions of step b) are satisfied. The radius of B_1 can be chosen (sufficiently small) in such a way that (1.5.9) is valid. For this purpose we choose the coordinate system in such a way that $h = 0$ and $\nabla h = 0$ in the origin. To φu we can now apply the inequality (1.5.16). This leads as above to the estimate (1.5.26).

Consider now any bounded subdomain $G_0 \subseteq \Omega$. Then we find finitely many balls B_0^j, B_1^j, and functions φ_j, $j = 1, \ldots, m$, which have the same properties as B_0, B_1, φ above, and which satisfy

$$G_0 \subseteq \bigcup_{j=1}^{m} (B_0^j \cap \Omega) \quad , \quad \sum_{j=1}^{m} \varphi_j(x) = 1 \quad \text{for all } x \in G_0.$$

For each $j = 1, \dots, m$ we can use the estimate (1.5.26), and we can take the sum on both sides of these inequalities. This yields the desired inequality (1.5.4) with

$$G_1 := \bigcup_{j=1}^{m} (B_1^j \cap \Omega).$$

This proves the theorem for $k = 0$. In order to prove the theorem for $k = 1$, we have to repeat the above arguments in each step. In step a) we now obtain instead of (1.5.8) the inequality

$$\|\nabla^3 u\|_2 + \|\nabla^2 p\|_2 \leq C\,(\|\nabla f\|_2 + \|\nabla^2 g\|_2 + \|\nabla g\|_2),$$

and in step b) we get

$$\begin{aligned} &\|\nabla^3 u\|_2 + \|\nabla^2 p\|_2 \\ &\quad \leq C\,(\|f\|_2 + \|\nabla f\|_2 + \|g\|_2 + \|\nabla g\|_2 + \|\nabla^2 g\|_2 + \|u\|_2 + \|\nabla u\|_2) \end{aligned}$$

instead of (1.5.16). The same arguments as in step c) then yield the result for $k = 1$, and so on. The case of general k follows by induction on k. The proof is complete. □

In the special case of a bounded domain we can prove the following formulation. It is a consequence of Theorem 1.5.1 and Theorem 1.4.1. See [Gal94a, IV.5] for similar results, see also [Catt61], [SoS73], and [GaM88].

1.5.3 Theorem *Let $k \in \mathbb{N}_0$, and let $\Omega \subseteq \mathbb{R}^n$, $n \geq 2$, be a bounded C^{k+2}-domain. Suppose*

$$f \in W^{-1,2}(\Omega)^n \quad , \quad g \in L^2(\Omega)$$

with $\int_\Omega g\,dx = 0$. Then there exists a unique pair $(u, p) \in W_0^{1,2}(\Omega)^n \times L^2(\Omega)$ satisfying $\int_\Omega p\,dx = 0$ and

$$-\nu\Delta u + \nabla p = f \quad , \quad \operatorname{div} u = g \tag{1.5.27}$$

in the sense of distributions. Moreover,

$$\|\nabla u\|_{L^2(\Omega)} + \nu^{-1}\|p\|_{L^2(\Omega)} \leq C\,(\nu^{-1}\|f\|_{W^{-1,2}(\Omega)} + \|g\|_{L^2(\Omega)}) \tag{1.5.28}$$

with some constant $C = C(\Omega) > 0$.

If additionally

$$f \in W^{k,2}(\Omega)^n \quad \textit{and} \quad g \in W^{k+1,2}(\Omega),$$

then $u \in W^{k+2,2}(\Omega)^n$, $p \in W^{k+1,2}(\Omega)$, *and the inequality*

$$\|u\|_{W^{k+2,2}(\Omega)} + \nu^{-1}\|p\|_{W^{k+1,2}(\Omega)} \leq C\left(\nu^{-1}\|f\|_{W^{k,2}(\Omega)} + \|g\|_{W^{k+1,2}(\Omega)}\right) \tag{1.5.29}$$

holds with some constant $C = C(k, \Omega) > 0$.

Proof. The first assertion follows from Theorem 1.4.1 if we write f in the form $f = \text{div } F$ with $F \in L^2(\Omega)^{n^2}$ satisfying

$$\|F\|_2 \leq \|f\|_{W^{-1,2}(\Omega)},$$

see Lemma 1.6.1, II. The estimate (1.5.28) follows from (1.4.6) and (1.4.7). The regularity result (1.5.29) follows from (1.5.4) after the following modifications:

We may apply (1.5.4) with $G_0 = G_1 = \Omega$. We may also set $\Omega_0 = \Omega$. The last two terms in (1.5.4) may be omitted because of (1.5.28).

It remains to investigate the constant C in (1.5.4). For this purpose we write (1.5.27) in the form

$$-\Delta u + \nabla(\nu^{-1}p) = \nu^{-1}f \ , \quad \text{div } u = g,$$

and apply (1.5.4) with $\nu = 1$, and with p, f replaced by $\nu^{-1}p, \nu^{-1}f$. This leads to (1.5.29) with $C = C(k, \Omega) > 0$. This proves the theorem. □

In the next theorem we consider the special case of a uniform C^2- domain, see the definition in Section 3.2, I. This result is only interesting for unbounded domains. Here we consider only the case $g = 0$. Since f in the theorem below has not necessarily the form $f =$ div F with $F \in L^2(\Omega)^{n^2}$, see Section 1.6, II, we have no existence result in this case.

1.5.4 Theorem *Let* $\Omega \subseteq \mathbb{R}^n$, $\Omega \neq \mathbb{R}^n$, $n \geq 2$, *be a uniform* C^2*-domain, and let* $f \in L^2(\Omega)^n$. *Suppose the pair*

$$(u, p) \in \widehat{W}^{1,2}_{0,\sigma}(\Omega) \times L^2_{loc}(\Omega)$$

solves the equation

$$-\nu\Delta u + \nabla p = f \tag{1.5.30}$$

in the sense of distributions, and suppose additionally that

$$u \in L^2(\Omega)^n .$$

Then

$$u \in W^{2,2}(\Omega)^n \ , \quad p \in L^2_{loc}(\overline{\Omega}) \ , \quad \nabla p \in L^2(\Omega)^n, \tag{1.5.31}$$

and

$$\|\nabla^2 u\|_{L^2(\Omega)} + \nu^{-1}\|\nabla p\|_{L^2(\Omega)} \qquad (1.5.32)$$
$$\leq C\,(\nu^{-1}\|f\|_{L^2(\Omega)} + \|\nabla u\|_{L^2(\Omega)} + \|u\|_{L^2(\Omega)})$$

with some constant $C = C(\Omega) > 0$.

Proof. See [Hey80, Lemma 1 and Theorem 1] and [GaM88, 3, Lemma 1] for a similar result. In fact, the constant C in (1.5.32) depends only on the constants which occur in the definition of a uniform C^2-domain in Section 3.2, I. To prove the theorem, we use the local estimate (1.5.26) with φ replaced by φ_j from (3.2.11), I, and ψ_j from (3.2.13), I, for all $j \in \mathbb{N}$. Note that Ω is also a uniform Lipschitz domain. The pressure term $\|\varphi \nabla p\|_2^2$ in (1.5.26) is now omitted. Since this term is finite with φ replaced by φ_j, ψ_j for all $j \in \mathbb{N}$, we can conclude that $p \in L^2_{loc}(\overline{\Omega})$.

Since Ω is a uniform C^2-domain, the constant C in (1.5.26) can be chosen independently of $j \in \mathbb{N}$. Therefore, taking the sum over $j \in \mathbb{N}$, we obtain with $g = 0$ the estimate

$$\|\nabla^2 u\|_{L^2(\Omega)} \;\leq\; C\,(\|f\|_{L^2(\Omega)} + \|\nabla u\|_{L^2(\Omega)} + \|u\|_{L^2(\Omega)})$$

with $C = C(\nu, \Omega) > 0$. Applying this inequality to the equation

$$-\Delta u + \nabla(\nu^{-1} p) = \nu^{-1} f$$

with $\nu = 1$, and with f replaced by $\nu^{-1} f$, we obtain

$$\|\nabla^2 u\|_{L^2(\Omega)} \;\leq\; C\,(\nu^{-1}\|f\|_{L^2(\Omega)} + \|\nabla u\|_{L^2(\Omega)} + \|u\|_{L^2(\Omega)})$$

where $C = C(\Omega) > 0$ does not depend on ν.

Writing $\nabla(\nu^{-1}p) = \nu^{-1} f + \Delta u$ we see that $\nabla p \in L^2(\Omega)$. From

$$\|u\|_2 + \|\nabla u\|_2 + \|\nabla^2 u\|_2 \;<\infty$$

we conclude that $u \in W^{2,2}(\Omega)^n$. This proves the theorem. □

The case $\Omega = \mathbb{R}^n$ is excluded in Theorem 1.5.1. If $n \geq 3$, the result of this theorem remains true for $\Omega = \mathbb{R}^n$, and the proof is the same with some obvious simplifications since $\partial\Omega = \emptyset$. However, the case $n = 2$, $\Omega = \mathbb{R}^2$ requires a modification. In this case, see the discussion in Section 1.1, the condition $u_0 \in \widehat{W}^{1,2}_{0,\sigma}(\mathbb{R}^2)$ is problematic. Therefore we suppose

$$u_0 \in W^{1,2}_{0,\sigma}(\mathbb{R}^2) \;:=\; \overline{C^\infty_{0,\sigma}(\mathbb{R}^2)}^{\|\cdot\|_{W^{1,2}(\mathbb{R}^2)^2}}$$

in the following theorem if $n = 2$.

Recall that

$$W^{1,2}_{0,\sigma}(\mathbb{R}^n) = \{u \in W^{1,2}(\mathbb{R}^n)^n;\ \operatorname{div} u = 0\} \tag{1.5.33}$$

for all $n \geq 2$, see (2.5.23), II, and that

$$W^{1,2}_0(\mathbb{R}^n)^n = W^{1,2}(\mathbb{R}^n)^n = \overline{C_0^\infty(\mathbb{R}^n)^n}^{\|\cdot\|_{W^{1,2}(\mathbb{R}^n)^n}} \tag{1.5.34}$$

for all $n \geq 2$, see (3.6.17), I.

1.5.5 Theorem *Let $k \in \mathbb{N}_0$, $n \in \mathbb{N}$, $n \geq 2$, and let $\Omega_0 \subseteq \mathbb{R}^n$, $\Omega_0 \neq \emptyset$, be a bounded subdomain.*

Suppose

$$\begin{aligned} &f \in W^{k,2}_{loc}(\mathbb{R}^n)^n \quad , \quad g \in W^{k+1,2}_{loc}(\mathbb{R}^n), \\ &u_0 \in \widehat{W}^{1,2}_{0,\sigma}(\mathbb{R}^n) \quad \text{if } n \geq 3 \quad , \quad u_0 \in W^{1,2}_{0,\sigma}(\mathbb{R}^2) \quad \text{if } n = 2\,, \\ &u_1 \in W^{1,2}(\mathbb{R}^n)^n \quad , \quad p \in L^2_{loc}(\mathbb{R}^n) \quad , \quad \int_{\Omega_0} p\, dx = 0, \end{aligned}$$

and assume that $u := u_0 + u_1$ and p solve the system

$$-\nu\Delta u + \nabla p = f \quad , \quad \operatorname{div} u = g$$

in the sense of distributions.

Then we get

$$u \in W^{k+2,2}_{loc}(\mathbb{R}^n)^n \quad , \quad p \in W^{k+1,2}_{loc}(\mathbb{R}^n), \tag{1.5.35}$$

and for each bounded subdomain $G_0 \subseteq \mathbb{R}^n$ there exists another bounded subdomain $G_1 \subseteq \mathbb{R}^n$ with $G_0 \subseteq G_1$ such that

$$\begin{aligned} &\|u\|_{W^{k+2,2}(G_0)} + \|p\|_{W^{k+1,2}(G_0)} \\ &\quad \leq C\big(\|f\|_{W^{k,2}(G_1)} + \|g\|_{W^{k+1,2}(G_1)} + \|\nabla u\|_{L^2(G_1)} + \|u\|_{L^2(G_1)}\big) \end{aligned} \tag{1.5.36}$$

with $C = C(\nu, k, n, \Omega_0, G_0, G_1) > 0$.

Proof. The proof is the same as the proof of Theorem 1.5.1 with the following modification: Since $\partial\Omega = \emptyset$, the local estimates (1.5.26) are now used only in the first case $\overline{B}_1 \subseteq \Omega$, see the proof of (1.5.26). This yields the result. □

In the same way we can extend Theorem 1.5.4 to the case $\Omega = \mathbb{R}^n$.

1.5.6 Theorem *Let $n \in \mathbb{N}$, $n \geq 2$, suppose $f \in L^2(\mathbb{R}^n)^n$,*

$$\begin{aligned} u \in \widehat{W}^{1,2}_{0,\sigma}(\mathbb{R}^n) \quad , \quad &u \in L^2(\mathbb{R}^n)^n \quad &&\text{if } n \geq 3, \\ &u \in W^{1,2}_{0,\sigma}(\mathbb{R}^2) &&\text{if } n = 2, \end{aligned}$$

and let $p \in L^2_{loc}(\mathbb{R}^n)$. *Suppose* u *and* p *solve the equation*

$$-\nu\Delta u + \nabla p = f$$

in the sense of distributions. Then

$$u \in W^{2,2}(\mathbb{R}^n)^n \ , \ \ \nabla p \in L^2(\mathbb{R}^n)^n, \tag{1.5.37}$$

and

$$\begin{aligned} &\|\nabla^2 u\|_{L^2(\mathbb{R}^n)} + \nu^{-1}\|\nabla p\|_{L^2(\mathbb{R}^n)} \\ &\quad \le C\left(\nu^{-1}\|f\|_{L^2(\mathbb{R}^n)} + \|\nabla u\|_{L^2(\mathbb{R}^n)} + \|u\|_{L^2(\mathbb{R}^n)}\right) \end{aligned} \tag{1.5.38}$$

with some constant $C = C(n) > 0$.

Proof. The arguments are the same as in the proof of Theorem 1.5.4 with obvious simplifications. □

There are several further regularity results for the $\mathbb{R}^n$. They will be given in the section on the Stokes operator, see Lemma 2.3.2.

2 The Stokes operator A

2.1 Definition and properties

The Stokes operator A is basic for our functional analytic approach to the Navier-Stokes system, see the discussion in Section 2, I. We need some elementary Hilbert space methods, see Section 3.2, II.

We develop only the L^2-theory for A. The advantage of this approach is that we can admit arbitrary domains. In particular we can include the interesting case of unbounded boundaries. For bounded and exterior domains we obtain more information on A in general L^q-spaces, see Varnhorn's book [Var94] and [Wie99], see also [Sol77], [Gig81], [vWa85], [Spe86], [Gig86], [GiRa86], [GiSo89], [FaS94a], [STh98], [SSp98].

The underlying domain $\Omega \subseteq \mathbb{R}^n$, $n \ge 2$, is completely general. We use the (real) Hilbert space

$$L^2_\sigma(\Omega) \ = \ \overline{C^\infty_{0,\sigma}(\Omega)}^{\|\cdot\|_{L^2(\Omega)}}$$

with scalar product

$$< u, v >_\Omega = < u, v > = \int_\Omega u \cdot v \, dx$$

and norm $\|u\|_2 = \|u\|_{L^2(\Omega)} = < u, u >^{\frac{1}{2}}$, see (2.5.1), II.

Further we need the Hilbert space

$$W^{1,2}_{0,\sigma}(\Omega) \;=\; \overline{C^\infty_{0,\sigma}(\Omega)}^{\|\cdot\|_{W^{1,2}(\Omega)}} \subseteq L^2_\sigma(\Omega) \tag{2.1.1}$$

with scalar product

$$< u, v > + < \nabla u, \nabla v > = \int_\Omega u \cdot v \, dx + \int_\Omega (\nabla u) \cdot (\nabla v) \, dx$$

and norm $(\|u\|_2^2 + \|\nabla u\|_2^2)^{\frac{1}{2}}$, see (1.2.1). Recall that the Hilbert space $\widehat{W}^{1,2}_{0,\sigma}(\Omega)$, with scalar product $< \nabla u, \nabla v >$ and norm $\|\nabla u\|_2$ is defined as the completion of $C^\infty_{0,\sigma}(\Omega)$ with respect to the norm $\|\nabla u\|_2$.

See Lemma 1.2.1 for embedding properties concerning these spaces. In particular we see that

$$W^{1,2}_{0,\sigma}(\Omega) \subseteq \widehat{W}^{1,2}_{0,\sigma}(\Omega) \tag{2.1.2}$$

with continuous embedding.

We define the operator $A : D(A) \to L^2_\sigma(\Omega)$ with domain $D(A) \subseteq L^2_\sigma(\Omega)$ and range $R(A) = \{Au;\ u \in D(A)\}$, as follows:

Let $D(A) \subseteq W^{1,2}_{0,\sigma}(\Omega)$ be the space of all those $u \in W^{1,2}_{0,\sigma}(\Omega)$ for which there exists some $f \in L^2_\sigma(\Omega)$ satisfying

$$\nu < \nabla u, \nabla v > = < f, v > \ , \quad v \in C^\infty_{0,\sigma}(\Omega). \tag{2.1.3}$$

Using the Riesz representation theorem (Section 3.2, II) we see that $D(A)$ is the space of all those $u \in W^{1,2}_{0,\sigma}(\Omega)$ such that the functional

$$v \mapsto \nu < \nabla u, \nabla v > \ , \quad v \in C^\infty_{0,\sigma}(\Omega)$$

is continuous in the norm $\|v\|_2$.

For all $u \in D(A)$, let $Au \in L^2_\sigma(\Omega)$ be defined by the relation

$$\nu < \nabla u, \nabla v > = < Au, v > \ , \quad v \in C^\infty_{0,\sigma}(\Omega). \tag{2.1.4}$$

Thus $Au = f$ with f in (2.1.3).

Then $A = A_\Omega$ is called the **Stokes operator** for the domain Ω. The following theorem collects some properties of A. Recall that $P : L^2(\Omega)^n \to L^2_\sigma(\Omega)$ means the Helmholtz projection, see Section 2.5, II.

2.1.1 Theorem *Let $\Omega \subseteq \mathbb{R}^n$, $n \geq 2$, be any domain, and let $A : D(A) \to L^2_\sigma(\Omega)$ be the Stokes operator for Ω. Then we have:*

a) *A is positive selfadjoint with dense domain $D(A) \subseteq L^2_\sigma(\Omega)$, $C^\infty_{0,\sigma}(\Omega) \subseteq D(A) \subseteq W^{1,2}_{0,\sigma}(\Omega)$. It holds $N(A) = \{u \in D(A);\ Au = 0\} = \{0\}$, and the* *inverse* $A^{-1} : D(A^{-1}) \to L^2_\sigma(\Omega)$ *with domain* $D(A^{-1}) = R(A)$ *is again positive selfadjoint.*

b) *Let* $u \in W^{1,2}_{0,\sigma}(\Omega)$, $f \in L^2_\sigma(\Omega)$. *Then* u *is a weak solution of the Stokes system* (1.1.1) *with force* f *iff*

$$u \in D(A) \quad \text{and} \quad Au = f\,, \tag{2.1.5}$$

and this holds iff there exists some $p \in L^2_{loc}(\Omega)$, *unique up to a constant, satisfying*

$$-\nu\Delta u + \nabla p = f \tag{2.1.6}$$

in the sense of distributions.

c) *If* Ω *is bounded, then* $D(A^{-1}) = R(A) = L^2_\sigma(\Omega)$, *and* A^{-1} *is a bounded operator with operator norm*

$$\|A^{-1}\| \;\le\; C^2\,\nu^{-1}\,, \tag{2.1.7}$$

where $C = C(\Omega) > 0$ *is the constant in Poincaré's inequality* $\|v\|_2 \le C\,\|\nabla v\|_2$, $v \in C^\infty_{0,\sigma}(\Omega)$.

d) *If* Ω *is a uniform* C^2*-domain or if* $\Omega = \mathbb{R}^n$, *then*

$$D(A) = W^{1,2}_{0,\sigma}(\Omega) \cap W^{2,2}(\Omega)^n \quad , \quad Au = -\nu P\Delta u\,, \tag{2.1.8}$$

and

$$\|\nabla^2 u\|_2 + \nu^{-1}\|\nabla p\|_2 \;\le\; C\,(\nu^{-1}\|Au\|_2 + \|\nabla u\|_2 + \|u\|_2) \tag{2.1.9}$$

for all $u \in D(A)$. *Here* $C = C(\Omega) > 0$ *is a constant, and* $p \in L^2_{loc}(\Omega)$ *is defined up to a constant by* (2.1.6).

e) *If* Ω *is a bounded* C^2*-domain, then*

$$D(A) = L^2_\sigma(\Omega) \cap W^{1,2}_0(\Omega)^n \cap W^{2,2}(\Omega)^n,$$

and

$$\|u\|_{W^{2,2}(\Omega)} + \nu^{-1}\|\nabla p\|_2 \;\le\; C\,\nu^{-1}\|Au\|_2 \tag{2.1.10}$$

for all $u \in D(A)$. *Here* $C = C(\Omega) > 0$ *is a constant, and* p *defined by* (2.1.6) *is contained in* $L^2(\Omega)$.

Proof. To prove a) we use the bilinear form S defined by

$$S(u,v) := \nu < \nabla u, \nabla v > \quad , \quad u, v \in W^{1,2}_{0,\sigma}(\Omega) \tag{2.1.11}$$

with domain $D(S) := W^{1,2}_{0,\sigma}(\Omega)$. Obviously, S is positive and symmetric. S is a closed form since $D(S)$ is complete with respect to the norm $(\|u\|_2^2 + \|\nabla u\|_2^2)^{\frac{1}{2}}$.

The relations (2.1.3), (2.1.4) can be extended by continuity to all $v \in D(S)$. Lemma 3.2.1, II, now shows that A defined by (2.1.4) is a positive selfadjoint operator.

If $u \in D(A)$, then we see that $\|\nabla u\|_2 = 0$ iff $u = 0$. Thus we conclude from (3.2.32), II, that $N(A) = \{0\}$. Therefore, see (3.2.26), II, the inverse A^{-1} : $D(A^{-1}) \to L^2_\sigma(\Omega)$, $D(A^{-1}) = R(A)$, exists, A^{-1} is selfadjoint and positive. Since $N(A) = \{0\}$ we conclude from the selfadjointness of A that $R(A) = D(A^{-1})$ is dense in $L^2_\sigma(\Omega)$.

Since A is a positive selfadjoint operator, there exists a uniquely determined resolution $\{E_\lambda;\ \lambda \geq 0\}$ of identity in the Hilbert space $L^2_\sigma(\Omega)$, see Section 3.2, II. A has the spectral representation

$$A = \int_0^\infty \lambda dE_\lambda\,, \tag{2.1.12}$$

and we get

$$D(A) = \{v \in L^2_\sigma(\Omega)\ ;\ \|Av\|_2^2 = \int_0^\infty \lambda^2 d\|E_\lambda v\|_2^2 < \infty\}.$$

A^{-1} possesses the representation

$$A^{-1} = \int_0^\infty \lambda^{-1} dE_\lambda \tag{2.1.13}$$

with

$$D(A^{-1}) = \{v \in L^2_\sigma(\Omega);\ \|A^{-1}v\|_2^2 = \int_0^\infty \lambda^{-2} d\|E_\lambda v\|_2^2 < \infty\}$$

see (3.2.27), II.

To prove b), let $u \in W^{1,2}_{0,\sigma}(\Omega)$ and $f \in L^2_\sigma(\Omega)$. If u is a weak solution with force f according to Definition 1.1.1, then we use

$$\nu < \nabla u, \nabla v > = < f, v > \quad , \quad v \in C^\infty_{0,\sigma}(\Omega), \tag{2.1.14}$$

and see that $u \in D(A)$ and $Au = f$. Conversely, if $Au = f$, this relation holds, and therefore u is a weak solution with force f. The result concerning p now follows from Lemma 1.1.2.

To prove c), let $u \in D(A)$, $f := Au$, and consider the functional

$$v \mapsto < f, v > \ , \quad v \in W^{1,2}_{0,\sigma}(\Omega).$$

Using Poincaré's inequality we get

$$|< f, v >| \ \leq\ \|f\|_2\, \|v\|_2 \ \leq\ C\, \|f\|_2\, \|\nabla v\|_2,$$

and this shows continuity with respect to the norm $\|\nabla v\|_2$. The Hahn-Banach theorem then yields some $F \in L^2(\Omega)^{n^2}$ satisfying

$$< f, v > = < F, \nabla v > \quad , \quad v \in W_{0,\sigma}^{1,2}(\Omega),$$

and $\|F\|_2 \leq C\,\|f\|_2$.

We may set $u = v$ in (2.1.14). This shows that $\nu\|\nabla u\|_2 \leq C\,\|f\|_2$ and therefore that

$$\|u\|_2 \leq C\,\|\nabla u\|_2 \leq \nu^{-1}C^2\,\|f\|_2 = \nu^{-1}C^2\,\|Au\|_2. \tag{2.1.15}$$

Setting $f = Au$, $u = A^{-1}f$, and using that $R(A)$ is dense, we obtain

$$\|A^{-1}f\|_2 \leq \nu^{-1}C^2\|f\|_2$$

for all $f \in L^2_\sigma(\Omega)$. This proves c).

To prove d) we consider some $u \in D(A)$ and set $f := Au$. Then we get (2.1.6) with some $p \in L^2_{loc}(\Omega)$ and from Theorem 1.5.4, we obtain $u \in W^{2,2}(\Omega)^n$, $p \in L^2_{loc}(\overline{\Omega})$, $\nabla p \in L^2(\Omega)^n$. Applying the Helmholtz projection P to the equation (2.1.6) we see that $-\nu P\Delta u = f$. See Theorem 1.5.6 concerning the case $\Omega = \mathbb{R}^n$.

Conversely, let $u \in W_{0,\sigma}^{1,2}(\Omega) \cap W^{2,2}(\Omega)^n$. Then $-\nu\Delta u \in L^2(\Omega)^n$ and

$$\nu < \nabla u, \nabla v > = -\nu < \Delta u, v > = -\nu < \Delta u, Pv > = -\nu < P\Delta u, v >$$

for all $v \in W_{0,\sigma}^{1,2}(\Omega)$. Indeed, these relations are obvious if $v \in C_{0,\sigma}^\infty(\Omega)$, and the closure argument yields the validity for all $v \in W_{0,\sigma}^{1,2}(\Omega)$. This shows that $u \in D(A)$ and $Au = -\nu P\Delta u$. The inequality (2.1.9) follows from (1.5.32), (1.5.38). This proves d).

To prove e) we use

$$\widehat{W}_{0,\sigma}^{1,2}(\Omega) = W_{0,\sigma}^{1,2}(\Omega) = \{u \in W_0^{1,2}(\Omega)^n;\ \operatorname{div} u = 0\}, \tag{2.1.16}$$

see (1.2.8) and (1.2.9), and we need that

$$L^2_\sigma(\Omega) = \{u \in L^2(\Omega)^n;\ \operatorname{div} u = 0,\ N \cdot u|_{\partial\Omega} = 0\},$$

see (2.5.9), II. This shows that

$$D(A) = W_{0,\sigma}^{1,2}(\Omega) \cap W^{2,2}(\Omega)^n = L^2_\sigma(\Omega) \cap W_0^{1,2}(\Omega)^n \cap W^{2,2}(\Omega)^n. \tag{2.1.17}$$

Using the estimates (2.1.9) and (2.1.15) we obtain inequality (2.1.10). Since $p \in L^2_{loc}(\overline{\Omega})$ we see that $p \in L^2(\Omega)$. This proves the theorem. □

The regularity theory for the Stokes system yields the following properties.

2.1.2 Lemma *Let $k \in \mathbb{N}_0$ and let $\Omega \subseteq \mathbb{R}^n$, $n \geq 2$, be a bounded C^{k+2}-domain. Suppose $u \in D(A)$ and $f := Au \in W^{k,2}(\Omega)^n$. Then $u \in W^{k+2}(\Omega)^n$ and the pressure p, determined up to a constant by $\nabla p = f + \nu \Delta u$, see (2.1.6), satisfies $p \in W^{k+1}(\Omega)$.*

Moreover,

$$\|u\|_{W^{k+2,2}(\Omega)} + \nu^{-1}\|\nabla p\|_{W^{k,2}(\Omega)} \leq C\,\nu^{-1}\|f\|_{W^{k,2}(\Omega)} \tag{2.1.18}$$

with $C = C(k, \Omega) > 0$.

Proof. This result is a consequence of Theorem 1.5.3. □

For smooth unbounded domains we obtain the following local result.

2.1.3 Lemma *Let $k \in \mathbb{N}_0$, let $\Omega = \mathbb{R}^n$ or let $\Omega \subseteq \mathbb{R}^n$, $n \geq 2$, be any unbounded C^{k+2}-domain.*

Suppose $u \in D(A)$ and $f := Au \in W^{k,2}_{loc}(\overline{\Omega})^n$. Then $u \in W^{k+2,2}_{loc}(\overline{\Omega})^n$, and the pressure p, determined up to a constant by $\nabla p = f + \nu \Delta u$, satisfies $p \in W^{k+1,2}_{loc}(\overline{\Omega})$.

Moreover, for each bounded subdomain $G_0 \subseteq \Omega$ there exists another bounded subdomain $G_1 \subseteq \Omega$ with $G_0 \subseteq G_1$ such that

$$\begin{aligned} &\|u\|_{W^{k+2,2}(G_0)} + \nu^{-1}\|\nabla p\|_{W^{k,2}(G_0)} \\ &\quad \leq C\,(\nu^{-1}\|f\|_{W^{k,2}(G_1)} + \|\nabla u\|_{L^2(G_1)} + \|u\|_{L^2(G_1)}) \end{aligned} \tag{2.1.19}$$

with $C = C(k, \Omega) > 0$.

Proof. This result follows from Theorem 1.5.1 if $\Omega \neq \mathbb{R}^n$, and from Theorem 1.5.5 if $\Omega = \mathbb{R}^n$. □

2.2 The square root $A^{\frac{1}{2}}$ of A

The **square root** $A^{\frac{1}{2}}$ of the Stokes operator A and its inverse $A^{-\frac{1}{2}}$ play an important role in the theory of weak solutions of the Navier-Stokes system, see Section 2, I. Later on, see Section 2.5, we study the completions of the domains $D(A^{\frac{1}{2}})$, $D(A^{-\frac{1}{2}})$ in the norms $\|A^{\frac{1}{2}}v\|$ and $\|A^{-\frac{1}{2}}v\|_2$, respectively, and extend $A^{\frac{1}{2}}$ and $A^{-\frac{1}{2}}$ to the corresponding larger domains obtained in this way.

The existence of the square root $A^{\frac{1}{2}}$ and some properties are given in the next lemma.

Here we use the notation $A^{\frac{1}{2}}D(A) := \{A^{\frac{1}{2}}u;\ u \in D(A)\}$.

2.2.1 Lemma *Let $\Omega \subseteq \mathbb{R}^n$, $n \geq 2$, be any domain, and let $A : D(A) \to L^2_\sigma(\Omega)$, $D(A) \subseteq L^2_\sigma(\Omega)$, be the Stokes operator for Ω.*

Then there exists a uniquely determined positive selfadjoint operator $A^{\frac{1}{2}}$: $D(A^{\frac{1}{2}}) \to L^2_\sigma(\Omega)$ *with domain* $D(A^{\frac{1}{2}}) \subseteq L^2_\sigma(\Omega)$ *satisfying* $D(A) \subseteq D(A^{\frac{1}{2}})$,

$$A^{\frac{1}{2}} D(A) = D(A^{\frac{1}{2}}) \ , \quad Au = A^{\frac{1}{2}} A^{\frac{1}{2}} u \quad \textit{for all } u \in D(A). \tag{2.2.1}$$

The operator $A^{\frac{1}{2}}$ *has the following properties:*

a) $D(A^{\frac{1}{2}}) = W^{1,2}_{0,\sigma}(\Omega)$ *and*

$$< A^{\frac{1}{2}} u, A^{\frac{1}{2}} v > = \nu < \nabla u, \nabla v > \ , \quad \|A^{\frac{1}{2}} u\|_2 = \nu^{\frac{1}{2}} \|\nabla u\|_2 \tag{2.2.2}$$

for all $u, v \in W^{1,2}_{0,\sigma}(\Omega)$.

b) $N(A^{\frac{1}{2}}) = \{u \in D(A^{\frac{1}{2}});\ A^{\frac{1}{2}} u = 0\} = \{0\}$, *and the inverse* $A^{-\frac{1}{2}} := (A^{\frac{1}{2}})^{-1}$ *with dense domain* $D(A^{-\frac{1}{2}}) = R(A^{\frac{1}{2}}) = \{A^{\frac{1}{2}} u;\ u \in D(A^{\frac{1}{2}})\}$ *and range* $R(A^{-\frac{1}{2}}) = D(A^{\frac{1}{2}})$ *is again positive selfadjoint.*

c) *If* Ω *is bounded, then* $D(A^{-\frac{1}{2}}) = R(A^{\frac{1}{2}}) = L^2_\sigma(\Omega)$, *and* $A^{-\frac{1}{2}}$ *is a bounded operator with operator norm*

$$\|A^{-\frac{1}{2}}\| \ \leq \ C \nu^{-\frac{1}{2}} \ , \tag{2.2.3}$$

where $C = C(\Omega) > 0$ *means the constant in Poincaré's inequality* $\|u\|_2 \leq C \|\nabla u\|_2$, $u \in W^{1,2}_0(\Omega)^n$.

The proof of this lemma rests on the spectral representation

$$A = \int_0^\infty \lambda \, dE_\lambda \tag{2.2.4}$$

with $D(A) = \{u \in L^2_\sigma(\Omega);\ \|Au\|_2^2 = \int_0^\infty \lambda^2 \, d\|E_\lambda u\|_2^2 < \infty\}$, see (3.2.17), II. $\{E_\lambda;\ \lambda \geq 0\}$ means the resolution of identity for A. Similarly we get the representations

$$A^{\frac{1}{2}} = \int_0^\infty \lambda^{\frac{1}{2}} dE_\lambda \ , \quad A^{-\frac{1}{2}} = \int_0^\infty \lambda^{-\frac{1}{2}} dE_\lambda \tag{2.2.5}$$

with

$$D(A^{\frac{1}{2}}) = \{u \in L^2_\sigma(\Omega);\ \|A^{\frac{1}{2}} u\|_2^2 = \int_0^\infty \lambda \, d\|E_\lambda u\|_2^2 < \infty\},$$

and

$$D(A^{-\frac{1}{2}}) = \{u \in L^2_\sigma(\Omega);\ \|A^{-\frac{1}{2}} u\|_2^2 = \int_0^\infty \lambda^{-1} \, d\|E_\lambda u\|_2^2 < \infty\}.$$

More generally, for $-1 \leq \alpha \leq 1$ we define the positive selfadjoint operator

$$A^\alpha := \int_0^\infty \lambda^\alpha \, dE_\lambda \tag{2.2.6}$$

with domain

$$D(A^\alpha) = \{u \in L^2_\sigma(\Omega);\ \|A^\alpha u\|_2^2 = \int_0^\infty \lambda^{2\alpha}\, d\|E_\lambda u\|_2^2 < \infty\},$$

see (3.2.21), II, and (3.2.28), II.

From $N(A) = \{0\}$, see Theorem 2.1.1, we obtain that $N(A^\alpha) = \{0\}$, and it holds that

$$A^{-\alpha} = (A^{-1})^\alpha = (A^\alpha)^{-1},\ D(A^\alpha) = R(A^{-\alpha}), \tag{2.2.7}$$

$A^\alpha A^{-\alpha} u = u$ for all $u \in D(A^{-\alpha}) = R(A^\alpha)$, see Section 3.2, II. If $0 \le \alpha \le 1$, then $D(A) \subseteq D(A^\alpha)$ and $D(A^{-1}) = R(A) \subseteq D(A^{-\alpha}) = R(A^\alpha)$.

An important property is the interpolation inequality

$$\|A^\alpha u\|_2 \ \le\ \|Au\|_2^\alpha\, \|u\|_2^{1-\alpha}\ ,\quad u \in D(A)\ ,\ \ 0 \le \alpha \le 1, \tag{2.2.8}$$

and correspondingly

$$\|A^{-\alpha} u\|_2 \ \le\ \|A^{-1}u\|_2^\alpha\, \|u\|_2^{1-\alpha}\ ,\quad u \in D(A^{-1})\ ,\ \ 0 \le \alpha \le 1, \tag{2.2.9}$$

see (3.2.33), II.

If Ω is a bounded domain, then A^{-1} is a bounded operator, see (2.1.7), and for $0 \le \alpha \le 1$ we obtain the following properties:

$D(A^{-\alpha}) = R(A^\alpha) = L^2_\sigma(\Omega)$, $A^{-\alpha}$ is a bounded operator, and

$$\|A^{-\alpha}\| \ \le\ \|A^{-1}\|^\alpha \ \le\ C^{2\alpha}\,\nu^{-\alpha} \tag{2.2.10}$$

with C as in (2.1.7). Indeed, from (2.2.9) we get

$$\|A^{-\alpha}u\|_2 \ \le\ \|A^{-1}\|^\alpha\, \|u\|_2^\alpha\, \|u\|_2^{1-\alpha} = \|A^{-1}\|^\alpha\, \|u\|_2$$

and using (2.1.7) yields (2.2.10).

Proof of Lemma 2.2.1. Since A is a selfadjoint operator and $N(A) = \{0\}$, see Theorem 2.1.1, the operators $A^{\frac12}$ and $A^{-\frac12}$ defined by (2.2.5) are also selfadjoint and positive. We obtain

$$\begin{aligned} Au = \int_0^\infty \lambda\, dE_\lambda u \ &=\ \int_0^\infty \lambda^{\frac12}\, d\Big(E_\lambda \int_0^\infty \lambda^{\frac12}\, dE_\lambda u\Big) = \int_0^\infty \lambda^{\frac12}\, dE_\lambda v \\ &=\ \Big(\int_0^\infty \lambda^{\frac12}\, dE_\lambda\Big)\Big(\int_0^\infty \lambda^{\frac12}\, dE_\lambda u\Big) = A^{\frac12}\, A^{\frac12}\, u \end{aligned}$$

for all $u \in D(A)$, $v = A^{\frac12} u$, and

$$\|A^{\frac12} v\|_2^2 = \int_0^\infty \lambda\, d\|E_\lambda v\|_2^2 < \infty \qquad \text{iff} \qquad \|Au\|_2^2 = \int_0^\infty \lambda^2\, d\|E_\lambda u\|_2^2 < \infty.$$

This follows using the Riemann sums in the definition of these integrals, see Section 3.2, II. This proves (2.2.1). In particular we get $C^\infty_{0,\sigma}(\Omega) \subseteq D(A) \subseteq D(A^{\frac{1}{2}})$; see [Kat66] for details.

To prove a) we use the definition of A, see (2.1.4), and get

$$\nu < \nabla u, \nabla v > = < Au, v > = < A^{\frac{1}{2}} A^{\frac{1}{2}} u, v > = < A^{\frac{1}{2}} u, A^{\frac{1}{2}} v > \qquad (2.2.11)$$

for all $u, v \in D(A)$. In particular,

$$\nu^{\frac{1}{2}} \, \|\nabla u\|_2 = \|A^{\frac{1}{2}} u\|_2 \quad \text{for} \quad u = v \in D(A). \qquad (2.2.12)$$

Since $C^\infty_{0,\sigma}(\Omega) \subseteq D(A)$ and since $A^{\frac{1}{2}}$ is selfadjoint and therefore closed, we conclude from (2.2.12) that $W^{1,2}_{0,\sigma}(\Omega) \subseteq D(A^{\frac{1}{2}})$. On the other hand, using the above integrals, we find for each $u \in D(A^{\frac{1}{2}})$ a sequence $(u_j)^\infty_{j=1}$ in $D(A)$ satisfying $u = s - \lim_{j\to\infty} u_j$ and $A^{\frac{1}{2}} u = s - \lim_{j\to\infty} A^{\frac{1}{2}} u_j$. From (2.2.12) we conclude that the norms

$$\|u\|_2 + \|\nabla u\|_2 \quad \text{and} \quad \|u\|_2 + \|A^{\frac{1}{2}} u\|_2 \qquad (2.2.13)$$

are equivalent on $D(A)$.

This shows, $u \in D(A^{\frac{1}{2}})$ implies $u \in W^{1,2}_{0,\sigma}(\Omega)$. It follows that $D(A^{\frac{1}{2}}) = W^{1,2}_{0,\sigma}(\Omega)$, the norms (2.2.13) are equivalent on $W^{1,2}_{0,\sigma}(\Omega)$, and (2.2.12) holds for all $u \in W^{1,2}_{0,\sigma}(\Omega)$.

To prove b) we observe that $A^{\frac{1}{2}} u = 0$ iff $\nabla u = 0$.

This shows that $N(A^{\frac{1}{2}}) = \{0\}$, and that $A^{-\frac{1}{2}}$ exists with $D(A^{-\frac{1}{2}}) = R(A^{\frac{1}{2}})$; $A^{\frac{1}{2}}$ is positive selfadjoint and $R(A^{\frac{1}{2}})$ is dense in $L^2_\sigma(\Omega)$.

To prove c) we use (2.2.12) and Poincaré's inequality with C from (1.1.1), II. This yields

$$\|u\|_2 \, \le \, C \, \|\nabla u\|_2 \, = \, C \nu^{-\frac{1}{2}} \|A^{\frac{1}{2}} u\|_2 \ , \quad u \in D(A^{\frac{1}{2}}),$$

and setting $f = A^{\frac{1}{2}} u$, $u = A^{-\frac{1}{2}} f$, we see that $\|A^{-\frac{1}{2}} f\|_2 \le C \nu^{-\frac{1}{2}} \|f\|_2$ for all $f \in R(A^{\frac{1}{2}})$. Since $D(A^{-\frac{1}{2}}) = R(A^{\frac{1}{2}}) \subseteq L^2_\sigma(\Omega)$ is dense and $A^{-\frac{1}{2}}$ is a closed operator, we conclude from this inequality that $D(A^{-\frac{1}{2}}) = R(A^{\frac{1}{2}}) = L^2_\sigma(\Omega)$. Thus $A^{-\frac{1}{2}}$ is bounded and (2.2.3) holds. The proof is complete. □

2.3 The Stokes operator A in $\mathbb{R}^n$

If $\Omega = \mathbb{R}^n$, the Stokes operator A has special properties and can be expressed completely by the Laplacian Δ and the Helmholtz projection. This enables us to apply the potential theoretic arguments in Section 3.3, II.

We already know the special characterizations

$$W_0^{k,2}(\mathbb{R}^n)^n := \overline{C_0^\infty(\mathbb{R}^n)^n}^{\|\cdot\|_{k,2}} = W^{k,2}(\mathbb{R}^n)^n\,, \tag{2.3.1}$$

see (3.6.17), I,

$$W_{0,\sigma}^{1,2}(\mathbb{R}^n) := \overline{C_{0,\sigma}^\infty(\mathbb{R}^n)}^{\|\cdot\|_{1,2}} = \{u \in W^{1,2}(\mathbb{R}^n)^n;\ \mathrm{div}\ u = 0\} \tag{2.3.2}$$

see (2.5.23), II,

$$L_\sigma^2(\mathbb{R}^n) := \overline{C_{0,\sigma}^\infty(\mathbb{R}^n)}^{\|\cdot\|_2} = \{u \in L^2(\mathbb{R}^n)^n;\ \mathrm{div}\ u = 0\} \tag{2.3.3}$$

see (2.5.11), II, and

$$G(\mathbb{R}^n) = \overline{\nabla C_0^\infty(\mathbb{R}^n)}^{\|\cdot\|_2} = \{\nabla h \in L^2(\mathbb{R}^n)^n;\ h \in L_{loc}^2(\mathbb{R}^n)\}, \tag{2.3.4}$$

see (2.5.2), II, and (2.5.13), II.

Further we need the Helmholtz projection

$$P : L^2(\mathbb{R}^n)^n \to L_\sigma^2(\mathbb{R}^n)$$

for the whole space $\mathbb{R}^n$, see Section 2.5, II. For each $g \in L^2(\mathbb{R}^n)^n$ we define $g_0 := Pg \in L_\sigma^2(\mathbb{R}^n)$ and $\nabla h := (I - P)g \in G(\mathbb{R}^n)$ such that

$$g = g_0 + \nabla h\ ,\quad \|g_0\|_2^2 + \|\nabla h\|_2^2 = \|g\|_2^2\,. \tag{2.3.5}$$

The Laplace operator

$$-\Delta : D(\Delta) \to L^2(\mathbb{R}^n)^n$$

with domain

$$D(-\Delta) = D(\Delta) = \{u \in W^{1,2}(\mathbb{R}^n)^n\ ;\ \Delta u \in L^2(\mathbb{R}^n)^n\} \subseteq L^2(\mathbb{R}^n)^n, \tag{2.3.6}$$

see Section 3.3, II, is selfadjoint and positive. In particular,

$$< \nabla u, \nabla v > = < (-\Delta)u, v >\ ,\quad u \in D(\Delta)\,,\ v \in W^{1,2}(\mathbb{R}^n)^n, \tag{2.3.7}$$

see (3.3.4), II, and we obtain the spectral representation

$$-\Delta = \int_0^\infty \lambda\, d\tilde{E}_\lambda \tag{2.3.8}$$

where $\{\tilde{E}_\lambda\,;\lambda \geq 0\}$ means the resolution of identity for $-\Delta$, see Section 3.2, II.

The spectral representation of the Stokes operator

$$A : D(A) \to L^2_\sigma(\mathbb{R}^n)$$

is written as before in the form

$$A = \int_0^\infty \lambda \, dE_\lambda \tag{2.3.9}$$

with $\{E_\lambda \, ; \, \lambda \geq 0\}$ for A.

In the following we use the method of difference quotients in the same way as in the proof of Theorem 1.5.1. We need the quotients

$$(D_j^\delta u)(x) := \delta^{-1}\big(u(x + \delta e_j) - u(x)\big) \quad , \quad x \in \mathbb{R}^n \, , \quad \delta > 0, \tag{2.3.10}$$

in the x_j-direction, $j = 1, \dots, n$, see (1.5.5).

First we collect some properties of P. Here we use the notation

$$\nabla^k := (D_{j_1} D_{j_2} \dots D_{j_k})_{j_1, \dots, j_k = 1}^n \quad , \quad k \in \mathbb{N}.$$

2.3.1 Lemma *Let $k \in \mathbb{N}$, $n \geq 2$, $g \in L^2(\mathbb{R}^n)^n$, and let*

$$g = g_0 + \nabla h \quad , \quad g_0 = Pg \in L^2_\sigma(\mathbb{R}^n) \quad , \quad \nabla h \in G(\mathbb{R}^n) \, , \tag{2.3.11}$$

be the Helmholtz decomposition of g.

Then

$$g \in W^{k,2}(\mathbb{R}^n)^n \quad \textit{implies} \quad Pg \in W^{k,2}(\mathbb{R}^n)^n \, , \tag{2.3.12}$$

$$\|Pg\|_{W^{k,2}(\mathbb{R}^n)} \; \leq \; C \, \|g\|_{W^{k,2}(\mathbb{R}^n)} \, , \tag{2.3.13}$$

$C = C(n) > 0$, and

$$P(\nabla^k g) \; = \; \nabla^k Pg \tag{2.3.14}$$

for all $g \in W^{k,2}(\mathbb{R}^n)^n$. In particular,

$$P \Delta g = \Delta P g \tag{2.3.15}$$

for all $g \in W^{2,2}(\mathbb{R}^n)^n$.

Proof. Let $k = 1$. Applying D_j^δ with $\delta > 0$, $j = 1, \dots, n$, to $g = g_0 + \nabla h$ we get the decomposition

$$D_j^\delta g \; = \; D_j^\delta \, g_0 + \nabla D_j^\delta h \, , \tag{2.3.16}$$

and from

$$\|D_j^\delta \, g_0\|_2^2 \, + \, \|\nabla D_j^\delta \, h\|_2^2 \; = \; \|D_j^\delta \, g\|_2^2 \, ,$$

we see letting $\delta \to 0$ that $D_j g_0$, $D_j \nabla h \in L^2(\mathbb{R}^n)^n$, $j = 1, \ldots, n$, and therefore that

$$\|\nabla g_0\|_2 \leq C \|\nabla g\|_2$$

with $C = C(n)$. This proves (2.3.12) and (2.3.13) for $k = 1$. Letting $\delta \to 0$ in (2.3.16), we conclude that

$$D_j g = D_j g_0 + \nabla D_j h \quad , \quad j = 1, \ldots, n .$$

This shows that $PD_j g = D_j Pg$ for $j = 1, \ldots, n$. Thus we get (2.3.14) for $k = 1$. We can repeat this procedure, and the general result follows by induction on k. The relation (2.3.15) means that P commutes with Δ. The proof is complete. □

The same method yields the following regularity properties of the Stokes operator A in $\mathbb{R}^n$.

2.3.2 Lemma *Let $A : D(A) \to L^2_\sigma(\mathbb{R}^n)$, $n \geq 2$, be the Stokes operator for $\mathbb{R}^n$. Then*

$$D(A) = W^{2,2}(\mathbb{R}^n)^n \cap L^2_\sigma(\mathbb{R}^n) = \{u \in W^{2,2}(\mathbb{R}^n)^n;\ \text{div } u = 0\}, \tag{2.3.17}$$

and

$$Au = -\nu \, P\Delta u = -\nu \Delta u \quad \textit{for all } u \in D(A). \tag{2.3.18}$$

Further let $k \in \mathbb{N}_0$, $u \in D(A)$, $f := Au$, $p \in L^2_{loc}(\mathbb{R}^n)$, and suppose that

$$-\nu \Delta u + \nabla p = f. \tag{2.3.19}$$

Then

$$f \in W^{k,2}(\mathbb{R}^n)^n \ \textit{implies} \ u \in W^{k+2,2}(\mathbb{R}^n)^n \quad , \quad \nabla p \in W^{k,2}(\mathbb{R}^n)^n ,$$

and

$$\|\nabla^{k+2} u\|_2 + \nu^{-1} \|\nabla^{k+1} p\|_2 \leq \nu^{-1} C \, \|\nabla^k f\|_2 \tag{2.3.20}$$

with $C = C(k, n) > 0$.

Proof. Note that the pressure p is determined only up to a constant, see Lemma 1.1.2.

Let $u \in D(A)$, $f := Au$, $p \in L^2_{loc}(\mathbb{R}^n)$, with $-\nu \Delta u + \nabla p = f$. Then we apply the difference operators D^δ_j as above for $\delta > 0$, $j = 1, \ldots, n$, and obtain

$$-\nu \Delta (D^\delta_j u) + \nabla (D^\delta_j p) = D^\delta_j f \quad , \quad D^\delta_j u \in W^{1,2}_{0,\sigma}(\mathbb{R}^n). \tag{2.3.21}$$

A calculation yields

$$(D_j^\delta f)(x) = \frac{1}{\delta}\int_0^\delta (D_j f)(x+\tau e_j)\,d\tau \;=\; D_j\Big(\frac{1}{\delta}\int_0^\delta f(x+\tau e_j)\,d\tau. \qquad (2.3.22)$$

This shows that $D_j^\delta f$ has the form $D_j^\delta f = \operatorname{div} F_{\delta,j}$ with some $F_{\delta,j} \in L^2(\mathbb{R}^n)^{n^2}$ defined by (2.3.22). See [Agm65, Sec. 3], [Fri69, I, 15] for details concerning difference quotients. Using Hölder's inequality we see that

$$\|F_{\delta,j}\|_2 \;\le\; C\,\|f\|_2$$

with some $C > 0$ not depending on δ and j. This enables us to apply Theorem 1.3.1. We obtain

$$\|D_j^\delta \nabla u\|_2 \;\le\; \nu^{-1}\,\|F_{\delta,j}\|_2 \;\le\; C\nu^{-1}\,\|f\|_2 \;\;,\;\; j = 1,\dots,n.$$

Letting $\delta \to 0$ yields $\|\nabla^2 u\|_2 < \infty$ and

$$\|\nabla^2 u\|_2 \;\le\; C\,\nu^{-1}\|f\|_2$$

with $C = C(n) > 0$. It follows that

$$D(A) \;\subseteq\; W^{2,2}(\mathbb{R}^n)^n \cap L^2_\sigma(\mathbb{R}^n).$$

Conversely, let $u \in W^{2,2}(\mathbb{R}^n)^n \cap L^2_\sigma(\mathbb{R}^n)$. Then from (2.3.3) and (2.3.2) we see that $u \in W^{1,2}_{0,\sigma}(\mathbb{R}^n)$, and

$$\nu < \nabla u, \nabla v > = < f, v > \qquad \text{for all } \; v \in C^\infty_{0,\sigma}(\mathbb{R}^n)$$

with some $f \in L^2_\sigma(\mathbb{R}^n)$. This shows that $u \in D(A)$ and (2.3.17) follows.

If $u \in D(A)$, $f = Au$, and $p \in L^2_{loc}(\mathbb{R}^n)$ with $-\nu\Delta u + \nabla p = f$, then $\nabla p \in L^2(\mathbb{R}^n)^n, \Delta u \in L^2(\mathbb{R}^n)^n$, and applying P yields

$$-\nu P\Delta u = f.$$

Using (2.3.15), we get $-\nu P\Delta u = -\nu\Delta P u = -\nu\Delta u = f$. This proves (2.3.18). From $-\nu\Delta u + \nabla p = f$ we obtain the estimate

$$\|\nabla p\|_2 \;\le\; C\,\|f\|_2$$

with $C = C(n) > 0$. This proves (2.3.20) for $k = 0$. Repeating the procedure with the difference operators D_j^δ, the result follows for $k = 1$, and so on. The general result follows by induction on k . This completes the proof. □

In the same way we obtain the regularity properties of the Laplace operator Δ in $\mathbb{R}^n$, see Section 3.3, II, concerning this operator.

If $u \in D(\Delta)$ and $f := -\Delta u$, then $u \in W^{2,2}(\mathbb{R}^n)^n$ and the estimate

$$\|\nabla^2 u\|_2 \;\le\; C\,\|f\|_2 \tag{2.3.23}$$

holds with $C = C(n) > 0$. Instead of (2.3.17) we now get

$$D(\Delta) = W^{2,2}(\mathbb{R}^n)^n. \tag{2.3.24}$$

If $u \in D(\Delta),\ f := -\Delta u$, then

$$f \in W^{k,2}(\mathbb{R}^n)^n \quad \text{implies} \quad u \in W^{k+2,2}(\mathbb{R}^n)^n,$$

and the estimate

$$\|\nabla^{k+2} u\|_2 \;\le\; C\,\|\nabla^k f\|_2 \tag{2.3.25}$$

holds for all $k \in \mathbb{N}_0$ with $C = C(k,n) > 0$, see [SiSo96].

Now we are able to express in $\mathbb{R}^n$ the fractional powers A^α in terms of the fractional powers $(-\Delta)^\alpha$ of $-\Delta$.

Let $\alpha \in \mathbb{R}$. Then we use the spectral representations (2.3.8), (2.3.9) and get

$$A^\alpha = \int_0^\infty \lambda^\alpha\, dE_\lambda \quad , \quad (-\Delta)^\alpha = \int_0^\infty \lambda^\alpha\, d\tilde{E}_\lambda, \tag{2.3.26}$$

see Section 3.2, II. Since P commutes with Δ by (2.3.15), we know, see [Yos80, XI, 12, Theorem 1], that P also commutes with each projection $\tilde{E}_\lambda$, $\lambda \ge 0$. Therefore, $\{\tilde{E}_\lambda P;\ \lambda \ge 0\}$ is a resolution of identity in the Hilbert space $PL^2(\mathbb{R}^n)^n = L^2_\sigma(\mathbb{R}^n)$. For each $u \in D(A)$, we see with $Pu = u$ and (2.3.18) that

$$\begin{aligned} Au &= \int_0^\infty \lambda\, dE_\lambda u \;=\; \nu(-\Delta)u \\ &= \nu(-\Delta)Pu \;=\; \nu \int_0^\infty \lambda\, d\tilde{E}_\lambda u \;=\; \nu \int_0^\infty \lambda\, d\tilde{E}_\lambda P u. \end{aligned}$$

If $\nu = 1$ we conclude that $E_\lambda = \tilde{E}_\lambda P$ for all $\lambda \ge 0$, since the resolution of identity is uniquely determined by A. In this case we see that

$$A^\alpha u = \int_0^\infty \lambda^\alpha\, dE_\lambda u = \int_0^\infty \lambda^\alpha\, d\tilde{E}_\lambda P u = \int_0^\infty \lambda^\alpha\, d\tilde{E}_\lambda u \;=\; (-\Delta)^\alpha u$$

holds for all $u \in D(A^\alpha)$, and that $D(A^\alpha) = D((-\Delta)^\alpha) \cap L^2_\sigma(\mathbb{R}^n)$. The general case $\nu > 0$ can be reduced to this case when we replace u by νu. This yields for arbitrary $\nu > 0$ the representation $Au = (-\Delta)\nu u$, $A^\alpha u = (-\Delta)^\alpha \nu^\alpha u$, and it follows that

$$A^\alpha u = (-\Delta)^\alpha \nu^\alpha u = \nu^\alpha \int_0^\infty \lambda^\alpha \, d\tilde{E}_\lambda u.$$

Thus we obtain the following result.

2.3.3 Lemma *Consider the Stokes operator $A : D(A) \to L^2_\sigma(\mathbb{R}^n)$ and the Laplace operator $\Delta : D(\Delta) \to L^2(\mathbb{R}^n)^n$ in $\mathbb{R}^n$, $n \geq 2$, and let $\alpha \in \mathbb{R}$. Then we obtain*

$$D(A^\alpha) \;=\; D((-\Delta)^\alpha) \cap L^2_\sigma(\mathbb{R}^n) \tag{2.3.27}$$

and

$$A^\alpha u = \nu^\alpha (-\Delta)^\alpha u \tag{2.3.28}$$

for all $u \in D(A^\alpha)$.

Proof. See above. □

2.4 Embedding properties of $D(A^\alpha)$

Our aim is to prove continuous embeddings of the domain $D(A^\alpha)$ into certain L^q-spaces. Such properties are needed to estimate the nonlinear term $u \cdot \nabla u$. First we consider the case $\Omega = \mathbb{R}^n$. In this case the problem can be reduced to the embedding properties of $D((-\Delta)^\alpha)$. These properties are developed in Lemma 3.3.1, II, and lead to the following result.

2.4.1 Lemma *Let $n \geq 2$, $0 \leq \alpha < \frac{n}{4}$, $2 \leq q < \infty$ such that*

$$2\alpha + \frac{n}{q} = \frac{n}{2}\,, \tag{2.4.1}$$

and let A be the Stokes operator for $\mathbb{R}^n$.

Suppose $u \in D(A^\alpha)$, then $u \in L^q(\mathbb{R}^n)^n$ and

$$\|u\|_{L^q(\mathbb{R}^n)^n} \;\leq\; C\,\nu^{-\alpha} \|A^\alpha u\|_{L^2(\mathbb{R}^n)^n} \tag{2.4.2}$$

where $C = C(\alpha, q) > 0$ is a constant.

Remark The condition (2.4.2) means that

$$D(A^\alpha) \;\subseteq\; L^q(\mathbb{R}^n)^n \tag{2.4.3}$$

is continuously embedded with respect to the norm $\|A^\alpha u\|_2$.

Setting $f = A^\alpha u$, $u = A^{-\alpha} f$, we get the following equivalent formulation of (2.4.2): If $f \in D(A^{-\alpha}) = R(A^\alpha)$, then $A^{-\alpha} f \in L^q(\mathbb{R}^n)^n$ and

$$\|A^{-\alpha} f\|_{L^q(\mathbb{R}^n)^n} \le C\nu^{-\alpha}\|f\|_{L^2(\mathbb{R}^n)^n}. \tag{2.4.4}$$

Proof. From Lemma 2.3.3 we obtain

$$D(A^\alpha) = D((-\Delta)^\alpha) \cap L^2_\sigma(\mathbb{R}^n),$$

and if $u \in D(A^\alpha)$ we get $\|A^\alpha u\|_2 = \nu^\alpha \|(-\Delta)^\alpha u\|_2$. Using Lemma 3.3.1, II, with $\frac{\alpha}{2}$ replaced by α, we see that $u \in L^q(\mathbb{R}^n)^n$ and that

$$\|u\|_q \le C\,\|(-\Delta)^\alpha u\|_2 = C\nu^{-\alpha}\|A^\alpha u\|_2$$

with some constant $C = C(\alpha, q) > 0$. This proves the lemma. □

In the next step we extend this embedding lemma to arbitrary domains Ω. For this purpose we use the Heinz inequality, see Lemma 3.2.3, II. This requires the additional restriction $0 \le \alpha \le \frac{1}{2}$.

2.4.2 Lemma *Let $\Omega \subseteq \mathbb{R}^n$, $n \ge 2$, be any domain, let $0 \le \alpha \le \frac{1}{2}$, $2 \le q < \infty$ with*

$$2\alpha + \frac{n}{q} = \frac{n}{2}, \tag{2.4.5}$$

and let A be the Stokes operator for Ω.

Then $u \in D(A^\alpha)$ implies $u \in L^q(\Omega)^n$ and

$$\|u\|_{L^q(\Omega)^n} \le C\nu^{-\alpha}\|A^\alpha u\|_{L^2(\Omega)^n} \tag{2.4.6}$$

where $C = C(\alpha, q) > 0$ is a constant.

Remarks Note that the constant C does not depend on Ω. In the special case $n = 2$, the result is contained in [BMi92, (3.6)] with a different proof. In this case, $\alpha = \frac{1}{2}$ is excluded since $2 \le q < \infty$. For bounded and exterior domains there is a more general result, see [BMi91] and [GiSo89].

From (2.4.6) we get the continuous embedding

$$D(A^\alpha) \subseteq L^q(\Omega)^n \tag{2.4.7}$$

with respect to the norm $\|A^\alpha u\|_2$ of $D(A^\alpha)$. Setting $f = A^\alpha u$, $u = A^{-\alpha} f$, we see, $f \in D(A^{-\alpha}) = R(A^\alpha)$ implies $A^{-\alpha} f \in L^q(\Omega)^n$ and

$$\|A^{-\alpha} f\|_{L^q(\Omega)^n} \le C\nu^{-\alpha}\|f\|_{L^2(\Omega)^n}. \tag{2.4.8}$$

Proof. Let $A_1 := A^{\frac{1}{2}}$ be the square root of the Stokes operator A for Ω in the Hilbert space $H_1 := L^2_\sigma(\Omega)$, let A_+ be the Stokes operator for $\mathbb{R}^n$, and let $A_2 = A_+^{\frac{1}{2}}$ be the square root of A_+ in the Hilbert space $H_2 := L^2_\sigma(\mathbb{R}^n)$. Define the operator B from $C^\infty_{0,\sigma}(\Omega)$ into $C^\infty_{0,\sigma}(\mathbb{R}^n)$ by setting $(Bu)(x) := u(x)$ if $x \in \Omega$, $(Bu)(x) := 0$ if $x \neq \Omega$ (extension by zero). Then B extends by closure to an isometric and therefore bounded linear operator, again denoted by B, from

$$H_1 = \overline{C^\infty_{0,\sigma}(\Omega)}^{\|\cdot\|_2} \quad \text{into} \quad H_2 = \overline{C^\infty_{0,\sigma}(\mathbb{R}^n)}^{\|\cdot\|_2}.$$

If $u \in D(A_1) = W^{1,2}_{0,\sigma}(\Omega)$, then $Bu \in D(A_2) = W^{1,2}_{0,\sigma}(\mathbb{R}^n)$. This follows since B is also isometric with respect to the norm $\|u\|_2 + \|\nabla u\|_2$, $u \in C^\infty_{0,\sigma}(\Omega)$. From Lemma 2.2.1 we get

$$\|A_2 Bu\|_{L^2(\mathbb{R}^n)} = \nu^{\frac{1}{2}} \|\nabla Bu\|_{L^2(\mathbb{R}^n)} = \nu^{\frac{1}{2}} \|\nabla u\|_{L^2(\Omega)} = \|A_1 u\|_{L^2(\Omega)}.$$

Therefore, H_1, H_2, A_1, A_2 and B, defined in this way, satisfy the assumptions of Lemma 3.2.3, II, and we may conclude for $0 \le \beta \le 1$, B maps $D(A_1^\beta)$ into $D(A_2^\beta)$ and

$$\|A_2^\beta Bu\|_{L^2(\mathbb{R}^n)^n} \le \|B\|^{1-\beta} \|A_1^\beta u\|_{L^2(\Omega)^n}$$

for all $u \in D(A_1^\beta)$.

Since B is isometric, we get $\|B\| = 1$ for the operator norm. Setting $\beta = 2\alpha$, and applying Lemma 2.4.1 to A_+ and Bu we get

$$\begin{aligned} \|u\|_{L^q(\Omega)^n} &= \|Bu\|_{L^q(\mathbb{R}^n)^n} \le C\nu^{-\alpha} \|A_+^\alpha Bu\|_{L^2(\mathbb{R}^n)^n} \\ &= C\nu^{-\alpha} \|A_2^\beta Bu\|_{L^2(\mathbb{R}^n)^n} \le C\nu^{-\alpha} \|A_1^\beta u\|_{L^2(\Omega)^n} \\ &= C\nu^{-\alpha} \|A^\alpha u\|_{L^2(\Omega)^n} \end{aligned}$$

with the same constant $C = C(\alpha, n) > 0$ as in Lemma 2.4.1. Since $0 \le \beta \le 1$, we get the restriction $0 \le \alpha \le \frac{1}{2}$. This proves the lemma. □

The next lemma yields the continuous embedding

$$D(A^\alpha) \subseteq W^{1,q}(\Omega)^n$$

for certain values α and q. Now the norm of $D(A^\alpha)$ must be the graph norm $\|u\|_2 + \|A^\alpha u\|_2$. In this case we need a smoothness property on Ω and we have to prepare some further facts.

Let $\Omega \subseteq \mathbb{R}^n$, $n \ge 2$, be a uniform C^2-domain, see Section 3.2, I. Then there exists a bounded linear operator $E : u \mapsto Eu$ from $W^{2,2}(\Omega)^n$ into $W^{2,2}(\mathbb{R}^n)^n$ satisfying $Eu|_\Omega = u$ on Ω. Thus

$$\|Eu\|_{W^{2,2}(\mathbb{R}^n)^n} \le C \|u\|_{W^{2,2}(\Omega)^n}, \quad u \in W^{2,2}(\Omega)^n \tag{2.4.9}$$

with some constant $C = C(\Omega) > 0$. Moreover, E can be chosen such that

$$\|Eu\|_{L^2(\mathbb{R}^n)^n} \le C\,\|u\|_{L^2(\Omega)^n}\ , \quad u \in W^{2,2}(\Omega)^n \tag{2.4.10}$$

with $C = C(\Omega) > 0$. E is called an **extension operator** from Ω to $\mathbb{R}^n$; see [Ada75, IV, 4.29]. The proof rests on the same localization method as in step c) of the proof of Theorem 1.5.1.

Further we use some regularity properties of A if Ω is a uniform C^2-domain, see Theorem 2.1.1, d). From (2.1.9) we obtain the inequality

$$\|\nabla^2 u\|_2 \le C\,(\nu^{-1}\|Au\|_2 + \|\nabla u\|_2 + \|u\|_2) \tag{2.4.11}$$

for all $u \in D(A) = W^{1,2}_{0,\sigma}(\Omega) \cap W^{2,2}(\Omega)^n$ with $C = C(\Omega) > 0$ and it holds that $Au = -\nu P\Delta u$.

From Lemma 2.2.1 we get

$$\|\nabla u\|_2 = \nu^{-\frac{1}{2}}\,\|A^{\frac{1}{2}}u\|_2\ , \quad u \in D(A^{\frac{1}{2}}) = W^{1,2}_{0,\sigma}(\Omega)\,, \tag{2.4.12}$$

and using the interpolation inequality (2.2.8) we see that

$$\begin{aligned}\|\nabla u\|_2 &= \nu^{-\frac{1}{2}}\,\|A^{\frac{1}{2}}u\|_2 \le \nu^{-\frac{1}{2}}\|Au\|_2^{\frac{1}{2}}\,\|u\|_2^{\frac{1}{2}}\\ &\le \frac{1}{2}\nu^{-1}\|Au\|_2 + \frac{1}{2}\|u\|_2.\end{aligned}$$

Therefore we get

$$\|\nabla^2 u\|_2 \le C\,(\nu^{-1}\|Au\|_2 + \|u\|_2) \tag{2.4.13}$$

for all $u \in D(A)$ with $C = C(\Omega) > 0$.

From the spectral representation $A = \int_0^\infty \lambda\, dE_\lambda$, see (2.1.12), we get

$$\begin{aligned}\nu^{-1}\|Au\|_2 + \|u\|_2 &= \nu^{-1}(\|Au\|_2 + \|\nu u\|_2)\\ &\le \sqrt{2}\,\nu^{-1}(\|Au\|_2^2 + \|\nu u\|_2^2)^{\frac{1}{2}} = \sqrt{2}\,\nu^{-1}\left(\int_0^\infty (\lambda^2+\nu^2)\, d\|E_\lambda u\|_2^2\right)^{\frac{1}{2}}\\ &\le 2\nu^{-1}\left(\int_0^\infty (\lambda+\nu)^2\, d\|E_\lambda u\|_2^2\right)^{\frac{1}{2}} = 2\nu^{-1}\|(\nu I + A)u\|_2\end{aligned}$$

for $u \in D(A)$.

If $\frac{1}{2} \le \alpha \le 1$, we obtain in the same way that

$$\nu^{-\alpha}\|A^\alpha u\|_2 + \|u\|_2 \le 2\nu^{-\alpha}\|(\nu I + A)^\alpha u\|_2, \tag{2.4.14}$$

and

$$\nu^{-\alpha}\|(\nu I + A)^\alpha u\|_2 \le C\,(\nu^{-\alpha}\|A^\alpha u\|_2 + \|u\|_2) \tag{2.4.15}$$

with $C = C(\alpha) > 0$. Using (2.4.13) and (2.4.14) with $\alpha = 1$ yields

$$\|u\|_{W^{2,2}(\Omega)^n} \leq C\nu^{-1}\|(\nu I + A)u\|_{L^2(\Omega)^n} \tag{2.4.16}$$

for all $u \in D(A)$ with $C = C(\Omega) > 0$. We get the following result.

2.4.3 Lemma *Let $\Omega \subseteq \mathbb{R}^n$, $n \geq 2$, be a uniform C^2-domain, let $\frac{1}{2} \leq \alpha \leq 1$, $2 \leq q < \infty$ such that*

$$2\alpha + \frac{n}{q} = 1 + \frac{n}{2}, \tag{2.4.17}$$

and let A be the Stokes operator for Ω.

If $u \in D(A^\alpha)$, then $u \in W^{1,q}(\Omega)^n$ and

$$\|u\|_{W^{1,q}(\Omega)^n} \leq C\,(\nu^{-\alpha}\|A^\alpha u\|_{L^2(\Omega)^n} + \|u\|_{L^2(\Omega)^n}) \tag{2.4.18}$$

where $C = C(\Omega, \alpha, n) > 0$ is a constant.

Proof. We use again the Heinz inequality (3.2.35), II. Let $H_1 := L^2_\sigma(\Omega)$, $A_1 := \nu I + A$ with $D(A_1) = D(A)$, $H_2 := L^2(\mathbb{R}^n)^n$, $A_2 := I - \Delta$ with $D(A_2) = D(\Delta) = W^{2,2}(\mathbb{R}^n)^n$, see (2.3.24), and let $B := E$ be the extension operator in (2.4.9).

Using (2.4.10) we see that B can be extended to a bounded operator from H_1 into H_2 with operator norm $\|B\|$. B maps $D(A_1)$ into $D(A_2)$, and from the inequalities (2.4.9), (2.4.16) we obtain

$$\begin{aligned}\|A_2 Bu\|_{L^2(\mathbb{R}^n)^n} &= \|(I-\Delta)Bu\|_{L^2(\mathbb{R}^n)^n} \leq C_1\,\|Bu\|_{W^{2,2}(\mathbb{R}^n)^n} \\ &\leq C_2\,\|u\|_{W^{2,2}(\Omega)^n} \leq C_3\,\nu^{-1}\|A_1 u\|_{L^2(\Omega)^n}\end{aligned}$$

for all $u \in D(A_1)$; $C_1, C_2, C_3 > 0$ are constants depending on Ω.

Therefore we may apply Lemma 3.2.3, II, and get

$$\|A_2^\alpha Bu\|_{L^2(\mathbb{R}^n)^n} \leq C_3^\alpha \nu^{-\alpha}\|B\|^{1-\alpha}\,\|A_1^\alpha u\|_{L^2(\Omega)^n}\,,$$

for $0 \leq \alpha \leq 1$, $u \in D(A_1^\alpha)$.

Now we can apply Lemma 3.3.3, II, replacing $\frac{\alpha}{2}$ by α, and obtain

$$\begin{aligned}\|u\|_{W^{1,q}(\Omega)^n} &\leq \|Bu\|_{W^{1,q}(\mathbb{R}^n)^n} \leq C_4\,\|(I-\Delta)^\alpha Bu\|_{L^2(\mathbb{R}^n)^n} \\ &= C_4\,\|A_2^\alpha Bu\|_{L^2(\mathbb{R}^n)^n} \leq C_4\,C_3^\alpha\,\nu^{-\alpha}\|B\|^{1-\alpha}\,\|A_1^\alpha u\|_{L^2(\Omega)^n} \\ &= C_4\,C_3^\alpha\,\nu^{-\alpha}\|B\|^{1-\alpha}\,\|(\nu I + A)^\alpha u\|_{L^2(\Omega)^n}\end{aligned}$$

with $C_4 = C_4(\alpha, n) > 0$. Using (2.4.15) leads now to the desired estimate (2.4.18). This proves the lemma. □

We conclude under the assumptions of Lemma 2.4.3, that the embedding

$$D(A^\alpha) \subseteq W^{1,q}(\Omega)^n \tag{2.4.19}$$

is continuous with respect to the norm $\|u\|_2 + \|A^\alpha u\|_2$ for $D(A^\alpha)$.

2.5 Completion of the space $D(A^\alpha)$

We consider the completion $\widehat{D}(A^\alpha)$ of the space $D(A^\alpha)$ with respect to the norm $\|A^\alpha u\|_2$, $-\frac{1}{2} \le \alpha \le \frac{1}{2}$, study some embedding properties, and extend the operator A^α by taking the closure from $D(A^\alpha)$ to the larger domain $\widehat{D}(A^\alpha)$. Here we need that these operators A^α are injective, which follows from the injectivity of A, see Theorem 2.1.1. This will enable us to characterize properties of weak solutions in terms of the Stokes operator. One important aim is to show that the weak solutions of the Stokes system are solutions of a certain operator equation, see Lemma 2.6.3. First we consider the most important case $\alpha = \frac{1}{2}$.

Let $\Omega \subseteq \mathbb{R}^n$, $n \ge 2$, be any domain, let A be the Stokes operator for Ω with domain $D(A) \subseteq L^2_\sigma(\Omega)$, and let $A^{\frac{1}{2}}$ be the square root of A with domain $D(A^{\frac{1}{2}}) \subseteq L^2_\sigma(\Omega)$. We know, see Lemma 2.2.1, that

$$D(A^{\frac{1}{2}}) = W^{1,2}_{0,\sigma}(\Omega) = \overline{C^\infty_{0,\sigma}(\Omega)}^{\|\cdot\|_{W^{1,2}(\Omega)}} \tag{2.5.1}$$

and that

$$\|A^{\frac{1}{2}}u\|_2 = \nu^{\frac{1}{2}}\|\nabla u\|_2 \ , \quad u \in D(A^{\frac{1}{2}}). \tag{2.5.2}$$

The norm

$$\|u\|_{D(A^{\frac{1}{2}})} := (\|u\|_2^2 + \|A^{\frac{1}{2}}u\|_2^2)^{\frac{1}{2}} = (\|u\|_2^2 + \nu\|\nabla u\|_2^2)^{\frac{1}{2}} \tag{2.5.3}$$

is the graph norm of $D(A^{\frac{1}{2}})$, and

$$\|u\|_{\widehat{D}(A^{\frac{1}{2}})} := \|A^{\frac{1}{2}}u\|_2 = \nu^{\frac{1}{2}}\|\nabla u\|_2 \tag{2.5.4}$$

is the **homogeneous graph norm** of $D(A^{\frac{1}{2}})$. Obviously, $D(A^{\frac{1}{2}})$ is complete and a Hilbert space with respect to the graph norm (2.5.3). But $D(A^{\frac{1}{2}})$ is complete with respect to the norm (2.5.4) and this norm is equivalent to (2.5.3), iff the estimate

$$\|u\|_2 \le C\,\|\nabla u\|_2 \ , \quad u \in D(A^{\frac{1}{2}})$$

holds with some $C > 0$ (Poincaré inequality). If this estimate does not hold for all $u \in D(A^{\frac{1}{2}})$, the completion of $D(A^{\frac{1}{2}})$ in the homogeneous norm (2.5.4) becomes a strictly larger space. This is an abstract space of (classes of) Cauchy sequences, and we have to study the concrete characterization (embeddings).

Let

$$\widehat{D}(A^{\frac{1}{2}}) := \overline{D(A^{\frac{1}{2}})}^{\|A^{\frac{1}{2}}u\|_2} \tag{2.5.5}$$

be the **completion** of $D(A^{\frac{1}{2}})$ with respect to the homogeneous norm (2.5.4). Because of (2.5.4) it is obvious that

$$\widehat{D}(A^{\frac{1}{2}}) = \widehat{W}^{1,2}_{0,\sigma}(\Omega) = \overline{C^\infty_{0,\sigma}(\Omega)}^{\|\nabla u\|_2} \tag{2.5.6}$$

where $\widehat{W}^{1,2}_{0,\sigma}(\Omega)$ is the weak solution space (1.1.2).

Therefore, $\widehat{D}(A^{\frac{1}{2}})$ has the same embedding properties as $\widehat{W}^{1,2}_{0,\sigma}(\Omega)$, see Lemma 1.2.1. To obtain the trivial embedding $D(A^{\frac{1}{2}}) \subseteq \widehat{D}(A^{\frac{1}{2}})$, we identify each $u \in D(A^{\frac{1}{2}})$ with the sequence $(u, u, \ldots) \in \widehat{D}(A^{\frac{1}{2}})$. See [Gal94a, II.5] concerning the homogeneous norm of $\widehat{W}^{1,2}_{0,\sigma}(\Omega)$.

By definition, $D(A^{\frac{1}{2}}) \subseteq \widehat{D}(A^{\frac{1}{2}})$ is a dense subspace with respect to the norm $\|A^{\frac{1}{2}}u\|_2$.
Therefore, by the usual closure procedure, we can extend the operator $u \mapsto A^{\frac{1}{2}}u$ from the original domain $D(A^{\frac{1}{2}})$ to the larger space $\widehat{D}(A^{\frac{1}{2}})$, keeping the same notation $A^{\frac{1}{2}}$. Thus we get the (well defined) extended operator

$$A^{\frac{1}{2}} : \widehat{D}(A^{\frac{1}{2}}) \to L^2_\sigma(\Omega) \tag{2.5.7}$$

which is called the **closure extension** of $u \mapsto A^{\frac{1}{2}}u$ from $D(A^{\frac{1}{2}})$ to $\widehat{D}(A^{\frac{1}{2}})$.

Let $u \in \widehat{D}(A^{\frac{1}{2}})$. Then, by definition, u is represented by a Cauchy sequence $u = (u_j)_{j=1}^\infty$, $u_j \in D(A^{\frac{1}{2}})$, with respect to the norm $\|A^{\frac{1}{2}}u\|_2$. This means that $(A^{\frac{1}{2}}u_j)_{j=1}^\infty$ is a Cauchy sequence in $L^2_\sigma(\Omega)$ with respect to the norm $\|\cdot\|_2$. Then we define $A^{\frac{1}{2}}u \in L^2_\sigma(\Omega)$ by setting

$$A^{\frac{1}{2}}u := s - \lim_{j\to\infty} A^{\frac{1}{2}}u_j \,. \tag{2.5.8}$$

This defines the extended operator (2.5.7).

Consider now the more general case $-\frac{1}{2} \le \alpha \le \frac{1}{2}$, and let

$$A^\alpha = \int_0^\infty \lambda^\alpha \, dE_\lambda \tag{2.5.9}$$

be the fractional power of A with domain

$$D(A^\alpha) = \{u \in L^2_\sigma(\Omega);\ \|A^\alpha u\|_2^2 = \int_0^\infty \lambda^{2\alpha}\, d\|E_\lambda u\|_2^2 < \infty\}, \tag{2.5.10}$$

see Section 3.2, II. Then we define the (abstract) **completion**

$$\widehat{D}(A^\alpha) := \overline{D(A^\alpha)}^{\|A^\alpha u\|_2} \tag{2.5.11}$$

of $D(A^\alpha)$ with respect to the homogeneous graph norm

$$\|u\|_{\widehat{D}(A^\alpha)} := \|A^\alpha u\|_2 \;, \quad u \in D(A^\alpha). \tag{2.5.12}$$

As above we extend the operator $u \mapsto A^\alpha u$ from $D(A^\alpha)$ to the larger domain $\widehat{D}(A^\alpha)$, and obtain the (well defined) extended operator

$$A^\alpha : \widehat{D}(A^\alpha) \to L^2_\sigma(\Omega) \tag{2.5.13}$$

as the closure extension. $\widehat{D}(A^\alpha)$ becomes a Hilbert space with scalar product

$$< A^\alpha u, A^\alpha v > = \int_\Omega (A^\alpha u) \cdot (A^\alpha v)\, dx \qquad u, v \in \widehat{D}(A^\alpha).$$

From the embedding estimate in Lemma 2.4.2 we obtain immediately the continuous embedding

$$\widehat{D}(A^\alpha) \subseteq L^q(\Omega)^n \tag{2.5.14}$$

with $0 \le \alpha \le \frac{1}{2}$, $2 \le q < \infty$, $2\alpha + \frac{n}{q} = \frac{n}{2}$. To obtain this embedding precisely, we have to identify each Cauchy sequence $(u_j)_{j=1}^\infty$ from $\widehat{D}(A^\alpha)$ with the (well defined) limit

$$u = s - \lim_{j \to \infty} u_j$$

in $L^q(\Omega)^n$. Note that the case $n = 2$, $\alpha = \frac{1}{2}$, $q = \infty$ is excluded.

In particular we are interested in the case $\alpha = -\frac{1}{2}$. Here we need some further preparations.

We know, see (2.2.7), that $A^{-\frac{1}{2}} : D(A^{-\frac{1}{2}}) \to L^2_\sigma(\Omega)$ is the inverse of $A^{\frac{1}{2}} : D(A^{\frac{1}{2}}) \to L^2_\sigma(\Omega)$. We get $D(A^{-\frac{1}{2}}) = R(A^{\frac{1}{2}})$, $D(A^{\frac{1}{2}}) = R(A^{-\frac{1}{2}})$,

$$A^{-\frac{1}{2}} = (A^{-1})^{\frac{1}{2}} = (A^{\frac{1}{2}})^{-1} = \int_0^\infty \lambda^{-\frac{1}{2}}\, dE_\lambda\,, \tag{2.5.15}$$

and

$$D(A^{-\frac{1}{2}}) = \{u \in L^2_\sigma(\Omega);\ \|A^{-\frac{1}{2}} u\|_2^2 = \int_0^\infty \lambda^{-1}\, d\|E_\lambda u\|_2^2 < \infty\}.$$

Consider now in particular the completion

$$\widehat{D}(A^{-\frac{1}{2}}) := \overline{D(A^{-\frac{1}{2}})}^{\|A^{-\frac{1}{2}} u\|_2}, \tag{2.5.16}$$

endowed with the homogeneous norm

$$\|u\|_{\widehat{D}(A^{-\frac{1}{2}})} := \|A^{-\frac{1}{2}} u\|_2\,, \tag{2.5.17}$$

and the closure extension

$$A^{-\frac{1}{2}} : \widehat{D}(A^{-\frac{1}{2}}) \to L^2_\sigma(\Omega)\,, \tag{2.5.18}$$

defined as in (2.5.8) with $A^{\frac{1}{2}}$ replaced by $A^{-\frac{1}{2}}$. $\widehat{D}(A^{-\frac{1}{2}})$ is a Hilbert space with scalar product

$$< A^{-\frac{1}{2}}u, A^{-\frac{1}{2}}v > = \int_\Omega (A^{-\frac{1}{2}}u)\cdot(A^{-\frac{1}{2}}v)\,dx \ , \quad u,v \in \widehat{D}(A^{-\frac{1}{2}}),$$

and the embedding $D(A^{-\frac{1}{2}}) \subseteq \widehat{D}(A^{-\frac{1}{2}})$ is obtained by identifying each $u \in D(A^{-\frac{1}{2}})$ with $(u, u, \dots)$.

Since $R(A^{-\frac{1}{2}}) = D(A^{\frac{1}{2}}) \subseteq L^2_\sigma(\Omega)$ is dense, we can show that the extended operator (2.5.18) is even surjective. Indeed, consider any $g \in L^2_\sigma(\Omega)$. Then we find $u_j \in D(A^{-\frac{1}{2}}) = R(A^{\frac{1}{2}})$, $j \in \mathbb{N}$, such that

$$g = s - \lim_{j\to\infty} A^{-\frac{1}{2}}u_j \ .$$

Thus $u := (u_j)_{j=1}^\infty \in \widehat{D}(A^{-\frac{1}{2}})$ holds in the abstract sense and we get

$$A^{-\frac{1}{2}}u = s - \lim_{j\to\infty} A^{-\frac{1}{2}}u_j = g.$$

Further we observe that $D(A^{-\frac{1}{2}})$ is never complete with respect to the norm $\|A^{-\frac{1}{2}}u\|_2$. Otherwise we conclude by the closed graph theorem [Yos80, II, 6] that $A^{-\frac{1}{2}}u \mapsto u$, $u \in D(A^{-\frac{1}{2}})$, is a bounded operator, and this means that $u \mapsto A^{\frac{1}{2}}u$, $u \in D(A^{\frac{1}{2}})$, is bounded, which is never true. Therefore, $\widehat{D}(A^{-\frac{1}{2}})$ is strictly larger than $D(A^{-\frac{1}{2}})$ for each domain Ω.

We need the following direct characterization of $\widehat{D}(A^{-\frac{1}{2}})$ and of the extended operator $A^{-\frac{1}{2}}$ in terms of functionals defined on $C^\infty_{0,\sigma}(\Omega)$. For this purpose let

$$u = (u_j)_{j=1}^\infty \ , \quad u_j \in D(A^{-\frac{1}{2}}) \ , \quad j \in \mathbb{N}, \tag{2.5.19}$$

be any element of $\widehat{D}(A^{-\frac{1}{2}})$ represented by a Cauchy sequence $(u_j)_{j=1}^\infty$ in the norm $\|\cdot\|_{\widehat{D}(A^{-\frac{1}{2}})}$. Then $(A^{-\frac{1}{2}}u_j)_{j=1}^\infty$ is a Cauchy sequence in $L^2_\sigma(\Omega)$, and

$$A^{-\frac{1}{2}}u = s - \lim_{j\to\infty} A^{-\frac{1}{2}}u_j \in L^2_\sigma(\Omega)\,.$$

We will identify u with the well defined functional

$$\begin{aligned} [u,\cdot\,] : v \mapsto [u,v] \ &:= \ \lim_{j\to\infty}[u_j,v] = \lim_{j\to\infty} < u_j, v > \\ &= \ \lim_{j\to\infty} < A^{-\frac{1}{2}}u_j, A^{\frac{1}{2}}v > = < A^{-\frac{1}{2}}u, A^{\frac{1}{2}}v >\,, \end{aligned}$$

$v \in C^\infty_{0,\sigma}(\Omega)$. Thus u is identified with the functional

$$[u,\cdot\,] : v \mapsto [u,v] = < A^{-\frac{1}{2}}u, A^{\frac{1}{2}}v > \ , \quad v \in C^\infty_{0,\sigma}(\Omega)\,,$$

which is an element of $C_{0,\sigma}^{\infty}(\Omega)'$, see (3.5.12), I. Since $A^{-\frac{1}{2}}u \in L^2_\sigma(\Omega)$, this functional is continuous in the norm $\|A^{\frac{1}{2}}v\|_2 = \nu^{\frac{1}{2}}\|\nabla v\|_2$. This yields the natural embedding

$$D(A^{-\frac{1}{2}}) \subseteq \widehat{D}(A^{-\frac{1}{2}}) \subseteq C_{0,\sigma}^{\infty}(\Omega)' . \tag{2.5.20}$$

Recall, $C_{0,\sigma}^{\infty}(\Omega)'$ is the space of restrictions of distributions in Ω to the test space $C_{0,\sigma}^{\infty}(\Omega)$. Another formulation: $C_{0,\sigma}^{\infty}(\Omega)'$ is the space of (classes of) distributions modulo gradients. Thus we may use the theory of distributions.

In particular, each $u \in D(A^{-\frac{1}{2}})$ is identified with the functional $[u, \cdot] = < u, \cdot >$ such that

$$v \mapsto [u,v] \; = \; < u, v > = < A^{-\frac{1}{2}}u, A^{\frac{1}{2}}v > \quad , \quad v \in C_{0,\sigma}^{\infty}(\Omega).$$

We write $u = [u, \cdot\,]$ for all $u \in \widehat{D}(A^{-\frac{1}{2}})$ and get the following easy characterization.

2.5.1 Lemma *Let $\Omega \subseteq \mathbb{R}^n$, $n \geq 2$, be any domain, let A be the Stokes operator of Ω, let $\widehat{D}(A^{-\frac{1}{2}})$ be the completion of $D(A^{-\frac{1}{2}})$ with respect to the norm $\|A^{-\frac{1}{2}}u\|_2$, and let $A^{-\frac{1}{2}} : \widehat{D}(A^{-\frac{1}{2}}) \to L^2_\sigma(\Omega)$ be the closure extension of $u \mapsto A^{-\frac{1}{2}}u$ from $D(A^{-\frac{1}{2}})$ to $\widehat{D}(A^{-\frac{1}{2}})$.*

Then we have:

a) *$\widehat{D}(A^{-\frac{1}{2}})$ is (identified with) the space of all functionals $[u, \cdot\,] : v \mapsto [u, v]$, $v \in C_{0,\sigma}^{\infty}(\Omega)$, which are continuous in the norm $\|\nabla v\|_2$. Here each (abstract) element $u = (u_j)_{j=1}^{\infty}$ from $\widehat{D}(A^{-\frac{1}{2}})$ is identified with the functional*

$$v \mapsto \; [u,v] \; = \; \lim_{j \to \infty} \; < A^{-\frac{1}{2}}u_j, A^{\frac{1}{2}}v > = < A^{-\frac{1}{2}}u, A^{\frac{1}{2}}v > \; .$$

b) *For each $u \in \widehat{D}(A^{-\frac{1}{2}})$, the element $A^{-\frac{1}{2}}u \in L^2_\sigma(\Omega)$ is uniquely determined by the relation*

$$[u, v] \; = < A^{-\frac{1}{2}}u, A^{\frac{1}{2}}v > \; , \quad v \in C_{0,\sigma}^{\infty}(\Omega). \tag{2.5.21}$$

Remarks Let $u \in C_{0,\sigma}^{\infty}(\Omega)'$ be any functional defined on the test space $C_{0,\sigma}^{\infty}(\Omega)$. Then we see from a) that $u \in \widehat{D}(A^{-\frac{1}{2}})$ iff there is some $C = C(u) > 0$ with

$$|[u, v]| \; \leq \; C\,\|\nabla v\|_2 \; , \quad v \in C_{0,\sigma}^{\infty}(\Omega). \tag{2.5.22}$$

From (2.5.21) we see that $\|A^{-\frac{1}{2}}u\|_2$ is the functional norm of $u \in \widehat{D}(A^{-\frac{1}{2}})$ with respect to the norm $\|A^{\frac{1}{2}}v\|_2$, and therefore we get

$$\|A^{-\frac{1}{2}}u\|_2 \; \leq \; C\,\nu^{-\frac{1}{2}} \tag{2.5.23}$$

with C from (2.5.22) if we use that $\|\nabla u\|_2 = \nu^{-\frac{1}{2}}\|A^{\frac{1}{2}}u\|_2$.

We will simply write

$$A^{-\frac{1}{2}}u \in L^2_\sigma(\Omega) \quad \text{iff} \quad u \in \widehat{D}(A^{-\frac{1}{2}}). \tag{2.5.24}$$

Proof of Lemma 2.5.1. Consider $u = (u_j)_{j=1}^\infty$, $u_j \in D(A^{-\frac{1}{2}})$, such that $(A^{-\frac{1}{2}}u_j)_{j=1}^\infty$ is a Cauchy sequence in $L^2_\sigma(\Omega)$. Then

$$\begin{aligned} |[u,v]| &= |\lim_{j\to\infty}[u_j,v]| = |\lim_{j\to\infty} < A^{-\frac{1}{2}}u_j, A^{\frac{1}{2}}v > | \\ &\le \|s-\lim_{j\to\infty} A^{-\frac{1}{2}}u_j\|_2 \, \|A^{\frac{1}{2}}v\|_2 = \nu^{\frac{1}{2}}\|s-\lim_{j\to\infty} A^{-\frac{1}{2}}u_j\|_2 \, \|\nabla v\|_2 \,, \end{aligned}$$

$v \in C^\infty_{0,\sigma}(\Omega)$, which means that the functional $v \mapsto [u,v]$ is continuous in the norm $\|\nabla v\|_2$. Setting

$$A^{-\frac{1}{2}}u = s - \lim_{j\to\infty} A^{-\frac{1}{2}}u_j$$

we see that (2.5.21) is satisfied, and that $A^{-\frac{1}{2}}u$ is uniquely determined by (2.5.21).

Conversely, let $v \mapsto [u,v]$, $v \in C^\infty_{0,\sigma}(\Omega)$, be any functional which is continuous in $\|\nabla v\|_2$. Then the Riesz representation theorem, see Section 3.2, II, yields a unique $\widehat{u} \in L^2_\sigma(\Omega)$ satisfying the relation

$$[u,v] = < \widehat{u}, A^{\frac{1}{2}}v > \quad , \quad v \in C^\infty_{0,\sigma}(\Omega).$$

Since $R(A^{-\frac{1}{2}}) = D(A^{\frac{1}{2}}) \subseteq L^2_\sigma(\Omega)$ is dense, we find a sequence $(u_j)_{j=1}^\infty$ in $D(A^{-\frac{1}{2}})$ satisfying

$$\widehat{u} = s - \lim_{j\to\infty} A^{-\frac{1}{2}}u_j.$$

This shows that $\widehat{u} = A^{-\frac{1}{2}}u$ in the extended sense, (2.5.21) is satisfied, and u is identified with the Cauchy sequence $(u_j)_{j=1}^\infty$ with respect to the norm $\|A^{-\frac{1}{2}}u\|_2$. This proves the lemma. □

The result of this lemma also holds with $A^{-\frac{1}{2}}$ replaced by $A^{-\alpha}$, $0 \le \alpha \le \frac{1}{2}$. This yields the characterization of

$$\widehat{D}(A^{-\alpha}) \subseteq C^\infty_{0,\sigma}(\Omega)'$$

as the space of all functionals $[u,\cdot\,] : v \mapsto [u,v]$, defined on $C^\infty_{0,\sigma}(\Omega)$, which are continuous with respect to the norm $\|A^\alpha v\|_2$. The closure extension

$$A^{-\alpha} : \widehat{D}(A^{-\alpha}) \to L^2_\sigma(\Omega)$$

from $D(A^{-\alpha})$ to the larger domain $\widehat{D}(A^{-\alpha})$ is determined by the relation

$$[u,v] = < A^{-\alpha}u, A^{\alpha}v > \quad , \quad v \in C^{\infty}_{0,\sigma}(\Omega)\,, \tag{2.5.25}$$

we write $u = [u,\cdot]$ for each $u \in \widehat{D}(A^{-\alpha})$, and we may use Lemma 2.5.1 with $A^{-\frac{1}{2}}$, $A^{\frac{1}{2}}$ replaced by $A^{-\alpha}$, A^{α}.

In the next lemma we use a notation which was already introduced in (3.5.17), I. Recall the natural embeddings

$$L^2_{\sigma}(\Omega) \subseteq C^{\infty}_{0,\sigma}(\Omega)' \quad , \quad L^2(\Omega)^n \subseteq C^{\infty}_0(\Omega)^{n\prime}$$

into the functional spaces $C^{\infty}_{0,\sigma}(\Omega)'$, $C^{\infty}_0(\Omega)^{n\prime}$, see (3.5.14), I, and (3.5.15), I, where we identify any element $f \in L^2(\Omega)^n$ with the functional

$$< f, \cdot > : v \mapsto < f, v >$$

defined on $C^{\infty}_{0,\sigma}(\Omega)$ or on $C^{\infty}_0(\Omega)^n$, respectively.

Then we define the natural extension of the Helmholtz projection

$$P : L^2(\Omega)^n \to L^2_{\sigma}(\Omega)$$

to get an operator from $C^{\infty}_0(\Omega)^{n\prime}$ into $C^{\infty}_{0,\sigma}(\Omega)'$, again denoted by P. We simply set

$$Pf := f|_{C^{\infty}_{0,\sigma}(\Omega)} \quad , \quad f \in C^{\infty}_0(\Omega)^{n\prime}. \tag{2.5.26}$$

Thus for arbitrary $f \in C^{\infty}_0(\Omega)^{n\prime}$, Pf means the restriction of the distribution $f \in C^{\infty}_0(\Omega)^{n\prime}$ to the test space $C^{\infty}_{0,\sigma}(\Omega)$. Using the embeddings above we see that Pf coincides with the Helmholtz projection if $f \in L^2(\Omega)^n$.

In particular we obtain

$$P(\nabla p) = 0$$

for each distribution $\nabla p \in C^{\infty}_0(\Omega)^{n\prime}$ with $p \in C^{\infty}_0(\Omega)'$, and

$$Pf \in \widehat{D}(A^{-\alpha})$$

in the following lemma is well defined.

2.5.2 Lemma *Let $\Omega \subseteq \mathbb{R}^n$, $n \geq 2$, be any domain, and let $0 \leq \alpha \leq \frac{1}{2}$, $1 < q \leq 2$ be given with*

$$2\alpha + \frac{n}{2} = \frac{n}{q}\,.$$

Then for each $f \in L^q(\Omega)^n$ we get $Pf \in \widehat{D}(A^{-\alpha})$ and

$$\|A^{-\alpha}Pf\|_2 \;\leq\; C\nu^{-\alpha}\|f\|_q \tag{2.5.27}$$

with some constant $C = C(\alpha, n) > 0$. Thus

$$A^{-\alpha}P \,:\, f \mapsto A^{-\alpha}Pf \tag{2.5.28}$$

is a bounded operator from $L^q(\Omega)^n$ to $L^2_{\sigma}(\Omega)$.

Proof. With $q' = \frac{q}{q-1}$, $\frac{1}{q'} + \frac{1}{q} = 1$, we get $2\alpha + \frac{n}{q'} = 2\alpha + n - \frac{n}{q} = 2\alpha + \frac{n}{2} + \frac{n}{2} - \frac{n}{q} = \frac{n}{2}$, and using (2.4.8) we obtain

$$\begin{aligned} |[f,v]| &= |[Pf,v]| = |< f, A^{-\alpha} A^{\alpha} v >| \leq \|f\|_q \, \|A^{-\alpha} A^{\alpha} v\|_{q'} \\ &= \|f\|_q \, \nu^{-\alpha} C \, \|A^{\alpha} v\|_2 \end{aligned}$$

for all $v \in C_{0,\sigma}^{\infty}(\Omega)$ with $C = C(\alpha, q') > 0$. Therefore, $Pf : v \mapsto [Pf, v]$ is continuous with respect to $\|A^{\alpha} v\|_2$. Using Lemma 2.5.1, a), with $A^{-\alpha}$ instead of $A^{-\frac{1}{2}}$, we see that $Pf \in \widehat{D}(A^{-\alpha})$. Correspondingly we get instead of (2.5.21) the relation

$$[Pf, v] = < A^{-\alpha} Pf, A^{\alpha} v > \ , \quad v \in C_{0,\sigma}^{\infty}(\Omega) .$$

Therefore, the last estimate leads to (2.5.27). This proves the lemma. □

2.6 The operator $A^{-\frac{1}{2}} P \operatorname{div}$

This operator plays a basic role in the theory of weak solutions in the next sections.

We explain two equivalent possibilities to define precisely the meaning of

$$A^{-\frac{1}{2}} P \operatorname{div} F \in L_{\sigma}^2(\Omega)$$

for each matrix field $F = (F_{kl})_{k,l=1}^n \in L^2(\Omega)^{n^2}$. Recall that the distribution

$$\operatorname{div} F = (D_1 F_{1l} + \cdots + D_n F_{nl})_{l=1}^n \in C_0^{\infty}(\Omega)^{n\prime}$$

is defined by applying div to the columns of F, see Section 1.2, I.

In the next lemma we see that

$$F \mapsto A^{-\frac{1}{2}} P \operatorname{div} F \tag{2.6.1}$$

is well defined as a bounded operator from $C_0^{\infty}(\Omega)^{n^2}$ into $L_{\sigma}^2(\Omega)$. This enables us to extend this operator by closure from $C_0^{\infty}(\Omega)^{n^2}$ to the larger domain $L^2(\Omega)^{n^2}$.

Another possibility is to use the extended meaning of P and $A^{-\frac{1}{2}}$, see (2.5.26), (2.5.18), and to show directly that (2.6.1) is a bounded operator from $L^2(\Omega)^{n^2}$ to $L_{\sigma}^2(\Omega)$, see Lemma 2.6.2.

2.6.1 Lemma *Let $\Omega \subseteq \mathbb{R}^n$, $n \geq 2$, be any domain. Then*

$$P \operatorname{div} F \in D(A^{-\frac{1}{2}})$$

and

$$\|A^{-\frac{1}{2}}P\text{div } F\|_2 \leq \nu^{-\frac{1}{2}}\|F\|_2 \tag{2.6.2}$$

for all $F \in C_0^\infty(\Omega)^{n^2}$. *Therefore,*

$$F \mapsto A^{-\frac{1}{2}}P\,\text{div } F \ , \quad F \in C_0^\infty(\Omega)^{n^2}$$

extends by closure to a bounded operator

$$A^{-\frac{1}{2}}P\text{div } : L^2(\Omega)^{n^2} \to L^2_\sigma(\Omega) \tag{2.6.3}$$

with operator norm

$$\|A^{-\frac{1}{2}}P\text{div }\| \leq \nu^{-\frac{1}{2}}. \tag{2.6.4}$$

For each $F \in L^2(\Omega)^{n^2}$, $A^{-\frac{1}{2}}P\text{div } F \in L^2_\sigma(\Omega)$ *is uniquely determined by the relation*

$$< A^{-\frac{1}{2}}P\text{div } F, A^{\frac{1}{2}}v > \; = \; - < F, \nabla v > \ , \quad v \in C^\infty_{0,\sigma}(\Omega). \tag{2.6.5}$$

Proof. Let $F \in C_0^\infty(\Omega)^{n^2}$, $v \in D(A^{\frac{1}{2}})$, and $w := A^{\frac{1}{2}}v$. Then div $F \in C_0^\infty(\Omega)^n \subseteq L^2(\Omega)^n$, P div $F \in L^2_\sigma(\Omega)$, and we get

$$< P \text{ div } F, v > = < P \text{ div } F, A^{-\frac{1}{2}}w > = - < F, \nabla v > .$$

Using

$$\|\nabla v\|_2 = \nu^{-\frac{1}{2}}\|A^{\frac{1}{2}}v\|_2 = \nu^{-\frac{1}{2}}\|w\|_2,$$

we see that

$$\begin{aligned} | < P \text{ div } F, A^{-\frac{1}{2}}w > | = | < F, \nabla v > | &\leq \|F\|_2\|\nabla v\|_2 \\ &\leq \nu^{-\frac{1}{2}}\|F\|_2\|w\|_2. \end{aligned} \tag{2.6.6}$$

Therefore, $< P$ div $F, A^{-\frac{1}{2}}w >$ is continuous in $\|w\|_2$, $w \in R(A^{\frac{1}{2}}) = D(A^{-\frac{1}{2}})$. Since $A^{-\frac{1}{2}}$ is selfadjoint we see that P div $F \in D(A^{-\frac{1}{2}})$, and (2.6.6) leads to (2.6.2).

Since $C_0^\infty(\Omega)^{n^2} \subseteq L^2(\Omega)^{n^2}$ is dense in the norm $\|\cdot\|_2$, we can extend the operator (2.6.1) by closure to a uniquely determined operator from $L^2(\Omega)^{n^2}$ to $L^2_\sigma(\Omega)$, keeping the notation $A^{-\frac{1}{2}}P$ div. The estimate (2.6.4) is a consequence of (2.6.2).

To prove (2.6.5) we choose any $F \in L^2(\Omega)^{n^2}$ and a sequence $(F_j)_{j=1}^{\infty}$ in $C_0^\infty(\Omega)^{n^2}$ satisfying

$$\lim_{j\to\infty} \|F - F_j\|_2 = 0.$$

Then the above definition by closure shows that

$$A^{-\frac{1}{2}}P \text{ div } F = s - \lim_{j\to\infty} A^{-\frac{1}{2}}P \text{ div } F_j.$$

Therefore we get

$$\begin{aligned} < A^{-\frac{1}{2}}P\text{div } F, A^{\frac{1}{2}}v > &= \lim_{j\to\infty} < A^{-\frac{1}{2}}P \text{ div } F_j, A^{\frac{1}{2}}v > \\ &= -\lim_{j\to\infty} < F_j, \nabla v > = - < F, \nabla v > \end{aligned}$$

in particular for all $v \in C_{0,\sigma}^\infty(\Omega)$. Since by definition $C_{0,\sigma}^\infty(\Omega)$ is dense in $W_{0,\sigma}^{1,2}(\Omega) = D(A^{\frac{1}{2}})$, the last relation can be extended to all $v \in D(A^{\frac{1}{2}})$. Thus (2.6.5) holds and $A^{-\frac{1}{2}}P$ div F is uniquely determined by this relation. This proves the lemma. □

The next lemma yields the direct characterization of $A^{-\frac{1}{2}}P$ div by means of functionals. Indeed, for each $F \in L^2(\Omega)^{n^2}$, div F is now only a distribution, and we can show that P div $F \in \widehat{D}(A^{-\frac{1}{2}})$ which leads to the same element $A^{-\frac{1}{2}}P$ div $F \in L^2_\sigma(\Omega)$ as above.

2.6.2 Lemma *Let $\Omega \subseteq \mathbb{R}^n$, $n \geq 2$, be any domain, and let $F \in L^2(\Omega)^{n^2}$. Then the functional*

$$P\text{div } F : v \mapsto [P\text{div } F, v] = [\text{div } F, v] = - < F, \nabla v >, \tag{2.6.7}$$

$v \in C_{0,\sigma}^\infty(\Omega)$, is contained in $\widehat{D}(A^{-\frac{1}{2}})$, and

$$A^{-\frac{1}{2}}P\text{div } F \in L^2_\sigma(\Omega)$$

in the extended sense is uniquely determined by

$$[\text{div } F, v] = - < F, \nabla v > = < A^{-\frac{1}{2}}P\text{div } F, A^{\frac{1}{2}}v >, \tag{2.6.8}$$

$v \in C_{0,\sigma}^\infty(\Omega)$.

Proof. For each $F \in L^2(\Omega)^{n^2}$ we get

$$\begin{aligned} |[\text{div } F, v]| = |[P \text{ div } F, v]| &= |< F, \nabla v >| \leq \|F\|_2 \, \|\nabla v\|_2 \\ &= \nu^{-\frac{1}{2}} \|F\|_2 \, \|A^{\frac{1}{2}}v\|_2, \end{aligned}$$

$v \in C_{0,\sigma}^\infty(\Omega)$. Therefore, the functional $v \mapsto [\text{ div } F, v]$ is continuous in $\|\nabla v\|_2$, and Lemma 2.5.1, a), yields P div $F \in \widehat{D}(A^{-\frac{1}{2}})$. The second assertion follows from (2.5.21) and the proof is complete. □

In other words, if we are interested in the element $A^{-\frac{1}{2}}P\text{div } F \in L^2_\sigma(\Omega)$, we have to apply test functions of the form $A^{\frac{1}{2}}v$, $v \in C^\infty_{0,\sigma}(\Omega)$, and get

$$\begin{aligned} < A^{-\frac{1}{2}}P\text{div } F, A^{\frac{1}{2}}v > &= [P \text{ div } F, v] = [\text{div } F, v] \\ &= - < F, \nabla v > . \end{aligned}$$

This yields

$$\begin{aligned} |< A^{-\frac{1}{2}}P\text{div } F, A^{\frac{1}{2}}v >| &= |< F, \nabla v >| \leq \|F\|_2 \, \|\nabla v\|_2 \\ &= \nu^{-\frac{1}{2}}\|F\|_2 \, \|A^{\frac{1}{2}}v\|_2 \end{aligned}$$

and

$$\begin{aligned} \|A^{-\frac{1}{2}}P\text{div } F\|_2 &\leq \nu^{-\frac{1}{2}}\|F\|_2 \, , \\ \|A^{-\frac{1}{2}}P \text{ div}\| &\leq \nu^{-\frac{1}{2}} . \end{aligned}$$

The operator $A^{-\frac{1}{2}}P$ div can be used to give a direct characterization of weak solutions of the Stokes system in terms of the extended operators $A^{\frac{1}{2}}$, $A^{-\frac{1}{2}}$, and P.

2.6.3 Lemma *Let $\Omega \subseteq \mathbb{R}^n$, $n \geq 2$, be any domain, and let $f = \text{div } F$ with $F \in L^2(\Omega)^{n^2}$.*

Then $u \in \widehat{W}^{1,2}_{0,\sigma}(\Omega) = \widehat{D}(A^{\frac{1}{2}})$ is a weak solution of the Stokes system

$$-\nu\Delta u + \nabla p = f \ \ , \quad \text{div } v = 0 \ \ , \quad u|_{\partial\Omega} = 0$$

iff the equation

$$A^{\frac{1}{2}}u = A^{-\frac{1}{2}}P\text{div } F \tag{2.6.9}$$

is satisfied in the extended sense.

Proof. From (2.5.6) we see that $\widehat{D}(A^{\frac{1}{2}}) = \widehat{W}^{1,2}_{0,\sigma}(\Omega)$. If $u \in \widehat{D}(A^{\frac{1}{2}})$ is a weak solution, we have

$$\nu < \nabla u, \nabla v > = [f, v] = - < F, \nabla v > \tag{2.6.10}$$

for all $v \in C^\infty_{0,\sigma}(\Omega)$, see Definition 1.1.1. If even $u \in W^{1,2}_{0,\sigma}(\Omega)$, we see by Lemma 2.2.1 that

$$\nu < \nabla u, \nabla v > = < A^{\frac{1}{2}}u, A^{\frac{1}{2}}v > \ \ , \quad v \in C^\infty_{0,\sigma}(\Omega). \tag{2.6.11}$$

The definition of $A^{\frac{1}{2}} : \widehat{D}(A^{\frac{1}{2}}) \to L^2_\sigma(\Omega)$ as operator closure, (2.5.7), shows that (2.6.11) also holds if only $u \in \widehat{W}^{1,2}_{0,\sigma}(\Omega)$. Using (2.6.5) we see that

$$- < F, \nabla v > = < A^{-\frac{1}{2}}P \text{ div } F, A^{\frac{1}{2}}v > \ \ , \quad v \in C^\infty_{0,\sigma}(\Omega).$$

Thus (2.6.10) implies the relation (2.6.9). Conversely, if $u \in \widehat{D}(A^{\frac{1}{2}})$ satisfies (2.6.9), we get (2.6.10) and u is a weak solution. This proves the lemma. □

3 The stationary Navier-Stokes equations

3.1 Weak solutions

In this section $\Omega \subseteq \mathbb{R}^n$ is any domain with $n = 2$ or $n = 3$. This restriction is caused by the structure of the nonlinear term. To estimate this term, we need Sobolev's embedding theorem which depends on the dimension n.

The stationary Navier-Stokes system has the form

$$\begin{aligned} -\nu\Delta u + u \cdot \nabla u + \nabla p &= f \, , \ \operatorname{div} u = 0 \, , \\ u|_{\partial\Omega} &= 0, \end{aligned} \tag{3.1.1}$$

where the latter condition is omitted if $\partial\Omega = \emptyset$, $\Omega = \mathbb{R}^n$.

Recall, $\nu > 0$ is a physical constant (viscosity), $f = (f_1, \dots, f_n)$ means the given exterior force, $u = (u_1, \dots, u_n)$ the unknown velocity field, and the scalar p is the unknown pressure.

First we develop the theory of weak solutions. Then we prove regularity properties and get also solutions in the classical sense, provided f and Ω are sufficiently smooth. Concerning weak solutions we refer to [Gal94a, VIII.3]. See also [Pil96].

To introduce weak solutions, we use as in the linear theory the completion

$$\widehat{W}_{0,\sigma}^{1,2}(\Omega) = \overline{C_{0,\sigma}^\infty(\Omega)}^{\|\nabla u\|_2}$$

equipped with scalar product $< \nabla u, \nabla v > = \int_\Omega (\nabla u) \cdot (\nabla v) \, dx$ and norm $\|\nabla u\|_2 = (\int_\Omega |\nabla u|^2 \, dx)^{\frac{1}{2}}$, where $\nabla u = (D_j u_l)_{j,l=1}^\infty$, $\nabla v = (D_j v_l)_{j,l=1}^\infty$, $(\nabla u) \cdot (\nabla v) = \sum_{j,l=1}^n (D_j u_l)(D_j v_l)$.

Further we need the closure

$$W_{0,\sigma}^{1,2}(\Omega) = \overline{C_{0,\sigma}^\infty(\Omega)}^{\|u\|_{W^{1,2}(\Omega)}},$$

$\|u\|_{W_{0,\sigma}^{1,2}(\Omega)} = (\|u\|_2^2 + \|\nabla u\|_2^2)^{\frac{1}{2}}$, and we use the embeddings

$$\begin{aligned} W_{0,\sigma}^{1,2}(\Omega) &\subseteq \widehat{W}_{0,\sigma}^{1,2}(\Omega) \, , \qquad \widehat{W}_{0,\sigma}^{1,2}(\Omega) \subseteq L^6(\Omega)^3 \quad \text{if } n = 3 \, , \\ \widehat{W}_{0,\sigma}^{1,2}(\Omega) &\subseteq L_{loc}^q(\overline{\Omega})^2 \qquad \text{if } n = 2, \ 1 < q < \infty, \ \overline{\Omega} \neq \mathbb{R}^2 \, , \end{aligned}$$

see Lemma 1.2.1.

We need these embeddings to estimate $u \cdot \nabla u$. $\widehat{W}_{0,\sigma}^{1,2}(\Omega)$ will be the space of weak solutions if $n = 3$, or if $n = 2$ and $\overline{\Omega} \neq \mathbb{R}^2$. In the exceptional case $\overline{\Omega} = \mathbb{R}^2$, we will replace $\widehat{W}_{0,\sigma}^{1,2}(\Omega)$ by the space $W_{0,\sigma}^{1,2}(\Omega)$.

In the definition below we admit exterior forces of the general form $f = f_0 + \operatorname{div} F$ where

$$f_0 \in L_{loc}^2(\Omega)^n \ , \quad F = (F_{kl})_{k,l=1}^n \in L^2(\Omega)^{n^2} . \tag{3.1.2}$$

The functional $v \mapsto [f, v]$, $v = (v_1, \ldots, v_n) \in C_0^\infty(\Omega)^n$, is defined by

$$\begin{aligned}[f,v] &= <f_0, v> + [\text{div } F, v] = <f_0, v> - <F, \nabla v> \\ &= \int_\Omega f_0 \cdot v \, dx - \int_\Omega F \cdot (\nabla v) \, dx\end{aligned}$$

as in the linear theory.

If $u \in \widehat{W}_{0,\sigma}^{1,2}(\Omega)$ with $\overline{\Omega} \neq \mathbb{R}^2$ for $n = 2$, then we will see by Lemma 3.2.1 that

$$\nabla u \in L^2(\Omega)^{n^2} \quad , \quad u \cdot \nabla u \in L_{\text{loc}}^{3/2}(\overline{\Omega})^n. \tag{3.1.3}$$

These properties justify the following elementary calculation. With div $u = 0$ we get

$$\begin{aligned} u \cdot \nabla u &= (u_1 D_1 + \ldots + u_n D_n) u = (u_1 D_1 u_l + \ldots + u_n D_n u_l)_{l=1}^n \\ &= (D_1(u_1 u_l) + \ldots + D_n(u_n u_l))_{l=1}^n - ((D_1 u_1) u_l + \ldots + (D_n u_n) u_l)_{l=1}^n \\ &= (D_1(u_1 u_l) + \ldots + D_n(u_n u_l))_{l=1}^n \\ &= D_1(u_1 u) + \ldots + D_n(u_n u) \\ &= \text{div } (u\, u) \end{aligned}$$

where $u\,u := (u_j u_l)_{j,l=1}^n$ is a matrix field. The operation div applies to the columns of $u\,u$. Recall, $u\,u = u \otimes u$ means the tensor product, see Section 1.2, I.

We will see that $u\,u \in L_{loc}^3(\overline{\Omega})^{n^2}$, see Lemma 3.2.1. Therefore, the functional

$$v \mapsto [u \cdot \nabla u, v] = [\text{ div } u\,u, v] \quad , \quad v \in C_0^\infty(\Omega)^n,$$

with

$$\begin{aligned}[\text{div } u\,u, v] &= - <u\,u, \nabla v> = - \int_\Omega (u\,u) \cdot (\nabla v) \, dx \\ &= - \sum_{j,k=1}^n \int_\Omega u_j u_k \, D_j v_k \, dx\end{aligned}$$

is well defined. We write div $u\,u =$ div $(u\,u)$ and

$$[\cdot, \cdot] = [\cdot, \cdot]_\Omega \quad , \quad <\cdot, \cdot> = <\cdot, \cdot>_\Omega$$

if necessary.

In order to motivate the notion of a weak solution in the definition below, we have to treat each term of the first equation of (3.1.1) as a functional defined on the test space $C_{0,\sigma}^\infty(\Omega)$. Then the gradient term ∇p vanishes. If u, p, f and Ω are sufficiently smooth, and if the system (3.1.1) is satisfied in the classical sense, then, taking in (3.1.1) the scalar product with $v \in C_{0,\sigma}^\infty(\Omega)$, we see that u is also a weak solution. This justifies the notion of a weak solution.

3.1.1 Definition *Let $\Omega \subseteq \mathbb{R}^n$, $n = 2, 3$, be any domain, let $f = f_0 + \text{div } F$ with $f_0 \in L^2_{loc}(\Omega)^n$, $F \in L^2(\Omega)^{n^2}$, let*

$$u \in \widehat{W}^{1,2}_{0,\sigma}(\Omega) \quad \text{if } n = 3 \,, \qquad \text{or if } n = 2 \text{ and } \overline{\Omega} \neq \mathbb{R}^2 \,,$$

and let

$$u \in W^{1,2}_{0,\sigma}(\Omega) \quad \text{if } n = 2 \,, \quad \overline{\Omega} = \mathbb{R}^2 .$$

Then u is called a **weak solution** *of the Navier-Stokes system* (3.1.1) *with force f iff*

$$\nu < \nabla u, \nabla v > - < u\,u, \nabla v > \; = \; [f, v] \tag{3.1.4}$$

holds for all $v \in C^\infty_{0,\sigma}(\Omega)$.

If u satisfies (3.1.4) *and if $p \in L^2_{loc}(\Omega)$ is given such that*

$$-\nu\Delta u + u \cdot \nabla u + \nabla p = f \tag{3.1.5}$$

holds in the sense of distributions in Ω, then (u, p) is called a **weak solution pair** *of the system* (3.1.1)*, and p is called an* **associated pressure** *of u.*

3.1.2 Remarks

a) If u is a weak solution of the system (3.1.1), then we always find an associated pressure $p \in L^2_{loc}(\Omega)$ such that (u, p) is a weak solution pair, see Lemma 3.3.1.

b) Let u be given as in the above definition, let $p \in L^2_{loc}(\Omega)$, and assume that u and p satisfy the equation (3.1.5) in the sense of distributions in Ω. Then u is obviously a weak solution of (3.1.1), (u, p) a weak solution pair, and p an associated pressure.

3.2 The nonlinear term $u \cdot \nabla u$

The next lemma yields some integrability properties of the nonlinear term

$$u \cdot \nabla u = (u_1 D_1 + \cdots + u_n D_n) u$$

if u is contained in the solution space $\widehat{W}^{1,2}_{0,\sigma}(\Omega)$ of the system (3.1.1). First we assume that $\overline{\Omega} \neq \mathbb{R}^2$ if $n = 2$. The exceptional case is treated briefly in Remark 3.2.2. See [Gal94b, VIII] concerning properties of $u \cdot \nabla u$.

From above we know that

$$u \cdot \nabla u = \; \text{div } u\,u$$

with $u\,u = (u_j u_l)_{j,l=1}^n$. Further we get

$$(u \cdot \nabla u) \cdot u = \sum_{j,l=1}^n (u_j\, D_j u_l) u_l = \sum_{j,l=1}^n \frac{1}{2} u_j\, D_j u_l^2 = \frac{1}{2} u \cdot \nabla |u|^2 ,$$

and if Ω is bounded, we obtain

$$\begin{aligned} < u \cdot \nabla u, u > &= < \operatorname{div} u\,u, u > = \frac{1}{2} < u, \nabla |u|^2 > \\ &= -\frac{1}{2} < \operatorname{div} u, |u|^2 > = 0. \end{aligned} \tag{3.2.1}$$

3.2.1 Lemma *Let $\Omega \subseteq \mathbb{R}^n$, $n = 2, 3$, be any domain with $\overline{\Omega} \neq \mathbb{R}^2$ if $n = 2$, and let $B_0 \subseteq \mathbb{R}^2$ be an open ball with $\overline{B}_0 \cap \overline{\Omega} = \emptyset$ if $n = 2$. Suppose $u \in \widehat{W}_{0,\sigma}^{1,2}(\Omega)$. Then we have :*

a)

$$u\,u \in L^3_{loc}(\overline{\Omega})^{n^2} \quad , \quad u \cdot \nabla u \in L^{3/2}_{loc}(\overline{\Omega})^n, \tag{3.2.2}$$

$$\nabla |u|^2 \in L^{3/2}_{loc}(\overline{\Omega})^n \quad , \quad (u \cdot \nabla u) \cdot u \in L^{6/5}_{loc}(\overline{\Omega}),$$

and

$$u \cdot \nabla u = \operatorname{div} u\,u \quad , \quad (u \cdot \nabla u) \cdot u = \frac{1}{2} u \cdot \nabla |u|^2. \tag{3.2.3}$$

b) *For each bounded subdomain $\Omega' \subseteq \Omega$ we get*

$$\|u\,u\|_{L^3(\Omega')} + \|u \cdot \nabla u\|_{L^{3/2}(\Omega')} \leq C \, \|\nabla u\|^2_{L^2(\Omega)} , \tag{3.2.4}$$

and

$$\|(u \cdot \nabla u) \cdot u\|_{L^{6/5}(\Omega')} \leq C \, \|\nabla u\|^3_{L^2(\Omega)} \tag{3.2.5}$$

where $C > 0$ depends on B_0, Ω' if $n = 2$.

c) *If $n = 3$, $1 < q \leq 6$, then for each bounded subdomain $\Omega' \subseteq \Omega$ we get*

$$\|u\|_{L^q(\Omega')} \leq C \, \|\nabla u\|_{L^2(\Omega)}, \tag{3.2.6}$$

where $C = C(q, \Omega') > 0$.

d) *If $n = 2$, $1 < q < \infty$, then for each bounded subdomain $\Omega' \subseteq \Omega$ we get*

$$\|u\|_{L^q(\Omega')} \leq C \, \|\nabla u\|_{L^2(\Omega)}, \tag{3.2.7}$$

where $C = C(q, \Omega', B_0) > 0$.

e) *If Ω is bounded, then*

$$u\,u \in L^3(\Omega)^{n^2} \quad , \quad u \cdot \nabla u \in L^{3/2}(\Omega)^n \quad , \quad (u \cdot \nabla u) \cdot u \in L^{6/5}(\Omega), \quad (3.2.8)$$

and

$$\begin{aligned} < u \cdot \nabla u, u > &= - < u\,u, \nabla u > = \frac{1}{2} < u, \nabla |u|^2 > \qquad (3.2.9) \\ &= -\frac{1}{2} < \text{div } u, |u|^2 > = 0. \end{aligned}$$

Proof. By the definition of $\widehat{W}^{1,2}_{0,\sigma}(\Omega)$ we find a sequence $(u_j)_{j=1}^{\infty}$ in $C^{\infty}_{0,\sigma}(\Omega)$ such that $(\nabla u_j)_{j=1}^{\infty}$ is a Cauchy sequence in $L^2(\Omega)^{n^2}$.

If $n = 3$ we apply the embedding property (1.2.4) with $q = 6$ and obtain $u \in L^6(\Omega)^3$, $u = s - \lim_{j\to\infty} u_j$ in $L^6(\Omega)^3$, and

$$\|u\|_{L^6(\Omega)} \le C\, \|\nabla u\|_{L^2(\Omega)}$$

with $C > 0$. For any bounded subdomain $\Omega' \subseteq \Omega$, we see, using Hölder's inequality, that

$$\|u\|_{L^q(\Omega')} \le C_1\, \|u\|_{L^6(\Omega)')} \le C_1\, C\, \|\nabla u\|_{L^2(\Omega)} \qquad (3.2.10)$$

with $C_1 = C_1(q, \Omega') > 0$. It follows that

$$u = s - \lim_{j\to\infty} u_j \quad \text{in } L^q(\Omega')^2,$$

with $1 < q \le 6$.

If $n = 2$, we use (1.2.7) instead of (1.2.4), and for each bounded subdomain $\Omega' \subseteq \Omega$ we get $u \in L^q(\Omega')^2$,

$$u = s - \lim_{j\to\infty} u_j \quad \text{in } L^q(\Omega')^2 \quad , \quad 1 < q < \infty,$$

and

$$\|u\|_{L^q(\Omega')} \le C\, \|\nabla u\|_{L^2(\Omega)} \qquad (3.2.11)$$

with $C = C(q, \Omega', B_0) > 0$.

In both cases we see with Hölder's inequality that

$$\begin{aligned} \|u\,u\|_{L^3(\Omega')} &\le C_1\, \|u\|^2_{L^6(\Omega')} \le C_2\, \|\nabla u\|^2_{L^2(\Omega)}\,, \qquad (3.2.12) \\ \|u \cdot \nabla u\|_{L^{3/2}(\Omega')} &\le C_1\, \|u\|_{L^6(\Omega')}\, \|\nabla u\|_{L^2(\Omega)} \qquad (3.2.13) \\ &\le C_2\, \|\nabla u\|^2_{L^2(\Omega)}\,, \end{aligned}$$

and that

$$\|(u\cdot\nabla u)\cdot u\|_{L^{6/5}(\Omega')} \;\leq\; C_1\,\|u\|_{L^6(\Omega')}\,\|\nabla u\|_{L^2(\Omega')}\|u\|_{L^6(\Omega')} \qquad (3.2.14)$$
$$\leq\; C_2\,\|\nabla u\|^3_{L^2(\Omega)}$$

with $C_1 > 0,\ C_2 > 0$. This leads to

$$\begin{aligned} u\,u &= s-\lim_{j\to\infty} u_j u && \text{in } L^3(\Omega'),\\ u\cdot\nabla u &= s-\lim_{j\to\infty} u_j\cdot\nabla u && \text{in } L^{3/2}(\Omega'),\end{aligned}$$

and to

$$(u\cdot\nabla u)\cdot u = s-\lim_{j\to\infty}(u_j\cdot\nabla u)\cdot u \quad \text{in } L^{6/5}(\Omega').$$

Since $u_j \in C^\infty_{0,\sigma}(\Omega)$ we see that div $(u_j u) = u_j\cdot\nabla u$, and letting $j\to\infty$ leads to

$$u\cdot\nabla u = \text{ div } (uu)\ ,\quad \nabla|u|^2 = 2|u|\,\nabla|u|\ ,\quad (u\cdot\nabla u)\cdot u = \frac{1}{2}u\cdot\nabla|u|^2\,.$$

Since $\Omega'\subseteq\Omega$ is an arbitrary bounded subdomain we thus obtain the properties (3.2.2) and (3.2.3). The inequalities (3.2.4), (3.2.5) follow from (3.2.12), (3.2.13) and (3.2.14). Inequality (3.2.6) follows from (3.2.10), and (3.2.7) is a consequence of (3.2.11).

It remains to prove e). If Ω is bounded, we may set $\Omega' = \Omega$ and obtain the properties (3.2.8). Since supp $u_j\subseteq\Omega$ we conclude that

$$\begin{aligned} < u\cdot\nabla u, u > &= \ < \text{ div } (u\,u), u > \;=\; \lim_{j\to\infty} < \text{ div } (u\,u), u_j > \\ &= -\lim_{j\to\infty} < u\,u, \nabla u_j > \;=\; - < u\,u, \nabla u > .\end{aligned}$$

In the same way we see with div $u_j = 0$ that

$$\begin{aligned} < u\cdot\nabla u, u > &= \frac{1}{2} < u, \nabla|u|^2 > = \frac{1}{2}\lim_{j\to\infty} < u_j, \nabla|u|^2 > \\ &= -\frac{1}{2}\lim_{j\to\infty} < \text{ div } u_j, |u|^2 > = 0.\end{aligned}$$

This proves the lemma. □

3.2.2 Remarks Consider the exceptional case $n = 2$, $\overline{\Omega} = \mathbb{R}^2$ in Definition 3.1.1. We may admit here that $\Omega\subseteq\mathbb{R}^2$ is an arbitrary domain.

Instead of $u \in \widehat{W}_{0,\sigma}^{1,2}(\Omega)$ in Lemma 3.2.1 we now suppose the stronger assumption $u \in W_{0,\sigma}^{1,2}(\Omega)$. Then by definition we find $u_j \in C_{0,\sigma}^{\infty}(\Omega)$, $j \in \mathbb{N}$, with

$$u = s - \lim_{j\to\infty} u_j \quad \text{in } L^2(\Omega)^n \ , \quad \nabla u = s - \lim_{j\to\infty} \nabla u_j \quad \text{in } L^2(\Omega)^{n^2} .$$

Applying Sobolev's inequality (1.3.2), II, with $r = \gamma = n = 2$, $0 \le \beta < 1$, $q = \frac{2}{1-\beta}$, we obtain

$$\begin{aligned} \|u\|_{L^q(\Omega)} &\le C \, \|\nabla u\|_{L^2(\Omega)}^{\beta} \, \|u\|_{L^2(\Omega)}^{1-\beta} \\ &\le C \, (\|\nabla u\|_{L^2(\Omega)} + \|u\|_{L^2(\Omega)}) \end{aligned} \tag{3.2.15}$$

with $C = C(\beta) > 0$. This holds for $2 \le q < \infty$, and also with u replaced by $u - u_j$. For each bounded subdomain $\Omega' \subseteq \Omega$, we thus obtain that

$$\|u\|_{L^q(\Omega')} \le C \, (\|\nabla u\|_{L^2(\Omega)} + \|u\|_{L^2(\Omega)}) \, , \tag{3.2.16}$$

$1 < q < \infty$ with $C = C(q) > 0$ where C also depends on Ω' if $1 < q < 2$.

We may replace inequality (3.2.11) in the above proof by (3.2.16) and obtain now for u, $u\,u$, $u \cdot \nabla u, \ldots$ the same properties as in Lemma 3.2.1, a)–e).

3.3 The associated pressure p

If u is a weak solution of the system (3.1.1), then the pressure term ∇p of this system can be constructed in the same way as in the linear theory; p is determined up to a constant and becomes unique under the additional condition $\int_{\Omega_0} p \, dx = 0$, see below. See [Gal94b, IX.1], [Tem77, Chap. II, Theorem 1.2] concerning the pressure of weak solutions.

We omit the exceptional case $n = 2$, $\overline{\Omega} = \mathbb{R}^2$ in Definition 3.1.1. Here we are not interested in this case.

3.3.1 Lemma *Let $\Omega \subseteq \mathbb{R}^n$, $n = 2, 3$, be any domain with $\overline{\Omega} \neq \mathbb{R}^2$ if $n = 2$, let $f = f_0 + \operatorname{div} F$ with $f_0 \in L^2_{loc}(\Omega)^n$, $F \in L^2(\Omega)^{n^2}$, and let $\Omega_0 \subseteq \Omega$ be a bounded subdomain with $\overline{\Omega_0} \subseteq \Omega$, $\Omega_0 \neq \emptyset$.*

Suppose $u \in \widehat{W}_{0,\sigma}^{1,2}(\Omega)$ is a weak solution of the Navier-Stokes system (3.1.1) *with force f. Then there exists a unique $p \in L^2_{loc}(\Omega)$ with*

$$\int_{\Omega_0} p \, dx = 0 \tag{3.3.1}$$

such that

$$-\nu \Delta u + u \cdot \nabla u + \nabla p = f \tag{3.3.2}$$

holds in the sense of distributions. Thus (u, p) is a weak solution pair of (3.1.1).

Conversely, let $(u,p) \in \widehat{W}^{1,2}_{0,\sigma}(\Omega) \times L^2_{loc}(\Omega)$ *satisfy* (3.3.2) *in the sense of distributions, then* u *is a weak solution,* (u,p) *a weak solution pair, and* p *an associated pressure for the system* (3.1.1).

Proof. Let $G : v \mapsto [G,v]$, $v \in C_0^\infty(\Omega)^n$, be the functional defined by

$$\begin{aligned} [G,v] &:= [f,v] - [u\cdot\nabla u, v] + [\nu\Delta u, v] \\ &= \; < f_0, v > - < F, \nabla v > + < u\,u, \nabla v > -\nu < \nabla u, \nabla v > . \end{aligned}$$

Let $\Omega' \subseteq \Omega$ be any bounded subdomain with $\overline{\Omega'} \subseteq \Omega$, let $C_1 = C_1(\Omega') > 0$ be the constant in the Poincaré inequality (1.1.1), II. Then, using (3.2.4), we obtain

$$\|u\,u\|_{L^2(\Omega')} \;\le\; C_2\,\|\nabla u\|^2_{L^2(\Omega)}$$

with $C_2 = C_2(\Omega') > 0$. Now we get

$$|[G,v]| \;\le\; \big(C_1\,\|f_0\|_{L^2(\Omega')} + \|F\|_2 + C_2\|\nabla u\|^2_{L^2(\Omega)} + \nu\|\nabla u\|_{L^2(\Omega)}\big)\,\|\nabla v\|_2,$$

$v \in C_0^\infty(\Omega')^n$, it follows that $G \in W^{-1,2}_{loc}(\Omega)^n$, from (3.1.4) we see that $[G,v] = 0$ for all $v \in C^\infty_{0,\sigma}(\Omega)$, and Lemma 2.2.1, II, yields a unique $p \in L^2_{loc}(\Omega)$ satisfying (3.3.1) and $G = \nabla p$. This yields (3.3.2).

If $(u,p) \in \widehat{W}^{1,2}_{0,\sigma}(\Omega) \times L^2_{loc}(\Omega)$ satisfies (3.3.2) in the sense of distributions, we get (3.1.4) and (u,p) is a weak solution pair. This proves the lemma. □

3.3.2 Remark If Ω in Lemma 3.3.1 is a bounded Lipschitz domain and if $f_0 \in L^2(\Omega)^n$, then

$$|[G,v]| \;\le\; C(\|f_0\|_2 + \|F\|_2 + \|\nabla u\|_2^2 + \nu\|\nabla u\|_2)\,\|\nabla v\|_2 \tag{3.3.3}$$

for all $v \in C_0^\infty(\Omega)^n$ with $C = C(\Omega) > 0$. It follows that $G \in W^{-1,2}(\Omega)^n$ and we may apply Lemma 2.2.2, II. This yields a unique

$$p \in L^2(\Omega) \;\text{ with }\; \int_\Omega p\,dx = 0 \tag{3.3.4}$$

satisfying $-\nu\Delta u + u\cdot\nabla u + \nabla p = f$ in the sense of distributions. Moreover we get

$$\|p\|_2 \;\le\; C\,(\|f_0\|_2 + \|F\|_2 + \|\nabla u\|_2^2 + \nu\|\nabla u\|_2) \tag{3.3.5}$$

where $C = C(\Omega) > 0$ is a constant.

3.4 Existence of weak solutions in bounded domains

The proof of the following existence result rests on the Leray-Schauder principle, see Lemma 3.1.1, II. A similar argument has been used by Ladyzhenskaya [Lad69, Chap. 5]. Another proof of this result is based on the Galerkin method, see [Hey80], [Tem77], [Gal94b, VIII].

We consider only exterior forces of the form $f = \text{div } F$ with $F \in L^2(\Omega)^{n^2}$; see Section 1.6, II, concerning this property. If $\Omega \subseteq \mathbb{R}^n$ is a bounded domain, then each $f \in W^{-1,2}(\Omega)^n$ can be written in the form $f = \text{div } F$, $F \in L^2(\Omega)^{n^2}$, with

$$\|F\|_2 \leq C\,\|f\|_{W^{-1,2}(\Omega)^n}\,, \tag{3.4.1}$$

$C = C(\Omega) > 0$, see Lemma 1.6.1, II.

3.4.1 Theorem *Let $\Omega \subseteq \mathbb{R}^n$, $n = 2, 3$, be any bounded domain and let $f = \text{div } F$ with $F \in L^2(\Omega)^{n^2}$.*

Then there exists at least one pair

$$(u, p) \in W^{1,2}_{0,\sigma}(\Omega) \times L^2_{loc}(\Omega)$$

satisfying

$$-\nu\Delta u + u \cdot \nabla u + \nabla p = f$$

in the sense of distributions; u is a weak solution, (u, p) a weak solution pair and p an associated pressure of the Navier-Stokes system (3.1.1) with force f.

Moreover,

$$\|\nabla u\|_2 \leq \nu^{-1}\|F\|_2. \tag{3.4.2}$$

If Ω is a bounded Lipschitz domain, then $p \in L^2(\Omega)$, p can be chosen with $\int_\Omega p\,dx = 0$ and satisfies the inequality

$$\|p\|_2 \leq C\,(\|F\|_2 + \nu^{-2}\|F\|_2^2) \tag{3.4.3}$$

with some constant $C = C(\Omega) > 0$.

Remark Let Ω be as above. Then with (3.4.1) we get the following result: For each $f \in W^{-1,2}(\Omega)^n$ there exists at least one weak solution $u \in W^{1,2}_{0,\sigma}(\Omega)$ satisfying

$$\|\nabla u\|_2 \leq C\,\nu^{-1}\|f\|_{W^{-1,2}(\Omega)^n} \tag{3.4.4}$$

with $C = C(\Omega) > 0$.

Proof. Let $u \in W^{1,2}_{0,\sigma}(\Omega)$. In the first step we show that u is a weak solution of (3.1.1) iff the equation

$$A^{\frac{1}{2}}u + A^{-\frac{1}{2}}P \text{ div } (u\,u) = A^{-\frac{1}{2}}P \text{ div } F \tag{3.4.5}$$

is satisfied. Here A means the Stokes operator and the operators $A^{-\frac{1}{2}}$, P have the extended meaning as in Lemma 2.6.2.

To show this we start with the definition of a weak solution u in (3.1.4) which means that

$$\nu < \nabla u, \nabla v > - < u\,u, \nabla v > = - < F, \nabla v > \tag{3.4.6}$$

for all $v \in W^{1,2}_{0,\sigma}(\Omega)$. Next we use $D(A^{\frac{1}{2}}) = W^{1,2}_{0,\sigma}(\Omega)$, $\|A^{\frac{1}{2}} v\|_2 = \nu^{\frac{1}{2}} \|\nabla v\|_2$, $\nu < \nabla u, \nabla v > = < A^{\frac{1}{2}} u, A^{\frac{1}{2}} v >$, see Lemma 2.2.1,

$$- < F, \nabla v > = < A^{-\frac{1}{2}} P \text{ div } F, A^{\frac{1}{2}} v > \ , \ \ \|A^{-\frac{1}{2}} P \text{ div } F\|_2 \ \le \ \nu^{-\frac{1}{2}} \|F\|_2,$$

see Lemma 2.6.1, and

$$- < u\,u, \nabla v > = < A^{-\frac{1}{2}} P \text{ div } (u\,u), A^{\frac{1}{2}} v > \ , \ \ \ v \in W^{1,2}_{0,\sigma}(\Omega)\ .$$

From Lemma 3.2.1 we get $u\,u \in L^3(\Omega)^{n^2}$, therefore $u\,u \in L^2(\Omega)^{n^2}$ and

$$\|u\,u\|_2 \ \le \ C\, \|\nabla u\|_2^2 \tag{3.4.7}$$

with $C = C(\Omega) > 0$. Since $R(A^{\frac{1}{2}}) = L^2_\sigma(\Omega)$, see Lemma 2.2.1, we conclude that (3.4.5) is equivalent with (3.4.6). Here the arguments are the same as in the proof of Lemma 2.6.3.

Setting $w = A^{\frac{1}{2}} u$, $u = A^{-\frac{1}{2}} w$ with $w \in D(A^{-\frac{1}{2}}) = L^2_\sigma(\Omega)$, (3.4.5) can be written in the form

$$w + A^{-\frac{1}{2}} P \text{ div } (A^{-\frac{1}{2}} w)(A^{-\frac{1}{2}} w) = A^{-\frac{1}{2}} P \text{ div } F. \tag{3.4.8}$$

Setting

$$Bw := A^{-\frac{1}{2}} P \text{ div } F - A^{-\frac{1}{2}} P \text{ div } (A^{-\frac{1}{2}} w)(A^{-\frac{1}{2}} w),$$

we obtain a well defined (nonlinear) operator B from $L^2_\sigma(\Omega)$ to $L^2_\sigma(\Omega)$ and (3.4.8) has the form

$$w = Bw.$$

In the next step we show that the Leray-Schauder principle, Lemma 3.1.1, II, is applicable to (3.4.8).

Since Ω is bounded and $\|\nabla v\|_2 = \nu^{-\frac{1}{2}} \|A^{\frac{1}{2}} v\|_2$, $v \in D(A^{\frac{1}{2}})$, we obtain from the compactness property in Lemma 1.5.1, II, the following fact:

For each bounded sequence $(w_j)_{j=1}^\infty$ in $L^2_\sigma(\Omega)$, $(A^{-\frac{1}{2}} w_j)_{j=1}^\infty$ contains a subsequence which converges in $L^2_\sigma(\Omega)$. Therefore, the operator $A^{-\frac{1}{2}} : L^2_\sigma(\Omega) \to L^2_\sigma(\Omega)$ is compact, see Section 3.1, II.

The same holds for the operator $A^{-\frac{\alpha}{2}} : L^2_\sigma(\Omega) \to L^2_\sigma(\Omega)$ with $0 < \alpha \le 1$. To show this we use the interpolation inequality, see Lemma 3.2.2, II, and obtain the estimate

$$\|A^{-\frac{\alpha}{2}}(w_j - w_l)\|_2 \le \|w_j - w_l\|_2^{1-\alpha} \, \|A^{-\frac{1}{2}}(w_j - w_l)\|_2^{\alpha} \ , \quad j, l = 1, 2, \ldots .$$

This shows that $(A^{-\frac{\alpha}{2}} w_j)_{j=1}^\infty$ contains a strongly convergent subsequence if $(w_j)_{j=1}^\infty$ is bounded in $L^2_\sigma(\Omega)$. Thus $A^{-\frac{\alpha}{2}}$ is a compact operator.

To apply Lemma 3.1.1, II, we have to show that B is completely continuous. For this purpose we use Hölder's inequality and the embedding property in Lemma 2.4.2. This leads with $\alpha = \frac{n}{8}$, $2\alpha + \frac{n}{4} = \frac{n}{2}$, to the estimate

$$\begin{aligned} \|(A^{-\frac{1}{2}} w)(A^{-\frac{1}{2}} w)\|_2 &\le C_1 \, \|A^{-\frac{1}{2}} w\|_4 \, \|A^{-\frac{1}{2}} w\|_4 \\ &\le C_2 \, \nu^{-2\alpha} \|A^{\alpha - \frac{1}{2}} w\|_2 \, \|A^{\alpha-\frac{1}{2}} w\|_2 \end{aligned} \tag{3.4.9}$$

where $C_1(n) > 0$, $C_2(n) > 0$ are constants.

To show the continuity of B, we consider any sequence $(w_j)_{j=1}^\infty$ in $L^2_\sigma\Omega)$ which converges strongly to $w \in L^2_\sigma(\Omega)$. Then, using (3.4.9) and estimate (2.6.2), we conclude that

$$\begin{aligned} \|Bw_j - Bw\|_2 &= \big\| A^{-\frac{1}{2}} P \text{ div } \big((A^{-\frac{1}{2}} w_j)(A^{-\frac{1}{2}} w - A^{-\frac{1}{2}} w_j)\big) \\ &\quad - A^{-\frac{1}{2}} P \text{ div } \big((A^{-\frac{1}{2}} w_j - A^{-\frac{1}{2}} w)(A^{-\frac{1}{2}} w)\big) \big\|_2 \\ &\le C \nu^{-\frac{1}{2} - 2\alpha} \, \|A^{\alpha - \frac{1}{2}}(w_j - w)\|_2 \, (\|A^{\alpha - \frac{1}{2}} w\|_2 + \|A^{\alpha-\frac{1}{2}} w_j\|_2) \end{aligned} \tag{3.4.10}$$

with $C = C(n) > 0$, $j \in \mathbb{N}$. Since $\alpha - \frac{1}{2} = \frac{n}{8} - \frac{1}{2} < 0$ for $n = 2, 3$, $A^{\alpha - \frac{1}{2}}$ is a bounded operator and we get

$$\lim_{j \to \infty} \|w - w_j\|_2 = 0 \quad \text{implies} \quad \lim_{j\to\infty} \|Bw - Bw_j\|_2 = 0.$$

This shows that B is continuous.

To show that B is completely continuous, we consider a bounded sequence $(w_j)_{j=1}^\infty$ in $L^2_\sigma(\Omega)$. Since $\alpha - \frac{1}{2} < 0$ we know that $A^{\alpha-\frac{1}{2}}$ is compact, see above. Therefore, $(A^{\alpha - \frac{1}{2}} w_j)_{j=1}^\infty$ contains a strongly convergent subsequence, and writing (3.4.10) with w replaced by w_l, $l \in \mathbb{N}$, we conclude that $(Bw_j)_{j=1}^\infty$ contains a strongly convergent subsequence. This shows that B is completely continuous.

To prove the property (3.1.11), II, we use Lemma 3.2.1, (3.2.9), and with $u = A^{-\frac{1}{2}} w$ we obtain

$$\begin{aligned} < A^{-\frac{1}{2}} P \text{ div } (A^{-\frac{1}{2}} w)(A^{-\frac{1}{2}} w), w > &= < \text{ div } (u\,u), A^{-\frac{1}{2}} w > \\ &= - < u\,u, \nabla u > \, = \, 0 \,. \end{aligned}$$

Therefore, the equation $w = \lambda Bw$, $0 \leq \lambda \leq 1$, $w \in L^2_\sigma(\Omega)$, leads to

$$\begin{aligned} \|w\|_2^2 &= \lambda < Bw, w > \\ &= \lambda < A^{-\frac{1}{2}} P \text{ div } F, w > - \lambda < A^{-\frac{1}{2}} P \text{ div } (A^{-\frac{1}{2}} w)(A^{-\frac{1}{2}} w), w > \\ &= \lambda < A^{-\frac{1}{2}} P \text{ div } F, w > \leq \lambda \|A^{-\frac{1}{2}} P \text{ div } F\|_2 \, \|w\|_2 \\ &\leq \lambda \nu^{-\frac{1}{2}} \|F\|_2 \, \|w\|_2. \end{aligned}$$

Setting $r := \nu^{-\frac{1}{2}} \|F\|_2$ we see that

$$w \in L^2_\sigma(\Omega)\,, \quad 0 \leq \lambda \leq 1\,, \quad w = \lambda Bw \;\text{ implies }\; \|w\|_2 \leq r.$$

We may assume that $F \neq 0$ and therefore that $r > 0$. The Leray-Schauder principle, Lemma 3.1.1, II, now yields at least one $w \in L^2_\sigma(\Omega)$ satisfying $w = Bw$ and

$$\|w\|_2 \;\leq\; \nu^{-\frac{1}{2}} \|F\|_2. \tag{3.4.11}$$

Setting $u := A^{-\frac{1}{2}} w$, we get a solution u of (3.4.5) which satisfies (3.4.6) and is therefore a weak solution of the system (3.1.1). From (3.4.11) we get

$$\|w\|_2 \;=\; \|A^{\frac{1}{2}} u\|_2 \;=\; \nu^{\frac{1}{2}} \|\nabla u\|_2 \;\leq\; \nu^{-\frac{1}{2}} \|F\|_2\,. \tag{3.4.12}$$

This proves (3.4.2).

The pressure $p \in L^2_{loc}(\Omega)$ is constructed by Lemma 3.3.1. If Ω is a Lipschitz domain, we may use (3.3.4) and see that $p \in L^2(\Omega)$. Subtracting a constant we can satisfy the condition $\int_\Omega p \, dx = 0$. From (3.3.5) we get

$$\|p\|_2 \;\leq\; C\,(\|F\|_2 + \|\nabla u\|_2^2 + \nu \|\nabla u\|_2),$$

$C = C(\Omega) > 0$, and together with (3.4.12) we see that

$$\|p\|_2 \;\leq\; C\,(\|F\|_2 + \nu^{-2} \|F\|_2^2).$$

This proves (3.4.3). The proof is complete. □

3.5 Existence of weak solutions in unbounded domains

The Leray-Schauder principle, see the proof of Theorem 3.4.1, is only applicable for a bounded domain Ω. Therefore, in order to prove an existence result of weak solutions for unbounded domains, we use an approximation argument which reduces the problem to bounded domains. In principle we follow here the argument in [Lad69, Chap. 5, 3].

As in the linear theory we can only treat exterior forces of the form $f =$ div F with $F \in L^2(\Omega)^{n^2}$. If Ω is unbounded, we know that not every $f \in$

$W^{-1,2}(\Omega)^n$ can be written in this form, see Section 1.6, II. The reason is that the Poincaré inequality does not always hold in unbounded domains. Lemma 1.6.2, II, yields sufficient conditions for the representation $f = \text{div } F$.

For example, if $n = 3$, $f \in L^{\frac{6}{5}}(\Omega)^3$ is a sufficient condition for the existence of some $F \in L^2(\Omega)^{3^2}$ with $f = \text{div } F$, and we get

$$\|F\|_2 \leq C \|f\|_{\frac{6}{5}}$$

with some constant $C > 0$.

If $n = 2$, we need the restriction $\overline{\Omega} \neq \mathbb{R}^2$ since we use the embedding properties of Lemma 1.3.5, II.

See [Gal94b] for further results in unbounded domains.

3.5.1 Theorem *Let $\Omega \subseteq \mathbb{R}^n$, $n = 2, 3$, be any unbounded domain with $\overline{\Omega} \neq \mathbb{R}^2$ if $n = 2$, and let $f = \text{div } F$ with $F \in L^2(\Omega)^{n^2}$.*

Then there exists at least one pair

$$(u, p) \in \widehat{W}^{1,2}_{0,\sigma}(\Omega) \times L^2_{loc}(\Omega)$$

satisfying

$$-\nu\Delta u + u \cdot \nabla u + \nabla p = f \tag{3.5.1}$$

in the sense of distributions; u is a weak solution, (u, p) a weak solution pair, and p an associated pressure of the Navier-Stokes system (3.1.1) *with force f.*

Moreover, u satisfies the inequality

$$\|\nabla u\|_2 \leq \nu^{-1} \|F\|_2 \, , \tag{3.5.2}$$

and if Ω is a Lipschitz domain, then $p \in L^2_{loc}(\overline{\Omega})$.

Proof. We use Lemma 1.4.1, II, and find a sequence $(\Omega_j)_{j=1}^{\infty}$ of bounded Lipschitz subdomains with the properties

$$\Omega = \bigcup_{j=1}^{\infty} \Omega_j \ , \quad \overline{\Omega}_j \subseteq \Omega_{j+1} \ , \quad j \in \mathbb{N}.$$

For each bounded subdomain $\Omega' \subseteq \Omega$ with $\overline{\Omega'} \subseteq \Omega$, there exists some $j_0 \in \mathbb{N}$ with $\overline{\Omega'} \subseteq \Omega_{j_0}$, see Remark 1.4.2, II. If $n = 2$, we choose an open ball $B_0 \subseteq \mathbb{R}^2$ with $\overline{B_0} \cap \overline{\Omega} = \emptyset$.

For each $u \in \widehat{W}^{1,2}_{0,\sigma}(\Omega)$ we obtain from Lemma 3.2.1, (3.2.6) and (3.2.7), the estimate

$$\|u\|_{L^2(\Omega_k)} \leq C_k \|\nabla u\|_{L^2(\Omega)} \ , \quad k \in \mathbb{N} \tag{3.5.3}$$

where $C_k > 0$ depends on k.

Applying Theorem 3.4.1 we find for each $j \in \mathbb{N}$ a pair

$$(u_j, p_j) \in W_{0,\sigma}^{1,2}(\Omega_j) \times L^2(\Omega_j)$$

solving

$$-\nu\Delta u_j + u_j \cdot \nabla u_j + \nabla p_j = \operatorname{div} F \tag{3.5.4}$$

in the sense of distributions in Ω_j. From (3.4.2) we get

$$\|\nabla u_j\|_{L^2(\Omega_j)} \leq \nu^{-1}\|F\|_{L^2(\Omega_j)} \leq \nu^{-1}\|F\|_{L^2(\Omega)} \tag{3.5.5}$$

with a bound not depending on $j \in \mathbb{N}$. Extending each element by zero we obtain the trivial continuous embeddings

$$W_{0,\sigma}^{1,2}(\Omega_j) \subseteq W_{0,\sigma}^{1,2}(\Omega_{j+1}) \subseteq \widehat{W}_{0,\sigma}^{1,2}(\Omega) \ , \quad j \in \mathbb{N}.$$

Therefore, we may treat $(u_j)_{j=1}^{\infty}$ as a sequence in $\widehat{W}_{0,\sigma}^{1,2}(\Omega)$.

Using (3.5.3), we obtain for fixed $k \in \mathbb{N}$ and all $j \geq k$ the estimate

$$\|u_j\|_{L^2(\Omega_k)} \leq C_k \|\nabla u_j\|_{L^2(\Omega)} \tag{3.5.6}$$

with a bound not depending on $j \geq k$.

The space $\widehat{W}_{0,\sigma}^{1,2}(\Omega)$ is reflexive. Therefore, using the uniform bound in (3.5.5), we find a subsequence of $(u_j)_{j=1}^{\infty}$ which converges weakly in $\widehat{W}_{0,\sigma}^{1,2}(\Omega)$ to some $u \in \widehat{W}_{0,\sigma}^{1,2}(\Omega)$, see Section 3.1, II. Using (3.5.3) we see that $(u_j)_{j=1}^{\infty}$ converges weakly to u in each fixed space $L^2(\Omega_k)^n$, $k \in \mathbb{N}$. Indeed, each functional $< \cdot, v >_{\Omega_k}$ is continuous on $\widehat{W}_{0,\sigma}^{1,2}(\Omega)$ for fixed $v \in L^2(\Omega_k)^n$. We also see that $(\nabla u_j)_{j=1}^{\infty}$ converges weakly to $\nabla u \in L^2(\Omega)^{n^2}$ in the space $L^2(\Omega)^{n^2}$.

For simplicity we may assume that the sequence $(u_j)_{j=1}^{\infty}$ itself has this property. From (3.5.6) we conclude that for each fixed $k \in \mathbb{N}$, the sequence $(u_j)_{j\geq k}$ is bounded in $L^2(\Omega_k)^n$. Together with (3.5.5) we see that this sequence is even bounded in $W^{1,2}(\Omega_k)^n$.

Since Ω_k is a bounded Lipschitz domain, the embedding

$$W^{1,2}(\Omega_k)^n \subseteq L^2(\Omega_k)^n \ , \quad k \in \mathbb{N},$$

is compact, see Lemma 1.5.3, II. Therefore, a subsequence of $(u_j)_{j\geq 1}$ converges strongly in $L^2(\Omega_1)^n$. It converges strongly to u in $L^2(\Omega_1)^n$ since $(u_j)_{j=1}^{\infty}$ converges weakly to u in $L^2(\Omega_1)^n$. Repeating this argument we can choose from this sequence a subsequence which converges strongly to u in $L^2(\Omega_2)^n$. Here we use that $(u_j)_{j\geq 2}$ is bounded in $W^{1,2}(\Omega_2)^n$.

In this way we find a sequence of subsequences, written as lines of a matrix, in such a way that each line is a subsequence of the previous one, and that the kth line converges to u strongly in $L^2(\Omega_k)^n$ for all $k \in \mathbb{N}$. The diagonal sequence of this matrix is a subsequence of $(u_j)_{j=1}^{\infty}$, which converges to u strongly in $L^2(\Omega_k)^n$ for each $k \in \mathbb{N}$. We may assume that the sequence $(u_j)_{j=1}^{\infty}$ itself has this property (**diagonal principle**).

In the next step we show that u is a weak solution of the system (3.1.1). For this purpose let $v \in C_{0,\sigma}^{\infty}(\Omega)$, and choose a fixed $k \in \mathbb{N}$ with supp $v \subseteq \Omega_k$. Since u_j is a weak solution in Ω_j we get

$$\nu < \nabla u_j, \nabla v > - < u_j u_j, \nabla v > = [f, v] = - < F, \nabla v > \tag{3.5.7}$$

for all $j \geq k$. Since $(\nabla u_j)_{j=1}^{\infty}$ converges to ∇u weakly in $L^2(\Omega)^{n^2}$ we get

$$< \nabla u, \nabla v > = \lim_{j \to \infty} < \nabla u_j, \nabla v > .$$

Since $(u_j)_{j \geq k}$ converges strongly to u in $L^2(\Omega_k)^n$, we see, using the representation

$$\begin{aligned} < u_j u_j, \nabla v > &- < uu, \nabla v > \\ &= < (u_j - u) \, u_j, \nabla v > + < u \, (u_j - u), \nabla v >, \end{aligned}$$

that

$$< u \, u, \nabla v > = \lim_{j \to \infty} < u_j u_j, \nabla v > .$$

Letting $j \to \infty$ in (3.5.7) we see that u is a weak solution of (3.1.1).

Lemma 3.3.1 yields an associated pressure $p \in L^2_{loc}(\Omega)$ so that (u, p) satisfies (3.5.1) in the sense of distributions. The estimate (3.5.2) is a consequence of (3.5.5).

Let Ω be a Lipschitz domain and let $\Omega' \subseteq \Omega$ be any bounded Lipschitz subdomain. We use the functional

$$G : v \mapsto [G, v] \, ,$$

defined in the proof of Lemma 3.3.1, but in this case only for all $v \in C_0^{\infty}(\Omega')^n$. Let G' be the restriction of G to $v \in C_0^{\infty}(\Omega')^n$. As in (3.3.3) we get the estimate

$$|[G', v]| \leq C \left(\|F\|_{L^2(\Omega)} + \|\nabla u\|^2_{L^2(\Omega)} + \nu \, \|\nabla u\|_{L^2(\Omega)} \right) \|\nabla v\|_{L^2(\Omega')} \, ,$$

$v \in C_0^{\infty}(\Omega')^n$, $C = C(\Omega') > 0$, and therefore $G' \in W^{-1,2}(\Omega')^n$. Lemma 2.2.2, II, shows that $p \in L^2(\Omega')$. Since $\Omega' \subseteq \Omega$ is an arbitrary bounded Lipschitz subdomain, we see that $p \in L^2_{loc}(\overline{\Omega})$. This proves the theorem. $\square$

In the same way as in the linear case, see Lemma 2.6.3, we can characterize any weak solution u completely in terms of the Stokes operator A. For this purpose we use the extended meaning of the operators $P, A^{\frac{1}{2}}, A^{-\frac{1}{2}}$ as in this lemma. This leads to the following result.

3.5.2 Lemma *Let $\Omega \subseteq \mathbb{R}^n, n = 2, 3$, be any domain with $\overline{\Omega} \neq \mathbb{R}^2$ if $n = 2$, and let $f = \text{div } F$ with $F \in L^2(\Omega)^{n^2}$.*

Then each weak solution $u \in \widehat{W}^{1,2}_{0,\sigma}(\Omega) = \widehat{D}(A^{\frac{1}{2}})$ of the Navier-Stokes system

$$-\nu\Delta u + u \cdot \nabla u + \nabla p = f \ , \quad \text{div } u = 0 \ , \quad u|_{\partial\Omega} = 0 \tag{3.5.8}$$

with force f satisfies $P \text{div } (u\, u) \in \widehat{D}(A^{-\frac{1}{2}})$ and is a solution of the equation

$$A^{\frac{1}{2}} u + A^{-\frac{1}{2}} P \text{div } (u\, u) = A^{-\frac{1}{2}} P \text{div } F \tag{3.5.9}$$

in the extended sense.

Conversely, each solution $u \in \widehat{W}^{1,2}_{0,\sigma}(\Omega)$ of the equation (3.5.9) is a weak solution of (3.5.8).

Proof. Let $u \in \widehat{W}^{1,2}_{0,\sigma}(\Omega)$ be a weak solution of (3.5.8). Then we use $\widehat{D}(A^{\frac{1}{2}}) = \widehat{W}^{1,2}_{0,\sigma}(\Omega)$,

$$\nu < \nabla u, \nabla v > = < A^{\frac{1}{2}} u, A^{\frac{1}{2}} v > \ , \quad v \in C^{\infty}_{0,\sigma}(\Omega), \tag{3.5.10}$$

as in the proof of Lemma 2.6.3, and with (3.1.4) we see that

$$\begin{aligned} [\text{div } (u\, u), v] &= [P \text{ div } (u\, u), v] \\ &= - < u\, u, \nabla v > \\ &= - < F, \nabla v > - < A^{\frac{1}{2}} u, A^{\frac{1}{2}} v > \end{aligned} \tag{3.5.11}$$

for all $v \in C^{\infty}_{0,\sigma}(\Omega)$. This functional is continuous in $\|\nabla v\|_2$, and Lemma 2.5.1 shows that $P \text{ div } (u\, u) \in \widehat{D}(A^{-\frac{1}{2}})$ and that

$$[P\text{div } (u\, u), v] = < A^{-\frac{1}{2}} P\text{div } (u\, u), A^{\frac{1}{2}} v > .$$

Together with

$$- < F, \nabla v > = < A^{-\frac{1}{2}} P \text{ div } F, A^{\frac{1}{2}} v >, \tag{3.5.12}$$

see (2.6.5), we conclude that

$$A^{-\frac{1}{2}} P \text{ div } (u\, u) = A^{-\frac{1}{2}} P \text{ div } F - A^{\frac{1}{2}} u.$$

This proves (3.5.9).

Conversely, let $u \in \widehat{D}(A^{\frac{1}{2}})$ be a solution of (3.5.9). Then (3.5.11) holds, and using (3.5.12) we see that u is a weak solution of (3.5.8). This proves the lemma. □

3.6 Regularity properties for the stationary nonlinear system

Our purpose is to prove smoothness properties of weak solutions u if the given exterior force f and the domain Ω are sufficiently smooth. The idea is to write the nonlinear system (3.1.1) in the form

$$-\nu\Delta u + \nabla p = f - u\cdot\nabla u \ , \quad \text{div}\ u = 0 \ , \quad u|_{\partial\Omega} = 0 \ ,$$

to use on the right side the information on the given weak solution u, and to improve these properties by applying the linear regularity theory. Using these properties again on the right side, we can repeat this argument, and so on. This enables us to reach classical solutions after a number of steps. The first step is the crucial one.

There are several other variants of regularity results, see [Gal94b, VIII.5], [Hey80, page 653], [Lad69, Chap. 5, 5], [Tem77, Chap. II, Prop 1.1].

The proof below is based on the fractional powers of the Stokes operator A. First we consider a bounded domain. For unbounded domains we prove only local regularity properties.

3.6.1 Theorem *Let $k \in \mathbb{N}_0$, let $\Omega \subseteq \mathbb{R}^n$, $n = 2, 3$, be a bounded C^{k+2}-domain, and let $f \in W^{k,2}(\Omega)^n$. Suppose the pair*

$$(u,p) \in W^{1,2}_{0,\sigma}(\Omega) \times L^2_{loc}(\Omega)$$

solves the equation

$$-\nu\Delta u + u\cdot\nabla u + \nabla p = f \tag{3.6.1}$$

in the sense of distributions. Then

$$u \in W^{k+2,2}(\Omega)^n \ , \quad p \in W^{k+1,2}(\Omega). \tag{3.6.2}$$

Proof. First let $k = 0$ and $f \in L^2(\Omega)^n$. Using Lemma 1.6.1, II, we see that f can be written in the form $f = \text{div}\ F$ with $F \in L^2(\Omega)^{n^2}$, and from Remark 3.3.2 we obtain that $p \in L^2(\Omega)$. From Lemma 3.3.1 and Lemma 3.5.2 we conclude that u satisfies the equation

$$A^{\frac{1}{2}}u = A^{-\frac{1}{2}}Pf - A^{-\frac{1}{2}}P\ \text{div}\ (u\,u). \tag{3.6.3}$$

See Lemma 2.6.2 for the properties of the operator $A^{-\frac{1}{2}}P$ div . In the first step we can improve the regularity of u only slightly, we show that

$$A^{\frac{1}{2}}u \in D(A^{\frac{1}{4}}) \text{ and therefore } u \in D(A^{\frac{3}{4}}). \tag{3.6.4}$$

For this purpose we choose $v \in C^\infty_{0,\sigma}(\Omega)$, set $w = A^{\frac{1}{4}}v$, and consider the functional

$$\begin{aligned} w \quad &\mapsto \quad < A^{\frac{1}{2}}u, A^{\frac{1}{4}}w > \\ &= \quad < A^{-\frac{1}{2}}Pf, A^{\frac{1}{4}}w > - < A^{-\frac{1}{2}}P\ \text{div}\ (u\,u), A^{\frac{1}{4}}w >, \end{aligned}$$

$v \in C_{0,\sigma}^{\infty}(\Omega)$. We show that this functional is continuous with respect to the norm $\|w\|_2$. Since $A^{\frac{1}{4}}$ is a selfadjoint operator, we can conclude that $A^{\frac{1}{2}} u \in D(A^{\frac{1}{4}})$. To prove this continuity property, we use the estimate

$$\|u \cdot \nabla u\|_{3/2} = \|\text{div}\,(u\,u)\|_{3/2} \leq C\,\|\nabla u\|_2^2$$

see Lemma 3.2.1, b), with $C = C(\Omega) > 0$, the inequality

$$\|A^{-\frac{1}{4}} w\|_3 \leq C\,\|w\|_2 \quad \text{if } n = 3,$$

see Lemma 2.4.2, (2.4.8), with $\alpha = \frac{1}{4}$, $q = 3$, $2\alpha + \frac{3}{q} = \frac{3}{2}$, $C > 0$, and the estimate

$$\|A^{-\frac{1}{4}} w\|_3 \leq C_1\,\|A^{-\frac{1}{4}} w\|_4 \leq C_2\,\|w\|_2 \quad \text{if } n = 2,$$

see the same lemma with $\alpha = \frac{1}{4}$, $q = 4$, $2\alpha + \frac{2}{q} = \frac{2}{2}$, $C_1 = C_1(\Omega) > 0$, $C_2 = C_2(\Omega) > 0$.

This shows with Hölder's inequality and with (2.5.21) that

$$\begin{aligned} |< A^{-\frac{1}{2}} P \,\text{div}\,(u\,u), A^{\frac{1}{4}} w >| &= |< \text{div}\,(u\,u), A^{-\frac{1}{4}} w >| \\ &\leq C_1\,\|\text{div}\,(u\,u)\|_{3/2}\,\|A^{-\frac{1}{4}} w\|_3 \leq C_2\,\|\nabla u\|_2^2\,\|w\|_2 \end{aligned}$$

with $C_1 = C_1(\Omega) > 0$, $C_2 = C_2(\Omega) > 0$. Similarly we get

$$\begin{aligned} |< A^{-\frac{1}{2}} P f, A^{\frac{1}{4}} w >| &= |< f, A^{-\frac{1}{4}} w >| \leq \|f\|_{3/2}\,\|A^{-\frac{1}{4}} w\|_3 \\ &\leq C\,\|f\|_2\,\|w\|_2 \end{aligned}$$

with $C = C(\Omega) > 0$. This proves the above continuity property, and (3.6.4) follows.

In the next step we show that $u \in D(A)$.

For this purpose we use the embedding property in Lemma 2.4.3 with $n = 3$, $\alpha = \frac{3}{4}$, $q = 3$, $2\alpha + \frac{3}{3} = 1 + \frac{3}{2}$, and get

$$\|u\|_{W^{1,3}(\Omega)} \leq C\,(\nu^{-3/4}\,\|A^{\frac{3}{4}} u\|_2 + \|u\|_2),$$

$C = C(\Omega) > 0$. Similarly, with $n = 2$, $\alpha = \frac{3}{4}$, $q = 4$, $2\alpha + \frac{2}{4} = 1 + \frac{2}{2}$, we obtain

$$\|u\|_{W^{1,4}(\Omega)} \leq C\,(\nu^{-3/4}\,\|A^{\frac{3}{4}} u\|_2 + \|u\|_2),$$

$C = C(\Omega) > 0$. The right sides are finite because of (3.6.4). In particular we obtain

$$\nabla u \in L^3(\Omega)^{n^2} \quad \text{if } n = 3\,, \quad \nabla u \in L^4(\Omega)^{n^2} \quad \text{if } n = 2. \tag{3.6.5}$$

Using (3.6.4) we get from (3.6.3) the equation

$$A^{\frac{3}{4}}u = A^{-\frac{1}{4}}Pf - A^{-\frac{1}{4}}P \operatorname{div}(u\,u).$$

Next we show that

$$\operatorname{div}(u\,u) \in L^2(\Omega)^n.$$

This follows with (3.2.6), (3.2.7), and Hölder's inequality. For $n = 2, 3$ we obtain

$$\|\operatorname{div}(u\,u)\|_2 = \|u \cdot \nabla u\|_2 \le C_1 \|u\|_6 \|\nabla u\|_3 \le C_2 \|\nabla u\|_2 \|\nabla u\|_3 < \infty$$

with $C_1 = C_1(n) > 0,\ C_2 = C_2(n, \Omega) > 0$.

In the same way as above, these properties show that the functional

$$w \mapsto < A^{\frac{1}{2}}u, A^{\frac{1}{2}}w > = < A^{-\frac{1}{2}}Pf, A^{\frac{1}{2}}w > - < A^{-\frac{1}{2}}P \operatorname{div}(u\,u), A^{\frac{1}{2}}w > ,$$

$w = A^{\frac{1}{2}}v,\ v \in C^\infty_{0,\sigma}(\Omega)$, is continuous in $\|w\|_2$. Since $A^{\frac{1}{2}}$ is selfadjoint, we conclude that $A^{\frac{1}{2}}u \in D(A^{\frac{1}{2}})$ and therefore that $u \in D(A)$.

Since $D(A) \subseteq W^{2,2}(\Omega)^n$, see Theorem 2.1.1, e), we conclude that $u \in W^{2,2}(\Omega)^n$.

Then we see that $f + \nu\Delta u - u \cdot \nabla u \in L^2(\Omega)^n$ and writing (3.6.1) in the form

$$\nabla p = f + \nu\Delta u - u \cdot \nabla u,$$

we conclude that $\nabla p \in L^2(\Omega)^n$ and $p \in W^{1,2}(\Omega)$. This proves the theorem for $k = 0$. In particular we see that u now satisfies the equation

$$Au = Pf - P \operatorname{div}(u\,u). \tag{3.6.6}$$

In the next steps $k = 1, \dots,$ we have only to apply the regularity theory of the linearized equations in Section 1.5. Consider the case $k = 1$. With $\tilde{f} := f - u \cdot \nabla u$, we write (3.6.1) in the form

$$-\nu\Delta u + \nabla p = \tilde{f}\,,$$

and apply Theorem 1.5.3 with $k = 1$. For this purpose we show that $\tilde{f} \in W^{1,2}(\Omega)^n$. Since $f \in W^{1,2}(\Omega)^n$, see the assumption for $k = 1$, we have only to show that $u \cdot \nabla u \in W^{1,2}(\Omega)^n$.

Since $u \in W^{2,2}(\Omega)^n$, we can apply Sobolev's embedding inequality, Lemma 1.3.3, II, (1.3.9) with u replaced by ∇u. With $r = \gamma = 2,\ 1 < q < \infty,\ 0 \le \beta \le 1,\ \beta(\frac{1}{2} - \frac{1}{n}) + (1 - \beta)\frac{1}{2} = \frac{1}{q}\,,\ \frac{\beta}{n} = \frac{1}{2} - \frac{1}{q}\,,\ \frac{1}{q} \ge \frac{1}{2} - \frac{1}{n}\,,$ we obtain

$$\|\nabla u\|_q \le C \|u\|_{W^{2,2}(\Omega)}\,,$$

$C = C(\Omega, q) > 0$. In particular, for $n = 2, 3$ we get

$$\|\nabla u\|_q \leq C \|u\|_{W^{2,2}(\Omega)} \ , \quad 1 < q \leq 6.$$

The embedding property in Lemma 1.3.2, II, (1.3.7) shows that

$$\|u\|_\infty \leq C \|\nabla u\|_6 \ ,$$

with $C = C(\Omega) > 0$. With Hölder's inequality this leads to

$$\begin{aligned} \|D_j(u \cdot \nabla u)\|_2 &= \|(D_j u) \cdot \nabla u + u \cdot \nabla(D_j u)\|_2 \\ &\leq \|(D_j u) \cdot \nabla u\|_2 + \|u \cdot \nabla(D_j u)\|_2 \\ &\leq C(\|\nabla u\|_4 \|\nabla u\|_4 + \|u\|_\infty \|\nabla^2 u\|_2) < \infty \, , \end{aligned}$$

$j = 1, \dots, n$, $C = C(\Omega) > 0$, and therefore we get $u \cdot \nabla u \in W^{1,2}(\Omega)^n$, $\tilde{f} \in W^{1,2}(\Omega)^n$.

Theorem 1.5.3 now yields

$$u \in W^{3,2}(\Omega)^n \ , \quad p \in W^{2,2}(\Omega).$$

This proves the theorem for $k = 1$.

This procedure can be repeated; differentiating $\nabla(u \cdot \nabla u)$ again, we obtain the result for $k = 2$, and so on. The general result follows by induction on k. This proves the theorem. □

The case of unbounded domains in the next theorem can be reduced to the case above by applying the localization method similarly as in the proof of Theorem 1.5.1. This yields only local regularity results. We can include the case $\Omega = \mathbb{R}^n$. In the exceptional $n = 2$, $\Omega = \mathbb{R}^2$, we have to suppose the stronger condition $u \in W^{1,2}_{0,\sigma}(\mathbb{R}^2)$ instead of $u \in \widehat{W}^{1,2}_{0,\sigma}(\mathbb{R}^2)$. The reason is, we need the embedding property of Lemma 1.3.5, II.

3.6.2 Theorem *Let $k \in \mathbb{N}_0$, $n = 2, 3$, let $\Omega = \mathbb{R}^n$ or let $\Omega \subseteq \mathbb{R}^n$ be any unbounded C^{k+2}-domain. Suppose $f \in W^{k,2}_{loc}(\overline{\Omega})^n$,*

$$\begin{aligned} (u, p) &\in \widehat{W}^{1,2}_{0,\sigma}(\Omega) \times L^2_{loc}(\overline{\Omega}) \quad \text{if } n = 3 \text{ or if } n = 2 \text{ and } \Omega \neq \mathbb{R}^2, \\ (u, p) &\in W^{1,2}_{0,\sigma}(\mathbb{R}^2) \times L^2_{loc}(\mathbb{R}^2) \quad \text{if } \Omega = \mathbb{R}^2, \end{aligned}$$

and assume that (u, p) satisfies the equation

$$-\nu \Delta u + u \cdot \nabla u + \nabla p = f \tag{3.6.7}$$

in the sense of distributions.

Then

$$u \in W^{k+2,2}_{loc}(\overline{\Omega})^n \ , \quad p \in W^{k+1,2}_{loc}(\overline{\Omega}). \tag{3.6.8}$$

3.6.3 Corollary *Suppose the assumptions of this theorem are satisfied for all $k \in \mathbb{N}$. Then, after a redefinition on a subset of Ω of measure zero, we obtain*

$$u \in C^{\infty}_{loc}(\overline{\Omega})^n \ , \quad p \in C^{\infty}_{loc}(\overline{\Omega}). \tag{3.6.9}$$

Proof of Theorem 3.6.2. Let $k = 0$, and consider open balls $B_0, B_1 \subseteq \mathbb{R}^n$ with $\overline{B}_0 \subseteq B_1$, $B_1 \cap \Omega \neq \emptyset$ as in step c) of the proof of Theorem 1.5.1. We choose a function $\varphi \in C_0^\infty(\mathbb{R}^n)$ satisfying

$$0 \le \varphi \le 1 \ , \quad \text{supp}\,\varphi \subseteq B_1 \ , \quad \text{and } \varphi(x) = 1 \quad \text{for all } x \in B_0 \,.$$

Since supp $\varphi \subseteq B_1$, we can choose a bounded C^{k+2}- domain $\Omega' \subseteq \Omega$ satisfying $\Omega' \subseteq \Omega \cap B_1$ and $\Omega \cap (\text{supp}\,\varphi) \subseteq \Omega'$. The multiplication of (3.6.7) with φ yields the **local equations** written in the form

$$-\nu\Delta(\varphi u) + \nabla(\varphi p) = \tilde{f} - \varphi(u\cdot\nabla u) \ , \quad \text{div}\,(\varphi u) = \tilde{g} \tag{3.6.10}$$

with $\tilde{f} := \varphi f - \nu 2(\nabla\varphi)(\nabla u) - \nu(\Delta\varphi)u + (\nabla\varphi)p$ and $\tilde{g} := (\nabla\varphi)\cdot u$, see (1.5.20).

We see that $\varphi u|_{\partial\Omega'} = 0$ in the sense of traces, and therefore, see (1.2.5), II, we get $\varphi u \in W_0^{1,2}(\Omega')^n$, $\tilde{g} \in W_0^{1,2}(\Omega')$. Green's formula (1.2.12), II, shows that

$$\int_{\Omega'} \text{div}\,\tilde{g}\,dx = \int_{\Omega'} \text{div}\,(\varphi u)\,dx = 0.$$

Now we can apply Lemma 2.3.1, II, with $k = 1$, and get some $u_0 \in W_0^{2,2}(\Omega')^n$ satisfying $\text{div}\,u_0 = \tilde{g}$ and

$$\|u_0\|_{W^{2,2}(\Omega')} \le C\,\|\tilde{g}\|_{W^{1,2}(\Omega')} \tag{3.6.11}$$

with $C = C(\Omega') > 0$.

Setting $\widehat{u} := \varphi u - u_0$ we get $\widehat{u} \in W_0^{1,2}(\Omega')^n$, div $\widehat{u} = 0$, and therefore, see (1.2.8), (1.2.9), we obtain $\widehat{u} \in W_{0,\sigma}^{1,2}(\Omega')$. From the assumptions we get $\varphi p \in L^2(\Omega')$. Further, the pair $(\widehat{u}, \varphi p) \in W_{0,\sigma}^{1,2}(\Omega') \times L^2(\Omega')$ satisfies the equation

$$-\nu\Delta\widehat{u} + \nabla(\varphi p) = \widehat{f} - \varphi(u\cdot\nabla u) \tag{3.6.12}$$

with $\widehat{f} := \tilde{f} + \nu\Delta u_0$ in the sense of distributions, and we see that $\widehat{f} \in L^2(\Omega')^n$. Therefore, this equation can be treated in the same way as equation (3.6.1) in the case $k = 0$. This yields (3.6.4), which means that $\widehat{u} \in D(A^{3/4})$. In the same way as in (3.6.5) we now obtain that

$$\nabla\widehat{u} \in L^3(\Omega')^{n^2} \ \text{ if } n = 3 \ , \quad \nabla\widehat{u} \in L^4(\Omega')^{n^2} \ \text{ if } n = 2.$$

Since the balls B_0, B_1 are arbitrary, and since $u_0 \in W_0^{2,2}(\Omega')^n$, we can conclude that

$$\nabla u \in L^3_{loc}(\overline{\Omega})^{n^2} \text{ if } n = 3 \quad \text{and} \quad \nabla u \in L^4_{loc}(\overline{\Omega})^{n^2} \text{ if } n = 2.$$

Using this information again on the right side of (3.6.12), we get in the same way as in (3.6.6) that $\widehat{u} \in D(A')$, where A' is the Stokes operator of Ω'. It follows that $\widehat{u} \in W^{2,2}(\Omega')^n$, $\varphi p \in W^{1,2}(\Omega')$, and as before we can conclude that

$$u \in W^{2,2}_{loc}(\overline{\Omega})^n \quad \text{and} \quad p \in W^{1,2}_{loc}(\overline{\Omega}). \tag{3.6.13}$$

This proves the theorem for $k = 0$.

As in the proof of Theorem 3.6.1 we conclude from (3.6.13) that $u \cdot \nabla u \in W^{1,2}_{loc}(\overline{\Omega})^n$. Therefore, in order to prove the case $k = 1$, we write the equation (3.6.7) in the form

$$-\nu \Delta u + \nabla p = f - u \cdot \nabla u \quad , \ \operatorname{div} u = 0,$$

and apply the linear theory in Theorem 1.5.1. This proves the result for $k = 1$. The next step $k = 2$ follows directly from the linear theory, and so on. The general result follows by induction on k. □

To prove Corollary 3.6.3, we have only to apply Sobolev's embedding results, see Lemma 1.3.4, II.

3.7 Some uniqueness results

The uniqueness of weak solutions of the stationary Navier-Stokes system (3.1.1) can be shown only under certain additional assumptions. See [Tem77, Chap. II, Theorem 1.3] and [Gal94b, VIII.3] concerning this problem. Theorem 3.7.3 below yields a uniqueness result for domains which have a finite width.

Let $\Omega \subseteq \mathbb{R}^n$, $n = 2, 3$, be any domain. Then Ω is said to have a **finite width** $d > 0$ iff Ω lies between two parallel hyperplanes having the distance d, see [Ada75, VI, 6.26]. This means, that after a translation and a rotation of the coordinate system the following holds:

$$\Omega \subseteq \{(x_1, \ldots, x_n) \in \mathbb{R}^n;\ 0 \le x_n \le d\}. \tag{3.7.1}$$

For uniqueness questions, see the proof below, it is important to determine explicitly the constants in some special embedding inequalities. Such results are given by the following lemmas.

3.7.1 Lemma *Let $\Omega \subseteq \mathbb{R}^n$, $n = 2, 3$, be a domain with finite width $d > 0$. Then*

$$\|u\|_{L^2(\Omega)} \le \frac{d}{\sqrt{2}} \|\nabla u\|_{L^2(\Omega)} \tag{3.7.2}$$

for all $u \in W_0^{1,2}(\Omega)^n$.

Proof. Without loss of generality we may assume that Ω has the special form (3.7.1). Then each $u \in W_0^{1,2}(\Omega)^n$ has the representation

$$u(x) = \int_0^{x_n} \frac{d}{dt} u(x', t)\, dt \ , \quad x = (x', x_n)\ , \quad x' = (x_1, \dots, x_{n-1}),$$

which leads with Hölder's inequality to

$$\begin{aligned} \|u\|_{L^2(\Omega)}^2 &\leq \int_{\mathbb{R}^{n-1}} \int_0^d \left(\int_0^{x_n} |D_n u(x', t)|\, dt \right)^2 dx_n\, dx' \\ &\leq \int_{\mathbb{R}^{n-1}} \int_0^d \left(\sqrt{x_n} \left(\int_0^{x_n} |D_n u|^2\, dt \right)^{\frac{1}{2}} \right)^2 dx_n\, dx' \\ &\leq \frac{d^2}{2} \|D_n u\|_{L^2(\Omega)}^2 . \end{aligned}$$

This proves the lemma. See [Ada75, VI, (31)] concerning this argument. □

The next lemma is essentially contained in [Lad69, Chap. 1.1, Lemma 1, Lemma 2].

3.7.2 Lemma *Let $\Omega \subseteq \mathbb{R}^n$, $n = 2, 3$, be an arbitrary domain. Then for all $u \in W_0^{1,2}(\Omega)^n$ we get*

$$\|u\|_{L^4(\Omega)} \leq 2^{\frac{1}{4}}\, \|u\|_{L^2(\Omega)}^{\frac{1}{2}}\, \|\nabla u\|_{L^2(\Omega)}^{\frac{1}{2}} \tag{3.7.3}$$

if $n = 2$, and

$$\|u\|_{L^4(\Omega)} \leq 4^{\frac{1}{4}}\, \|u\|_{L^2(\Omega)}^{\frac{1}{4}}\, \|\nabla u\|_{L^2(\Omega)}^{\frac{3}{4}} \tag{3.7.4}$$

if $n = 3$.

Proof. We follow [Lad69, Chap. 1.1]. Arguing by closure we may assume that $u \in C_0^\infty(\Omega)^n$.

An elementary calculation yields

$$|u(x)|^2 = u(x) \cdot u(x) = 2 \int_{-\infty}^{x_n} u(x', t) \cdot (D_n\, u(x', t))\, dt\ ,$$

$x = (x', x_n)$, and

$$\sup_{x_n} |u(x', x_n)|^2 \leq 2 \int_{-\infty}^{+\infty} |u \cdot D_n\, u|\, dt\ .$$

The corresponding estimates hold with D_n replaced by $D_1, \dots, D_{n-1}$.

If $n = 2$ we conclude with Hölder's inequality:

$$\begin{aligned}
\int_{\mathbb{R}^2} |u(x_1,x_2)|^4\, dx_1\, dx_2 &= \int_{\mathbb{R}^2} |u(x_1,x_2)|^2\, |u(x_1,x_2)|^2 dx_1\, dx_2 \\
&\le \int_{-\infty}^{+\infty} \big(\sup_{x_2} |u(x_1,x_2)|^2\big)\, dx_1 \int_{-\infty}^{+\infty} \big(\sup_{x_1} |u(x_1,x_2)|^2\big)\, dx_2 \\
&\le 4 \int_{\mathbb{R}^2} |u\cdot D_2 u|\, dx_1\, dx_2 \int_{\mathbb{R}^2} |u\cdot D_1 u|\, dx_1\, dx_2 \\
&\le 4 \Big(\int_{\mathbb{R}^2} |u|^2\, dx_1\, dx_2\Big) \Big(\int_{\mathbb{R}^2} |D_1 u|^2\, dx_1\, dx_2\Big)^{\frac{1}{2}} \Big(\int_{\mathbb{R}^2} |D_2 u|^2\, dx_1\, dx_2\Big)^{\frac{1}{2}} \\
&= 4\|u\|_2^2\, \|D_1 u\|_2\, \|D_2 u\|_2 \\
&\le 4\|u\|_2^2\, \big(\frac{1}{2}\|D_1 u\|_2^2 + \frac{1}{2}\|D_2 u\|_2^2\big) \\
&= 2\|u\|_2^2\, \|\nabla u\|_2^2 .
\end{aligned}$$

This proves (3.7.3).

If $n = 3$ we use the above estimates with fixed x_3 and obtain

$$\begin{aligned}
&\int_{\mathbb{R}^2} |u(x_1,x_2,x_3)|^4\, dx_1\, dx_2\, dx_3 \\
&\le 2\int_{-\infty}^{+\infty} \Big(\int_{\mathbb{R}^2} |u(x_1,x_2,x_3)|^2\, dx_1\, dx_2\Big)\Big(\int_{\mathbb{R}^2} (|D_1 u|^2 + |D_2 u|^2) dx_1\, dx_2\Big)\, dx_3 \\
&\le 2\Big(\sup_{x_3} \int_{\mathbb{R}^2} |u|^2\, dx_1\, dx_2\Big) \Big(\int_{\mathbb{R}^3} (|D_1 u|^2 + |D_2 u|^2)\, dx_1\, dx_2\, dx_3\Big) \\
&\le 4\Big(\int_{\mathbb{R}^3} |u\cdot D_3 u|\, dx_1\, dx_2\, dx_3\Big) \Big(\int_{\mathbb{R}^3} (|D_1 u|^2 + |D_2 u|^2)\, dx_1\, dx_2\, dx_3\Big) \\
&\le 4\Big(\int_{\mathbb{R}^3} |u|^2\, dx_1\, dx_2\, dx_3\Big)^{\frac{1}{2}} \Big(\int_{\mathbb{R}^3} (|D_1 u|^2 + |D_2 u|^2 + |D_3 u|^2) dx_1\, dx_2\, dx_3\Big)^{\frac{3}{2}} \\
&= 4\|u\|_2\, \|\nabla u\|_2^3 .
\end{aligned}$$

This proves (3.7.4), and the proof is complete. □

Next we consider some consequences of these inequalities. Let $\Omega \subseteq \mathbb{R}^n$, $n = 2,3$, be a domain with finite width $d > 0$. Then from (3.7.2) we conclude that the norms

$$\|\nabla u\|_2 \quad \text{and} \quad (\|u\|_2^2 + \|\nabla u\|_2^2)^{\frac{1}{2}}$$

are equivalent on $C_{0,\sigma}^\infty(\Omega)$. This shows that

$$\widehat{W}_{0,\sigma}^{1,2}(\Omega) = W_{0,\sigma}^{1,2}(\Omega) = \overline{C_{0,\sigma}^\infty(\Omega)}^{\|\cdot\|_{W^{1,2}(\Omega)}} \tag{3.7.5}$$

holds for the solution space $\widehat{W}_{0,\sigma}^{1,2}(\Omega)$ of weak solutions.

Let $u, v \in W^{1,2}_{0,\sigma}(\Omega)$ with $u = (u_1, \dots, u_n), v = (v_1, \dots v_n)$. Using the notation $uv = (u_j\, v_l)_{j,l=1}^n$ and Hölder's inequality we obtain

$$\begin{aligned} \|u\,v\|_2 &= \left(\sum_{j,l=1}^n \|u_j\,v_l\|_2^2\right)^{\frac{1}{2}} \le \left(\sum_{j,l=1}^n \|u_j\|_4^2\,\|v_l\|_4^2\right)^{\frac{1}{2}} \qquad (3.7.6) \\ &= \left(\sum_{j=1}^n \|u_j\|_4^2\right)^{\frac{1}{2}} \left(\sum_{l=1}^n \|v_l\|_4^2\right)^{\frac{1}{2}} \le n^{\frac{1}{2}}\|u\|_4\,\|v\|_4 . \end{aligned}$$

Inserting (3.7.3), (3.7.4), (3.7.2) leads to

$$\|u\,v\|_2 \le 2^{\frac{1}{2}} d\, \|\nabla u\|_2\, \|\nabla v\|_2 \qquad (3.7.7)$$

if $n = 2$, and to

$$\|u\,v\|_2 \le 3^{\frac{1}{2}} 2^{\frac{3}{4}} d^{\frac{1}{2}}\, \|\nabla u\|_2\, \|\nabla v\|_2 \qquad (3.7.8)$$

if $n = 3$.

Consider the trilinear form

$$(u, v, w) \mapsto < uv, \nabla w > := \sum_{j,l=1}^n (u_j v_l) D_j w_l\,, \qquad (3.7.9)$$

$u, v, w \in \widehat{W}^{1,2}_{0,\sigma}(\Omega)$, as a mapping from $\widehat{W}^{1,2}_{0,\sigma}(\Omega) \times \widehat{W}^{1,2}_{0,\sigma}(\Omega) \times \widehat{W}^{1,2}_{0,\sigma}(\Omega)$ to $\mathbb{R}$. In the same way as in (3.2.1) we obtain the relations

$$< u \cdot \nabla v, w > = < \operatorname{div}\,(u\,v), w > = - < u\,v, \nabla w > \qquad (3.7.10)$$

for $u, v, w \in \widehat{W}^{1,2}_{0,\sigma}(\Omega)$, and

$$< u \cdot \nabla v, v > = \frac{1}{2} < u, \nabla |v|^2 > \qquad (3.7.11)$$

for $u, v \in \widehat{W}^{1,2}_{0,\sigma}(\Omega)$.

From (3.7.7), (3.7.8) we get

$$|< u\,v, \nabla w >| \le \|u\,v\|_2\, \|\nabla w\|_2 \le K\, \|\nabla u\|_2\, \|\nabla v\|_2\, \|\nabla w\|_2\,, \qquad (3.7.12)$$

$u, v, w \in \widehat{W}^{1,2}_{0,\sigma}(\Omega)$, with $K := 2^{\frac{1}{2}} d$ if $n = 2$, and $K := 3^{\frac{1}{2}} 2^{\frac{3}{4}} d^{\frac{1}{2}}$ if $n = 3$.

Let $u, v \in \widehat{W}^{1,2}_{0,\sigma}(\Omega)$, and let $(u_j)_{j=1}^\infty$ be a sequence in $C^\infty_{0,\sigma}(\Omega)$ satisfying $u = \lim_{j\to\infty} u_j$ with respect to $\|\cdot\|_{W^{1,2}(\Omega)}$. The last estimate shows that the following limits exist and that

$$\begin{aligned} < uv, \nabla v > &= \lim_{j \to \infty} < u_j v, \nabla v > \\ &= -\lim_{j \to \infty} < u_j \cdot \nabla v, v > \\ &= -\frac{1}{2} \lim_{j \to \infty} < u_j, \nabla |v|^2 > \\ &= \frac{1}{2} \lim_{j \to \infty} < \operatorname{div} u_j, |v|^2 > = 0. \end{aligned} \tag{3.7.13}$$

Now we can prove the following result

3.7.3 Theorem *Let $\Omega \subseteq \mathbb{R}^n$, $n = 2, 3$, be any domain having a finite width $d > 0$, and let $\Omega_0 \subseteq \Omega$ be a bounded subdomain with $\overline{\Omega}_0 \subseteq \Omega$, $\Omega_0 \neq \emptyset$. Suppose $f = \operatorname{div} F$, $F \in L^2(\Omega)^{n^2}$, satisfies*

$$\|F\|_2 < \nu^2 K^{-1} \tag{3.7.14}$$

with $K = 2^{\frac{1}{2}} d$ if $n = 2$ and $K = 3^{\frac{1}{2}} 2^{\frac{3}{4}} d^{\frac{1}{2}}$ if $n = 3$.

Then there exists one and only one pair

$$(u, p) \in W^{1,2}_{0,\sigma}(\Omega) \times L^2_{loc}(\Omega)$$

satisfying $\int_{\Omega_0} p \, dx = 0$ and

$$-\nu \Delta u + u \cdot \nabla u + \nabla p = f \tag{3.7.15}$$

in the sense of distributions.

Proof. The existence follows from Theorem 3.5.1. To prove the uniqueness we consider two pairs

$$(u, p), \ (\widehat{u}, \widehat{p}) \in W^{1,2}_{0,\sigma}(\Omega) \times L^2_{loc}(\Omega)$$

satisfying $\int_{\Omega_0} p \, dx = \int_{\Omega_0} \widehat{p} \, dx = 0$ and (3.7.15) in the sense of distributions. Then

$$\begin{aligned} &-\nu \Delta (u - \widehat{u}) + \nabla (p - \widehat{p}) + (u - \widehat{u}) \cdot \nabla u + \widehat{u} \cdot \nabla (u - \widehat{u}) \\ &\qquad = -\nu \Delta u + u \cdot \nabla u + \nabla p - (-\nu \Delta \widehat{u} + \widehat{u} \cdot \nabla \widehat{u} + \nabla \widehat{p}) = 0 \end{aligned}$$

and for each $v \in C^\infty_{0,\sigma}(\Omega)$ we get

$$\begin{aligned} \nu < \nabla (u - \widehat{u}), \nabla v > &= - < (u - \widehat{u}) \cdot \nabla u, v > - < \widehat{u} \cdot \nabla (u - \widehat{u}), v > \\ &= < (u - \widehat{u}) u, \nabla v > + < \widehat{u} (u - \widehat{u}), \nabla v > . \end{aligned}$$

Since $(u-\widehat{u})u,\ \widehat{u}(u-\widehat{u}) \in L^2(\Omega)^{n^2}$, see (3.7.7), (3.7.8), we see, using a sequence as in (3.7.13), that we may insert $v = u - \widehat{u}$ in the last equation. This yields $< u(u-\widehat{u}), \nabla(u-\widehat{u}) > = 0$ by (3.7.13) and we obtain

$$\nu\|\nabla(u-\widehat{u})\|_2^2 = < (u-\widehat{u})u, \nabla(u-\widehat{u}) > . \tag{3.7.16}$$

In the same way we get from (3.7.15) that

$$\nu\|\nabla u\|_2^2 = - < F, \nabla u > . \tag{3.7.17}$$

Assume that $u \neq 0$ and that $u \neq \widehat{u}$. Then from (3.7.12) we obtain

$$\nu\|\nabla(u-\widehat{u})\|_2^2 \ \leq\ K\,\|\nabla(u-\widehat{u})\|_2^2\,\|\nabla u\|_2\,, \tag{3.7.18}$$

and therefore

$$\nu \ \leq\ K\,\|\nabla u\|_2\,.$$

From (3.7.17) we get

$$\nu\|\nabla u\|_2^2 \ \leq\ \|F\|_2\,\|\nabla u\|_2\ , \quad \nu\|\nabla u\|_2 \ \leq\ \|F\|_2.$$

This leads to $\nu^2 \leq K\,\|F\|_2$ and with (3.7.14) we get the contradiction

$$\nu^2 < \nu^2\,.$$

Therefore, if $u \neq 0$ we see that $u = \widehat{u}$. If $u = 0$ we conclude from (3.7.16) that $\widehat{u} = u = 0$. It follows $\nabla(p-\widehat{p}) = 0$. Thus $p - \widehat{p}$ is a constant which is zero since $\int_{\Omega_0}(p-\widehat{p})\,dx = 0$. This proves the theorem. □

3.7.4 Corollary *Let $\Omega \subseteq \mathbb{R}^n$, $n = 2,3$, be any domain having a finite width $d > 0$, and let $\Omega_0 \subseteq \Omega$ be a bounded subdomain with $\overline{\Omega}_0 \subseteq \Omega$, $\Omega_0 \neq \emptyset$. Let $K = 2^{\frac{1}{2}}d$ if $n = 2$ and $K = 3^{\frac{1}{2}}\,2^{\frac{3}{4}}d^{\frac{1}{2}}$ if $n = 3$.*

Suppose f satisfies one of the following conditions:

a) *$f \in W^{-1,2}(\Omega)^n$ and*

$$\|f\|_{W^{-1,2}(\Omega)^n} \ <\ \nu^2\,(1+\frac{d^2}{2})^{-\frac{1}{2}}\,K^{-1} \tag{3.7.19}$$

or

b) *$f \in L^2(\Omega)^n$ and*

$$\|f\|_{L^2(\Omega)^n} \ <\ \nu^2\,d^{-1}2^{\frac{1}{2}}K^{-1}. \tag{3.7.20}$$

Then there exists one and only one pair

$$(u,p) \in W^{1,2}_{0,\sigma}(\Omega) \times L^2_{loc}(\Omega)$$

satisfying $\int_{\Omega_0} p\,dx = 0$ *and*

$$-\nu\,\Delta u + u \cdot \nabla u + \nabla p = f$$

in the sense of distributions.

Proof. Using (3.7.2) we see in case a) that

$$\begin{aligned} |[f, v]| &\le \|f\|_{W^{-1,2}(\Omega)} \, (\|v\|_2^2 + \|\nabla v\|_2^2)^{\frac{1}{2}} \\ &\le (1 + \frac{d^2}{2})^{\frac{1}{2}} \, \|f\|_{W^{-1,2}(\Omega)} \, \|\nabla v\|_2 \, , \end{aligned}$$

$v \in C_{0,\sigma}^\infty(\Omega)$. Then, in the same way as in Lemma 1.6.1, II, we find some $F \in L^2(\Omega)^{n^2}$ satisfying $f = \operatorname{div} F$,

$$\|F\|_2 \le (1 + \frac{d^2}{2})^{\frac{1}{2}} \, \|f\|_{W^{-1,2}(\Omega)},$$

and the assertion follows from Theorem 3.7.3. In case b) we find in the same way as above some $F \in L^2(\Omega)^{n^2}$ satisfying $f = \operatorname{div} F$,

$$\|F\|_2 \le \frac{d}{\sqrt{2}} \, \|f\|_2 \, ,$$

and the assertion follows again from Theorem 3.7.3. This proves the corollary.

$\square$

Chapter IV

The Linearized Nonstationary Theory

1 Preliminaries for the time dependent linear theory

1.1 The nonstationary Stokes system

Let $\Omega \subseteq \mathbb{R}^n$ be an arbitrary domain with $n \geq 2$ and boundary $\partial\Omega$. In the linear time dependent theory we admit arbitrary dimensions $n \geq 2$. Let $0 < T \leq \infty$. Then $[0,T)$ is called the time interval. The case $T = \infty$ is admitted. We call $t \in [0,T)$ the time variable and $x = (x_1, \ldots, x_n) \in \Omega$ the space variables. For each scalar or vector function

$$v : (t,x) \mapsto v(t,x) \quad , \quad t \in [0,T),\ x \in \Omega$$

let $v(t) = v(t,\cdot)$ be the function $x \mapsto v(t,x)$ only in the space variables with fixed t.

On the cylinder $[0,T) \times \Omega$ the nonstationary Stokes system has the form

$$\begin{aligned} u_t - \nu\Delta u + \nabla p = f \quad &, \quad \mathrm{div}\ u = 0, \\ u|_{\partial\Omega} = 0 \quad &, \quad u(0) = u_0, \end{aligned} \tag{1.1.1}$$

where $u|_{\partial\Omega} = 0$ means the boundary condition, and $u(0) = u_0$ the initial condition at $t = 0$. If $\Omega = \mathbb{R}^n$, $\partial\Omega = \emptyset$, the boundary condition is omitted.

Here $u_t = u' = \frac{du}{dt}$ means the time derivative, $\nu > 0$ as before the viscosity constant, $f = (f_1, \ldots, f_n)$ the given exterior force, $u = (u_1, \ldots, u_n)$ the unknown velocity field, and p the unknown pressure; $u_0 = u(0)$ is the given velocity field at $t = 0$ (initial value).

In the following we will give these equations a precise meaning and develop the theory of existence, uniqueness and regularity of solutions. The linear theory considered in this chapter is basic for the full nonlinear equations in Chapter V. Here we refer to [Tem77], [Tem83], [Sol77], [vWa85], [Ama95], [Wie99].

1.2 Basic spaces for the time dependent theory

In the following we introduce some special notations for the time dependent theory, see, e.g., [Ama95, Chap. III, 1.1]. Let $0 < T \leq \infty$ and let X be any Banach space with norm $\|\cdot\|_X$.

Consider any function $u : t \mapsto u(t)$, $t \in [0, T)$, with values in X. Then u is called finitely valued or a **step function** iff there are finitely many points

$$0 \leq a_1 < b_1 \leq a_2 < b_2 \leq \cdots \leq a_m < b_m < T \ , \quad m \in \mathbb{N},$$

and values $C_1, \ldots, C_m \in X$ such that

$$u(t) = C_j \ \text{ for } \ t \in [a_j, b_j) \ , \quad j = 1, \ldots, m,$$

and $u(t) = 0$ for t outside of these intervals. We define the elementary integral

$$\int_0^T u(t)\, dt \ = \ \int_0^T u\, dt := \sum_{j=1}^{m} (b_j - a_j)\, C_j \ \ \in X$$

for such a step function. A function $u : [0, T) \to X$ is called **(Bochner-) measurable** iff there exists a sequence $(u_j)_{j=1}^{\infty}$ of step functions such that

$$\lim_{j \to \infty} \|u(t) - u_j(t)\|_X = 0$$

holds for almost all (a.a.) $t \in [0, T)$. If additionally

$$\lim_{j,l \to \infty} \int_0^T \|u_j - u_l\|_X \, dt = 0 \, ,$$

which means that $(u_j)_{j=1}^{\infty}$ is a Cauchy sequence with respect to the norm

$$\int_0^T \|\cdot\|_X \, dt \, ,$$

then u is called **(Bochner-) integrable**, and the (well defined) limit

$$\int_0^T u(t)\, dt \ = \ \int_0^T u\, dt \ := \ \lim_{j \to \infty} \int_0^T u_j\, dt \quad \in X \tag{1.2.1}$$

(with respect to the norm $\|\cdot\|_X$) is called the **(Bochner-) integral** of u on $[0, T)$, see [Yos80, V, 5] or [HiPh57, Sec. 3.7]. Indeed, we see that $(\int_0^T u_j dt)_{j=1}^{\infty}$ is a Cauchy sequence with respect to $\|\cdot\|_X$.

We know, see Bochner's theorem [Yos80, V, 5, Theorem 1], that a measurable function $u : [0,T) \to X$ is integrable iff the function $t \mapsto \|u(t)\|_X$, $t \in [0,T)$, is integrable in the scalar-valued usual sense. We simply write

$$\int_0^T \|u(t)\|_X \, dt < \infty$$

in this case, and we get the estimate

$$\| \int_0^T u \, dt \|_X \leq \int_0^T \|u(t)\|_X \, dt. \tag{1.2.2}$$

Let $1 \leq s < \infty$. Then $L^s(0,T;X)$ denotes the Banach space of all measurable (classes of) functions $u : [0,T) \to X$ such that $t \mapsto \|u(t)\|_X^s$ is integrable on $[0,T)$; we simply write

$$\int_0^T \|u\|_X^s \, dt < \infty.$$

The norm in $L^s(0,T;X)$ is defined by

$$\|u\|_{X,s;T} := \left(\int_0^T \|u\|_X^s \, dt \right)^{\frac{1}{s}}. \tag{1.2.3}$$

Similarly, $L^\infty(0,T;X)$ means the Banach space of all measurable (classes of) functions $u : [0,T) \to X$ with finite norm

$$\|u\|_{X,\infty;T} := \operatorname*{ess\,-\,sup}_{t \in [0,T)} \|u(t)\|_X .$$

The Banach spaces $L^s(a,b;X)$ with $-\infty < a < b \leq \infty$ are defined in the same way with $[0,T)$ replaced by $[a,b)$.

Let $1 \leq s \leq \infty$. Then the vector space $L^s_{loc}([0,T);X)$ is the space of all measurable functions $u : [0,T) \to X$ with

$$u \in L^s(0,T';X) \ \text{ for all } \ T' \ \text{ with } \ 0 < T' < T. \tag{1.2.4}$$

Let X' be the dual space of X, and let $[f,u]$ be the value of the functional $f \in X'$ at $u \in X$. Then

$$\|f\|_{X'} = \sup_{0 \neq u \in X} (|[f,u]| / \|u\|_X) \tag{1.2.5}$$

is the norm of f in X'; $\|f\|_{X'}$ is the infimum of all $C = C(f) \geq 0$ such that

$$|[f,u]| \leq C \, \|u\|_X \ , \quad u \in X. \tag{1.2.6}$$

If X is a reflexive Banach space and if $1 < s < \infty$, then $L^s(0,T;X)$ is reflexive too, see [HiPh57, Sec. 3.8]. We get the important relation

$$L^{s'}(0,T;X') = L^s(0,T;X)' \ , \quad s' := \frac{s}{s-1} \ , \quad 1 = \frac{1}{s} + \frac{1}{s'} \tag{1.2.7}$$

in the sense that each $f \in L^{s'}(0,T;X')$ is considered as (identified with) the functional $[f,\cdot] : u \mapsto [f,u]$, $u \in L^s(0,T;X)$, defined by

$$[f,u] := [f,u]_{X,T} := \int_0^T [f(t),u(t)]\, dt. \tag{1.2.8}$$

We obtain the relations

$$|[f(t),u(t)]| \ \leq \ \|f(t)\|_{X'} \, \|u(t)\|_X \quad \text{for a. a. } t \in [0,T) \tag{1.2.9}$$

and

$$|[f,u]| \ \leq \ \|f\|_{X',s';T} \, \|u\|_{X,s;T} \ . \tag{1.2.10}$$

If in particular $X = H$ is a Hilbert space with scalar product $< \cdot,\cdot >_H$ and norm $\| \cdot \|_H$, then $L^2(0,T;H)$ is a Hilbert space with scalar product

$$< u,v >_{H,T} := \int_0^T < u(t),v(t) >_H \ dt \tag{1.2.11}$$

and norm $\|u\|_{H,2;T} = (\int_0^T \|u(t)\|_H^2\, dt)^{\frac{1}{2}}$.

Let $B : D(B) \to X$ be a closed linear operator with dense domain $D(B) \subseteq X$. We consider $D(B)$ as a Banach space with the graph norm

$$\|u\|_{D(B)} := \|u\|_X + \|Bu\|_X, \tag{1.2.12}$$

and we obtain the continuous embedding $D(B) \subseteq X$. Then for $1 \leq s < \infty$ we consider the Banach space $L^s(0,T;D(B))$ with norm

$$\|u\|_{D(B),s;T} := (\int_0^T \|u\|_{D(B)}^s \, dt)^{\frac{1}{s}} \ . \tag{1.2.13}$$

The subspace $L^s(0,T;D(B)) \subseteq L^s(0,T;X)$ is dense in $L^s(0,T;X)$ with respect to the norm $\| \cdot \|_{X,s;T}$.

We define the operator $\widehat{B} : D(\widehat{B}) \to L^s(0,T;X)$ with domain $D(\widehat{B}) := L^s(0,T;D(B)) \subseteq L^s(0,T;X)$ by setting

$$(\widehat{B}u)(t) := Bu(t)$$

for almost all $t \in (0,T)$ and all $u \in D(\widehat{B})$. $\widehat{B}$ is a closed densely defined operator. We write $\widehat{B} = B$ if there is no confusion.

Thus we get $\widehat{B}u = Bu \in L^s(0,T;X)$ for each $u \in L^s(0,T;D(B))$, and we simply write

$$Bu \in L^s(0,T;X) \quad \text{or} \quad \|Bu\|_{X,s;T} < \infty \tag{1.2.14}$$

iff $u \in L^s(0,T;D(B))$.

If $0 < T < \infty$, $u \in L^s(0,T;D(B))$, then $\int_0^T u\,dt$ and $\int_0^T Bu\,dt$ are well defined, and we get

$$B\int_0^T u\,dt = \int_0^T Bu\,dt \ , \quad \int_0^T u\,dt \in D(B). \tag{1.2.15}$$

This can be shown by going back to the step functions and using the closedness of B; see [Fri69, Part 2, Lemma 1.2] or [HiPh57, Chap. III, Theorem 3.7.12].

If in particular $D(B) = X$ and if B is a bounded operator with operator norm $\|B\|$, then the graph norm $\|u\|_{D(B)}$ is equivalent to $\|u\|_X$ and we get $D(\widehat{B}) = L^s(0,T;X)$. In this case $\widehat{B}$ is a bounded operator.

We need the following important fact. Let X be a reflexive Banach space and let $1 < s < \infty$. Suppose $(u_j)_{j=1}^{\infty}$ is a bounded sequence in $L^s(0,T;D(B))$ and assume that u_j converges weakly in $L^s(0,T;X)$ to some $u \in L^s(0,T;X)$ as $j \to \infty$. Then we get

$$u \in L^s(0,T;D(B)) \ , \quad \|u\|_{D(B),s;T} \le \liminf_{j\to\infty} \|u_j\|_{D(B),s;T}. \tag{1.2.16}$$

To prove this we use that $L^s(0,T;D(B))$ is a reflexive Banach space. Therefore, the given bounded sequence $(u_j)_{j=1}^{\infty}$ contains a subsequence which converges weakly in $L^s(0,T;D(B))$ to some $\widehat{u} \in L^s(0,T;D(B))$. Then we see that $u = \widehat{u}$ and (1.2.16) follows using (1.2.10). See the similar argument in (3.1.3), II.

Let $\Omega \subseteq \mathbb{R}^n$, $n \ge 1$, be any domain, let $1 \le q \le \infty$, $m \in \mathbb{N}$, $0 < T \le \infty$, $1 \le s \le \infty$, and set $X := L^q(\Omega)^m$. In this case we introduce some special notations.

The norm in the Banach space $L^s(0,T;L^q(\Omega)^m)$ will be denoted by

$$\|u\|_{q,s;T} := \left(\int_0^T \|u(t)\|_q^s\,dt\right)^{\frac{1}{s}} \ , \quad u \in L^s(0,T;L^q(\Omega)^m), \tag{1.2.17}$$

with the obvious modification if $s = \infty$.

We write

$$\|u\|_{q,s;T} = \|u\|_{L^q(\Omega),s;T} = \|u\|_{L^q(\Omega)^m,s;T}.$$

In the case $1 < q < \infty$, $1 < s < \infty$, the Banach space $L^s(0,T;L^q(\Omega)^m)$ is reflexive, and the dual space is given by

$$L^{s'}(0,T;L^{q'}(\Omega)^m) = L^s(0,T;L^q(\Omega)^m)' \,, \tag{1.2.18}$$

where $s' = \frac{s}{s-1}$, $q' = \frac{q}{q-1}$ such that $1 = \frac{1}{s} + \frac{1}{s'}$, $1 = \frac{1}{q} + \frac{1}{q'}$. The value of some $f \in L^{s'}(0,T;L^{q'}(\Omega)^m)$ at $u \in L^s(0,T;L^q(\Omega)^m)$ is given by

$$\begin{aligned} < f, u >_{\Omega,T} &= \int_0^T < f(t), u(t) >_\Omega \, dt \\ &= \int_0^T \int_\Omega f(t,x) \cdot u(t,x) \, dx \, dt \end{aligned} \tag{1.2.19}$$

where $f = (f_j)_{j=1}^m$, $u = (u_j)_{j=1}^m$, $f \cdot u = \sum_{j=1}^m f_j u_j$. Thus we identify f with the functional $< f, \cdot >_{\Omega,T}$ defined by (1.2.19).

If in particular $s = q = 2$, we get the Hilbert space $L^2(0,T;L^2(\Omega)^m)$ with scalar product

$$< u, v >_{\Omega,T} = \int_0^T < u(t), v(t) >_\Omega \, dt = \int_0^T \int_\Omega u(t,x) \cdot v(t,x) \, dx \, dt, \tag{1.2.20}$$

and norm $\|u\|_{2,2;T} = (\int_0^T \|u(t)\|_2^2 \, dt)^{\frac{1}{2}}$, where $u = (u_j)_{j=1}^m$, $v = (v_j)_{j=1}^m$, $u \cdot v = \sum_{j=1}^m u_j v_j$.

We conclude this subsection with some general remarks, see [Yos80, V, 4-5] and [HiPh57, Chap. III] for further information.

A function $u : [0,T) \to X$ is integrable iff $u \in L^1(0,T;X)$.

Consider a sequence $(u_j)_{j=1}^\infty$ in $L^1(0,T;X)$ and a function $u : [0,T) \to X$ such that $u(t) = s - \lim_{j\to\infty} u_j(t)$ holds for almost all $t \in [0,T)$. This means, the sequence converges **pointwise** for almost all $t \in [0,T)$ in the strong sense.

Further assume that there is some $g \in L^1(0,T;\mathbb{R})$ satisfying

$$\|u_j(t)\|_X \leq |g(t)| \quad \text{for a.a. } t \in [0,T) \text{ and all } j \in \mathbb{N}\,. \tag{1.2.21}$$

Then **Lebesgue's dominated convergence lemma** yields

$$u \in L^1(0,T;X) \quad \text{and} \quad \int_0^T u \, dt = s - \lim_{j\to\infty} \int_0^T u_j \, dt. \tag{1.2.22}$$

This vector valued version of Lebesgue's theorem can be reduced to the usual scalar valued case by considering the integrals $\int_0^T \|u_j - u_l\|_X \, dt$, $j, l \in \mathbb{N}$, see [HiPh57, Chap. III, Theorem 3.7.9].

Next we consider a sequence $(u_j)_{j=1}^{\infty}$ in $L^s(0,T;X)$, $1 \le s < \infty$, which converges to $u \in L^s(0,T;X)$ in the sense that

$$\lim_{j\to\infty} \|u - u_j\|_{X,s;T} = 0. \tag{1.2.23}$$

Then the **Fisher-Riesz theorem** shows that there exists a subsequence which converges pointwise to u in the strong sense for almost all $t \in [0,T)$, see [Apo74, Note in Sec. 10.25].

In the same way we also obtain a vector valued version of **Fubini's theorem**, see [HiPh57, Chap. III, Theorem 3.7.13].

All spaces considered here for the interval $[0,T)$ can be defined in the same way for any other interval $I \subseteq \mathbb{R}$. This leads to the spaces

$$L^s(I,X),\ L^s(I;L^q(\Omega)^m),\ \dots. \tag{1.2.24}$$

1.3 The vector valued operator $\frac{d}{dt}$

This operator is the first part of the nonstationary Stokes system (1.1.1). Our purpose is to define this operator precisely for X-valued functions and to investigate its properties. Here we follow essentially the arguments in [Tem77, Chap. III, Lemma 1.1]. See also [Ama95, Chap. III, 1.2].

Let X be a Banach space with norm $\|\cdot\|_X$, let $0 < T \le \infty$ and $1 \le s < \infty$. Recall, $C_0^\infty((0,T))$ means the space of all scalar valued test functions φ with support supp $\varphi \subseteq (0,T)$, see Section 3.1, I.

Let $u \in L^s(0,T;X)$. Then we call a function $u' \in L^s(0,T;X)$ the (weak) **derivative** of u iff

$$-\int_0^T u\varphi'\,dt = \int_0^T u'\varphi\,dt \quad \text{for all } \varphi \in C_0^\infty((0,T)). \tag{1.3.1}$$

The short notation $u' \in L^s(0,T;X)$ always means that u possesses a weak derivative u' in $L^s(0,T;X)$.

The next lemma yields important equivalent conditions for the property $u' \in L^s(0,T;X)$. In particular it shows that $u' \in L^s(0,T;X)$ is uniquely determined by the condition (1.3.1). We use the notations

$$u' = \dot{u} = \frac{d}{dt}u = du/dt. \tag{1.3.2}$$

Thus u' is defined in the sense of (vector valued) distributions in $(0,T)$ with scalar valued test functions φ.

The vector valued **Sobolev space** of **first order** is defined as

$$W^{1,s}(0,T;X) := \{u \in L^s(0,T;X);\ u' \in L^s(0,T;X)\} \tag{1.3.3}$$

with norm

$$\|u\|_{W^{1,s}(0,T;X)} := \|u\|_{L^s(0,T;X)} + \|u'\|_{L^s(0,T;X)} = \|u\|_{X,s;T} + \|u'\|_{X,s;T}\,.$$

Further we define the vector space $W^{1,s}_{loc}([0,T);X)$ by setting

$$u \in W^{1,s}_{loc}([0,T);X) \;\text{ iff }\; u \in W^{1,s}(0,T';X) \;\text{ for all }\; T' \text{ with } 0 < T' < T.$$

The time derivative $\frac{d}{dt}$ is a well defined linear operator

$$\frac{d}{dt} : u \mapsto \frac{d}{dt}u = u' \tag{1.3.4}$$

with domain

$$D(\frac{d}{dt}) := W^{1,s}(0,T;X) \subseteq L^s(0,T;X)$$

and range

$$R(\frac{d}{dt}) := \{u' \in L^s(0,T;X);\; u \in D(\frac{d}{dt})\} \subseteq L^s(0,T;X).$$

We will show that $\frac{d}{dt}$ is a closed densely defined operator, see [Ama95, Chap. III, Theorem 1.2.2].

Let $C^0([0,T);X)$ be the space of all strongly (that is in $\|\cdot\|_X$) continuous functions $u : [0,T) \to X$, and let $C^1([0,T);X)$ be the space of all $u \in C^0([0,T);X)$ such that

$$u'(t) := \lim_{\delta\to 0} \frac{1}{\delta}(u(t+\delta) - u(t)) \in X \quad (\delta > 0) \tag{1.3.5}$$

exists (strongly) for all $t \in [0,T)$ with $u' \in C^0([0,T);X)$. Let $C^1_0([0,T);X)$ be the space of all $u \in C^1([0,T);X)$ having a compact support contained in $[0,T)$.

If $u \in C^1_0([0,T);X)$, the derivative u' defined in (1.3.5) coincides with (1.3.2), and we get

$$C^1_0([0,T);X) \subseteq W^{1,s}(0,T;X) \tag{1.3.6}$$

for $1 \le s < \infty$.

More generally, let $k \in \mathbb{N}$, $1 \le s < \infty$. Then $W^{k,s}(0,T;X)$ means the space of all $u \in L^s(0,T;X)$ such that all derivatives

$$u',\; u'' = (u')',\ldots, u^{(k)} = (u^{(k-1)})'$$

exist in the weak sense (1.3.1) and are contained in $L^s(0,T;X)$. We set $u^{(k)} = (\frac{d}{dt})^k u$.

$C^k([0,T);X)$ means the space of all $u \in C^0([0,T);X)$ with $u', u'', \dots, u^{(k)} \in C^0([0,T);X)$. $C_0^k([0,T);X)$ is the space of all $u \in C^k([0,T);X)$ having a compact support contained in $[0,T)$. We get

$$C_0^k([0,T);X) \subseteq W^{k,s}(0,T;X). \tag{1.3.7}$$

Further we set

$$C^\infty([0,T);X) := \bigcap_{k \in \mathbb{N}} C^k([0,T);X) \tag{1.3.8}$$

$$C_0^\infty([0,T);X) := \bigcap_{k \in \mathbb{N}} C_0^k([0,T);X).$$

The next lemma is essentially contained in [Tem77, Chap. III, 1.1]. It yields several characterizations of the space $W^{1,s}(0,T;X)$, see also [Ama95, Chap. III, 1.2.2]. Recall, X' means the dual space of X, and $[f,v]$ means the value of $f \in X'$ at $v \in X$.

1.3.1 Lemma *Let X be a Banach space with dual space X', and let $0 < T \le \infty$, $1 \le s < \infty$. Then the following conditions are equivalent:*

a) $u \in W^{1,s}(0,T;X)$.

b) $u \in L^s(0,T;X)$ *and there exist* $g \in L^s(0,T;X)$ *and* $u_0 \in X$ *such that*

$$u(t) = u_0 + \int_0^t g(\tau)\,d\tau \quad \textit{for almost all} \quad t \in [0,T). \tag{1.3.9}$$

c) $u \in L^s(0,T;X)$ *and there exist* $g \in L^s(0,T;X)$ *and a dense subspace* $D \subseteq X'$ *such that*

$$\frac{d}{dt}[f, u(t)] = [f, g(t)] \;, \quad t \in (0,T) \tag{1.3.10}$$

holds for all $f \in D$ in the (usual) sense of distributions in $(0,T)$.

Before we prove this lemma, we mention some further properties.

Each of the conditions b) and c) yields a characterization of the weak derivative $g = u'$ of u. Indeed, the proof shows that $g = u'$.

The function $t \mapsto u_0 + \int_0^t g(\tau)d\tau$, $t \in [0,T)$, is obviously continuous in the norm $\|\cdot\|_X$. Since (1.3.9) holds for almost all $t \in [0,T)$, we see that each $u \in W^{1,s}(0,T;X)$ is continuous in $\|\cdot\|_X$ after a redefinition on a subset of $[0,T)$ of measure zero. We obtain from (1.3.9) that

$$u_0 = u(0)$$

is well defined; u_0 is called the **initial value** of u.

Moreover, the representation in (1.3.9) enables us to apply the Bochner theorem [Yos80, V, 5, Theorem 2], which yields the following result:

For each $u \in W^{1,s}(0,T;X)$, the classical derivative $u'(t)$ in the sense of (1.3.5) exists with respect to the norm $\|\cdot\|_X$ and $u'(t) = g(t)$ for almost all $t \in [0,T)$. In particular we see that u' is uniquely determined for each $u \in W^{1,s}(0,T;X)$.

We know that the function $t \mapsto u_0 + \int_0^t g(\tau)\,d\tau$, $t \in [0,T)$, is absolutely continuous on each finite subinterval $[0,T']$, $0 < T' < T$, see [Ama95, Chap. III, 1.2] for this notion. It has been shown by [Kom67], see also [Ama95, Chap. III, 1.2], that if X is reflexive, the following condition is equivalent to $u \in W^{1,s}(0,T;X)$:

$u : [0,T) \to X$ is absolutely continuous (after redefinition on a null set) on each subinterval $[0,T']$, $0 < T' < T$, and the derivative u', which exists for almost all $t \in [0,T)$ in the sense of (1.3.5), is contained in the space $L^s(0,T;X)$.

Using (1.3.6) we see that $D(\frac{d}{dt}) = W^{1,s}(0,T;X)$ is dense in $L^s(0,T;X)$ with respect to the norm $\|u\|_{X,s;T} = (\int_0^T \|u\|_X^s\,dt)^{\frac{1}{s}}$. Therefore, the operator (1.3.4) is densely defined.

The condition (1.3.1) immediately shows that d/dt in (1.3.4) is a closed operator. This is equivalent to the fact that $W^{1,s}(0,T;X)$ is a Banach space with respect to the Sobolev space norm $\|u\|_{X,s;T} + \|u'\|_{X,s;T}$.

Proof of Lemma 1.3.1. In the following proof we slightly modify the arguments in [Tem77, Chap. III, 1.1].

First we show that a) implies b). Let $u \in W^{1,s}(0,T;X)$, $g := u' \in L^s(0,T;X)$, and $\varphi \in C_0^\infty((0,T))$. We consider any subinterval $(\delta, T') \subseteq [0,T)$ with $0 < \delta < T' < T$, and use the mollification method in Section 1.7, II, with the function $\mathcal{F}_\varepsilon \in C_0^\infty(\mathbb{R})$, see (1.7.3), II, (for $n = 1$), $0 < \varepsilon < \varepsilon_0 < \delta$, $\varepsilon_0 < T - T'$. As in (1.7.5), II, we define the mollified function u^ε by

$$u^\varepsilon(t) := (\mathcal{F}_\varepsilon \star u)(t) := \int_{\mathbb{R}} \mathcal{F}_\varepsilon(t-\tau)u(\tau)\,d\tau \quad , \quad t \in \mathbb{R}.$$

Here we set $u = 0$ outside of $[0,T)$. In the same way we define g^ε, φ^ε, and $(\frac{d}{dt}\varphi)^\varepsilon$.

The convergence properties as $\varepsilon \to 0$ are the same as in the scalar valued case, see Lemma 1.7.1, II. Thus

$$\begin{aligned} \lim_{\varepsilon\to 0} \|u - u^\varepsilon\|_{X,s;T} &= 0 \quad , \quad \lim_{\varepsilon\to 0} \|g - g^\varepsilon\|_{X,s;T} = 0, \\ \lim_{\varepsilon\to 0} \|\varphi - \varphi^\varepsilon\|_{X,s;T} &= 0. \end{aligned} \tag{1.3.11}$$

If $u : [0,T) \to X$ is continuous we get from the representation (1.7.7), II, that

$$\|u^\varepsilon(t)\|_X \leq C\,\|u(t)\|_X \quad \text{for all } t \in [0,T) \tag{1.3.12}$$

with some constant $C > 0$ depending on u, ε_0.

We use the relation (1.3.1) with φ replaced by φ^ε where now supp $\varphi \subseteq (\delta, T')$. Then we get $\frac{d}{dt}\varphi^\varepsilon = (\frac{d}{dt}\varphi)^\varepsilon$, and an elementary calculation yields

$$\begin{aligned} -\int_0^T u\frac{d}{dt}\varphi^\varepsilon\,dt &= \int_0^T g\varphi^\varepsilon\,dt = \int_0^T g^\varepsilon\varphi\,dt \\ &= -\int_0^T u(\frac{d}{dt}\varphi)^\varepsilon\,dt = -\int_0^T u^\varepsilon\frac{d}{dt}\varphi\,dt. \end{aligned}$$

The functions u^ε, g^ε are contained in $C^\infty([0,T];X)$, see (1.7.16), II. Therefore, in the last relation

$$\int_0^T g^\varepsilon\varphi\,dt = -\int_0^T u^\varepsilon\frac{d}{dt}\varphi\,dt \;, \quad \varphi \in C_0^\infty((\delta,T'))$$

we may apply the elementary rule of integration by parts in the same way as in the usual scalar valued case. This leads to

$$\frac{d}{dt}u^\varepsilon(t) = g^\varepsilon(t) \;, \quad u^\varepsilon(t) = u^\varepsilon(t_0) + \int_{t_0}^t g^\varepsilon(\tau)\,dt \tag{1.3.13}$$

for all $t, t_0 \in (\delta, T')$, $t_0 < t$, $0 < \varepsilon < \varepsilon_0$.

In the next step we choose any sequence $(\varepsilon_j)_{j=1}^\infty$ with $0 < \varepsilon_j < \varepsilon_0$, $\lim_{j\to\infty}\varepsilon_j = 0$, and set $u_j := u^{\varepsilon_j}$, $g_j := g^{\varepsilon_j}$, $j \in \mathbb{N}$. Using (1.3.11) and the Fischer-Riesz theorem, see (1.2.23), we obtain a subsequence of $(u_j)_{j=1}^\infty$ which converges pointwise to u in the norm $\|\cdot\|_X$ for almost all $t \in (\delta, T')$. Inserting this subsequence in the second equation of (1.3.13) and taking the limit $j \to \infty$, we see that

$$u(t) = u(t_0) + \int_{t_0}^t g(\tau)\,d\tau \;, \quad u(t) - u(t_0) = \int_{t_0}^t g(\tau)\,d\tau \tag{1.3.14}$$

for almost all $t_0, t \in (\delta, T')$, $t_0 < t$. Since $0 < \delta < T' < T$ are arbitrary, we may conclude that (1.3.14) holds for almost all $t_0, t \in (0,T)$, $t_0 < t$.

Let $0 < T' < T$ be fixed. Then the function $t_0, t \mapsto \int_{t_0}^t g(\tau)\,d\tau$ with t_0, $t \in (0, T']$ is uniformly continuous in the norm $\|\cdot\|_X$. Therefore, redefining $u : [0,T'] \mapsto X$ on a null set, we may assume that (1.3.14) holds for all $0 \le t_0 \le t \le T'$. Since T' is arbitrary, this holds for $0 \le t_0 \le t < T$. Setting $u_0 = u(0)$ we get in particular the property (1.3.9). This proves b).

In the next step we show that b) implies a). Let $\varphi \in C_0^\infty((0,T))$. We may use Fubini's theorem, see [Apo74, Theorem 15.6], in the same way as in the scalar valued case. This yields with (1.3.9) that

$$-\int_0^T u\varphi'\,dt = -\int_0^T \left(u_0 + \int_0^t g(\tau)\,d\tau\right)\varphi'(t)\,dt$$

$$\begin{aligned}
&= -\Big(\int_0^T \varphi'(t)\,dt\Big)u_0 - \int_0^T \Big(\int_0^t \varphi'(t)g(\tau)\,d\tau\Big)\,dt \\
&= -\int_0^T \Big(\int_\tau^T \varphi'(t)g(\tau)\,dt\Big)\,d\tau = -\int_0^T g(\tau)\,(\varphi(T)-\varphi(\tau))\,d\tau \\
&= \int_0^T g(\tau)\varphi(\tau)\,d\tau.
\end{aligned}$$

Thus we get (1.3.1) and we see, b) implies a). In particular it holds $g = u'$.

Next we prove that a) implies c). Suppose $u \in W^{1,s}(0,T;X)$ and let $g := u'$, $D := X'$. Then we use (1.3.1) and get

$$\begin{aligned}
-[f, \int_0^T u\varphi'\,dt] &= -\int_0^T [f,u(t)]\,\varphi'(t)\,dt \\
&= [f, \int_0^T g\,\varphi\,dt] = \int_0^T [f,g(t)]\,\varphi(t)\,dt
\end{aligned}$$

for all $f \in D$, $\varphi \in C_0^\infty((0,T))$. This proves (1.3.10) and we see, a) implies c). Suppose now that c) is satisfied with $D \subseteq X'$ and with $u, g \in L^s(0,T;X)$. Then for all $\varphi \in C_0^\infty((0,T))$, $f \in D$, we get

$$\begin{aligned}
-\int_0^T [f,u(t)]\varphi'(t)\,dt &= -[f, \int_0^T u\varphi'\,dt] \\
&= \int_0^T [f,g(t)]\,\varphi(t)\,dt = [f, \int_0^T g\,\varphi\,dt].
\end{aligned}$$

Since D is dense in X', this relation holds as well for all $f \in X'$. This proves (1.3.1) and we see, c) implies a). Now we conclude that a), b) and c) are equivalent conditions. This proves the lemma. □

The properties above enable us to prove a formula which can be understood as the rule of **integration by parts**, see (1.3.15). Here we assume that $X = H$ is a Hilbert space.

1.3.2 Lemma *Let H be a Hilbert space with scalar product $<\cdot,\cdot>_H$ and norm $\|\cdot\|_H$, and let $0 < T \le \infty$, $1 \le s < \infty$. Suppose $u, v \in W^{1,s}(0,T;H)$. Then, after a redefinition on a null set, u and v are continuous in the norm $\|\cdot\|_H$ and*

$$\begin{aligned}
<u(t),v(t)>_H &= <u(0),v(0)>_H \\
&\quad + \int_0^t (<u',v>_H + <u,v'>_H)\,d\tau
\end{aligned} \tag{1.3.15}$$

for all $t \in [0,T)$.

Proof. The continuity of u, v, after a corresponding redefinition, follows from Lemma 1.3.1, b). Let $g := u'$, $h := v'$, and use the mollified functions u^ε, v^ε, g^ε, h^ε, $0 < \varepsilon < \varepsilon_0$, see the previous proof. These functions are contained in $C^\infty([0,T);H)$, and for $0 < \delta < t_0 < t < T' < T$, $\varepsilon_0 < \delta$, $\varepsilon_0 < T - T'$, we get by an elementary calculation that

$$
\begin{aligned}
< u^\varepsilon(t), v^\varepsilon(t) >_H \; - \; & < u^\varepsilon(t_0), v^\varepsilon(t_0) >_H \\
&= \int_{t_0}^t \frac{d}{d\tau} < u^\varepsilon(\tau), v^\varepsilon(\tau) >_H \, d\tau \\
&= \int_{t_0}^t \Big(< \frac{d}{d\tau} u^\varepsilon, v^\varepsilon >_H + < u^\varepsilon, \frac{d}{d\tau} v^\varepsilon >_H \Big) \, d\tau \\
&= \int_{t_0}^t (< g^\varepsilon, v^\varepsilon >_H + < u^\varepsilon, h^\varepsilon >_H) \, d\tau .
\end{aligned}
$$

Since u, v are continuous we get

$$
u(t) = \lim_{\varepsilon \to 0} u^\varepsilon(t) \quad , \quad v(t) = \lim_{\varepsilon \to 0} v^\varepsilon(t) \tag{1.3.16}
$$

in the norm $\| \cdot \|_H$, see (1.7.8), II. The same holds with t replaced by t_0.

To treat the limit as $\varepsilon \to 0$, we write

$$
< g^\varepsilon, v^\varepsilon >_H \; = \; < g^\varepsilon - g, v^\varepsilon >_H \; + \; < g, v^\varepsilon >_H ,
$$

use the properties (1.3.11), (1.3.12), and the estimate

$$
\left| \int_{t_0}^t < g^\varepsilon - g, v^\varepsilon > \, d\tau \right| \; \le \; C \, \| g^\varepsilon - g \|_{H,s;t} \, \| v \|_{H,\infty;t} \, ,
$$

with some $C > 0$. This yields

$$
\lim_{\varepsilon \to 0} \int_{t_0}^t < g^\varepsilon - g, v^\varepsilon >_H \, d\tau = 0
$$

and

$$
\lim_{\varepsilon \to 0} \int_{t_0}^t < g, v^\varepsilon >_H \, d\tau = \int_{t_0}^t < g, v > \, d\tau .
$$

For the last limit we need the estimate

$$
| < g(\tau), v^\varepsilon(\tau) >_H | \; \le \; C \, \| g(\tau) \|_H \, \| v(\tau) \|_H \quad , \quad 0 \le \tau \le t,
$$

with $C > 0$, use Lebesgue's dominated convergence theorem, see (1.2.22), and the continuity of v. This yields

$$\lim_{\varepsilon \to 0} \int_{t_0}^{t} < g^\varepsilon, v^\varepsilon >_H \, d\tau = \int_{t_0}^{t} < g, v >_H \, d\tau$$

and correspondingly

$$\lim_{\varepsilon \to 0} \int_{t_0}^{t} < u^\varepsilon, h^\varepsilon >_H \, d\tau = \int_{t_0}^{t} < u, h >_H \, d\tau .$$

Thus letting $\varepsilon \to 0$ we obtain

$$\begin{aligned} < u(t), v(t) >_H \; - \; & < u(t_0), v(t_0) >_H \\ &= \int_{t_0}^{t} (< g(\tau), v(\tau) >_H + < u(\tau), h(\tau) >_H) \, d\tau . \end{aligned}$$

Now we let $t_0 \to 0$ and obtain the desired rule (1.3.15). □

1.4 Time dependent gradients ∇p

The time dependent gradient ∇p is another part of the Stokes system (1.1.1) which we have to investigate in this preliminary section. For this purpose we need a time dependent version of the results on stationary gradients, see Section 2.2, II. Recall that ∇ and div only concern the space variables.

Let $\Omega \subseteq \mathbb{R}^n$, $n \geq 2$, be any domain and let $0 < T \leq \infty$. First we recall some test spaces (without norm) introduced in Section 3.1, I.

In particular we need the space $C_{0,\sigma}^\infty(\Omega) := \{v \in C_0^\infty(\Omega)^n ; \; \text{div } v = 0\}$ of smooth solenoidal test functions, and we use the test spaces

$$C_0^\infty((0,T); C_{0,\sigma}^\infty(\Omega)) := \{v \in C_0^\infty((0,T) \times \Omega)^n \, ; \text{div } v = 0\}, \tag{1.4.1}$$

and

$$\begin{aligned} & C_0^\infty([0,T); C_{0,\sigma}^\infty(\Omega)) \\ & \quad := \{v|_{[0,T)\times\Omega} \, ; \; v \in C_0^\infty((-1,T) \times \Omega)^n ; \; \text{div } v = 0\}. \end{aligned} \tag{1.4.2}$$

Thus $C_0^\infty([0,T); C_{0,\sigma}^\infty(\Omega))$ consists of all restrictions to $[0,T) \times \Omega$ of functions $v \in C_0^\infty((-1,T) \times \Omega)^n$ with $\text{div } v = D_1 v_1 + \cdots + D_n v_n = 0$.

We always write $v(t) = v(t, \cdot)$, $t \in [0,T)$, for these functions, and for each $v \in C_0^\infty([0,T); C_{0,\sigma}^\infty(\Omega))$ we get the well defined initial value

$$v_0 := v(0) = v|_{t=0} . \tag{1.4.3}$$

Further we need some special spaces. Let $1 < s < \infty$, $1 < q < \infty$, and let $s' = \frac{s}{s-1}$, $q' = \frac{q}{q-1}$ so that $1 = \frac{1}{q} + \frac{1}{q'}$, $1 = \frac{1}{s} + \frac{1}{s'}$.

The space $W_0^{1,q'}(\Omega)^n$, see Section 3.6, I, is reflexive, the space $L^{s'}(0,T; W_0^{1,q'}(\Omega)^n)$ is reflexive too, and its dual space is given by

$$L^s(0,T;W^{-1,q}(\Omega)^n) = L^{s'}(0,T;W_0^{1,q'}(\Omega)^n)' \tag{1.4.4}$$

where $W^{-1,q}(\Omega)^n = W_0^{1,q'}(\Omega)^{n\prime}$, see (3.6.5), I, and (1.2.7).

This means that each $f \in L^s(0,T;W^{-1,q}(\Omega)^n)$ is identified with the functional defined on $L^{s'}(0,T;W_0^{1,q'}(\Omega)^n)$ which is given by

$$[f;\,.\,] : v \mapsto [f,v]_{\Omega,T} := \int_0^T [f(t),v(t)]_\Omega \, dt \quad , \quad v \in L^{s'}(0,T;W_0^{1,q'}(\Omega)^n)\,.$$

Here $[f(t),v(t)]_\Omega$ means the value of $f(t) \in W^{-1,q}(\Omega)^n$ at $v(t) \in W_0^{1,q'}(\Omega)^n$ for almost all $t \in [0,T)$.

The following lemma is the time dependent version of Lemma 2.2.2, II.

1.4.1 Lemma *Let $\Omega \subseteq \mathbb{R}^n$, $n \geq 2$, be a bounded Lipschitz domain, let $\Omega_0 \subseteq \Omega$, $\Omega_0 \neq \emptyset$, be a subdomain, and let $0 < T \leq \infty$, $1 < q < \infty$, $1 < s < \infty$. Suppose $f \in L^s(0,T;W^{-1,q}(\Omega)^n)$ satisfies*

$$[f,v]_{\Omega,T} = 0 \quad \textit{for all} \quad v \in C_0^\infty((0,T);C_{0,\sigma}^\infty(\Omega)). \tag{1.4.5}$$

Then there exists a unique $p \in L^s(0,T;L^q(\Omega))$ satisfying

$$f = \nabla p$$

in the sense of distributions in $(0,T) \times \Omega$, and

$$\int_{\Omega_0} p(t)\, dx = 0$$

for almost all $t \in [0,T)$. We obtain

$$f(t) = \nabla p(t) \quad \textit{for almost all} \quad t \in [0,T), \tag{1.4.6}$$

and

$$\|p\|_{L^s(0,T;L^q(\Omega))} \;\leq\; C\,\|f\|_{L^s(0,T;W^{-1,q}(\Omega)^n)} \tag{1.4.7}$$

with some constant $C = C(q,\Omega_0,\Omega) > 0$.

Proof. We use the same duality argument as in the proof of Lemma 2.1.1, II, now for the time dependent case. For this purpose let $L^q_0(\Omega) \subseteq L^q(\Omega)$ be as in (2.1.3), II, and consider the operator

$$\operatorname{div} \,:\, L^{s'}(0,T;W_0^{1,q'}(\Omega)^n) \to L^{s'}(0,T;L_0^{q'}(\Omega)), \tag{1.4.8}$$

defined by $v \mapsto \operatorname{div} v$, the dual operator

$$-\nabla \,:\, L^s(0,T;L^q_0(\Omega)) \to L^s(0,T;W^{-1,q}(\Omega)^n) \tag{1.4.9}$$

defined by $p \mapsto -\nabla p$, and the relation

$$[-\nabla p, v]_{\Omega,T} \;=\; <p, \operatorname{div}\, v>_{\Omega,T} = \int_0^T <p, \operatorname{div}\, v>_\Omega \, dt.$$

These operators are bounded, the range spaces $R(-\nabla) = R(\nabla)$, $R(\operatorname{div})$, and the null spaces $N(-\nabla) = N(\nabla)$, $N(\operatorname{div})$ have the same properties as in the stationary case, see the proof of Lemma 2.1.1, II.

First let $\Omega_0 = \Omega$. From (2.1.2), II, we conclude that

$$\left(\int_0^T \|p(t)\|_q^s \, dt\right)^{\frac{1}{s}} \;\leq C \left(\int_0^T \|\nabla p(t)\|_{-1,q}^s \, dt\right)^{\frac{1}{s}} \tag{1.4.10}$$

for all $p \in R(-\nabla)$, $C = C(q, \Omega_0, \Omega) > 0$, and we get $\int_\Omega p(t)\, dx = 0$ for almost all $t \in [0,T)$. This shows that $R(-\nabla)$ is closed in $L^s(0,T;W^{-1,q}(\Omega)^n)$. It follows that $R(\operatorname{div})$ is closed in $L^{s'}(0,T;L_0^{q'}(\Omega))$ and that

$$f \in R(-\nabla) \quad \text{iff} \quad [f,v]_{\Omega,T} = 0 \;\text{ for all } v \in N(\operatorname{div}), \tag{1.4.11}$$

see the closed range theorem [Yos80, VII, 5].

Using Lemma 2.2.3, II, we conclude that

$$C_0^\infty((0,T);C_{0,\sigma}^\infty(\Omega)) \;\subseteq\; N(\operatorname{div})$$

is dense with respect to the norm of $L^{s'}(0,T;W_0^{1,q'}(\Omega)^n)$. To prove this property, we consider step functions with values in $C_{0,\sigma}^\infty(\Omega)$ and apply the mollification procedure in the time direction, see Section 1.7, II. Thus we get from (1.4.5) that $[f,v]_{\Omega,T} = 0$ holds for all $v \in N(\operatorname{div})$. Now from (1.4.11) we obtain the existence of some $p \in L^s(0,T;L^q_0(\Omega))$ with $f = \nabla p$. The uniqueness follows from $N(-\nabla) = \{0\}$. Indeed, if $\nabla p = \nabla p'$ with $p, p' \in L^s(0,T;L^q_0(\Omega))$, then $\nabla(p - p') = 0$, $p(t) - p'(t) = C(t)$ for almost all $t \in [0,T)$ with constants $C(t)$, and from $\int_\Omega C(t)\, dx = 0$ we get $C(t) = 0$ for almost all $t \in [0,T)$. This shows that $p = p'$. The relation (1.4.6) is obvious and (1.4.7) is a consequence of (1.4.10).

This proves the lemma for the case $\Omega_0 = \Omega$. To treat the general case $\Omega_0 \subseteq \Omega$ we let p be as above, subtract from p the term

$$p_0(t) := |\Omega_0|^{-1} \int_{\Omega_0} p(t)\, dx \ , \quad t \in [0,T) ,$$

and argue in the same way as in the stationary case, see (2.1.20), II. This completes the proof. □

The next lemma is the time dependent version of Lemma 2.2.1, II, for general domains.

We define the space $L^s(0,T; W_{loc}^{-1,q}(\Omega)^n)$ of functionals

$$f : v \mapsto [f, v]_{\Omega,T} \ , \quad v \in C_0^\infty((0,T) \times \Omega)^n$$

by the following condition:

$$f \in L^s(0,T; W_{loc}^{-1,q}(\Omega)^n) \quad \text{iff} \quad f \in L^s(0,T; W^{-1,q}(\Omega')^n) \tag{1.4.12}$$

for all bounded subdomains $\Omega' \subseteq \Omega$ with $\overline{\Omega'} \subseteq \Omega$.

We see that for each $f \in L^s(0,T; W_{loc}^{-1,q}(\Omega)^n)$ there exist functionals $f(t) \in W_{loc}^{-1,q}(\Omega)^n$ for almost all $t \in [0,T)$ such that

$$[f, v]_{\Omega,T} = \int_0^T [f(t), v(t)]_\Omega \, dt \tag{1.4.13}$$

for all $v \in C_0^\infty((0,T) \times \Omega)^n$; $[f(t), v(t)]_\Omega$ means the value of $f(t)$ at $v(t)$. Further we get

$$|[f, v]_{\Omega,T}| \leq \|f\|_{L^s(0,T; W^{-1,q}(\Omega')^n)} \|v\|_{L^{s'}(0,T; W_0^{1,q'}(\Omega')^n)} \tag{1.4.14}$$

for each such subdomain $\Omega' \subseteq \Omega$.

Similarly, the space $L^s(0,T; L_{loc}^q(\Omega))$ in the next lemma is defined by the condition

$$p \in L^s(0,T; L_{loc}^q(\Omega)) \quad \text{iff} \quad p \in L^s(0,T; L^q(\Omega')) \tag{1.4.15}$$

for all bounded subdomains $\Omega' \subseteq \Omega$ with $\overline{\Omega'} \subseteq \Omega$. Thus for each such Ω' the norm

$$\|p\|_{L^s(0,T; L^q(\Omega'))} < \infty$$

is well defined.

1.4.2 Lemma *Let $\Omega \subseteq \mathbb{R}^n$, $n \geq 2$, be a general domain, let $\Omega_0 \subseteq \Omega$, $\Omega_0 \neq \emptyset$, be a bounded subdomain with $\overline{\Omega}_0 \subseteq \Omega$, and let $0 < T \leq \infty$, $1 < q < \infty$, $1 < s < \infty$. Suppose $f \in L^s(0,T; W_{loc}^{-1,q}(\Omega)^n)$ satisfies*

$$[f, v]_{\Omega,T} = 0 \quad \textit{for all} \quad v \in C_0^\infty((0,T); C_{0,\sigma}^\infty(\Omega)). \tag{1.4.16}$$

Then there exists a unique $p \in L^s(0,T; L^q_{loc}(\Omega))$ satisfying

$$f = \nabla p$$

in the sense of distributions in $(0,T) \times \Omega$, and

$$\int_{\Omega_0} p(t)\, dx = 0$$

for almost all $t \in [0,T)$.

Moreover, for each bounded Lipschitz subdomain $\Omega' \subseteq \Omega$ with $\Omega_0 \subseteq \Omega'$, $\overline{\Omega'} \subseteq \Omega$, the estimate

$$\|p\|_{L^s(0,T;L^q(\Omega'))} \leq C \, \|f\|_{L^s(0,T;W^{-1,q}(\Omega')^n)} \tag{1.4.17}$$

holds with some constant $C = C(q, \Omega_0, \Omega') > 0$.

Proof. Consider the Lipschitz subdomains $\Omega_j \subseteq \Omega$, $j \in \mathbb{N}$, as in Lemma 1.4.1, II. We may assume that $\Omega_0 \subseteq \Omega_1$. Applying Lemma 1.4.1 above to each Ω_j, we get a unique $p_j \in L^s(0,T; L^q(\Omega_j))$ satisfying $f = \nabla p_j$ in $(0,T) \times \Omega_j$ and $\int_{\Omega_0} p_j(t)\, dx = 0$ for almost all $t \in [0,T)$. Since $\overline{\Omega}_j \subseteq \Omega_{j+1}$, we deduce from the uniqueness property in this lemma that

$$p_{j+1}|_{(0,T)\times\Omega_j} = p_j$$

holds for all $j \in \mathbb{N}$. Since $\Omega = \bigcup_{j=1}^\infty \Omega_j$, this yields a well-defined

$$p \in L^s(0,T; L^q_{loc}(\Omega))$$

satisfying $f = \nabla p$, and $\int_{\Omega_0} p(t)\, dx = 0$ for almost all $t \in [0,T)$.

If $\Omega' \subseteq \Omega$ is a subdomain as in (1.4.17), we apply again Lemma 1.4.1, obtain some $\tilde{p} \in L^s(0,T; L^q(\Omega'))$ satisfying $f = \nabla \tilde{p}$ and $\int_{\Omega_0} \tilde{p}(t)\, dx = 0$ for almost all $t \in [0,T)$. Further, (1.4.7) implies (1.4.17) for $\tilde{p}$. The uniqueness property as above yields $\tilde{p} = p$. This proves the lemma. □

1.5 A special solution class of the homogeneous system

In this and the next subsection we consider special solutions of the nonstationary Stokes system

$$u_t - \nu\Delta u + \nabla p = f \quad , \quad \text{div}\ u = 0, \qquad (1.5.1)$$
$$u|_{\partial\Omega} = 0 \quad , \quad u(0) = u_0$$

which are (first formally) obtained as follows. Applying the Helmholtz projection P and using the Stokes operator in the form $A = -\nu P A$, we obtain the **evolution system**

$$u_t + Au = f \ , \quad u(0) = u_0 \ . \qquad (1.5.2)$$

The first equation is called an **evolution equation**, and $u(0) = u_0$ is called an initial condition, see [Ama95, Chap. II, 1.2] or [Fri69, Part 2] concerning this notion. See [BuBe67], [LiMa72], [Kre71] concerning evolution equations and [Gig81], [FaS94b] concerning resolvents of A.

Our purpose is to investigate a very general solution class of (1.5.2). An important fact will be the variation of constant formula

$$u(t) = S(t)u_0 + \int_0^t S(t-\tau)\, f(\tau)\, d\tau \ , \quad t \geq 0 \, , \qquad (1.5.3)$$

where $S(t) = e^{-tA}$ is defined by the spectral representation, see Section 3.2, II.

The function u defined by $u(t) := S(t)u_0$ satisfies the homogeneous system

$$u_t + Au = 0 \ , \quad u(0) = u_0 \qquad (1.5.4)$$

which we treat in this subsection.

The more general theory of weak solutions of the system (1.5.1) can be reduced completely to the treatment of the evolution system (1.5.2), see the next section. In particular we will prove a representation formula for weak solutions which is based on the formula (1.5.3).

In the following, $\Omega \subseteq \mathbb{R}^n$, $n \geq 2$, means an arbitrary domain and

$$A = \int_0^\infty \lambda\, dE_\lambda, \qquad (1.5.5)$$

the Stokes operator of Ω, see (2.1.12), III. The fractional powers

$$A^\alpha = \int_0^\infty \lambda^\alpha\, dE_\lambda \ , \quad -1 \leq \alpha \leq 1 \qquad (1.5.6)$$

are positive selfadjoint operators, see (2.2.6), III, and (3.2.18), (3.2.28), II.

Similarly, using the spectral representation (3.2.10), II, we define for each $t \geq 0$ the operator

$$S(t) := e^{-tA} := \int_0^\infty e^{-t\lambda} \, dE_\lambda \, . \tag{1.5.7}$$

Since $\lambda \mapsto e^{-t\lambda}$, $\lambda \geq 0$, is a bounded positive function defined on $[0, \infty)$, each $S(t)$ is a bounded everywhere defined and positive selfadjoint operator in the Hilbert space $L^2_\sigma(\Omega)$.

From (3.2.15), II, we know that the operator norm $\|S(t)\|$ of $S(t)$ satisfies the estimate

$$\|S(t)\| \leq \sup_{\lambda \geq 0} e^{-t\lambda} \leq 1 \tag{1.5.8}$$

for all $t \geq 0$.

The representation (1.5.7) yields

$$S(t)\, S(\tau) = \int_0^\infty e^{-t\lambda} e^{-\tau\lambda} \, dE_\lambda = \int_0^\infty e^{-(t+\tau)\lambda} \, dE_\lambda \tag{1.5.9}$$

and therefore we get

$$S(t)\, S(\tau) = S(t + \tau) \tag{1.5.10}$$

for all t, $\tau \geq 0$. Further we have

$$S(0) = \int_0^\infty dE_\lambda = I$$

where I means the identity, see (3.2.14), II.

The operator family $\{S(t); t \geq 0\}$ is called the **Stokes semigroup** of Ω, see [BuBe67], [Kre71], [Yos80], [Gig81], [Gig85], [BSo87] concerning semigroups.

In the next lemma we consider the solution class of the homogeneous equation (1.5.4) which is given by $u(t) := S(t)u_0$, $t \geq 0$. Formally we see that

$$u' = \frac{d}{dt} u(t) = \frac{d}{dt} e^{-tA} u_0 = -Au(t) \;\; , \quad u' + Au = 0 \, ,$$

and we have to justify this calculation.

1.5.1 Lemma *Let $\Omega \subseteq \mathbb{R}^n$, $n \geq 2$, be any domain, let $\{S(t); t \geq 0\}$ be the Stokes semigroup of Ω, and let $u_0 \in L^2_\sigma(\Omega)$.*

Then the function $u : [0, \infty) \to L^2_\sigma(\Omega)$ defined by

$$u(t) := S(t) u_0 \;\; , \quad t \geq 0, \tag{1.5.11}$$

has the following properties:

a) *u is strongly continuous for $t \geq 0$, $u(0) = u_0$, for each $t > 0$*

$$u'(t) = \frac{d}{dt}u(t) = s - \lim_{\delta \to 0} \frac{1}{\delta}(u(t+\delta) - u(t)) \ , \quad \delta > 0 \, , \qquad (1.5.12)$$

exists in the strong sense and $t \mapsto u'(t)$ is strongly continuous for $t > 0$.

b) *For all $t > 0$ we get $u(t) \in D(A)$ and*

$$u'(t) + Au(t) = 0. \qquad (1.5.13)$$

If $u_0 \in D(A)$, then $u'(0) = s - \lim_{\delta \to 0} \frac{1}{\delta}(u(\delta) - u(0))$, $\delta > 0$, exists and $u'(0) + Au(0) = 0$.

c) *For all $t \geq 0$ we get $\|u(t)\|_2 \leq \|u_0\|_2$ and*

$$s - \lim_{t \to \infty} u(t) = 0. \qquad (1.5.14)$$

Proof. The important property (1.5.14) has been observed by Masuda [Mas84, (5.2)]. Let $t, t' \geq 0$. We use the representation

$$\|u(t) - u(t')\|_2^2 = \int_0^\infty (e^{-t\lambda} - e^{-t'\lambda})^2 \, d\|E_\lambda u_0\|_2^2,$$

see (3.2.12), II. Since $\lim_{t \to t'}(e^{-t\lambda} - e^{-t'\lambda})^2 = 0$ for $\lambda \geq 0$, and $(e^{-t\lambda} - e^{-t'\lambda})^2 \leq 1$, we may apply Lebesgue's dominated convergence lemma, see (1.2.22), and obtain

$$\lim_{t \to t'} \|u(t) - u(t')\|_2^2 = \int_0^\infty \lim_{t \to t'} (e^{-t\lambda} - e^{-t'\lambda})^2 \, d\|E_\lambda u_0\|_2^2 = 0.$$

This yields the strong continuity of u. If $t > 0$, $t + \delta \geq \delta_0 > 0$, δ_0 fixed, we use the same argument and obtain

$$\begin{aligned} &\lim_{\delta \to 0} \| \, [\frac{1}{\delta}(u(t+\delta) - u(t)) + Au(t)] \, \|_2^2 \\ &\quad = \lim_{\delta \to 0} \int_0^\infty [\frac{1}{\delta}(e^{-(t+\delta)\lambda} - e^{-t\lambda}) + \lambda e^{-t\lambda}]^2 \, d\|E_\lambda u_0\|_2^2 \\ &\quad = \int_0^\infty \lim_{\delta \to 0} [\frac{1}{\delta}(e^{-(t+\delta)\lambda} - e^{-t\lambda}) + \lambda e^{-t\lambda}]^2 \, d\|E_\lambda u_0\|_2^2 = 0, \end{aligned}$$

and

$$\|Au(t)\|_2^2 = \int_0^\infty \lambda^2 e^{-2t\lambda} \, d\|E_\lambda u_0\|_2^2 < \infty.$$

Thus we get $u(t) \in D(A)$ and (1.5.13) for $t > 0$. The strong continuity of $t \mapsto Au(t)$ and $t \mapsto u'(t)$, $t > 0$, follows as above. If $u_0 \in D(A)$, then

$$\|Au_0\|_2^2 = \int_0^\infty \lambda^2 \, d\|E_\lambda u_0\|_2^2 < \infty,$$

and the last calculation also holds for $t = 0$, $\delta > 0$. This proves the properties a) and b).

To prove c) we use (3.2.12), II, and get

$$\|u(t)\|_2^2 = \int_0^\infty e^{-2t\lambda} \, d\|E_\lambda u_0\|_2^2 \le \int_0^\infty d\|E_\lambda u_0\|_2^2 = \|u_0\|_2^2 \, .$$

Since $0 \le e^{-2t\lambda} \le 1$, $\lim_{t\to\infty} e^{-2t\lambda} = 0$ for all $\lambda > 0$, we can again use Lebesgue's theorem as above. Because of $N(A) = \{0\}$, see Theorem 2.1.1, III, a), and the argument after (3.2.25), II, the point $\lambda = 0$ is a continuity point of $\lambda \mapsto \|E_\lambda u_0\|_2^2$, $\lambda \ge 0$.

Therefore, $\{0\}$ is a null set concerning this measure, and using Lebesgue's theorem as above, we only need that $\lim_{t\to\infty} e^{-2t\lambda} = 0$ holds for each $\lambda > 0$ in order to prove that

$$\lim_{t\to\infty} \|u(t)\|_2^2 = \int_0^\infty (\lim_{t\to\infty} e^{-2t\lambda}) \, d\|E_\lambda u_0\|_2^2 = 0.$$

This yields (1.5.14), see [Mas84, p. 641]. The proof of the lemma is complete.

□

We mention some further properties of the operators $S(t)$.

Let $0 \le \alpha \le 1$ and $t > 0$. Then we get $\sup_{\lambda \ge 0} \lambda^\alpha e^{-t\lambda} \le t^{-\alpha}$. This shows, see (3.2.20), II, and (3.2.15), II, that

$$A^\alpha e^{-tA} = A^\alpha S(t) = \int_0^\infty \lambda^\alpha e^{-t\lambda} \, dE_\lambda$$

is a bounded operator with operator norm

$$\|A^\alpha e^{-tA}\| \le t^{-\alpha}. \tag{1.5.15}$$

We also see that $e^{-tA} v \in D(A^\alpha)$ for all $v \in L^2_\sigma(\Omega)$ and that

$$A^\alpha e^{-tA} v = e^{-tA} A^\alpha v \tag{1.5.16}$$

for all $v \in D(A^\alpha)$ and $t > 0$. This means, e^{-tA} commutes with A^α, see (3.2.19), II.

For $\mu > 0$ we consider the resolvent

$$(\mu I + A)^{-1} = \int_0^\infty (\mu + \lambda)^{-1} \, dE_\lambda$$

of the Stokes operator, see (3.2.22), II, and obtain

$$\|(\mu I + A)^{-1}\| \leq \sup_{\lambda \geq 0} (\mu + \lambda)^{-1} \leq \frac{1}{\mu}. \tag{1.5.17}$$

If $0 \leq \alpha \leq 1$, $\mu > 0$, $t \geq 0$, we get as above

$$\begin{aligned} A^\alpha (\mu I + A)^{-1} S(t) &= A^\alpha S(t) (\mu I + A)^{-1} \\ &= \int_0^\infty \lambda^\alpha (\mu + \lambda)^{-1} e^{-t\lambda} \, dE_\lambda \, , \end{aligned} \tag{1.5.18}$$

and

$$\|A^\alpha (\mu I + A)^{-1} S(t)\| \leq \sup_{\lambda \geq 0} \lambda^\alpha (\mu + \lambda)^{-1} e^{-t\lambda} \leq \mu^{\alpha - 1} . \tag{1.5.19}$$

Here we use that

$$\mu^{1-\alpha} \lambda^\alpha \leq \mu + \lambda \ , \quad e^{-t\lambda} \leq 1 \, .$$

Let $u_0 \in L^2_\sigma(\Omega)$, $k \in \mathbb{N}$. Then we can show in the same way as above that the kth derivative

$$\left(\frac{d}{dt} \right)^k u = (-1)^k A^k u \tag{1.5.20}$$

of the function $t \mapsto u(t) = S(t) u_0$ exists in the strong sense and is strongly continuous for $t > 0$.

We have to investigate further properties of the solutions

$$u := S(\cdot) u_0 \ , \quad u_0 \in L^2_\sigma(\Omega)$$

of the homogeneous system (1.5.4). In particular we are interested in sufficient conditions on the initial value u_0 that

$$u \in L^s(0, T; D(A)) \ \text{ and } \ u', Au \in L^s(0, T; L^2_\sigma(\Omega))$$

where $1 \leq s \leq \infty$, $0 < T \leq \infty$. See Section 1.2 for these spaces. Here $D(A)$ means the Banach space endowed with the graph norm

$$\|v\|_{D(A)} = \|v\|_2 + \|Av\|_2 \ , \quad v \in D(A). \tag{1.5.21}$$

The following theorem yields such properties. The case $1 < s < 2$ is critical in the method which we use here. In this case we cannot discuss the optimal assumptions on the initial value u_0. Recall the notation (1.2.14) here for $A = B$.

1.5.2 Theorem *Let $\Omega \subseteq \mathbb{R}^n$, $n \geq 2$, be an arbitrary domain, let $\{S(t);\, t \geq 0\}$ be the Stokes semigroup of Ω, and let $0 < T \leq \infty$, $1 < s < \infty$. Suppose*

$$u_0 \in D(A^{1-\frac{1}{s}}) \quad \text{if} \quad 2 \leq s < \infty,$$
$$u_0 \in D(A^{1-\frac{1}{s}+\varepsilon}) \quad \text{if} \quad 1 < s < 2, \ \text{where} \ 0 < \varepsilon < \frac{1}{s} - \frac{1}{2},$$

and let $u : [0, T) \to L^2_\sigma(\Omega)$ be defined by

$$u(t) := S(t)u_0 \ , \quad t \in [0, T).$$

Then u', $Au \in L^s(0, T; L^2_\sigma(\Omega))$, u is strongly continuous with $u(0) = u_0$, and

$$u' + Au = 0$$

is satisfied as an equation in $L^s(0, T; L^2_\sigma(\Omega))$.

Further we get

$$\|u'\|_{2,s;T} + \|Au\|_{2,s;T} \ \leq \ 2\, \|A^{1-\frac{1}{s}} u_0\|_2 \tag{1.5.22}$$

if $2 \leq s < \infty$, and

$$\|u'\|_{2,s;T} + \|Au\|_{2,s;T} \ \leq \ C\, (\|u_0\|_2 + \|A^{1-\frac{1}{s}+\varepsilon} u_0\|_2) \tag{1.5.23}$$

if $1 < s < 2$, $C = C(s, \varepsilon) > 0$ is a constant.

For the proof of this theorem we need the following technical result.

1.5.3 Lemma *Let $\Omega \subseteq \mathbb{R}^n$, $n \geq 2$, be any domain with the Stokes semigroup $\{S(t);\, t \geq 0\}$, let $0 < T \leq \infty$, $2 \leq s < \infty$, and let $u_0 \in L^2_\sigma(\Omega)$. Then $A^{\frac{1}{s}} S(\cdot) u_0 \in L^s(0, T; L^2_\sigma(\Omega))$ and*

$$\|A^{\frac{1}{s}} S(\cdot) u_0\|_{2,s;T} \ \leq \ \|u_0\|_2. \tag{1.5.24}$$

Proof of Lemma 1.5.3. Using the interpolation inequality (2.2.8), III, we get

$$\begin{aligned}
\|A^{\frac{1}{s}} S(\cdot) u_0\|_{2,s;T} \ &= \ \left(\int_0^T \|A^{\frac{1}{s}}\, S(t) u_0\|_2^s \, dt \right)^{\frac{1}{s}} \\
&= \ \left(\int_0^T \|(A^{\frac{1}{2}})^{2/s}\, S(t) u_0\|_2^s \, dt \right)^{\frac{1}{s}} \\
&\leq \ \left(\int_0^T \|A^{\frac{1}{2}}\, S(t) u_0\|_2^2 \, \|S(t) u_0\|_2^{(1-\frac{2}{s})s} \, dt \right)^{\frac{1}{s}}
\end{aligned}$$

$$
\begin{aligned}
&\leq \|u_0\|_2^{1-\frac{2}{s}} \left(\int_0^T \|A^{\frac{1}{2}} S(t) u_0\|_2^2 \, dt \right)^{\frac{1}{s}} \\
&= \|u_0\|_2^{1-\frac{2}{s}} \left(\int_0^T < AS(t)^2 u_0, u_0 > dt \right)^{\frac{1}{s}} \\
&= \|u_0\|_2^{1-\frac{2}{s}} \left(-\frac{1}{2} \int_0^T \frac{d}{dt} < S(2t) u_0, \, u_0 > dt \right)^{\frac{1}{s}} \\
&= \|u_0\|_2^{1-\frac{2}{s}} \left(\frac{1}{2} \|u_0\|_2^2 - \frac{1}{2} < S(2T) u_0, u_0 > \right)^{\frac{1}{s}} \\
&\leq \|u_0\|_2^{1-\frac{2}{s}} \, \|u_0\|_2^{\frac{2}{s}} = \|u_0\|_2 .
\end{aligned}
$$

This calculation is carried out first in the case $0 < T < \infty$. To get this estimate for $T = \infty$, we let $T \to \infty$, and use (1.5.14). Further we use the relation (1.5.13) and the following calculation:

$$
\begin{aligned}
\int_0^T \|A^{\frac{1}{2}} S(t) u_0\|_2^2 \, dt &= \int_0^T < AS(t)^2 u_0, u_0 > dt \\
&= -\frac{1}{2} \int_0^T \frac{d}{dt} < S(2t) u_0, u_0 > dt \\
&= \frac{1}{2} \|u_0\|_2^2 - \frac{1}{2} < S(2T) u_0, u_0 > \\
&\leq \frac{1}{2} \|u_0\|_2^2 + \frac{1}{2} \|S(2T)\| \, \|u_0\|_2 \, \|u_0\|_2 \\
&\leq \|u_0\|_2^2 .
\end{aligned}
$$

This proves the lemma. Similar calculations are carried out in [AsSo94]. □

Proof of Theorem 1.5.2. Let $2 \leq s < \infty$. Then we apply (1.5.24) with u_0 replaced by $A^{1-\frac{1}{s}} u_0$. This yields

$$
\|AS(\cdot) u_0\|_{2,s;T} = \|A^{\frac{1}{s}} S(\cdot) A^{1-\frac{1}{s}} u_0\|_{2,s;T} \leq \|A^{1-\frac{1}{s}} u_0\|_2 \, ,
$$

and using (1.5.13) we get (1.5.22). If $1 < s < 2$, we use (1.5.15) and obtain

$$
\begin{aligned}
\|AS(\cdot) u_0\|_{2,s;T} &= \left(\int_0^T \|AS(t) u_0\|_2^s \, dt \right)^{\frac{1}{s}} \\
&\leq \left(\int_0^1 \|AS(t) u_0\|_2^s \, dt \right)^{\frac{1}{s}} + \left(\int_1^\infty \|AS(t) u_0\|_2^s \, dt \right)^{\frac{1}{s}}
\end{aligned}
$$

$$
\begin{aligned}
&= \left(\int_0^1 \|A^{-\varepsilon+\frac{1}{s}}S(t)A^{1+\varepsilon-\frac{1}{s}}u_0\|_2^s\,dt\right)^{\frac{1}{s}} + \left(\int_1^\infty \|AS(t)u_0\|_2^s\,dt\right)^{\frac{1}{s}} \\
&\le \left(\int_0^1 t^{-(1-s\varepsilon)}\,dt\right)^{\frac{1}{s}} \|A^{1+\varepsilon-\frac{1}{s}}u_0\|_2 + \left(\int_1^\infty t^{-s}\,dt\right)^{\frac{1}{s}} \|u_0\|_2.
\end{aligned}
$$

This proves (1.5.23) since $s > 1,\ s\varepsilon < 1$.

From (1.5.22), (1.5.23) we conclude that $Au \in L^s(0,T;L^2_\sigma(\Omega))$. Then from (1.5.13) we obtain $u' = -Au \in L^s(0,T;L^2_\sigma(\Omega)),\ u' + Au = 0$. From Lemma 1.5.1 we know that u is strongly continuous and that $u(0) = u_0$. This proves the result. □

Note that $T = \infty$ is admitted in Theorem 1.5.2. This yields

$$
\|Au\|_{2,s;\infty} = \left(\int_0^\infty \|AS(t)u_0\|_2^s\,dt\right)^{\frac{1}{s}} \le \|A^{1-\frac{1}{s}}u_0\|_2 \tag{1.5.25}
$$

if $2 \le s < \infty$, and

$$
\|Au\|_{2,s;\infty} = \left(\int_0^\infty \|AS(t)u_0\|_2^s\,dt\right)^{\frac{1}{s}} \le C\left(\|u_0\|_2 + \|A^{1+\varepsilon-\frac{1}{s}}u_0\|_2\right) \tag{1.5.26}
$$

if $1 < s < 2$.

The next lemma yields a useful estimate of $A^\alpha S(t)u_0$.

1.5.4 Lemma *Let $\Omega \subseteq \mathbb{R}^n$, $n \ge 2$, be any domain with the Stokes semigroup $\{S(t); t \ge 0\}$, let $0 < T \le \infty$, $1 < s < \infty$, $0 \le \alpha \le \frac{1}{2}$ such that*

$$
\alpha \le \frac{1}{s} < \alpha + \frac{1}{2}, \tag{1.5.27}
$$

and let $u_0 \in D(A^{-\frac{1}{2}})$.

Then $A^\alpha S(\cdot)u_0 \in L^s(0,T;L^2_\sigma(\Omega))$ and

$$
\|A^\alpha S(\cdot)u_0\|_{2,s;T} \le C\left(\|u_0\|_2 + \|A^{-\frac{1}{2}}u_0\|_2\right) \tag{1.5.28}
$$

where $C = C(\alpha, s) > 0$ is a constant.

Proof. We obtain

$$
\begin{aligned}
\|A^\alpha S(\cdot)u_0\|_{2,s;T} &= \left(\int_0^T \|A^\alpha S(t)u_0\|_2^s\,dt\right)^{\frac{1}{s}} \\
&\le \left(\int_0^\infty \|A^\alpha S(t)u_0\|_2^s\,dt\right)^{\frac{1}{s}} \\
&\le \left(\int_0^1 \|A^\alpha S(t)u_0\|_2^s\,dt\right)^{\frac{1}{s}} + \left(\int_1^\infty \|A^\alpha S(t)u_0\|_2^s\,dt\right)^{\frac{1}{s}}.
\end{aligned}
$$

Applying the interpolation inequality (2.2.8), III, yields

$$\begin{aligned}\left(\int_0^1 \|A^\alpha S(t)u_0\|_2^s\,dt\right)^{\frac{1}{s}} &= \left(\int_0^1 \|(A^{\frac{1}{2}})^{2\alpha} S(t)u_0\|_2^s\,dt\right)^{\frac{1}{s}} \\ &\le \left(\int_0^1 \|A^{\frac{1}{2}} S(t)u_0\|_2^{2\alpha s}\,\|S(t)u_0\|_2^{(1-2\alpha)s}\,dt\right)^{\frac{1}{s}}.\end{aligned}$$

Since $\|S(t)u_0\|_2 \le \|u_0\|_2$ and $s\alpha \le 1$, we get with Hölder's inequality that the last expression is

$$\begin{aligned}&\le \|u_0\|_2^{1-2\alpha}\Big(\int_0^1 \|A^{\frac{1}{2}} S(t)u_0\|_2^{2\alpha s}\,dt\Big)^{\frac{1}{s}} \\ &\le \|u_0\|_2^{1-2\alpha}\Big(\int_0^1 \|A^{\frac{1}{2}} S(t)u_0\|_2^{2}\,dt\Big)^{\frac{1}{2}\cdot 2\alpha}.\end{aligned}$$

Using

$$\begin{aligned}\|A^{\frac{1}{2}} S(t)u_0\|_2^2 &= \langle A^{\frac{1}{2}} S(t)u_0, A^{\frac{1}{2}} S(t)u_0 \rangle \\ &= \langle AS(2t)u_0, u_0 \rangle = -\frac{1}{2}\frac{d}{dt} \langle S(2t)u_0, u_0 \rangle\end{aligned}$$

for $t > 0$, and

$$\begin{aligned}-\frac{1}{2}\int_0^1 \frac{d}{dt} \langle S(2t)u_0, u_0 \rangle\,dt &= \frac{1}{2} \langle u_0, u_0 \rangle - \frac{1}{2} \langle S(2)u_0, u_0 \rangle \\ &\le \frac{1}{2}\|u_0\|_2^2 + \frac{1}{2}\|S(2)\|\,\|u_0\|_2^2 \\ &\le \|u_0\|_2^2,\end{aligned}$$

we obtain

$$\begin{aligned}\left(\int_0^1 \|A^\alpha S(t)u_0\|_2^s\,dt\right)^{\frac{1}{s}} &\le \|u_0\|_2^{1-2\alpha}\left(-\frac{1}{2}\int_0^1 \frac{d}{dt} \langle S(2t)u_0, u_0 \rangle\,dt\right)^{\alpha} \\ &\le \|u_0\|_2^{1-2\alpha}\,\|u_0\|_2^{2\alpha} = \|u_0\|_2.\end{aligned}$$

Using again (1.5.15) we get

$$\begin{aligned}\left(\int_1^\infty \|A^\alpha S(t)u_0\|_2^s\,dt\right)^{\frac{1}{s}} &= \left(\int_1^\infty \|A^{\alpha+\frac{1}{2}}\, S(t)A^{-\frac{1}{2}}u_0\|_2^s\,dt\right)^{\frac{1}{s}} \\ \le \left(\int_1^\infty t^{-(\alpha+\frac{1}{2})s}\,dt\right)^{\frac{1}{s}} &\|A^{-\frac{1}{2}}u_0\|_2 \le C\,\|A^{-\frac{1}{2}}u_0\|_2,\end{aligned}$$

with some constant $C = C(\alpha, s) > 0$, since $(\alpha+\frac{1}{2})s > 1$. This proves the lemma.

□

Note that the case $T = \infty$ is admitted in (1.5.28). This yields the inequality

$$\begin{aligned} \|A^\alpha S(\cdot)u_0\|_{2,s;\infty} &= \Big(\int_0^\infty \|A^\alpha S(t)u_0\|_2^s \, dt\Big)^{\frac{1}{s}} \\ &\le C\,(\|u_0\|_2 + \|A^{-\frac{1}{2}}u_0\|_2) \end{aligned} \tag{1.5.29}$$

with some constant $C = C(\alpha, s) > 0$.

1.6 The inhomogeneous evolution equation $u' + Au = f$

In this subsection we investigate the inhomogeneous evolution system

$$u' + Au = f \ , \quad u(0) = u_0, \tag{1.6.1}$$

and consider the solution class given by the formula

$$u(t) = S(t)u_0 + \int_0^t S(t-\tau)f(\tau)\, d\tau \ , \quad t \ge 0, \tag{1.6.2}$$

see [Ama95, Chap. III, 1.5].

The general theory of weak solutions of the nonstationary Stokes system (1.5.1) is based on this formula, see the next section.

Since the first term $S(t)u_0$ has been already treated in the previous subsection, we may restrict ourselves now to the second term. This means we may treat now only the case $u_0 = 0$.

The second term in (1.6.2) determines an integral operator $\mathcal{J}$ defined by

$$(\mathcal{J}f)(t) := \int_0^t S(t-\tau)f(\tau)\, d\tau \ , \quad t \ge 0.$$

Our aim is to study the properties of this integral operator

$$\mathcal{J} : f \mapsto \mathcal{J}f$$

in several Banach spaces. The following lemma yields first results. For simplicity we first consider the case $0 < T < \infty$.

1.6.1 Lemma *Let $\Omega \subseteq \mathbb{R}^n$, $n \ge 2$, be any domain, and let $0 < T < \infty$, $1 \le s < \infty$, $f \in L^s(0,T; L^2_\sigma(\Omega))$. Then $u = \mathcal{J}f$ defined by*

$$u(t) = (\mathcal{J}f)(t) = \int_0^t S(t-\tau)f(\tau)\, d\tau \ , \quad t \in [0,T) \tag{1.6.3}$$

has the following properties:

a) $u : [0,T] \to L^2_\sigma(\Omega)$ *is strongly continuous,* $u(0) = 0$, *and*

$$\|u\|_{2,\infty;T} \leq \|f\|_{2,1;T}\,. \tag{1.6.4}$$

b) *Under the additional assumption*

$$Au \in L^s(0,T;L^2_\sigma(\Omega)), \tag{1.6.5}$$

we get

$$u \in W^{1,s}(0,T;L^2_\sigma(\Omega)) \;,\quad u' + Au = f \tag{1.6.6}$$

and

$$u'(t) + Au(t) = f(t) \tag{1.6.7}$$

for almost all $t \in [0,T)$.

Proof. Using (1.5.8), (1.5.10) we get $\|S(t-\tau)\| \leq 1$, and

$$\|S(t-\tau)f(\tau)\|_2 \leq \|f(\tau)\|_2 \;,\quad 0 \leq \tau \leq t \leq T. \tag{1.6.8}$$

This shows that $\mathcal{J}f : [0,T] \to L^2_\sigma(\Omega)$ is well defined, the strong continuity is obvious, and

$$\|(\mathcal{J}f)(t)\|_2 \leq \int_0^t \|f(\tau)\|_2\, d\tau \;,\quad 0 \leq t \leq T.$$

Thus we get (1.6.4) and $(\mathcal{J}f)(0) = 0$.

Let now (1.6.5) be satisfied, and let $w \in D(A)$, $\varphi \in C_0^\infty((0,T))$. Then, using Fubini's theorem [Apo74, 15.7], integration by parts, and

$$\frac{d}{dt} S(t-\tau)w = -A\,S(t-\tau)w\,,$$

see (1.5.13), we obtain

$$\begin{aligned}
-\int_0^T < u(t), w > \varphi'(t)\, dt &= -\int_0^T \int_0^t < S(t-\tau)f(\tau), w > \varphi'(t)\, d\tau\, dt \\
&= -\int_0^T \Big(\int_\tau^T < S(t-\tau)f(\tau), w > \varphi'(t)\, dt\Big)\, d\tau \\
&= -\int_0^T \Big(< S(T-\tau)f(\tau), w > \varphi(T) - < f(\tau), w > \varphi(\tau) \\
&\quad - \int_\tau^T < \frac{d}{dt} S(t-\tau)f(\tau), w > \varphi(t)\, dt \Big)\, d\tau \\
&= \int_0^T < f(\tau), w > \varphi(\tau)\, d\tau - \int_0^T \Big(\int_\tau^T < S(t-\tau)f(\tau), Aw > \varphi(t)\, dt\Big)\, d\tau
\end{aligned}$$

$$= \int_0^T < f(t), w > \varphi(t)\, dt - \int_0^T (\int_0^t < S(t-\tau) f(\tau), Aw > d\tau) \varphi(t)\, dt$$

$$= \int_0^T < f(t), w > \varphi(t)\, dt - \int_0^T < (\mathcal{J} f)(t), Aw > \varphi(t)\, dt$$

$$= \int_0^T < f(t) - A(\mathcal{J} f)(t), w > \varphi(t)\, dt.$$

This shows that

$$\frac{d}{dt} < u(t), w > = < f(t) - A(\mathcal{J} f)(t), w >$$

in the sense of distributions in $(0,T)$. Lemma 1.3.1, c), now yields that $u \in W^{1,s}(0,T; L^2_\sigma(\Omega))$ and that $u' = f - A(\mathcal{J} f)$. Thus we get (1.6.6), (1.6.7), and the proof of the lemma is complete. $\square$

Now it is easy to include the case $T = \infty$. Let Ω be as in the above lemma, and suppose that

$$f \in L^1_{loc}([0,\infty); L^2_\sigma(\Omega)), \tag{1.6.9}$$

see (1.2.4). Then we may apply the above lemma with $s = 1$, $0 < T' < \infty$, and we see that $u = \mathcal{J} f$ defined by (1.6.3) is strongly continuous in $[0,\infty)$, $u(0) = 0$, and (1.6.4) holds for each subinterval $[0,T') \subseteq [0,\infty)$.

Under the assumption

$$Au \in L^1_{loc}([0,\infty); L^2_\sigma(\Omega)) \tag{1.6.10}$$

we conclude that

$$u \in W^{1,1}_{loc}([0,\infty); L^2_\sigma(\Omega)), \tag{1.6.11}$$

see Section 1.3, that $u' + Au = f$ holds in each space $L^1(0,T'; L^2_\sigma(\Omega))$, $0 < T' < \infty$, and that

$$u'(t) + Au(t) = f(t) \tag{1.6.12}$$

for almost all $t \in [0,\infty)$.

Our next aim is to remove the critical assumption (1.6.5) in Lemma 1.6.1. First we consider some conditions which are sufficient for (1.6.5). Let Ω, T, s and f be as in this lemma. If

$$Af \in L^s(0,T; L^2_\sigma(\Omega)), \tag{1.6.13}$$

then we can use (1.2.15) and obtain

$$Au(t) = A(\mathcal{J} f)(t) = A \int_0^t S(t-\tau) f(\tau)\, d\tau = \int_0^t S(t-\tau) Af(\tau)\, d\tau\ ,$$

$0 \le t \le T$. This yields the validity of (1.6.5). Thus (1.6.13) is sufficient for (1.6.5).

A more general criterion for (1.6.5) is obtained as follows:

Let Ω be as above, let $0 < T \le \infty$, $0 < \alpha < 1$, $1 < r < s < \infty$ with

$$1 - \alpha + \frac{1}{s} = \frac{1}{r},$$

and suppose

$$A^{1-\alpha} f \in L^r(0,T; L^2_\sigma(\Omega)). \tag{1.6.14}$$

Then $u = \mathcal{J} f$ in (1.6.3) is well defined and strongly continuous in $[0,T)$. Using (1.5.15) we get

$$\|A^\alpha S(t-\tau)\| \le (t-\tau)^{-\alpha} \ , \quad 0 \le \tau \le t < \infty, \tag{1.6.15}$$

and this yields

$$\begin{aligned} \|Au(t)\|_2 &= \| \int_0^t A^\alpha S(t-\tau) A^{1-\alpha} f(\tau)\, d\tau \|_2 \\ &\le \int_0^t (t-\tau)^{-\alpha} \|A^{1-\alpha} f(\tau)\|_2 \, d\tau \\ &\le \int_0^T |t-\tau|^{-\alpha} \, \|A^{1-\alpha} f(\tau)\|_2 \, d\tau \, . \end{aligned}$$

The integral on the right side can be estimated by Lemma 3.3.2, II. This yields the estimate

$$\|Au\|_{2,s;T} \le C \, \|A^{1-\alpha} f\|_{2,r;T} \tag{1.6.16}$$

with some constant $C = C(\alpha, s) > 0$, and therefore

$$Au \in L^s(0,T; L^2_\sigma(\Omega)). \tag{1.6.17}$$

Thus (1.6.14) is sufficient for (1.6.5).

The case $\alpha = 1$, $r = s$, $A^{1-\alpha} f = f$ is not admitted in (1.6.14). In this case the integral kernel (1.6.15) is called **strongly singular**.

If we are able to include in (1.6.16) this singular case, we can set $r = s$, and (1.6.14) would already follow from the assumption on f. Then we get rid of the critical condition (1.6.5).

To prove (1.6.16) in the strongly singular case $\alpha = 1$, we need a new non-elementary argument. There are several approaches in the literature. Here we use without proof a result given by de Simon [deS64]. Note that the cases $s = 1$, $s = \infty$ in the following lemma are excluded.

1.6.2 Lemma (de Simon) *Let $\Omega \subseteq \mathbb{R}^n$, $n \geq 2$, be any domain and let $0 < T \leq \infty$, $1 < s < \infty$. Suppose*

$$f \in L^s(0,T;L^2_\sigma(\Omega))$$

and let $u = \mathcal{J}f$ be defined by (1.6.3). *Then $Au \in L^s(0,T;L^2_\sigma(\Omega))$ and*

$$\|Au\|_{2,s;T} \;\leq\; C\,\|f\|_{2,s;T} \tag{1.6.18}$$

with some constant $C = C(s) > 0$ not depending on T.

Remarks on the proof. The first proof was given by de Simon [deS64, Theorem 4.4]. An immediate consequence of (1.6.18) is the estimate

$$\|u_t\|_{2,s;T} + \|Au\|_{2,s;T} \;\leq\; (1+2C)\,\|f\|_{2,s;T}$$

which is called the estimate of **maximal regularity** for the evolution system $u_t + Au = f$, $u(0) = 0$, see [Ama95, Chap. III, 4.10]. Of course, in general, the regularity of u_t and Au cannot be better than that of f.

Recall that $Au \in L^s(0,T;L^2_\sigma(\Omega))$ means the following, see (1.2.14):

$$u(t) \in D(A) \quad \text{for a.a. } t \in [0,T), \tag{1.6.19}$$
$$\|Au\|^s_{2,s;T} \;=\; \int_0^T \|Au\|_2^s\,dt \;<\infty.$$

The result of this lemma is contained as a special case in the more general theory of Cannarsa-Vespri [CaV86]. The proof uses in particular the following estimates of the operator norms:

$$\|(\mu+A)^{-1}\| \;\leq\; \mu^{-1}\;,\quad \mu > 0, \tag{1.6.20}$$
$$\|e^{-tA}\| \;\leq\; 1\;,\quad t \geq 0, \tag{1.6.21}$$
$$\|Ae^{-tA}\| \;\leq\; t^{-1}\;,\quad t > 0. \tag{1.6.22}$$

If $0 < T < \infty$, the result of Lemma 1.6.2 is contained as a special case in the theory of Dore-Venni [DoVe87], see [Ama95, Chap. III, Theorem 4.10.8, (4.10.28)]. The case $T = \infty$ is included in the extension of the Dore-Venni theory given by Prüss-Sohr [PrS90] and Giga-Sohr [GiSo91], see also [Ama95, Chap. III, (4.10.33)], [Monn99]. A completely different (potential theoretic) proof for the Stokes operator A in bounded and exterior domains has been given by Solonnikov [Sol77, Theorem 4.1] for $0 < T < \infty$, and by Maremonti-Solonnikov [MSol97, Theorem 1.4] for the general case $0 < T \leq \infty$. □

The following result is essentially a combination of Lemma 1.6.1 with Lemma 1.6.2.

1.6.3 Theorem *Let $\Omega \subseteq \mathbb{R}^n$, $n \geq 2$, be any domain, let $0 < T \leq \infty$, $0 \leq \alpha \leq 1$, $1 < s < \infty$, $f \in L^s(0,T; L^2_\sigma(\Omega))$, and let $u = \mathcal{J} f$ be defined by (1.6.3).*

Then $u : [0,T) \to L^2_\sigma(\Omega)$ is strongly continuous, $u(0) = 0$, and we get the following properties:

a) *$Au,\ u' \in L^s(0,T; L^2_\sigma(\Omega))$,*

$$u'(t) + Au(t) = f(t) \ \text{ for almost all } \ t \in [0,T), \tag{1.6.23}$$

and

$$\|u'\|_{2,s;T} + \|Au\|_{2,s;T} \ \leq \ C \, \|f\|_{2,s;T} \tag{1.6.24}$$

with some constant $C = C(s) > 0$ not depending on T.

b) *$u \in W^{1,s}(0,T'; L^2_\sigma(\Omega))$, $u \in L^s(0,T'; D(A))$, $u \in L^\infty(0,T'; L^2_\sigma(\Omega))$, $A^\alpha u \in L^s(0,T'; L^2_\sigma(\Omega))$, and*

$$\begin{aligned} \|u\|_{2,\infty;T'} + \|u\|_{2,s;T'} + \|u'\|_{2,s;T'} + \|A^\alpha u\|_{2,s;T'} + \|Au\|_{2,s;T'} \\ \leq C\,(1+T')\,\|f\|_{2,s;T'} \end{aligned} \tag{1.6.25}$$

for all finite T' with $0 < T' \leq T$, where $C = C(s) > 0$.

c) *$u \in W^{1,s}_{loc}([0,\infty); L^2_\sigma(\Omega))$, $u \in L^s_{loc}([0,\infty); D(A))$, $u \in L^\infty_{loc}([0,\infty); L^2_\sigma(\Omega))$, $A^\alpha u \in L^s_{loc}([0,\infty); L^2_\sigma(\Omega))$ if $T = \infty$.*

Proof. See (1.2.4), (1.3.3) concerning the loc-spaces. Lemma 1.6.2 yields $Au \in L^s(0,T; L^2_\sigma(\Omega))$ for $0 < T \leq \infty$. First let $0 < T < \infty$. Then Lemma 1.6.1 shows that $u \in W^{1,s}(0,T; L^2_\sigma(\Omega))$. Therefore, $u' \in L^s(0,T; L^2_\sigma(\Omega))$, $u' + Au = f$, and (1.6.23) follows. Using (1.6.18) we get

$$\begin{aligned} \|u'\|_{2,s;T} + \|Au\|_{2,s;T} \ &\leq \ \|f - Au\|_{2,s;T} \ + \ C\,\|f\|_{2,s;T} \\ &\leq \ \|f\|_{2,s;T} \ + \ 2C\,\|f\|_{2,s;T} \ \leq \ (1+2C)\,\|f\|_{2,s;T} \end{aligned}$$

which leads to (1.6.24). The last inequality also holds if $T = \infty$. This proves a).

To prove b) we need some embedding inequalities. First let $0 < T < \infty$. Writing

$$u(t) = \int_0^t u'(\tau)\,d\tau \ , \quad 0 \ \leq \ t \leq T \, ,$$

we obtain

$$\|u(t)\|_2 \ \leq \ \int_0^t \|u'(\tau)\|_2 \, d\tau \ ,$$

and Hölder's inequality yields

$$\|u\|_{2,\infty;T} \ \leq \ T^{1/s'}\,\|u'\|_{2,s;T} \ \leq \ (1+T)\,\|u'\|_{2,s;T} \ , \tag{1.6.26}$$

$s' = \frac{s}{s-1}$. Using the interpolation inequality (2.2.8), III, we obtain

$$\|A^\alpha u(t)\|_2 \le \|Au(t)\|_2^\alpha \, \|u(t)\|_2^{1-\alpha} \le \alpha \, \|Au(t)\|_2 + (1-\alpha)\, \|u(t)\|_2 \quad (1.6.27)$$

for almost all $t \in [0,T)$, and with Hölder's inequality we get

$$\begin{aligned} \|A^\alpha u\|_{2,s;T} &= \Big(\int_0^T \|A^\alpha u\|_2^s \, dt\Big)^{\frac{1}{s}} \le \Big(\int_0^T \|Au\|_2^{\alpha s} \, \|u\|_2^{(1-\alpha)s} \, dt\Big)^{\frac{1}{s}} \quad (1.6.28) \\ &\le \Big(\int_0^T \|Au\|_2^s \, dt\Big)^{\frac{\alpha}{s}} \Big(\int_0^T \|u\|_2^s \, dt\Big)^{\frac{1-\alpha}{s}} = \|Au\|_{2,s;T}^\alpha \, \|u\|_{2,s;T}^{1-\alpha} \\ &\le \alpha \, \|Au\|_{2,s;T} + (1-\alpha)\, \|u\|_{2,s;T} \end{aligned}$$

with $0 \le \alpha \le 1$. Further we use the inequality

$$\begin{aligned} \|u\|_{2,s;T} &= \Big(\int_0^T \|u\|_2^s \, dt\Big)^{\frac{1}{s}} \le T^{\frac{1}{s}} \|u\|_{2,\infty;T} \\ &\le T^{\frac{1}{s'}} T^{\frac{1}{s}} \|u'\|_{2,s;T} \le TC\, \|f\|_{2,s;T} \, . \end{aligned}$$

Combining this with (1.6.26), (1.6.28), (1.6.24), we obtain the desired inequality (1.6.25) if $T < \infty$.

If $T = \infty$, the properties under b) hold with T replaced by T', $0 < T' < \infty$. This proves b), and c) is a consequence. The proof of the theorem is complete. □

In the case $T = \infty$ we obtain from (1.6.24) the estimate

$$\|u'\|_{2,s;\infty} + \|Au\|_{2,s;\infty} \le C\, \|f\|_{2,s;\infty} \quad (1.6.29)$$

with $C = C(s) > 0$. Note that u need not satisfy the condition $\|u\|_{2,s;\infty} < \infty$ in this case.

Finally we mention the following embedding property.

1.6.4 Lemma *Let $\Omega \subseteq \mathbb{R}^n$, $n \ge 2$, be any domain, let $0 < T \le \infty$, $f \in L^s(0,T;L^2_\sigma(\Omega))$, $1 < s < \rho < \infty$, $0 < \alpha < 1$ such that*

$$1 - \alpha + \frac{1}{\rho} = \frac{1}{s} \, , \quad (1.6.30)$$

and let $u = \mathcal{J} f$ be defined by (1.6.3).

Then $A^\alpha u \in L^\rho(0,T;L^2_\sigma(\Omega))$ and

$$\|A^\alpha u\|_{2,\rho;T} \le C\, \|f\|_{2,s;T} \quad (1.6.31)$$

with some constant $C = C(\rho, s) > 0$.

Proof. Using the inequality (1.6.15) we obtain

$$\begin{aligned} \|A^\alpha u(t)\|_2 &\le \int_0^t \|A^\alpha S(t-\tau)f(\tau)\|_2\, d\tau \\ &\le \int_0^t (t-\tau)^{-\alpha}\, \|f(\tau)\|_2\, d\tau \le \int_0^T |t-\tau|^{-\alpha}\, \|f(\tau)\|_2\, d\tau. \end{aligned}$$

Let $g : \mathbb{R} \to \mathbb{R}$ be defined by $g(\tau) = \|f(\tau)\|_2$ if $\tau \in [0,T)$, and by $g(\tau) = 0$ if $\tau \notin [0,T)$. Then $g \in L^s(\mathbb{R})$, and Lemma 3.3.2, II, shows, the integral

$$v(t) := \int_{\mathbb{R}} |t-\tau|^{(1-\alpha)-1}\, g(\tau)$$

converges absolutely for almost all $t \in \mathbb{R}$, and

$$\|v\|_{L^\rho(\mathbb{R})} \le C\, \|g\|_{L^s(\mathbb{R})}$$

with some constant $C = C(\rho, s) > 0$. This yields

$$\|A^\alpha u\|_{2,\rho;T} \le \|v\|_{L^\rho(\mathbb{R})} \le C\, \|g\|_{L^s(\mathbb{R})} = C\, \|f\|_{2,s;T}$$

and the proof is complete. □

2 Theory of weak solutions in the linearized case

2.1 Weak solutions

The definition below yields a very general solution class of the nonstationary Stokes system. The exterior force f is a distribution of the form $f = f_0 + \operatorname{div} F$ with $f_0 \in L^1_{loc}([0,T); L^2(\Omega)^n)$ and $F \in L^1_{loc}([0,T); L^2(\Omega)^{n^2})$. This means, see (1.2.4), that

$$f_0 \in L^1(0,T'; L^2(\Omega)^n) \;\text{ and }\; F \in L^1(0,T'; L^2(\Omega)^{n^2})$$

for all T' with $0 < T' < T$. f is considered as a functional defined by

$$\begin{aligned} [f,v]_{\Omega,T} &= < f_0, v >_{\Omega,T} + [\operatorname{div} F, v]_{\Omega,T} \\ &= \int_0^T < f_0, v >_\Omega\, dt + \int_0^T [\operatorname{div} F, v]_\Omega\, dt \\ &= \int_0^T < f_0, v >_\Omega\, dt - \int_0^T < F, \nabla v >_\Omega\, dt \\ &= \int_0^T \int_\Omega f_0 \cdot v\, dx\, dt - \int_0^T \int_\Omega F \cdot \nabla v\, dx\, dt \end{aligned}$$

for all $v \in C_0^\infty((0,T) \times \Omega)^n$.

Recall that $f_0 = (f_{01}, \dots, f_{0n})$, $F = (F_{jl})_{j,l=1}^n$, $v = (v_1, \dots, v_n)$, $\nabla v = (D_j v_l)_{j,l=1}^n$, $f_0 \cdot v = f_{01}v_1 + \dots + f_{0n}v_n$, and $F \cdot \nabla v = \sum_{j,l=1}^n F_{jl} D_j v_l$, depending on the variables $t \in [0,T)$ and $x = (x_1, \dots, x_n) \in \Omega$.

Further recall, (2.1.2) below means that

$$u \in L^1(0,T'; W_{0,\sigma}^{1,2}(\Omega)) \quad \text{for } 0 < T' < T,$$

where

$$W_{0,\sigma}^{1,2}(\Omega) = \overline{C_{0,\sigma}^\infty(\Omega)}^{\|\cdot\|_{W^{1,2}}} \subseteq L_\sigma^2(\Omega) = \overline{C_{0,\sigma}^\infty(\Omega)}^{\|\cdot\|_2} .$$

The condition (2.1.4) below is motivated formally if we consider each term of $u_t - \nu\Delta u + \nabla p = f$ as a functional applied to a test function

$$v \in C_0^\infty([0,T);\, C_{0,\sigma}^\infty(\Omega)),$$

and use the rule of integration by parts. This calculation becomes precise under some smoothness conditions. In this way the notion of a weak solution is justified. See (1.4.2) for the definition of the test space $C_0^\infty([0,T); C_{0,\sigma}^\infty(\Omega))$.

2.1.1 Definition *Let $\Omega \subseteq \mathbb{R}^n$, $n \geq 2$, be any domain, let $0 < T \leq \infty$, $u_0 \in L_\sigma^2(\Omega)$, and let $f = f_0 + \operatorname{div} F$ with*

$$f_0 \in L_{loc}^1([0,T); L^2(\Omega)^n) \ , \quad F \in L_{loc}^1([0,T); L^2(\Omega)^{n^2}). \tag{2.1.1}$$

Then a function

$$u \in L_{loc}^1([0,T); W_{0,\sigma}^{1,2}(\Omega)) \tag{2.1.2}$$

is called a **weak solution** *of the Stokes system*

$$u_t - \nu\Delta u + \nabla p = f \ , \quad \operatorname{div} u = 0 \ , \quad u|_{\partial\Omega} = 0 \ , \quad u(0) = u_0 \tag{2.1.3}$$

with data f, u_0, iff the condition

$$- < u, v_t >_{\Omega,T} + \nu < \nabla u, \nabla v >_{\Omega,T} = < u_0, v(0) >_\Omega + [f, v]_{\Omega,T} \tag{2.1.4}$$

is satisfied for all $v \in C_0^\infty([0,T); C_{0,\sigma}^\infty(\Omega))$. A distribution p in $(0,T) \times \Omega$ is called an **associated pressure** *of a weak solution u iff the equation*

$$u_t - \nu\Delta u + \nabla p = f \tag{2.1.5}$$

is satisfied in the sense of distributions.

Concerning weak solutions we refer to [Lad69, Chap. 4] and [Tem77, Chap. III, 1].

Our aim is to develop the theory of existence, uniqueness, and regularity of weak solutions. Important facts in this theory are the energy equality and the explicit representation formula of weak solutions. This formula is based on the Stokes operator A, see Section 2, III, and the Stokes semigroup $\{S(t); t \geq 0\}$, $S(t) = e^{-tA}$ of Ω, see (1.5.7).

In particular we use the square root $A^{\frac{1}{2}}$ of A with domain $D(A^{\frac{1}{2}}) = W^{1,2}_{0,\sigma}(\Omega)$ and with

$$< A^{\frac{1}{2}}u, A^{\frac{1}{2}}v >_\Omega = \nu < \nabla u, \nabla v >_\Omega \ , \quad \|A^{\frac{1}{2}}u\|_2 = \nu^{\frac{1}{2}}\|\nabla u\|_2, \tag{2.1.6}$$

$u, v \in D(A^{\frac{1}{2}})$, and we use the operator $A^{-\frac{1}{2}}$ with domain $D(A^{-\frac{1}{2}}) = R(A^{\frac{1}{2}})$, see Lemma 2.2.1, III.

An important role plays the natural extension of the operator $A^{-\frac{1}{2}}$ from $D(A^{-\frac{1}{2}})$ to the completion $\widehat{D}(A^{-\frac{1}{2}})$ of $D(A^{-\frac{1}{2}})$ in the norm $\|A^{-\frac{1}{2}}u\|_2$, see Lemma 2.6.2, III.

Further we use the notation

$$PF(t) = Pf_0(t) + P \text{ div } F(t)$$

for almost all $t \in [0, T)$, see (2.5.26), III. Here $Pf_0(t) \in L^2_\sigma(\Omega)$ means the Helmholtz projection applied to $f_0(t) \in L^2(\Omega)^n$, and P div $F(t)$ simply means (extending this operator in a natural way) the restriction of the distribution div $F(t)$ to test functions $v \in C^\infty_{0,\sigma}(\Omega)$, see (2.5.26), III.

With F as above, $A^{-\frac{1}{2}}P \text{ div } F(t) \in L^2_\sigma(\Omega)$ is determined by the relation

$$< A^{-\frac{1}{2}}P \text{ div } F(t), A^{\frac{1}{2}}v >_\Omega \ = \ - < F(t), \nabla v >_\Omega \ , \quad v \in W^{1,2}_{0,\sigma}(\Omega), \tag{2.1.7}$$

and it holds that

$$\|A^{-\frac{1}{2}}P \text{ div } F(t)\|_2 \ \leq \ \nu^{-\frac{1}{2}}\|F(t)\|_2 \tag{2.1.8}$$

for almost all $t \in [0, T)$; see Lemma 2.6.1, III.

The Yosida approximation of a weak solution u is defined by

$$u_k := J_k u \ , \quad J_k = (I + k^{-1}A^{\frac{1}{2}})^{-1} \ , \quad k \in \mathbb{N}, \tag{2.1.9}$$

and we use the properties in Section 3.4, II.

2.2 Equivalent formulation and approximation

In order to analyse the above formulation of weak solutions, it is convenient to use simple test functions $v \in C^\infty_0([0, T); C^\infty_{0,\sigma}(\Omega))$ of the special form

$$v(t, x) \ = \ \varphi(t)w(x) \ , \quad w \in C^\infty_{0,\sigma}(\Omega) \ , \quad \varphi \in C^\infty_0([0, T))$$

where

$$C_0^\infty([0,T)) = C_0^\infty([0,T);\mathbb{R}) := \{\varphi|_{[0,T)};\ \varphi \in C_0^\infty((-1,T))\} \tag{2.2.1}$$

means the space of restrictions to $[0,T)$ of the scalar test functions

$$\varphi \in C_0^\infty((-1,T)).$$

The following lemma yields an equivalent condition for weak solutions using these special test functions. Further it shows that Yosida's approximation, Section 3.4, II, can be used as a smoothing procedure. We will see that each

$$u_k = J_k u$$

satisfies an evolution equation in the strong sense for almost all $t \in [0,T)$.

2.2.1 Lemma *Let $\Omega \subseteq \mathbb{R}^n$, $n \geq 2$, be any domain. Let $0 < T \leq \infty$, $u_0 \in L^2_\sigma(\Omega)$, $f = f_0 + \operatorname{div} F$ with*

$$f_0 \in L^1_{loc}([0,T); L^2(\Omega)^n) \ , \quad F \in L^1_{loc}([0,T); L^2(\Omega)^{n^2}),$$

and let

$$u \in L^1_{loc}([0,T); W^{1,2}_{0,\sigma}(\Omega)) .$$

Then we have:

a) *u is a weak solution of the Stokes system (2.1.3) with data f, u_0 iff there exists a dense subspace $D \subseteq W^{1,2}_{0,\sigma}(\Omega)$ such that*

$$-\int_0^T < u, w >_\Omega \varphi_t \, dt + \nu \int_0^T < \nabla u, \nabla w >_\Omega \varphi \, dt \tag{2.2.2}$$
$$= < u_0, w >_\Omega \varphi(0) + \int_0^T [f, w]_\Omega \varphi \, dt$$

holds for all $w \in D$, $\varphi \in C_0^\infty([0,T))$.

b) *If u is a weak solution of (2.1.3) with data f, u_0, then, after a redefinition of u on a null set of $[0,T)$, each $u_k := J_k u$, $k \in \mathbb{N}$, is strongly continuous, and satisfies $u_k(0) = J_k u_0$,*

$$u_k' \in L^1(0,T'; L^2_\sigma(\Omega)) \quad , \quad Au_k \in L^1(0,T'; L^2_\sigma(\Omega)), \tag{2.2.3}$$
$$u_k' + Au_k = f_k \tag{2.2.4}$$

in each subinterval $[0,T')$, $0 < T' < T$, where

$$f_k := J_k P f_0 + A^{\frac{1}{2}} J_k A^{-\frac{1}{2}} P \operatorname{div} F.$$

Proof. Using (2.1.7), (2.1.8) we see that the last expression is well defined. Let u be a weak solution. Then in the first step we prove that (2.2.2) holds with $D = W^{1,2}_{0,\sigma}(\Omega)$ and that b) is satisfied. Set $v = \varphi w$ as above. Then from (2.1.4) we get the relation (2.2.2) for all $w \in C^\infty_{0,\sigma}(\Omega)$, $\varphi \in C^\infty_0([0,T))$.

Let $\varphi \in C^\infty_0([0,T))$ be fixed. Since $C^\infty_{0,\sigma}(\Omega)$ is dense in $W^{1,2}_{0,\sigma}(\Omega)$ by definition, see (1.2.1), III, we can extend (2.2.2) by closure to all $w \in W^{1,2}_{0,\sigma}(\Omega)$. For this purpose let $w \in W^{1,2}_{0,\sigma}(\Omega)$, choose a sequence $(w_j)_{j=1}^\infty$ in $C^\infty_{0,\sigma}(\Omega)$ with

$$w = s - \lim_{j\to\infty} w_j \ \text{ in } L^2_\sigma(\Omega) \ , \quad \nabla w = s - \lim_{j\to\infty} \nabla w_j \ \text{ in } L^2(\Omega)^{n^2} \ , \tag{2.2.5}$$

insert w_j in (2.2.2), let $j \to \infty$, and use Lebesgue's dominated convergence lemma, see (1.2.22). Thus we obtain (2.2.2) with $D = W^{1,2}_{0,\sigma}(\Omega)$.

Since $R(J_k) = D(A^{\frac{1}{2}}) = W^{1,2}_{0,\sigma}(\Omega)$, see Section 3.4, II, we may insert $w = J_k h$, $h \in L^2_\sigma(\Omega)$, in (2.2.2). Then we use the relations:

$$\begin{aligned} < u(t), J_k h >_\Omega &= < J_k u(t), h >_\Omega = < u_k(t), h >_\Omega \, , \\ \nu < \nabla u(t), \nabla J_k h >_\Omega &= < A^{\frac{1}{2}} u(t), A^{\frac{1}{2}} J_k h >_\Omega = < A^{\frac{1}{2}} J_k A^{\frac{1}{2}} u(t), h >_\Omega \, , \end{aligned}$$

see (2.2.2), III, and

$$\begin{aligned} [f(t), J_k h]_\Omega &= < f_0(t), J_k h >_\Omega - < F(t), \nabla J_k h >_\Omega \\ &= < J_k P f_0(t), h >_\Omega + < A^{-\frac{1}{2}} P \operatorname{div} F(t), A^{\frac{1}{2}} J_k h >_\Omega \\ &= < f_k(t), h >_\Omega \, , \end{aligned} \tag{2.2.6}$$

see (2.1.7). Since $A^{\frac{1}{2}} J_k$ is a bounded operator, see (3.4.5), II, we see that

$$f_k \in L^1(0, T'; L^2_\sigma(\Omega)) \, .$$

Further we see that for almost all $t \in [0,T)$ the functional

$$h \mapsto < A^{\frac{1}{2}} J_k u(t), A^{\frac{1}{2}} h >_\Omega = < A^{\frac{1}{2}} J_k A^{\frac{1}{2}} u(t), h >_\Omega \ , \quad h \in D(A^{\frac{1}{2}})$$

is continuous in the norm $\|h\|_2$. Since $A^{\frac{1}{2}}$ is selfadjoint, we obtain

$$A^{\frac{1}{2}} J_k u(t) = A^{\frac{1}{2}} u_k(t) \in D(A^{\frac{1}{2}}),$$

therefore $u_k(t) \in D(A)$ and

$$< A u_k(t), h >_\Omega = < A^{\frac{1}{2}} J_k A^{\frac{1}{2}} u(t), h >_\Omega = \nu < \nabla u(t), \nabla J_k h >_\Omega \, .$$

Since $u \in L^1(0,T'; W^{1,2}_{0,\sigma}(\Omega))$ and $\|\nabla u(t)\|_2 = \nu^{-\frac{1}{2}}\|A^{\frac{1}{2}}u(t)\|_2$, we get $A^{\frac{1}{2}}u \in L^1(0,T'; L^2_\sigma(\Omega))$, and since $A^{\frac{1}{2}}J_k$ is a bounded operator it follows that $Au_k \in L^1(0,T'; L^2_\sigma(\Omega))$ for $0 < T' < T$. This yields

$$\begin{aligned} &- \int_0^T < u_k, h >_\Omega \varphi' \, dt \\ &= \int_0^T < f_k - Au_k, h >_\Omega \varphi \, dt + < J_k u_0, h >_\Omega \varphi(0) \,. \end{aligned} \tag{2.2.7}$$

From Lemma 1.3.1, c), we conclude, choosing $\varphi \in C_0^\infty((0,T'))$, that

$$u_k \in W^{1,1}(0,T'; L^2_\sigma(\Omega)) \,,$$

and that

$$u_k' = f_k - Au_k$$

holds in $L^1(0,T'; L^2_\sigma(\Omega))$, for $0 < T' < T$.

Let $k = 1$. Then we see, after redefining u on a null set of $[0,T)$, that $u_1 = J_1 u$ is strongly continuous. We may write

$$J_k u = (I + A^{\frac{1}{2}})\, J_k\, J_1 u,$$

and since $(I + A^{\frac{1}{2}})J_k$ is a bounded operator, we see that $J_k u$ is strongly continuous for all $k \in \mathbb{N}$. Thus, $u_k(0)$ is well defined, and we may apply (1.3.15) with u, v replaced by $< u_k, h >_\Omega$, φ. This yields (rule of integration by parts) that

$$0 = < u_k(0), h >_\Omega \varphi(0) + \int_0^T (< u_k', h >_\Omega \varphi + < u_k, h > \varphi')\, dt. \tag{2.2.8}$$

for all $\varphi \in C_0^\infty([0,T))$. Inserting this in (2.2.7) and using $u_k' = f_k - Au_k$, we see that

$$< u_k(0), h >_\Omega \; = \; < J_k u_0, h >_\Omega$$

for all $h \in L^2_\sigma(\Omega)$ and therefore that $u_k(0) = J_k\, u_0$, $k \in \mathbb{N}$. This proves the property b).

Next we show the converse direction in a). For this purpose let $D \subseteq W^{1,2}_{0,\sigma}(\Omega)$ be any dense subspace and let (2.2.2) be satisfied for all $w \in D$. Then the same density argument as in (2.2.5) shows that (2.2.2) holds even with $D = W^{1,2}_{0,\sigma}(\Omega)$. But this was used to prove (2.2.3) and (2.2.4). Then we use (2.2.4), the rule (1.3.15), and (2.2.6). This yields

$$\int_0^T < u'_k, v >_\Omega dt + \int_0^T < Au_k, v >_\Omega dt = \int_0^T < f_k, v >_\Omega dt,$$

$$-\int_0^T < u_k, v' >_\Omega dt - < u_k(0), v(0) >_\Omega + \int_0^T < A^{\frac{1}{2}} u_k, A^{\frac{1}{2}} v >_\Omega dt$$
$$= \int_0^T [f, J_k v]_\Omega \, dt$$

for all $v \in C_0^\infty([0,T); C^\infty_{0,\sigma}(\Omega))$, and we get

$$-\int_0^T < u, J_k v' >_\Omega dt + \int_0^T < A^{\frac{1}{2}} u, J_k A^{\frac{1}{2}} v >_\Omega dt$$
$$= < u_k(0), v(0) >_\Omega + \int_0^T [f, J_k v]_\Omega \, dt.$$

Using the strong convergence property of J_k in Lemma 3.4.1, II, and Lebesgue's dominated convergence lemma, we may let $k \to \infty$ in each term of the last equation. This yields the condition (2.1.4) and therefore, u is a weak solution. The proof of the lemma is complete. □

2.3 Energy equality and strong continuity

Our next aim is to study important properties of weak solutions of the nonstationary Stokes system (2.1.3). In particular we are interested in the equality (2.3.3) below which is called the **energy equality**. To motivate this equality we consider (first formally) the scalar product

$$< u_t - \nu \Delta u + \nabla p, u >_{\Omega,t} = < f, u >_{\Omega,t} , \quad 0 \le t < T,$$

use integration by parts and the rule

$$\begin{aligned} \frac{d}{dt} \|u(t)\|_2^2 &= \frac{d}{dt} < u(t), u(t) >_\Omega \\ &= < u'(t), u(t) >_\Omega + < u(t), u'(t) >_\Omega \\ &= 2 < u'(t), u(t) >_\Omega . \end{aligned}$$

This yields

$$\frac{1}{2}\|u(t)\|_2^2 - \frac{1}{2}\|u_0\|_2^2 + \nu \int_0^t \|\nabla u\|_2^2 \, d\tau$$
$$= \int_0^t < f_0, u >_\Omega d\tau - \int_0^t < F, \nabla u >_\Omega d\tau.$$

The definition of a weak solution u however does not contain enough regularity in order to justify this calculation directly. Therefore, we use Yosida's smoothing procedure in Lemma 2.2.1. This enables us to give a precise proof.

2.3.1 Theorem *Let $\Omega \subseteq \mathbb{R}^n$, $n \geq 2$, be any domain, let $0 < T \leq \infty$, $u_0 \in L^2_\sigma(\Omega)$, $f = f_0 + \operatorname{div} F$ with*

$$f_0 \in L^1(0,T;L^2(\Omega)^n) \ , \quad F \in L^2(0,T;L^2(\Omega)^{n^2}) \, , \tag{2.3.1}$$

and let

$$u \in L^1_{loc}([0,T); W^{1,2}_{0,\sigma}(\Omega))$$

be a weak solution of the Stokes system (2.1.3) with data f, u_0.

Then u has the following properties:

a)
$$u \in L^\infty(0,T;L^2_\sigma(\Omega)), \quad \nabla u \in L^2(0,T;L^2(\Omega)^{n^2}) \, . \tag{2.3.2}$$

b) *$u : [0,T) \to L^2_\sigma(\Omega)$ is strongly continuous, after a redefinition on a null set of $[0,T)$, $u(0) = u_0$, and the energy equality*

$$\frac{1}{2}\|u(t)\|_2^2 + \nu \int_0^t \|\nabla u\|_2^2 \, d\tau = \frac{1}{2}\|u_0\|_2^2 + \int_0^t < f_0, u >_\Omega \, d\tau - \int_0^t < F, \nabla u >_\Omega \, d\tau \tag{2.3.3}$$

holds for all $t \in [0,T)$.

c)
$$\frac{1}{2}\|u\|^2_{2,\infty;T} + \nu\|\nabla u\|^2_{2,2,T} \leq 2\,\|u_0\|_2^2 + 8\,\|f_0\|^2_{2,1;T} + 4\nu^{-1}\|F\|^2_{2,2;T} \, . \tag{2.3.4}$$

Proof. First we assume that u satisfies the following additional properties:

$u : [0,T) \to L^2_\sigma(\Omega)$ is strongly continuous, $u(0) = u_0$,

$$u \in W^{1,1}(0,T';L^2_\sigma(\Omega)) \ , \quad Au \in L^1(0,T';L^2_\sigma(\Omega)) \ , \quad f \in L^1(0,T';L^2_\sigma(\Omega))$$

and $u' + Au = f$ in $[0,T')$ for all T' with $0 < T' < T$.

In this case we may use the rule in Lemma 1.3.2 and obtain

$$\|u(t)\|_2^2 - \|u_0\|_2^2 = 2\int_0^t < u', u >_\Omega \, d\tau. \tag{2.3.5}$$

Taking in $u' + Au = f$ the scalar product with u, we get

$$\frac{1}{2}\|u(t)\|_2^2 - \frac{1}{2}\|u_0\|_2^2 + \int_0^t < A^{\frac{1}{2}}u, A^{\frac{1}{2}}u >_\Omega \, d\tau = \int_0^t < f, u >_\Omega \, d\tau, \qquad (2.3.6)$$

and

$$\frac{1}{2}\|u(t)\|_2^2 + \nu \int_0^t \|\nabla u\|_2^2 \, d\tau = \frac{1}{2}\|u_0\|_2^2 + \int_0^t < f, u >_\Omega \, d\tau.$$

It follows that

$$\sup_{0 \le \tau \le t} \left(\frac{1}{2}\|u(\tau)\|_2^2 + \nu \int_0^\tau \|\nabla u\|_2^2 \, d\rho \right) \le \frac{1}{2}\|u_0\|_2^2 + \int_0^t | < f, u >_\Omega | \, d\rho,$$

and therefore we get

$$\frac{1}{2}(\sup_{0 \le \tau \le t} \|u(\tau)\|_2^2) + \nu \int_0^t \|\nabla u\|_2^2 \, d\tau \le \|u_0\|_2^2 + 2 \int_0^t | < f, u >_\Omega | \, d\tau. \qquad (2.3.7)$$

The additional properties above are always satisfied if u, f are replaced by u_k, f_k, $k \in \mathbb{N}$, see (2.2.3), (2.2.4). Thus we may use (2.3.6), (2.3.7) for u_k, f_k. Further we use the calculation (2.2.6), and get with $\|J_k\| \le 1$ and with (2.1.8) that

$$\begin{aligned} | < f_k, u_k >_\Omega | &= | < J_k P f_0, u_k >_\Omega + < A^{\frac{1}{2}} J_k A^{-\frac{1}{2}} P \operatorname{div} F, u_k >_\Omega | \\ &= | < f_0, J_k u_k >_\Omega + < J_k A^{-\frac{1}{2}} P \operatorname{div} F, A^{\frac{1}{2}} u_k >_\Omega | \\ &\le \|f_0\|_2 \, \|J_k u_k\|_2 + \|J_k A^{-\frac{1}{2}} P \operatorname{div} F\|_2 \, \|A^{\frac{1}{2}} u_k\|_2 \\ &\le \|f_0\|_2 \, \|u_k\|_2 + \|A^{-\frac{1}{2}} P \operatorname{div} F\|_2 \, \|A^{\frac{1}{2}} u_k\|_2 \\ &\le \|f_0\|_2 \, \|u_k\|_2 + \nu^{-\frac{1}{2}} \|F\|_2 \, \|A^{\frac{1}{2}} u_k\|_2 \\ &= \|f_0\|_2 \, \|u_k\|_2 + \|F\|_2 \, \|\nabla u_k\|_2 \end{aligned}$$

for almost all $\tau \in [0, t]$. Using

$$\sup_{0 \le \tau \le t} \|u_k(\tau)\|_2^2 = \|u_k\|_{2,\infty;t}^2 ,$$

and inserting u_k, f_k, $J_k u_0$ in (2.3.6), (2.3.7), we now obtain

$$\begin{aligned} \frac{1}{2} \|u_k\|_{2,\infty;t}^2 + \nu \|\nabla u_k\|_{2,2;t}^2 \le{} & \|J_k u_0\|_2^2 \\ & + 2 \int_0^t \|f_0\|_2 \, \|u_k\|_2 \, d\tau + 2 \int_0^t \|F\|_2 \, \|\nabla u_k\|_2 \, d\tau . \end{aligned} \qquad (2.3.8)$$

Using Young's inequality (3.3.8), I, we conclude that

$$\begin{aligned}\int_0^t \|f_0\|_2 \, \|u_k\|_2 \, d\tau \ + \ \int_0^t \|F\|_2 \, \|\nabla u_k\|_2 \, d\tau \\ &\le \ \|f_0\|_{2,1;t} \, \|u_k\|_{2,\infty;t} \ + \ \|F\|_{2,2;t} \, \|\nabla u_k\|_{2,2;t} \\ &\le \ 2\, \|f_0\|_{2,1;t}^2 \ + \ \frac{1}{8} \|u_k\|_{2,\infty;t}^2 \ + \ \nu^{-1} \|F\|_{2,2;t}^2 \ + \ \frac{\nu}{4} \|\nabla u_k\|_{2,2;t}^2 \ .\end{aligned}$$

Inserting this in (2.3.8), we finally obtain

$$\frac{1}{2} \, \|u_k\|_{2,\infty;t}^2 \ + \ \nu \, \|\nabla u_k\|_{2,2;t}^2 \ \le \ 2\, \|u_0\|_2^2 \ + \ 8 \|f_0\|_{2,1;T}^2 \ + \ 4\nu^{-1} \|F\|_{2,2;T}^2 \tag{2.3.9}$$

for $0 \le t < T$, $k \in \mathbb{N}$. In particular,

$$\frac{1}{2} \, \|J_k u(\tau)\|_2^2 + \nu \|\nabla u_k\|_{2,2;t}^2 \ \le \ 2\, \|u_0\|_2^2 \ + \ 8 \|f_0\|_{2,1;T}^2 \ + \ 4\nu^{-1} \|F\|_{2,2;T}^2$$

for each $\tau \in [0,t)$ and all $k \in \mathbb{N}$. Using the convergence property (3.4.8), II, we get

$$\lim_{k \to \infty} \ \|u(\tau) - J_k \, u(\tau)\|_2 \ = \ 0 \, ,$$

and

$$\frac{1}{2} \|u\|_{2,\infty;t}^2 + \lim_{k \to \infty} \inf \, (\nu \|\nabla u_k\|_{2,2;t}^2) \le 2\, \|u_0\|_2^2 + 8\, \|f_0\|_{2,1;T}^2 + 4\nu^{-1} \|F\|_{2,2;T}^2 .$$

Using the argument in (3.1.8), (3.1.9), II, and letting $t \to T$ we get the inequality (2.3.4) and obtain (2.3.2).

To prove b) we use (2.3.7) with u, u_0, f replaced by $u_k - u_l$, $(J_k - J_l) u_0$, $f_k - f_l$ where $k, l \in \mathbb{N}$. Similarly as above we conclude with $\tilde{f} := A^{-\frac{1}{2}} P$ div F that

$$\begin{aligned}& |< f_k - f_l, u_k - u_l >| \\ &\quad = \ |< (J_k - J_l) P f_0, u_k - u_l >_\Omega \\ &\qquad\quad + < (J_k - J_l) A^{-\frac{1}{2}} P \operatorname{div} F, A^{\frac{1}{2}} (u_k - u_l) >_\Omega | \\ &\quad \le \ \|(J_k - J_l) P f_0\|_2 \, \|u_k - u_l\|_2 + \nu^{\frac{1}{2}} \|(J_k - J_l) \tilde{f}\|_2 \, \|\nabla (u_k - u_l)\|_2 .\end{aligned}$$

Instead of (2.3.8) we now obtain

$$\begin{aligned}\frac{1}{2} \|u_k - u_l\|_{2,\infty;t}^2 \ + \ \nu \|\nabla (u_k - u_l)\|_{2,2;t}^2 \\ &\le \ \|(J_k - J_l) u_0\|_2^2 \ + 2 \int_0^t \|(J_k - J_l) P f_0\|_2 \, \|u_k - u_l\|_2 \, d\tau \\ &\quad + 2\nu^{\frac{1}{2}} \int_0^t \|(J_k - J_l) \tilde{f}\|_2 \, \|\nabla (u_k - u_l)\|_2 \, d\tau.\end{aligned}$$

In the same way as above we get instead of (2.3.9) the following inequality, we may set $t = T$ on the left side of (2.3.9).

$$\frac{1}{2}\|u_k - u_l\|^2_{2,\infty;T} + \nu\|\nabla(u_k - u_l)\|^2_{2,2;T}$$
$$\leq 2\|(J_k - J_l)u_0\|_2^2 + 8\,\|(J_k - J_l)Pf_0\|^2_{2,1;T} + 4\|(J_k - J_l)\tilde{f}\|^2_{2,2;T}\,.$$

We know, see (3.4.8), II, that

$$\lim_{k,l\to\infty} \|(J_k - J_l)Pf_0(t)\|_2 = 0$$

for almost all $t \in [0, T)$. Further, see (3.4.6), II, we get

$$\|(J_k - J_l)Pf_0(t)\|_2 \leq 2\|Pf_0(t)\|_2 \leq 2\,\|f_0(t)\|_2.$$

Since $f_0 \in L^1(0, T; L^2(\Omega)^n)$, we may use Lebesgue's dominated convergence lemma and see that

$$\lim_{k,l\to\infty} \|(J_k - J_l)Pf_0\|^2_{2,1;T} = 0.$$

Similarly, using (2.1.8) we obtain

$$\|\tilde{f}(t)\|_2 = \|A^{-\frac{1}{2}}P \text{ div } F(t)\|_2 \leq \nu^{-\frac{1}{2}}\|F(t)\|_2$$

for almost all $t \in [0, T)$, therefore $\tilde{f} \in L^2(0, T; L^2(\Omega)^n)$, and

$$\lim_{k,l\to\infty} \|(J_k - J_l)\tilde{f}\|^2_{2,2;T} = 0.$$

This shows that

$$\lim_{k,l\to\infty} (\frac{1}{2}\,\|u_k - u_l\|^2_{2,\infty;T} + \nu\,\|\nabla(u_k - u_l)\|^2_{2,2;T}) = 0.$$

In particular we conclude that

$$\lim_{k\to\infty} \|u(t) - J_k u(t)\|_2 = 0$$

holds uniformly for all $t \in [0, T)$. Since each $J_k\, u$, $k \in \mathbb{N}$, is strongly continuous, see Lemma 2.2.1, we see that the limit function $u : [0, T) \to L^2_\sigma(\Omega)$ is also strongly continuous. We get $u_0 = \lim_{k\to\infty} u_k(0)$.

To prove equality (2.3.3) for fixed $t \in [0, T)$, we insert u_k, f_k, $J_k\, u_0$ in (2.3.6), use the convergence properties

$$\lim_{k\to\infty} < A^{\frac{1}{2}}u_k, A^{\frac{1}{2}}u_k >_\Omega = \lim_{k\to\infty} \|J_k A^{\frac{1}{2}}u\|_2^2 = \|A^{\frac{1}{2}}u\|_2^2\,,$$

$$\begin{aligned}\lim_{k\to\infty} < f_k, u_k >_\Omega &= \lim_{k\to\infty} (< f_0, J_k u >_\Omega + < A^{-\frac{1}{2}} P \operatorname{div} F, J_k^2 A^{\frac{1}{2}} u >_\Omega) \\ &= < f_0, u >_\Omega + < A^{-\frac{1}{2}} P \operatorname{div} F, A^{\frac{1}{2}} u >_\Omega \\ &= < f_0, u >_\Omega - < F, \nabla u >_\Omega ,\end{aligned}$$

observe (2.1.7), and apply Lebesgue's dominated convergence lemma. This enables us to let $k \to \infty$ in each term and we get the desired equality (2.3.3). This proves the theorem. □

2.4 Representation formula for weak solutions

The next important step in the theory of weak solutions u is to prove the explicit representation formula

$$\begin{aligned}u(t) &= S(t)u_0 + \int_0^t S(t-\tau) P f_0(\tau)\, d\tau \\ &\quad + A^{\frac{1}{2}} \int_0^t S(t-\tau) A^{-\frac{1}{2}} P \operatorname{div} F(\tau)\, d\tau ,\end{aligned} \tag{2.4.1}$$

$0 \leq t < T$, see below. This formula is basic for the functional analytic approach to the Navier-Stokes equations. It characterizes completely the weak solutions u in terms of the Stokes operator A.

See (1.5.7)–(1.5.10) concerning the operators $S(t) = e^{-tA}$, $t \geq 0$. Recall that $A^{-\frac{1}{2}} P \operatorname{div}$ is a bounded operator satisfying

$$\|A^{-\frac{1}{2}} P \operatorname{div} F(t)\|_2 \leq \nu^{-\frac{1}{2}} \|F(t)\|_2, \tag{2.4.2}$$

and that $A^{-\frac{1}{2}}$ and P have an extended meaning, see Lemma 2.6.1, III. In particular we see that $A^{-\frac{1}{2}} P \operatorname{div} F(t) \in L^2_\sigma(\Omega)$ is well defined in the theorem below, for almost all $t \in [0, T)$.

Using the integral operator $\mathcal{J}$, see Lemma 1.6.1, we can write (2.4.1) in the form

$$u(t) = S(t)u_0 + (\mathcal{J} P f_0)(t) + A^{\frac{1}{2}} (\mathcal{J} A^{-\frac{1}{2}} P \operatorname{div} F)(t),$$

$0 \leq t < T$. The next theorem shows that each weak solution u can be expressed in this way. Note that it is not possible, in general, to write $A^{\frac{1}{2}} \mathcal{J} A^{-\frac{1}{2}} P \operatorname{div} F = \mathcal{J} P \operatorname{div} F$.

2.4.1 Theorem *Let $\Omega \subseteq \mathbb{R}^n$, $n \geq 2$, be any domain, let $0 < T \leq \infty$, $1 < s < \infty$, $u_0 \in L^2_\sigma(\Omega)$, and let $f = f_0 + \operatorname{div} F$ with*

$$f_0 \in L^1(0, T; L^2(\Omega)^n) \;,\quad F \in L^s(0, T; L^2(\Omega)^{n^2}) \,. \tag{2.4.3}$$

Then $(\mathcal{J} A^{-\frac{1}{2}} P\mathrm{div}\ F)(t) \in D(A^{\frac{1}{2}})$ *for almost all* $t \in [0,T)$, *and the function* $u : [0,T) \to L^2_\sigma(\Omega)$, *well defined by*

$$u(t) := S(t)u_0 + (\mathcal{J} P f_0)(t) + A^{\frac{1}{2}}(\mathcal{J} A^{-\frac{1}{2}} P\mathrm{div}\ F)(t) \tag{2.4.4}$$

for almost all $t \in [0,T)$, *satisfies*

$$u \in L^1_{loc}([0,T); W^{1,2}_{0,\sigma}(\Omega)), \tag{2.4.5}$$

and is a weak solution of the Stokes system (2.1.3) *with data* f, u_0.

Conversely, each weak solution

$$u \in L^1_{loc}([0,T); W^{1,2}_{0,\sigma}(\Omega))$$

of the Stokes system (2.1.3) *with data* f, u_0 *satisfies the representation* (2.4.4) *for almost all* $t \in [0,T)$.

Proof. We need several steps.

a) First we show that

$$S(\cdot)u_0 \in L^2(0,T; W^{1,2}_{0,\sigma}(\Omega)) \subseteq L^1_{loc}([0,T); W^{1,2}_{0,\sigma}(\Omega)), \tag{2.4.6}$$

and that $S(\cdot)u_0$ is a weak solution of (2.1.3) with data $f = 0$ and u_0.

The property (2.4.6) follows from Lemma 1.5.3 with $s = 2$. From (1.5.24) we get

$$\|A^{\frac{1}{2}} S(\cdot)u_0\|_{2,2;\infty} = \nu^{\frac{1}{2}} \|\nabla S(\cdot)u_0\|_{2,2;\infty} \le \|u_0\|_2. \tag{2.4.7}$$

Next we use Lemma 2.2.1, a), with $D = C^\infty_{0,\sigma}(\Omega)$, $w \in D$, $\varphi \in C_0^\infty([0,T))$, and we use Theorem 1.5.2. We obtain

$$\begin{aligned}
\int_0^T \frac{d}{dt}(\varphi(t) < S(t)u_0, w >_\Omega)\, dt &= \int_0^T \frac{d}{dt}(\varphi(t) < u_0, S(t)w >_\Omega)\, dt \\
&= - < u_0, w >_\Omega \varphi(0) \\
&= \int_0^T < S(t)u_0, w >_\Omega \varphi'\, dt - \int_0^T < u_0, S(t)Aw >_\Omega \varphi(t)\, dt,
\end{aligned}$$

and therefore

$$\begin{aligned}
&- \int_0^T < S(t)u_0, w >_\Omega \varphi'(t)\, dt + \int_0^T < A^{\frac{1}{2}} S(t)u_0, A^{\frac{1}{2}} w >_\Omega \varphi\, dt \\
&= - \int_0^T < S(t)u_0, w > \varphi'(t)\, dt + \nu \int_0^T < \nabla S(t)u_0, \nabla w >_\Omega \varphi\, dt \\
&= < u_0, w >_\Omega \varphi(0).
\end{aligned}$$

Thus (2.2.2) is satisfied, and from Lemma 2.2.1, a), we see that $S(\cdot)u_0$ has the desired property.

b) Next we consider the part $\mathcal{J}Pf_0$ in (2.4.4). We show that

$$\begin{aligned} \mathcal{J}Pf_0 &\in L^\infty(0,T;L^2_\sigma(\Omega))\,, \\ \nabla\mathcal{J}Pf_0 &\in L^2(0,T;L^2(\Omega)^{n^2}), \\ \mathcal{J}Pf_0 &\in L^1_{loc}([0,T);W^{1,2}_{0,\sigma}(\Omega))\,, \end{aligned}$$

$$\frac{1}{2}\,\|\mathcal{J}Pf_0\|^2_{2,\infty;T} + \nu\,\|\nabla\mathcal{J}Pf_0\|^2_{2,2;T} \;\le\; 8\,\|f_0\|^2_{2,1;T}\,, \tag{2.4.8}$$

and that $\mathcal{J}Pf_0$ is a weak solution of (2.1.3) with data f_0 and $u_0 = 0$.

First we assume that $0 < T < \infty$, and we suppose the (stronger) condition $f_0 \in L^s(0,T;L^2(\Omega)^n)$. We use Lemma 2.2.1 with $D = W^{1,2}_{0,\sigma}(\Omega)$, $w \in D$, $\varphi \in C_0^\infty([0,T))$, and we use the properties in Theorem 1.6.3. We obtain

$$\begin{aligned} 0 &= \int_0^T \frac{d}{dt}\,(\varphi < \mathcal{J}Pf_0, w >_\Omega)\,dt \\ &= \int_0^T \varphi' < \mathcal{J}Pf_0, w >_\Omega\,dt \;-\; \int_0^T \varphi < A\mathcal{J}Pf_0, w >_\Omega\,dt \\ &\quad + \int_0^T \varphi < Pf_0, w >_\Omega\,dt\,, \end{aligned}$$

and using

$$< A\mathcal{J}Pf_0, w >_\Omega = < A^{\frac{1}{2}}\mathcal{J}Pf_0, A^{\frac{1}{2}}w >_\Omega = \nu < \nabla\mathcal{J}Pf_0, \nabla w >_\Omega$$

it follows that

$$\begin{aligned} -\int_0^T < \mathcal{J}Pf_0, w >_\Omega \varphi'\,dt + \nu\int_0^T < \nabla\mathcal{J}Pf_0, \nabla w >_\Omega \varphi dt \\ = \int_0^T < f_0, w >_\Omega \varphi\,dt. \end{aligned} \tag{2.4.9}$$

This shows with Lemma 2.2.1, a), that $\mathcal{J}Pf_0$ is a weak solution with data f_0 and $u_0 = 0$. From (1.6.24) we get in particular that

$$\mathcal{J}Pf_0 \in W^{1,s}(0,T;L^2_\sigma(\Omega))\;,\quad A\mathcal{J}Pf_0 \in L^s(0,T;L^2_\sigma(\Omega))$$

with $0 < T < \infty$. From Lemma 1.6.1 we know that $\mathcal{J}Pf_0$ is strongly continuous with $(\mathcal{J}Pf_0)(0) = 0$, and that

$$\|\mathcal{J}Pf_0\|_{2,\infty;T} \;\le\; \|f_0\|_{2,1;T}\,, \tag{2.4.10}$$

see (1.6.4).

In the case $f_0 \in L^1(0,T;L^2(\Omega)^n)$, $0 < T < \infty$, we find a sequence $(f_j)_{j=1}^\infty$ in $L^s(0,T;L^2(\Omega)^n)$ such that

$$\lim_{j\to\infty} \|f_0 - f_j\|_{2,1;T} = 0.$$

To construct this sequence, we use the mollification method, see Lemma 1.7.1, II. Then $\mathcal{J}Pf_j$ is a weak solution with data $f_j, u_0 = 0$, and using (2.4.10) with f_0 replaced by $f_j - f_0$, we see that

$$(\mathcal{J}Pf_0)(t) = \lim_{j\to\infty} (\mathcal{J}Pf_j)(t) \quad , \quad t \in [0,T).$$

Using (2.3.4) we see that

$$\nu\,\|\nabla(\mathcal{J}Pf_0 - \mathcal{J}Pf_j)\|_{2,2;T}^2 \;\le\; 8\,\|f_0 - f_j\|_{2,1;T}^2.$$

This shows that $\nabla\mathcal{J}Pf_0 \in L^2(0,T;L^2(\Omega)^{n^2})$, and (2.4.8) is a consequence of (2.3.4). Using (2.4.9) with f_0 replaced by f_j and letting $j \to \infty$, we see that $\mathcal{J}Pf_0$ is a weak solution with data f_0 and $u_0 = 0$. If $T = \infty$, this result holds with T replaced by T', $0 < T' < \infty$. In particular, (2.4.8) and (2.4.9) hold with T replaced by T'. Letting $T' \to \infty$, we obtain the desired result for $T = \infty$. Therefore, (2.4.9) also holds in this case.

c) Finally we consider the last part in (2.4.4) which has the form $A^{\frac12}\mathcal{J}\tilde f$ with $\tilde f := A^{-\frac12}P \operatorname{div} F \in L^s(0,T;L^2_\sigma(\Omega))$, see (2.4.2). From above we know that $\mathcal{J}\tilde f$ is a weak solution with data $\tilde f$ and $u_0 = 0$. Therefore, we may use (2.4.9) with $\mathcal{J}Pf_0, f_0$ replaced by $\mathcal{J}\tilde f$, $\tilde f$ for all $w \in W^{1,2}_{0,\sigma}(\Omega)$, $\varphi \in C_0^\infty([0,T))$. In particular, we may set $w = A^{\frac12}h$ with $h \in C^\infty_{0,\sigma}(\Omega)$. Further we use (2.1.6). This yields

$$-\int_0^T <\mathcal{J}\tilde f, A^{\frac12}h>_\Omega \varphi'\,dt + \int_0^T < A^{\frac12}\mathcal{J}\tilde f, A^{\frac12}A^{\frac12}h>_\Omega \varphi\,dt = \int_0^T <\tilde f, A^{\frac12}h>_\Omega \varphi\,dt. \tag{2.4.11}$$

Next we show that $\tilde u := A^{\frac12}\mathcal{J}\tilde f$ is a weak solution with data $\operatorname{div} F$ and $u_0 = 0$. From Theorem 1.6.3, (1.6.24), we obtain that $(\mathcal{J}\tilde f)(t) \in D(A)$ and therefore that $\tilde u(t) \in D(A^{\frac12})$ for almost all $t \in [0,T)$. Further we get

$$\|\nabla\tilde u\|_{2,s;T} \;=\; \nu^{-\frac12}\,\|A^{\frac12}\tilde u\|_{2,s;T} \;=\; \nu^{-\frac12}\,\|A\mathcal{J}\tilde f\|_{2,s;T} \;\le\; C\,\nu^{-\frac12}\|\tilde f\|_{2,s;T}$$

with $C = C(s) > 0$. Using (2.4.2) we obtain

$$\|\nabla\tilde u\|_{2,s;T} \;\le\; C\,\nu^{-1}\|F\|_{2,s;T}\,.$$

For each finite T' with $0 < T' \leq T$, we see that

$$\tilde{u} \in L^s(0,T'; L^2_\sigma(\Omega)) \ , \quad \nabla \tilde{u} \in L^s(0,T'; L^2(\Omega)^{n^2}) \,, \tag{2.4.12}$$

which shows that $\tilde{u} \in L^1_{loc}([0,T); W^{1,2}_{0,\sigma}(\Omega))$.

Using

$$< \tilde{u}, A^{\frac{1}{2}} A^{\frac{1}{2}} h >_\Omega = < A^{\frac{1}{2}} \tilde{u}, A^{\frac{1}{2}} h >_\Omega = \ \nu < \nabla \tilde{u}, \nabla h >_\Omega \,,$$

and

$$< \tilde{f}, A^{\frac{1}{2}} h >_\Omega = < A^{-\frac{1}{2}} P \ \mathrm{div}\ F, A^{\frac{1}{2}} h >_\Omega = \ - < F, \nabla h >_\Omega \,,$$

see (2.1.7), we obtain from (2.4.11) that

$$- \int_0^T < \tilde{u}, h >_\Omega \varphi' \, dt + \nu \int_0^T < \nabla \tilde{u}, \nabla h >_\Omega \varphi \, dt \ = \ - \int_0^T < F, \nabla h > \varphi \, dt$$

for all $h \in C^\infty_{0,\sigma}(\Omega)$, $\varphi \in C^\infty_0([0,T))$. This shows that $\tilde{u}$ is a weak solution with data $\mathrm{div}\ F$ and $u_0 = 0$.

We conclude from the steps a), b) and c) above that $S(\cdot)u_0 + \mathcal{J} P f_0 + A^{\frac{1}{2}} \mathcal{J} A^{-\frac{1}{2}} P \ \mathrm{div}\ F$ is a weak solution with data f, u_0.

Let $\hat{u} \in L^1_{loc}([0,T); W^{1,2}_{0,\sigma}(\Omega))$ be any weak solution with data f, u_0. Then $\hat{u} - u$ with u from (2.4.4) is a weak solution with $f = 0$, $u_0 = 0$. From Theorem 2.3.1, (2.3.4), we deduce that $u - \hat{u} = 0$ and $u = \hat{u}$. This completes the proof.

□

The following lemma yields uniqueness and continuity properties of weak solutions.

2.4.2 Lemma *Consider $\Omega \subseteq \mathbb{R}^n$, $0 < T \leq \infty$, $1 < s < \infty$, $u_0 \in L^2_\sigma(\Omega)$, and $f = f_0 + \mathrm{div}\ F$ as in Theorem 2.4.1, and let $u : [0,T) \to L^2_\sigma(\Omega)$ be defined by*

$$u(t) = S(t) u_0 + (\mathcal{J} P f_0)(t) + A^{\frac{1}{2}} (\mathcal{J} A^{-\frac{1}{2}} P \,\mathrm{div}\ F)(t) \tag{2.4.13}$$

for almost all $t \in [0,T)$.

Then we have:

a) *u is the only weak solution of the Stokes system (2.1.3) with data f, u_0 within the space $L^1_{loc}([0,T); W^{1,2}_{0,\sigma}(\Omega))$.*

b) *Under the additional assumption*

$$u \in L^\infty(0,T; L^2_\sigma(\Omega)), \tag{2.4.14}$$

$u : [0,T) \to L^2_\sigma(\Omega)$ is weakly continuous after a redefinition on a null set of $[0,T)$, and $u(0) = u_0$.

c) *If $u : [0,T) \to L^2_\sigma(\Omega)$ is weakly continuous, then, after a redefinition on a null set of $[0,T)$, we get*

$$(\mathcal{J} A^{-\frac{1}{2}} P \operatorname{div} F)(t) \in D(A^{\frac{1}{2}})$$

and

$$u(t) = S(t)u_0 + (\mathcal{J} P f_0)(t) + A^{\frac{1}{2}} (\mathcal{J} A^{-\frac{1}{2}} P \operatorname{div} F)(t) \tag{2.4.15}$$

for all $t \in [0,T)$.

d) *If $F \in L^2(0,T; L^2(\Omega)^{n^2})$, then $u : [0,T) \to L^2_\sigma(\Omega)$ is strongly continuous, after a redefinition on a null set of $[0,T)$, we obtain*

$$\frac{1}{2} \|u\|^2_{2,\infty;T} + \nu \|\nabla u\|^2_{2,2;T} \le 2 \|u_0\|^2_2 + 8 \|f_0\|^2_{2,1;T} + 4\nu^{-1} \|F\|^2_{2,2;T} \tag{2.4.16}$$

and the energy equality

$$\begin{aligned} \frac{1}{2} \|u(t)\|^2_2 + \nu \int_0^t \|\nabla u\|^2_2 \, d\tau &= \frac{1}{2} \|u_0\|^2_2 + \int_0^t < f_0, u >_\Omega \, d\tau \\ &\quad - \int_0^t < F, \nabla u >_\Omega \, d\tau \end{aligned} \tag{2.4.17}$$

for all $t \in [0,T]$.

Proof. To prove a) let $\tilde{u} \in L^1_{loc}([0,T); W^{1,2}_{0,\sigma}(\Omega))$ be another weak solution of (2.1.3) with data f, u_0. Then from (2.4.5) we know that

$$u \in L^1_{loc}([0,T); W^{1,2}_{0,\sigma}(\Omega)) .$$

Therefore we get $u - \tilde{u} \in L^1_{loc}([0,T); W^{1,2}_{0,\sigma}(\Omega))$, and $u - \tilde{u}$ is a weak solution of (2.1.3) with data $f = 0$, $u_0 = 0$. Then inequality (2.3.4) holds with u replaced by $u - \tilde{u}$, and with $u_0 = 0$, $f_0 = 0$, $F = 0$. This shows that $u = \tilde{u}$.

To prove b) we suppose (2.4.14). From Lemma 1.5.1 and Lemma 1.6.1 we know that $S(\cdot)u_0$ and $\mathcal{J} P f_0$ are strongly continuous with $S(0)u_0 = u_0$ and $(\mathcal{J} P f_0)(0) = 0$. Therefore, $S(\cdot)u_0$, $\mathcal{J} P f_0$ are also weakly continuous. It remains to show the weak continuity of $A^{\frac{1}{2}} \mathcal{J} A^{-\frac{1}{2}} P \operatorname{div} F$.

Let $\tilde{f} := A^{-\frac{1}{2}} P \operatorname{div} F$. From (1.5.14) we get $S(\cdot)u_0 \in L^\infty(0,T; L^2_\sigma(\Omega))$, and from (1.6.4) we obtain that

$$\mathcal{J} P f_0 \in L^\infty(0,T; L^2_\sigma(\Omega))$$

since $f_0 \in L^1(0,T;L^2(\Omega)^n)$. Therefore, using (2.4.13) and (2.4.14) we see that

$$A^{\frac{1}{2}}\mathcal{J}\tilde{f} \in L^\infty(0,T;L^2_\sigma(\Omega)),$$

and we find a constant $C > 0$, and a null set $N \subseteq [0,T)$ such that

$$\sup_{t\in[0,T)\backslash N} \|A^{\frac{1}{2}}(\mathcal{J}\tilde{f})(t)\|_2 \leq C. \tag{2.4.18}$$

Let $h \in L^2_\sigma(\Omega)$. Since $J_k h \in D(A^{\frac{1}{2}})$, see (2.1.9), and since $\mathcal{J}\tilde{f}$ is strongly continuous, the function g_k defined on $[0,T)\backslash N$ by

$$g_k(t) := < A^{\frac{1}{2}}(\mathcal{J}\tilde{f})(t), J_k h >_\Omega = < (\mathcal{J}\tilde{f})(t), A^{\frac{1}{2}} J_k h >_\Omega ,$$

$t \in [0,T)\backslash N$, $k \in \mathbb{N}$, is continuous. We have

$$\begin{aligned} |g_k(t)| &\leq \|A^{\frac{1}{2}}(\mathcal{J}\tilde{f})(t)\|_2 \, \|J_k h\|_2 \leq C\,\|h\|_2 , \\ |g_k(t) - g_l(t)| &\leq C\,\|(J_k - J_l)h\|_2, \end{aligned}$$

and $\lim_{k\to\infty} J_k h = h$ in $L^2_\sigma(\Omega)$. This shows that $(g_k)_{k=1}^\infty$ converges uniformly on $[0,T)\backslash N$ to the function g defined by $g(t) := < A^{\frac{1}{2}}(\mathcal{J}\tilde{f})(t), h >_\Omega$. Thus we see,

$$t \mapsto < A^{\frac{1}{2}}(\mathcal{J}\tilde{f})(t), h >_\Omega \ , \quad t \in [0,T)\backslash N$$

is continuous on $[0,T)\backslash N$ for each $h \in L^2_\sigma(\Omega)$. This means that

$$A^{\frac{1}{2}}\mathcal{J}\tilde{f} : [0,T)\backslash N \to L^2_\sigma(\Omega) \tag{2.4.19}$$

is weakly continuous. Because of (2.4.18) and since the set $[0,T)\backslash N$ is dense in $[0,T)$, the function (2.4.19) has a unique weakly continuous extension from $[0,T)\backslash N$ to the whole interval $[0,T)$, with the same norm bound C from (2.4.18). Consider any $t_0 \in N$, and let $(t_j)_{j=1}^\infty$ be a sequence in $[0,T)\backslash N$ such that $t_0 = \lim_{j\to\infty} t_j$. Then we get

$$\lim_{j\to\infty} \|(\mathcal{J}\tilde{f})(t_0) - (\mathcal{J}\tilde{f})(t_j)\|_2 = 0 \ , \quad \sup_j \|A^{\frac{1}{2}}(\mathcal{J}\tilde{f})(t_j)\|_2 \leq C$$

and $A^{\frac{1}{2}}(\mathcal{J}\tilde{f})(t_j)$ converges weakly. Since $A^{\frac{1}{2}}$ is a closed operator, its graph is strongly and also weakly closed. Therefore, we get $(\mathcal{J}\tilde{f})(t_0) \in D(A^{\frac{1}{2}})$, and $A^{\frac{1}{2}}(\mathcal{J}\tilde{f})(t_0)$ is the value obtained by the extension of (2.4.19) to t_0. Thus we get $(\mathcal{J}\tilde{f})(t) \in D(A^{\frac{1}{2}})$ for all $t \in [0,T)$, $t \mapsto A^{\frac{1}{2}}(\mathcal{J}\tilde{f})(t)$ is well defined for all $t \in [0,T)$. We can redefine u on a null set of $[0,T)$, so that (2.4.13) holds for all $t \in [0,T)$ and that u is weakly continuous on $[0,T)$. Since $(\mathcal{J}Pf_0)(0) = 0$, $(A^{\frac{1}{2}}\mathcal{J}\tilde{f})(0) = 0$, we get $u(0) = u_0$.

This proves b). To prove c) we observe that each term of (2.4.13) is (at least) weakly continuous after a corresponding redefinition. The same argument as for b) now shows that $(\mathcal{J}\tilde{f})(t) \in D(A^{\frac{1}{2}})$ and (2.4.15) hold for all $t \in [0,T)$. To prove d) we use Theorem 2.3.1 and obtain the strong continuity of u, the inequality (2.4.16) and the energy equality (2.4.17). This proves the lemma. □

2.5 Basic estimates of weak solutions

The representation formula in the preceding subsection enables us to reduce the theory of weak solutions of the Stokes system

$$\begin{aligned} u_t - \nu\Delta u + \nabla p &= f \;,\quad \operatorname{div} u = 0, \\ u|_{\partial\Omega} &= 0 \;,\quad u(0) = u_0 \end{aligned} \tag{2.5.1}$$

completely to the theory of the evolution system

$$u_t + Au = f \;,\quad u(0) = u_0, \tag{2.5.2}$$

developed in Subsections 1.5 and 1.6. This theory rests only on properties of the operators

$$A,\; A^\alpha \text{ with } -1 \le \alpha \le 1 \;,\quad S(t) = e^{-tA} \text{ with } t \ge 0$$

where A means the Stokes operator.

It is important to recall that the operators P and $A^{-\frac{1}{2}}$ have an extended meaning, see Lemma 2.6.1, III. See also (2.5.26), III, and (2.5.18), III.

The representation formula (2.4.4) can be written in the form

$$u = S(\cdot)u_0 + \mathcal{J}Pf_0 + A^{\frac{1}{2}}\mathcal{J}A^{-\frac{1}{2}}P \operatorname{div} F, \tag{2.5.3}$$

$A^{-\frac{1}{2}}P$ div means the operator in Lemma 2.6.1, III, and $f = f_0 + \operatorname{div} F$.

In the following we may treat these three parts separately. The first part $S(\cdot)u_0$ is a weak solution of (2.5.1) with data $f = 0$ and u_0, the second part $\mathcal{J}Pf_0$ is a weak solution with data f_0 and $u_0 = 0$, and the last part $A^{\frac{1}{2}}\mathcal{J}A^{-\frac{1}{2}}P$ div F is a weak solution with data div F and $u_0 = 0$.

The first theorem below concerns the part $S(\cdot)u_0$.

To simplify the following formulations we will write

$$\|u\|_{q,s;T} = \left(\int_0^T \|u(t)\|_q^s \, dt\right)^{\frac{1}{s}} < \infty,$$

$1 \leq q < \infty$, $1 \leq s < \infty$, iff $u \in L^s(0,T; L^q(\Omega)^n)$. Similarly,

$$\|A^\alpha u\|_{2,s;T} = \left(\int_0^T \|A^\alpha u(t)\|_2^s \, dt \right)^{\frac{1}{s}} < \infty \ , \quad -1 \leq \alpha \leq 1$$

means that $u(t) \in D(A^\alpha)$ for almost all $t \in [0,T)$ and that

$$A^\alpha u \in L^s(0,T; L^2_\sigma(\Omega)) .$$

Further,

$$\|u'\|_{2,s;T} = \|\frac{d}{dt} u\|_{2,s;T} = \left(\int_0^T \|u'(t)\|_2^s \, dt \right)^{\frac{1}{s}} < \infty$$

means in the case $0 < T < \infty$ that $u \in W^{1,s}(0,T; L^2_\sigma(\Omega))$, see Section 1.3. However,

$$\|u'\|_{2,s;\infty} = \|\frac{d}{dt} u\|_{2,s;\infty} = \left(\int_0^\infty \|u'(t)\|_2^s \, dt \right)^{\frac{1}{s}} < \infty$$

means only that $u' \in L^s(0,\infty; L^2_\sigma(\Omega))$ and $u \in L^s_{loc}([0,\infty); L^2_\sigma(\Omega))$, i.e., $u \in L^s(0,T'; L^2_\sigma(\Omega))$ for $0 < T' < \infty$, see (1.6.26). See Subsection 1.2 for these spaces.

The restriction $\frac{1}{2} + \frac{1}{\rho} \geq \frac{1}{s}$ in the next theorem concerns only the case $1 < s < 2$. It is used to simplify the assumptions in this critical case. This condition means that ρ cannot be "too large" if $1 < s < 2$.

2.5.1 Theorem *Let $\Omega \subseteq \mathbb{R}^n$, $n \geq 2$, be any domain, let $0 < T \leq \infty$, $1 < s \leq \rho < \infty$, $\frac{1}{2} + \frac{1}{\rho} \geq \frac{1}{s}$ and let*

$$u_0 \in D(A^{1-\frac{1}{s}}) \ \textit{if} \ s \geq 2 \ , \quad u_0 \in D(A^{\frac{1}{2}}) = W^{1,2}_{0,\sigma}(\Omega) \ \textit{if} \ 1 < s < 2.$$

Let

$$u \in L^1_{loc}([0,T); W^{1,2}_{0,\sigma}(\Omega))$$

be a weak solution of the Stokes system (2.5.1) with data $f = 0$ and u_0.

Then, after redefining u on a null set of $[0,T)$, $u : [0,T) \to L^2_\sigma(\Omega)$ is strongly continuous, $u(0) = u_0$, and

$$u(t) = S(t)u_0 \tag{2.5.4}$$

for all $t \in [0,T)$.

Moreover, u has the following properties:

a) $$\|u'\|_{2,s;T} + \|Au\|_{2,s;T} \leq \begin{cases} 2\,\|A^{1-\frac{1}{s}}u_0\|_2 & \text{if } s \geq 2, \\ C\,(\|u_0\|_2 + \|A^{\frac{1}{2}}u_0\|_2) & \text{if } 1 < s < 2 \end{cases} \tag{2.5.5}$$

with $C = C(s) > 0$, $u' + Au = 0$ in $L^s(0,T;L^2_\sigma(\Omega))$ and therefore

$$u'(t) + Au(t) = 0 \tag{2.5.6}$$

for almost all $t \in [0,T)$.

b) $$\|A^\alpha u\|_{2,\rho;T} \leq \begin{cases} \|A^{1-\frac{1}{s}}u_0\|_2 & \text{if } s \geq 2, \\ C\,(\|u_0\|_2 + \|A^{\frac{1}{2}}u_0\|_2) & \text{if } 1 < s < 2 \end{cases} \tag{2.5.7}$$

with $\alpha = 1 + \frac{1}{\rho} - \frac{1}{s}$, $C = C(\alpha,\rho) > 0$.

c) $$\|u\|_{q,\rho;T} \leq \begin{cases} C\,\nu^{-\alpha}\,\|A^{1-\frac{1}{s}}u_0\|_2 & \text{if } s \geq 2, \\ C\,\nu^{-\alpha}(\|u_0\|_2 + \|A^{\frac{1}{2}}u_0\|_2) & \text{if } 1 < s < 2 \end{cases} \tag{2.5.8}$$

with $\alpha := 1 + \frac{1}{\rho} - \frac{1}{s}$, $C = C(\alpha,\rho,q) > 0$, and with $2 \leq q \leq \infty$ determined by

$$2 + \frac{n}{q} + \frac{2}{\rho} = \frac{n}{2} + \frac{2}{s}\,, \qquad \frac{1}{q} \geq \frac{1}{2} - \frac{1}{n}\,.$$

d) $$\|A^{\frac{1}{2}}u\|_{q,\rho;T} \leq \begin{cases} C\,\nu^{-\alpha}\,\|A^{1-\frac{1}{s}}u_0\|_2 & \text{if } s \geq 2, \\ C\,\nu^{-\alpha}(\|u_0\|_2 + \|A^{\frac{1}{2}}u_0\|_2) & \text{if } 1 < s < 2 \end{cases} \tag{2.5.9}$$

with $\alpha := \frac{1}{2} + \frac{1}{\rho} - \frac{1}{s}$, $C = C(\alpha,\rho,q) > 0$, and with $2 \leq q < \infty$ determined by

$$1 + \frac{n}{q} + \frac{2}{\rho} = \frac{n}{2} + \frac{2}{s}\,.$$

e) $$\frac{1}{2}\,\|u\|^2_{2,\infty;T} + \nu\,\|\nabla u\|^2_{2,2;T} \leq 2\,\|u_0\|_2^2\,. \tag{2.5.10}$$

Proof. Theorem 2.4.1 yields the representation $u(t) = S(t)u_0$ for almost all $t \in [0,T)$. Redefining u leads to (2.5.4). Now Theorem 1.5.2 with (1.5.22) yields (2.5.5) if $s \geq 2$. If $1 < s < 2$, we use (1.5.23) with $\varepsilon > 0$ so that $0 < \varepsilon < \frac{1}{s} - \frac{1}{2}$. Since $1 - \frac{1}{s} + \varepsilon < \frac{1}{2}$ we may apply to (1.5.23) the interpolation inequality (2.2.8), III, and Young's inequality (3.3.8), I. Setting $\delta := 1 - \frac{1}{s} + \varepsilon$ we can choose $0 < \beta < 1$ with $\delta = \beta\frac{1}{2}$ and obtain

$$\|A^\delta u_0\|_2 = \|A^{\frac{1}{2}\beta}u_0\|_2 \leq \|A^{\frac{1}{2}}u_0\|_2^\beta\,\|u_0\|_2^{1-\beta} \leq \|A^{\frac{1}{2}}u_0\|_2 + \|u_0\|_2\,.$$

Therefore, (1.5.23) implies (2.5.5) if $1 < s < 2$. Theorem 1.5.2 shows that u is strongly continuous, that $u(0) = u_0$, and that $u' + Au = 0$. This proves a).

To prove b) we use Lemma 1.5.3, (1.5.24), and get with $2 \leq s \leq \rho$ that

$$\begin{aligned} \|A^\alpha u\|_{2,\rho;T} &= \|A^{1+\frac{1}{\rho}-\frac{1}{s}} S(\cdot)u_0\|_{2,\rho;T} = \|A^{\frac{1}{\rho}} S(\cdot)A^{1-\frac{1}{s}} u_0\|_{2,\rho;T} \\ &\leq \|A^{1-\frac{1}{s}} u_0\|_2. \end{aligned}$$

In the case $1 < s < 2$, we use Lemma 1.5.4 with $0 \leq \frac{1}{2} + \frac{1}{\rho} - \frac{1}{s} < \frac{1}{\rho} < 1 + \frac{1}{\rho} - \frac{1}{s}$, and get from (1.5.28) that

$$\begin{aligned} \|A^\alpha u\|_{2,\rho;T} &= \|A^{1+\frac{1}{\rho}-\frac{1}{s}} S(\cdot)u_0\|_{2,\rho;T} = \|A^{\frac{1}{2}+\frac{1}{\rho}-\frac{1}{s}} S(\cdot)A^{\frac{1}{2}} u_0\|_{2,\rho;T} \\ &\leq C\,(\|u_0\|_2 + \|A^{\frac{1}{2}} u_0\|_2) \end{aligned}$$

with $C = C(\alpha, \rho) > 0$. This proves b).

To prove c), we use the embedding inequality (2.4.6), III, and Lemma 2.4.2 with $2\alpha + \frac{n}{q} = \frac{n}{2}$, $\alpha = 1 + \frac{1}{\rho} - \frac{1}{s}$, $0 < \alpha \leq \frac{1}{2}(\frac{n}{2} - \frac{n}{q}) \leq \frac{1}{2}$, $\frac{1}{q} \geq \frac{1}{2} - \frac{1}{n}$. This yields the estimate

$$\|u\|_{q,\rho;T} \leq C\nu^{-\alpha}\, \|A^\alpha u\|_{2,\rho;T} \tag{2.5.11}$$

with $C = C(\alpha, \rho, q) > 0$, and (2.5.8) follows from (2.5.7).

To prove d), we use the embedding inequality (2.4.6), III, now with $2\alpha + \frac{n}{q} = \frac{n}{2}$, $\alpha = \frac{1}{2} + \frac{1}{\rho} - \frac{1}{s}$, $0 \leq \alpha = \frac{1}{2}(\frac{n}{2} - \frac{n}{q}) = \frac{1}{2} - (\frac{1}{s} - \frac{1}{\rho}) \leq \frac{1}{2}$. This yields

$$\|A^{\frac{1}{2}} u\|_{q,\rho;T} \leq C\nu^{-\alpha} \|A^{\alpha+\frac{1}{2}} u\|_{2,\rho;T} = C\nu^{-\alpha} \|A^{1+\frac{1}{\rho}-\frac{1}{s}} u\|_{2,\rho;T}$$

with $C = C(\alpha, \rho, q) > 0$, and (2.5.9) follows from (2.5.7).

Property e) is a consequence of the energy estimate (2.3.4) in Theorem 2.3.1. The proof of the theorem is complete. □

In the next theorem we investigate the second part $\mathcal{J}Pf_0$ of the representation formula (2.5.3). Here we write $f = f_0$.

2.5.2 Theorem *Let $\Omega \subseteq \mathbb{R}^n$, $n \geq 2$, be any domain, let $0 < T \leq \infty$, $1 \leq s < \infty$, and let $f \in L^s(0, T; L^2(\Omega)^n)$.*

Suppose

$$u \in L^1_{loc}([0, T); W^{1,2}_{0,\sigma}(\Omega))$$

is a weak solution of the Stokes system (2.5.1) with data f and $u_0 = 0$.

Then, after redefining on a null set of $[0, T)$, we get

$$u(t) = (\mathcal{J}Pf)(t) = \int_0^t S(t - \tau)Pf(\tau)\, d\tau \tag{2.5.12}$$

for all $t \in [0, T)$, $u : [0, T) \to L^2_\sigma(\Omega)$ is strongly continuous and $u(0) = 0$.

Moreover, u has the following properties:

a)
$$\|u'\|_{2,s;T} + \|Au\|_{2,s;T} \le C\,\|f\|_{2,s;T} \tag{2.5.13}$$

with $1 < s < \infty$, $C = C(s) > 0$, *and*

$$u'(t) + Au(t) = Pf(t) \tag{2.5.14}$$

holds for almost all $t \in [0,T)$.

b)
$$\|A^\alpha u\|_{2,\rho;T} \le C\,\|f\|_{2,s;T} \tag{2.5.15}$$

with $\alpha := 1 + \frac{1}{\rho} - \frac{1}{s}$, $1 < s \le \rho < \infty$, $C = C(\alpha,\rho) > 0$.

c)
$$\|u\|_{q,\rho;T} \le C\,\nu^{-\alpha}\,\|f\|_{2,s;T} \tag{2.5.16}$$

with $\alpha = 1 + \frac{1}{\rho} - \frac{1}{s}$, $1 < s \le \rho < \infty$, $C = C(\alpha,\rho,q) > 0$, *and with* $2 \le q < \infty$ *determined by the conditions*

$$2 + \frac{n}{q} + \frac{2}{\rho} = \frac{n}{2} + \frac{2}{s}\,, \quad \frac{1}{q} \ge \frac{1}{2} - \frac{1}{n}\,.$$

d)
$$\frac{1}{2}\,\|u\|^2_{2,\infty;T} + \nu\,\|\nabla u\|^2_{2,2;T} \le 8\,\|f\|^2_{2,1;T} \tag{2.5.17}$$

if $s = 1$.

e)
$$\|u'\|^2_{2,2;T} + \frac{1}{2}\,\|A^{\frac{1}{2}}u\|^2_{2,\infty;T} + \|Au\|^2_{2,2;T} \le 14\,\|f\|^2_{2,2;T} \tag{2.5.18}$$

if $s = 2$.

Proof. The representation formula (2.4.4) implies (2.5.12) for almost all $t \in [0,T)$. Here we apply Theorem 2.4.1 to finite intervals $[0,T')$, $0 < T' \le T$. After redefining of u on a null set, (2.5.12) holds for all $t \in [0,T)$.

If $1 < s < \infty$, the properties in a) are a consequence of Theorem 1.6.3. Further properties are given in b) and c) of this theorem. In particular, u is strongly continuous, and $u(0) = 0$.

Inequality (2.5.15) is a consequence of Lemma 1.6.4, (1.6.31).

To prove c) we use the embedding inequality (2.4.6), III, and obtain with

$$\alpha = 1 + \frac{1}{\rho} - \frac{1}{s}\,, \quad 2\alpha + \frac{n}{q} = \frac{n}{2}\,, \quad 0 \le \alpha = \frac{1}{2}(\frac{n}{2} - \frac{n}{q}) \le \frac{1}{2}$$

that

$$\|u\|_{q,\rho;T} \le C\,\nu^{-\alpha}\,\|A^\alpha u\|_{2,\rho;T}, \tag{2.5.19}$$

where $C = C(\alpha,\rho,q) > 0$. Inequality (2.5.16) now follows from (2.5.15).

Inequality (2.5.17) is a consequence of the energy estimate (2.3.4), Theorem 2.3.1.

To prove e), we first show that Pf can be written in the form $Pf = A^{-\frac{1}{2}}P$ div F with some $F \in L^2(0,T; L^2(\Omega)^{n^2})$ satisfying

$$\|F\|_{2,2;T} \leq \nu^{\frac{1}{2}} \|f\|_{2,2;T} .$$

To prove this property we use the Hahn-Banach theorem, see [Yos80, IV]. Since

$$|< Pf, A^{\frac{1}{2}}v >_{\Omega,T}| \leq \|f\|_{2,2;T} \|A^{\frac{1}{2}}v\|_{2,2;T} = \nu^{\frac{1}{2}} \|f\|_{2,2;T} \|\nabla v\|_{2,2;T}$$

for all $v \in L^2(0,T; W^{1,2}_{0,\sigma}(\Omega))$, this theorem yields some $F \in L^2(0,T; L^2(\Omega)^{n^2})$ satisfying the estimate above and the relation

$$< Pf, A^{\frac{1}{2}}v >_{\Omega,T} = - < F, \nabla v >_{\Omega,T} ,$$

see Section 1.6, II, for similar arguments. The relation (2.6.5), III, shows that

$$- < F, \nabla v >_{\Omega,T} = < A^{-\frac{1}{2}}P\text{div } F, A^{\frac{1}{2}}v >_{\Omega,T}$$

and therefore that $Pf = A^{-\frac{1}{2}}P$ div F.

Inserting $Pf = A^{-\frac{1}{2}}P$ div F in (2.5.12), and using Theorem 2.4.1, we see that $A^{\frac{1}{2}}u$ is a weak solution of (2.5.1) with data div F and $u_0 = 0$. We conclude that $A^{\frac{1}{2}}u = A^{\frac{1}{2}}\mathcal{J}A^{-\frac{1}{2}}P$ div F has the form of the last term in (2.4.4). Therefore, $A^{\frac{1}{2}}u$ has the same properties as u in the estimate (2.3.4) for the case $u_0 = 0,\ f_0 = 0$. Thus we apply (2.3.4) with u replaced by $A^{\frac{1}{2}}u$, use

$$\nu \|\nabla u\|^2_{2,2;T} = \|A^{\frac{1}{2}}u\|^2_{2,2;T}$$

and obtain the estimate

$$\frac{1}{2} \|A^{\frac{1}{2}}u\|^2_{2,\infty;T} + \|Au\|^2_{2,2;T} \leq 4\nu^{-1}\|F\|^2_{2,2;T} \leq 4 \|f\|^2_{2,2;T} .$$

Using (2.5.14) we get

$$\begin{aligned} \|u'\|^2_{2,2;T} &\leq (\|f\|_{2,2;T} + \|Au\|_{2,2;T})^2 \\ &\leq 2(\|f\|^2_{2,2;T} + \|Au\|^2_{2,2;T}) \\ &\leq 10 \|f\|^2_{2,2;T} \end{aligned}$$

and this yields (2.5.18). The proof of the theorem is complete. □

In the next theorem we consider the last part $A^{\frac{1}{2}}\mathcal{J}A^{-\frac{1}{2}}P$ div F of the representation formula (2.5.3).

Setting $\tilde{f} := A^{-\frac{1}{2}}P$ div F in the theorem below we see that $\mathcal{J}\tilde{f}$ has the same properties as $\mathcal{J}Pf$ in the preceding theorem. Therefore, $A^{-\frac{1}{2}}u$ below has the properties of u in Theorem 2.5.2. This enables us to treat the next theorem as a corollary of the previous one.

2.5.3 Theorem *Let $\Omega \subseteq \mathbb{R}^n$, $n \geq 2$, be any domain, let $0 < T \leq \infty$, $1 < s \leq \rho < \infty$, and let $F \in L^s(0,T; L^2(\Omega)^{n^2})$.*

Suppose

$$u \in L^1_{loc}([0,T); W^{1,2}_{0,\sigma}(\Omega))$$

is a weak solution of the Stokes system (2.5.1) with data $f = \text{div } F$ and $u_0 = 0$.

Then we get

$$u(t) = A^{\frac{1}{2}} \int_0^t S(t-\tau) A^{-\frac{1}{2}} P\text{div } F(\tau)\, d\tau \;, \quad u(t) \in D(A^{-\frac{1}{2}}) \tag{2.5.20}$$

for almost all $t \in [0,T)$. After redefining u on a null set of $[0,T)$, $A^{-\frac{1}{2}}u : [0,T) \to L^2_\sigma(\Omega)$ is strongly continuous and $(A^{-\frac{1}{2}}u)(0) = 0$.

Moreover, u has the following properties:

a)
$$\|(A^{-\frac{1}{2}}u)'\|_{2,s;T} + \|A^{\frac{1}{2}}u\|_{2,s;T} \leq C\,\nu^{-\frac{1}{s}}\,\|F\|_{2,s;T} \tag{2.5.21}$$

with $C = C(s) > 0$,

$$(A^{-\frac{1}{2}}u)'(t) + A^{\frac{1}{2}}u(t) = A^{-\frac{1}{2}}P\,\text{div } F(t) \tag{2.5.22}$$

for almost all $t \in [0,T)$, and $A^{-\frac{1}{2}}u$ is a weak solution of the Stokes system (2.5.1) with data $A^{-\frac{1}{2}}P\,\text{div } F$ and $u_0 = 0$.

b)
$$\|A^{\alpha-\frac{1}{2}}u\|_{2,\rho;T} \leq C\,\nu^{-\frac{1}{s}}\,\|F\|_{2,s;T} \tag{2.5.23}$$

with $\alpha := 1 + \frac{1}{\rho} - \frac{1}{s}$, $C = C(\alpha,\rho) > 0$.

c)
$$\|u\|_{q,\rho;T} \leq C\,\nu^{-\alpha}\,\|F\|_{2,s;T} \tag{2.5.24}$$

with $\alpha := 1 + \frac{1}{\rho} - \frac{1}{s}$, $C = C(\alpha,\rho,q) > 0$, and with $2 \leq q < \infty$ determined by the condition

$$1 + \frac{n}{q} + \frac{2}{\rho} = \frac{n}{2} + \frac{2}{s}\,.$$

d)
$$\frac{1}{2}\,\|A^{-\frac{1}{2}}u\|^2_{2,\infty;T'} + \|u\|^2_{2,2;T'} \leq 8\,\nu^{-1}\,\|F\|^2_{2,1;T'} \tag{2.5.25}$$

for all finite T' with $0 < T' \leq T$.

e)
$$\frac{1}{2}\,\|u\|^2_{2,\infty;T} + \|A^{\frac{1}{2}}u\|^2_{2,2;T} \leq 4\,\nu^{-1}\,\|F\|^2_{2,2;T} \tag{2.5.26}$$

if $s = 2$.

Proof. The representation (2.5.20) follows from Theorem 2.4.1, (2.4.4). In particular, $u(t) \in R(A^{\frac{1}{2}}) = D(A^{-\frac{1}{2}})$ for almost all $t \in [0,T)$, and with $\tilde{f} := A^{-\frac{1}{2}} P \operatorname{div} F$ we get

$$A^{-\frac{1}{2}} u(t) = \int_0^t S(t-\tau)\tilde{f}(\tau)\, d\tau = (\mathcal{J}\tilde{f})(t) \tag{2.5.27}$$

for almost all $t \in [0,T)$.

Since $\tilde{f} \in L^s(0,T; L^2_\sigma(\Omega))$, see (2.4.2), we see that $\mathcal{J}\tilde{f}$ satisfies the properties of $\mathcal{J}Pf$ in Theorem 2.5.2. In particular, $\mathcal{J}\tilde{f}$ is strongly continuous and $(\mathcal{J}\tilde{f})(0) = 0$. Therefore, modifying u on a null set, we see that $A^{-\frac{1}{2}}u$ is strongly continuous with $(A^{-\frac{1}{2}}u)(0) = 0$. The above properties a), b) and c) are now consequences of Theorem 2.5.2 with $\mathcal{J}Pf$ replaced by $A^{-\frac{1}{2}}u = \mathcal{J}\tilde{f}$. The inequality (2.5.24) follows directly from (2.5.23) using the embedding (2.4.6), III, now with α replaced by $\alpha - \frac{1}{2}$, and with

$$\alpha - \frac{1}{2} = \frac{1}{2} + \frac{1}{\rho} - \frac{1}{s}\ , \quad 2(\alpha - \frac{1}{2}) + \frac{n}{q} = \frac{n}{2}\ , \quad 0 \le \alpha - \frac{1}{2} \le \frac{1}{2}\,.$$

The inequality (2.5.17) can be used with f replaced by $\tilde{f}$ only if T is finite, since we do not know from the assumptions that $\tilde{f} \in L^1(0,T; L^2_\sigma(\Omega))$. Therefore, we can apply (2.5.17) only for all finite T' with $0 < T' \le T$, and using

$$\nu\, \|\nabla(A^{-\frac{1}{2}}u)\|^2_{2,2;T'} = \|A^{\frac{1}{2}} A^{-\frac{1}{2}} u\|^2_{2,2;T'} = \|u\|^2_{2,2:T'}$$

we get (2.5.25). However, if $T = \infty$ and $F \in L^1(0,\infty; L^2(\Omega)^{n^2})$, then we obtain (2.5.25) also with $T' = \infty$.

To prove e) we use Theorem 2.3.1, c). Then (2.5.26) is a consequence of (2.3.4) with $u_0 = 0$ and $f_0 = 0$. The proof is complete. □

If Ω is a uniform C^2-domain or if $\Omega = \mathbb{R}^n$, we obtain more regularity properties on the weak solution u if $f \in L^s(0,T; L^2(\Omega)^n)$.

In this case we know from Theorem 2.1.1, III, d), (2.1.8), that $D(A) = W^{1,2}_{0,\sigma}(\Omega) \cap W^{2,2}(\Omega)^n$, and that

$$\|\nabla^2 u(t)\|_2 \;\le\; C\,(\nu^{-1}\|Au(t)\|_2 + \|\nabla u(t)\|_2 + \|u(t)\|_2)$$

for almost all $t \in [0,T)$ with $C = C(\Omega) > 0$. Using the relation $\nu^{\frac{1}{2}}\, \|\nabla u\|_2 = \|A^{\frac{1}{2}}u\|_2$ and the interpolation inequality, see (2.2.8), III, we obtain

$$\begin{aligned}\|\nabla u(t)\|_2 \;&=\; \nu^{-\frac{1}{2}}\, \|A^{\frac{1}{2}}u(t)\|_2 \;\le\; \nu^{-\frac{1}{2}}\, \|Au(t)\|_2^{\frac{1}{2}}\, \|u(t)\|_2^{\frac{1}{2}} \\ &=\; \left\|\nu^{-1}Au(t)\right\|_2^{\frac{1}{2}}\, \|u(t)\|_2^{\frac{1}{2}} \;\le\; \nu^{-1}\, \|Au(t)\|_2 + \|u(t)\|_2\,,\end{aligned}$$

and therefore

$$\|\nabla^2 u\|_{2,s;T} \leq C\,(\nu^{-1}\|Au\|_{2,s;T} + \|u\|_{2,s;T}) \tag{2.5.28}$$

for all $u \in L^s(0,T;D(A))$ with $1 < s < \infty$; $C = C(\Omega)$ is a constant depending on Ω.

This estimate, combined with Lemma 2.4.3, III, and Theorem 1.6.3, leads to the following result.

2.5.4 Theorem *Let $\Omega \subseteq \mathbb{R}^n$, $n \geq 2$, be any uniform C^2-domain or let $\Omega = \mathbb{R}^n$, let $0 < T < \infty$, $1 < s < \infty$, $2 \leq q < \infty$ such that*

$$\frac{1}{q} + \frac{1}{n} \geq \frac{1}{2}\,,$$

and let $u_0 \in D(A^{1-\frac{1}{s}}) \cap D(A^{\frac{1}{2}})$, $f \in L^s(0,T;L^2(\Omega)^n)$.

Suppose

$$u \in L^1_{loc}([0,T);W^{1,2}_{0,\sigma}(\Omega))$$

is a weak solution of the Stokes system (2.5.1) with data f, u_0.

Then $u : [0,T) \to L^s_\sigma(\Omega)$ is strongly continuous, after redefinition on a null set of $[0,T)$, and has the following properties:

$$u_t\,,\ \ Au \in L^s(0,T;L^2_\sigma(\Omega))\ \ ,\ \ \ u_t + Au = Pf\ \ ,\ \ \ u(0) = u_0, \tag{2.5.29}$$

and

$$\begin{aligned}\|u_t\|_{2,s;T} \ &+\ \|u\|_{q,s;T} + \|\nabla u\|_{q,s;T} + \|u\|_{2,s;T} + \|\nabla^2 u\|_{2,s;T} \\ &\leq\ C\,(1+T)(\|u_0\|_2 + \|A^{\frac{1}{2}}u_0\|_2 + \|A^{1-\frac{1}{s}}u_0\|_2 + \|f\|_{2,s;T})\end{aligned} \tag{2.5.30}$$

with $C = C(s,\nu,\Omega) > 0$.

Proof. Note that $D(A^{1-\frac{1}{s}}) \subseteq D(A^{\frac{1}{2}})$ if $\frac{1}{2} \leq 1 - \frac{1}{s}$, $s \geq 2$, see (3.2.30), II. Therefore, the assumption $u_0 \in D(A^{\frac{1}{2}})$ is needed only if $1 < s < 2$.

Using Theorem 2.4.1 we obtain the representation $u = S(\cdot)u_0 + \mathcal{J}Pf$, and applying Theorem 2.5.1 and Theorem 2.5.2 we obtain (2.5.29) and the inequality

$$\begin{aligned}\|u'\|_{2,s;T} + \|Au\|_{2,s;T} \ &\leq\ C\,(\|u_0\|_2 + \|A^{\frac{1}{2}}u_0\|_2 + \|A^{1-\frac{1}{s}}u_0\|_2 \\ &\qquad + \|f\|_{2,s;T})\end{aligned} \tag{2.5.31}$$

with $C = C(s) > 0$. Using (1.6.28) with $0 \leq \alpha \leq 1$ we get

$$\|A^\alpha u\|_{2,s;T} \leq \|Au\|_{2,s;T} + \|u\|_{2,s;T}, \tag{2.5.32}$$

and from

$$u(t) = u_0 + \int_0^t u'(\tau)\, d\tau \ , \quad \|u(t)\|_2 \ \leq \ \|u_0\|_2 + \int_0^t \|u'(\tau)\|_2 \, d\tau$$

we conclude that

$$\|u\|_{2,s;T} \ \leq \ T^{\frac{1}{s}} T^{\frac{1}{s'}} \|u'\|_{2,s;T} \ + T^{\frac{1}{s}} \|u_0\|_2 \tag{2.5.33}$$

with $s' := \frac{s}{s-1}$. This leads to

$$\|u\|_{2,s;T} \ \leq \ (1+T)(\|u_0\|_2 + \|u'\|_{2,s;T}). \tag{2.5.34}$$

From the embedding inequality (2.4.18), III, we get

$$\|u(t)\|_q + \|\nabla u(t)\|_q \ \leq \ C\,(\nu^{-\alpha}\|A^\alpha u(t)\|_2 + \|u(t)\|_2)$$

with $\frac{1}{2} \leq \alpha \leq 1$, $2 \leq q < \infty$, $2\,(\alpha - \frac{1}{2}) + \frac{n}{q} = \frac{n}{2}$. For each q with $2 \leq q < \infty$, $\frac{1}{q} \geq \frac{1}{2} - \frac{1}{n}$, we find some α satisfying these properties. The last inequality now leads to

$$\|u\|_{q,s;T} + \|\nabla u\|_{q,s;T} \ \leq \ C\,(\nu^{-\alpha}\|A^\alpha u\|_{2,s;T} + \|u\|_{2,s;T}). \tag{2.5.35}$$

Combining (2.5.31) with (2.5.34), (2.5.28), (2.5.32) and (2.5.35), we obtain the desired inequality (2.5.30). This proves the theorem. □

Note that the norms

$$\|u\|_{q,s;T} + \|\nabla u\|_{q,s;T} \ \text{ and } \ \|u\|_{L^s(0,T;W^{1,q}(\Omega)^n)} \tag{2.5.36}$$

are equivalent. The same holds for the norms

$$\|u\|_{2,s;T} + \|\nabla^2 u\|_{2,s;T} \ \text{ and } \ \|u\|_{L^s(0,T;W^{2,2}(\Omega)^n)}, \tag{2.5.37}$$

see [Ada75] for a proof.

2.6 Associated pressure of weak solutions

Theorem 2.4.1 yields the existence of a weak solution u of the Stokes system

$$u_t - \nu\Delta u + \nabla p = f \ , \quad \text{div } u = 0 \ , \quad u|_{\partial\Omega} = 0, \quad u(0) = u_0, \tag{2.6.1}$$

with data $f = f_0 + \text{div } F$, u_0, satisfying $u_0 \in L^2_\sigma(\Omega)$ and (2.4.3), and we obtain the representation formula

$$u = S(\cdot)u_0 + \mathcal{J}Pf_0 + A^{\frac{1}{2}}\mathcal{J}A^{-\frac{1}{2}}P \text{ div } F. \tag{2.6.2}$$

Our next purpose is to construct an associated pressure p, see Definition 2.1.1. This means, p is a distribution in $(0,T) \times \Omega$ satisfying

$$u_t - \nu \Delta u + \nabla p = f\,.$$

According to the representation (2.6.2) we can divide p into three parts

$$p = p^{(1)} + p^{(2)} + p^{(3)} \tag{2.6.3}$$

in the following way: Setting $u^{(1)} := S(\cdot)u_0$, $u^{(2)} := \mathcal{J}Pf_0$, $u^{(3)} := A^{\frac{1}{2}}\mathcal{J}A^{-\frac{1}{2}}P$ div F, we choose $p^{(j)}$, $j = 1,2,3$, such that

$$\begin{aligned} u_t^{(1)} - \nu\Delta u^{(1)} + \nabla p^{(1)} &= 0, \\ u_t^{(2)} - \nu\Delta u^{(2)} + \nabla p^{(2)} &= f_0, \\ u_t^{(3)} - \nu\Delta u^{(3)} + \nabla p^{(3)} &= \operatorname{div} F \end{aligned} \tag{2.6.4}$$

in $(0,T) \times \Omega$ in the sense of distributions.

The third term $p^{(3)}$ is problematic. From the equation $u_t^{(3)} - \nu\Delta u^{(3)} + \nabla p^{(3)} = \operatorname{div} F$ we could expect that $p^{(3)}$ has the same regularity as F. However we cannot prove this property, see the counter-example given by Heywood-Walsh [HeW94]. See also [Tem77, Chap. III, Prop. 1.1] concerning properties of p. This is the reason that we can show only a very weak regularity property of the total pressure p, see the next theorem. The method is to integrate the first equation in (2.6.1) over the interval $[0,t)$ in the time direction. Then we get rid of the term u_t, obtain an equation of the form

$$u - \nu\Delta\widehat{u} + \nabla\widehat{p} = \widehat{f} + u_0$$

and can apply the theorem of the stationary system to get some $\widehat{p}$. The time derivative $p = \widehat{p}_t$ yields the pressure p in the sense of distributions and leads to a lack of regularity.

If $F = 0$ in (2.6.1) we get $p = p^{(1)} + p^{(2)}$, and in this case we can improve the regularity of p, see Theorem 2.6.3.

In the nonlinear theory, see Chapter V, we cannot avoid the problematic pressure term $p^{(3)}$. To see this we write the nonlinear equation in the form

$$u_t - \nu\Delta u + \nabla p = f - \operatorname{div}(u\,u),$$

where $u\,u$ now plays the role of F in the linear system (2.6.1). This term will not vanish in general. Therefore, the next theorem will be important for the nonlinear theory.

The condition $\int_{\Omega_0} \widehat{p}(t)\,dx = 0$ is needed only in order to get the uniqueness of $\widehat{p}$. First we treat only the case $0 < T < \infty$.

2.6.1 Theorem *Let $\Omega \subseteq \mathbb{R}^n$, $n \geq 2$, be any domain, let $\Omega_0 \subseteq \Omega$ be a bounded subdomain with $\overline{\Omega}_0 \subseteq \Omega$, $\Omega_0 \neq \emptyset$, let $0 < T < \infty$, $1 < s < \infty$, and let $u_0 \in L^2_\sigma(\Omega)$, $f = f_0 + \operatorname{div} F$ with*

$$f_0 \in L^1(0,T; L^2(\Omega)^n) \ , \quad F \in L^s(0,T; L^2(\Omega)^{n^2}) \, .$$

Suppose

$$u \ \in \ L^1(0,T; W^{1,2}_{0,\sigma}(\Omega))$$

is a weak solution of the Stokes system (2.6.1) with data f, u_0, and let $\widehat{u}$, $\widehat{f} = \widehat{f}_0 + \operatorname{div} \widehat{F}$ be defined by

$$\widehat{u}(t) := \int_0^t u(\tau)\, d\tau \ , \quad \widehat{f}_0(t) := \int_0^t f_0(\tau)\, d\tau \ , \quad \widehat{F}(t) := \int_0^t F(\tau)\, d\tau \, , \tag{2.6.5}$$

$t \in [0,T)$.

Then there exists a unique

$$\widehat{p} \ \in \ L^s(0,T; L^2_{loc}(\Omega))$$

satisfying $\int_{\Omega_0} \widehat{p}(t)\, dx = 0$ for almost all $t \in [0,T)$, and

$$u - \nu \Delta \widehat{u} + \nabla \widehat{p} = \widehat{f} + u_0 \tag{2.6.6}$$

in the sense of distributions. The distributional derivative

$$p \ := \ \widehat{p}_t \ = \ \frac{\partial}{\partial t} \widehat{p}$$

satisfies

$$u_t - \nu \Delta u + \nabla p = f \tag{2.6.7}$$

and is an associated pressure of u.

Proof. The construction of $\widehat{p}$ rests on Lemma 1.4.2. We define the functional $G : v \mapsto [G, v]_{\Omega,T}$, $v \in C_0^\infty((0,T) \times \Omega)^n$, by setting

$$\begin{aligned} [G, v]_{\Omega,T} \ &:= \ [\widehat{f} + u_0 - u + \nu \Delta \widehat{u}, \ v]_{\Omega,T} \\ &:= \ < \widehat{f}_0, v >_{\Omega,T} - < \widehat{F}, \nabla v >_{\Omega,T} \\ &\quad + < u_0 - u, v >_{\Omega,T} - \nu < \nabla \widehat{u}, \nabla v >_{\Omega,T} \, . \end{aligned}$$

The representation formula (2.4.4) yields the decomposition

$$u = S(\cdot) u_0 + \mathcal{J} P f_0 + A^{\frac{1}{2}} \mathcal{J} \tilde{f} \tag{2.6.8}$$

with $\tilde{f} = A^{-\frac{1}{2}} P \operatorname{div} F$. From (1.5.8), (2.4.8) and (2.4.12) we conclude that $u \in L^s(0,T; L^2_\sigma(\Omega))$.

Let $\Omega' \subseteq \Omega$ be any bounded subdomain with $\overline{\Omega'} \subseteq \Omega$. Then a calculation yields for $v \in C_0^\infty((0,T) \times \Omega')^n$ the estimate

$$\begin{aligned} |[G,v]_{\Omega,T}| \leq\; & C([\|f_0\|_{2,1;T} + \|F\|_{2,s;T} + \|\nabla u\|_{2,1;T} \\ & + \|u - u_0\|_{2,s;T})\|\nabla v\|_{2,s';T} \end{aligned}$$

with $s' = \frac{s}{s-1}$, $C = C(\nu, s, \Omega', T) > 0$. This shows that

$$G \in L^s(0,T; W^{-1,2}_{loc}(\Omega)^n). \tag{2.6.9}$$

Consider now the special test function $v \in C_0^\infty((0,T); C_{0,\sigma}^\infty(\Omega))$ needed in (1.4.16). Then, setting $w(t) := \int_t^T v(\tau)\, d\tau$, $t \in [0,T)$, we see that $w \in C_0^\infty([0,T); C_{0,\sigma}^\infty(\Omega))$, $w(0) = \int_0^T v(\tau)\, d\tau$, and using the definition of a weak solution in (2.1.4) we conclude that

$$\begin{aligned} 0 &= - < u, w_t >_{\Omega,T} + \nu < \nabla u, \nabla w >_{\Omega,T} - < u_0, w(0) >_\Omega - [f,w]_{\Omega,T} \\ &= < u, v >_{\Omega,T} + \nu < \nabla \widehat{u}, \nabla v >_{\Omega,T} - \int_0^T < u_0, v >_\Omega \, d\tau - [\widehat{f}, v]_{\Omega,T} \\ &= - [\widehat{f} + u_0 - u + \nu \Delta \widehat{u}, v]_{\Omega,T}. \end{aligned}$$

Now the assumptions of Lemma 1.4.2 are satisfied, and we get a unique $\widehat{p} \in L^s(0,T; L^2_{loc}(\Omega))$ satisfying $\int_{\Omega_0} \widehat{p}(t)\, dx = 0$ for almost all $t \in [0,T)$ and $G = \nabla \widehat{p}$ in the sense of distributions. From the above estimate and from (1.4.17), we conclude that

$$\|\widehat{p}\|_{L^s(0,T;L^2(\Omega'))} \leq C\,(\|f_0\|_{2,1;T} + \|F\|_{2,s;T} + \|\nabla u\|_{2,1;T} + \|u - u_0\|_{2,s;T}) \tag{2.6.10}$$

with some constant $C = C(\nu, s, T, \Omega_0, \Omega') > 0$ depending on T and $\Omega' \subseteq \Omega$.

$G = \nabla \widehat{p}$ means that (2.6.6) is satisfied. Since $\widehat{u}' = u$, $\widehat{f}' = f$, we see that $p := \widehat{p}' = \widehat{p}_t$ satisfies (2.6.7) in the sense of distributions. This completes the proof. $\square$

If Ω in Theorem 2.6.1 is a Lipschitz domain, bounded or not, then we may consider any bounded Lipschitz subdomain $\Omega' \subseteq \Omega$ with $\Omega' \supseteq \Omega_0$. The estimate (2.6.10) remains valid for Ω', and we conclude that $\widehat{p} \in L^s(0,T; L^2(\Omega'))$, see Lemma 1.4.1. This leads to the property

$$\widehat{p} \in L^s(0,T; L^2_{loc}(\overline{\Omega})). \tag{2.6.11}$$

The case $T = \infty$ can be treated as a corollary of the above theorem.

2.6.2 Corollary *Let $\Omega \subseteq \mathbb{R}^n$, $n \geq 2$, be any domain, let $\Omega_0 \subseteq \Omega$ be a bounded subdomain with $\overline{\Omega}_0 \subseteq \Omega$, $\Omega_0 \neq \emptyset$, let $T = \infty$, $1 < s < \infty$, $u_0 \in L^2_\sigma(\Omega)$, and let $f = f_0 + \operatorname{div} F$ with*

$$f_0 \in L^1_{loc}([0,\infty); L^2(\Omega)^n) \ , \quad F \in L^s_{loc}([0,\infty); L^2(\Omega)^{n^2}). \tag{2.6.12}$$

Suppose

$$u \in L^1_{loc}([0,\infty); W^{1,2}_{0,\sigma}(\Omega))$$

is a weak solution of the Stokes system (2.6.1) with data f, u_0, and let $\widehat{u}$, $\widehat{f}_0, \widehat{F}$ be defined by (2.6.5).

Then there exists a unique

$$\widehat{p} \in L^s_{loc}([0,\infty); L^2_{loc}(\Omega)) \,, \tag{2.6.13}$$

satisfying $\int_{\Omega_0} \widehat{p}(t)\,dx = 0$ for almost all $t \in [0,\infty)$, and equation (2.6.6) in the sense of distributions. The derivative

$$p := \widehat{p}_t = \frac{\partial}{\partial t}\widehat{p}$$

satisfies (2.6.7) and is an associated pressure of u.

Proof. We apply Theorem 2.6.1 for all T' with $0 < T' < \infty$. Then the uniqueness assertion in this theorem enables us to construct $\widehat{p}$ as an extension of the corresponding functions in $[0, T')$. This yields the result. □

Recall, (2.6.13) means that

$$\widehat{p} \in L^s(0,T'; L^2_{loc}(\Omega)) \text{ for all } T' \text{ with } 0 < T' < \infty .$$

In the case $F = 0$, the given weak solution u has the form $u = S(\cdot)u_0 + \mathcal{J}Pf$, and this enables us to improve the regularity of the corresponding associated pressure p. For this purpose we need a smoothness property on u_0.

2.6.3 Theorem *Let $\Omega \subseteq \mathbb{R}^n$, $n \geq 2$, be any domain, let $\Omega_0 \subseteq \Omega$ be a bounded subdomain with $\overline{\Omega}_0 \subseteq \Omega$, $\Omega_0 \neq \emptyset$, let $0 < T < \infty$, $1 < s < \infty$, $u_0 \in D(A^{1-\frac{1}{s}}) \cap D(A^{\frac{1}{2}})$, and let*

$$f \in L^s(0,T; L^2(\Omega)^n). \tag{2.6.14}$$

Suppose

$$u \in L^1([0,T); W^{1,2}_{0,\sigma}(\Omega))$$

is a weak solution of the Stokes system (2.6.1) with data f, u_0.

Then there exists a unique

$$p \in L^s(0,T;L^2_{loc}(\Omega))$$

satisfying $\int_{\Omega_0} p(t)\,dx = 0$ *for almost all* $t \in [0,T)$ *and*

$$u_t - \nu\Delta u + \nabla p = f \tag{2.6.15}$$

in the sense of distributions in $(0,T)\times\Omega$.

If Ω *is a Lipschitz domain, then*

$$p \in L^s(0,T;L^2_{loc}(\overline{\Omega})), \tag{2.6.16}$$

and if Ω *is a uniform* C^2*-domain or if* $\Omega = \mathbb{R}^n$, *then*

$$\nabla p \in L^s(0,T;L^2(\Omega)^n)$$

and

$$\|\nabla p\|_{L^s(0,T;L^2(\Omega)^n)} \le C\,(\|u_0\|_2 + \|A^{\frac{1}{2}}u_0\|_2 + \|A^{1-\frac{1}{s}}u_0\|_2 + \|f\|_{2,s;T}) \tag{2.6.17}$$

with $C = C(\nu,s,\Omega,T) > 0$.

Proof. From Theorem 2.4.1 we obtain the representation $u = S(\cdot)u_0 + \mathcal{J}Pf$, and we may use the properties in Theorem 2.5.1 and Theorem 2.5.2. Combining (2.5.5) with (2.5.13) we get the inequality

$$\|u'\|_{2,s;T} + \|Au\|_{2,s;T} \le C\,(\|u_0\|_2 + \|A^{\frac{1}{2}}u_0\|_2 + \|A^{1-\frac{1}{s}}u_0\|_2 + \|f\|_{2,s;T}) \tag{2.6.18}$$

with $C = C(s) > 0$. Using the interpolation inequality (2.2.8), III, we obtain $\nu\|\nabla u\|_2 = \nu^{\frac{1}{2}}\,\|A^{\frac{1}{2}}u\|_2 \le \nu^{\frac{1}{2}}\,\|Au\|_2^{\frac{1}{2}}\,\|u\|_2^{\frac{1}{2}} \le \nu^{\frac{1}{2}}(\frac{1}{2}\|Au\|_2 + \frac{1}{2}\|u\|_2)$, and therefore

$$\begin{aligned} \nu\,\|\nabla u\|_{2,s;T} &= \nu^{\frac{1}{2}}\,\|A^{\frac{1}{2}}u\|_{2,s;T} \\ &\le \nu^{\frac{1}{2}}\,\Big(\frac{1}{2}\|Au\|_{2,s;T} + \frac{1}{2}\,\|u\|_{2,s;T}\Big). \end{aligned} \tag{2.6.19}$$

To find p we define the functional $G : v \mapsto [G,v]_{\Omega,T},\ v \in C_0^\infty((0,T)\times\Omega)^n$ by setting

$$\begin{aligned} [G,v]_{\Omega,T} &:= [f - u_t + \nu\Delta u, v]_{\Omega,T} \\ &= < f,v >_{\Omega,T} - < u_t, v >_{\Omega,T} - \nu < \nabla u, \nabla v >_{\Omega,T}\ . \end{aligned}$$

Let $\Omega' \subseteq \Omega$ be a bounded subdomain with $\overline{\Omega'} \subseteq \Omega$, and let

$$v \in C_0^\infty((0,T) \times \Omega')^n .$$

Then, using Poincaré's inequality and (2.6.18), (2.6.19), we get the estimate

$$\begin{aligned} |[G,v]|_{\Omega,T} &\leq C\,(\|f\|_{2,s;T} + \|u_t\|_{2,s;T} + \nu\,\|\nabla u\|_{2,s;T})\,\|\nabla v\|_{2,s';T} \\ &\leq C'\,(\|u_0\|_2 + \|A^{\frac{1}{2}}u_0\|_2 \\ &\quad + \|A^{1-\frac{1}{s}}u_0\|_2 + \|f\|_{2,s;T} + \|u\|_{2,s;T})\,\|\nabla v\|_{2,s';T} \end{aligned}$$

with $s' = \frac{s}{s-1}$, where C, C' depend on Ω'. Using

$$\|u\|_{2,s;T} \leq (1+T)\,(\|u_0\|_2 + \|u_t\|_{2,s;T}),$$

see (2.5.34), we obtain

$$|[G,v]_{\Omega,T}| \leq C\,(\|u_0\|_2 + \|A^{\frac{1}{2}}u_0\|_2 + \|A^{1-\frac{1}{s}}u_0\|_2 + \|f\|_{2,s;T})\,\|\nabla v\|_{2,s';T} \quad (2.6.20)$$

with $C = C(s,\nu,\Omega',T) > 0$.

Since u is a weak solution we get with $v \in C_0^\infty([0,T); C_{0,\sigma}^\infty(\Omega))$ that

$$\begin{aligned} [G,v]_{\Omega,T} &= \; < f,v >_{\Omega,T} - < u_t, v >_{\Omega,T} - \nu < \nabla u, \nabla v >_{\Omega,T} \\ &= \; < f,v >_{\Omega,T} + < u_0, v(0) >_\Omega + < u, v_t >_{\Omega,T} - \nu < \nabla u, \nabla v >_{\Omega,T} \\ &= \; 0. \end{aligned}$$

From (2.6.20) we see that $G \in L^s(0,T;W_{loc}^{-1,2}(\Omega)^n)$, and Lemma 1.4.2 yields a unique $p \in L^s(0,T;L_{loc}^2(\Omega))$ satisfying $G = f - u_t + \nu\Delta u = \nabla p$, $\int_{\Omega_0} p(t)\,dx = 0$ for a.a. $t \in [0,T)$. This proves (2.6.15).

From (2.6.20), (1.4.17) we get

$$\|p\|_{L^s(0,T;L^2(\Omega'))} \leq C\,(\|u_0\|_2 + \|A^{\frac{1}{2}}u_0\|_2 + \|A^{1-\frac{1}{s}}u_0\|_2 + \|f\|_{2,s;T}) \quad (2.6.21)$$

for each bounded subdomain $\Omega' \subseteq \Omega$ with $\Omega_0 \subseteq \Omega'$, $\overline{\Omega'} \subseteq \Omega$, and $C = C(s,\nu,\Omega_0,\Omega',T) > 0$.

If Ω is a uniform C^2-domain or if $\Omega = \mathbb{R}^n$, we apply Theorem 2.5.4 and see that $\|\nabla^2 u\|_{2,s;T} < \infty$. Thus it follows that $\nabla p = f - u_t + \nu\Delta u \in L^s(0,T;L^2(\Omega)^n)$, and using (2.5.30) we get the desired inequality (2.6.17).

If Ω is a Lipschitz domain, the same argument as used for (2.6.11) also yields the property (2.6.16). This proves the theorem. □

If the assumptions of Theorem 2.6.3 are satisfied with $T = \infty$, where now $f \in L_{loc}^s([0,\infty); L^2(\Omega)^n)$ and

$$u \in L_{loc}^1([0,\infty); W_{0,\sigma}^{1,2}(\Omega)), \quad (2.6.22)$$

then we can apply the above result for all T' with $0 < T' < \infty$, and using the uniqueness assertion we get a uniquely determined

$$p \in L^s_{loc}([0,\infty); L^2_{loc}(\Omega)) \tag{2.6.23}$$

satisfying $u_t - \nu\Delta u + \nabla p = f$ and $\int_{\Omega_0} p(t)\,dx = 0$ for almost all $t \in [0,\infty)$. On each finite interval $[0,T')$ we obtain all the properties of Theorem 2.6.3.

2.7 Regularity properties of weak solutions

Our aim is to prove regularity properties of weak solutions u of the Stokes system

$$u_t - \nu\Delta u + \nabla p = f \ , \quad \operatorname{div} u = 0 \ , \quad u|_{\partial\Omega} = 0 \ , \quad u(0) = u_0, \tag{2.7.1}$$

see [Lad69, Chap. 4,2], [Tem77, Chap. III, 3.5], [Sol68], [Hey80], [Gig86], [GaM88] concerning such results.

The following theorem concerns only the first regularity step. We obtain it combining the results of Theorem 2.5.4 and Theorem 2.6.3.

2.7.1 Theorem *Let $\Omega \subseteq \mathbb{R}^n$, $n \geq 2$, be a uniform C^2-domain or let $\Omega = \mathbb{R}^n$, let $0 < T \leq \infty$, $1 < s < \infty$, and let*

$$u_0 \in D(A^{1-\frac{1}{s}}) \cap D(A^{\frac{1}{2}}) \ , \quad f \in L^s_{loc}([0,T); L^2(\Omega)^n). \tag{2.7.2}$$

Suppose

$$u \in L^1_{loc}([0,T); W^{1,2}_{0,\sigma}(\Omega))$$

is a weak solution of the Stokes system (2.7.1) with data f, u_0.

Then $u : [0,T) \to L^2_\sigma(\Omega)$ is strongly continuous after a redefinition on a null set of $[0,T)$, it holds that $u(0) = u_0$,

$$u', \ Au \in L^s_{loc}([0,T); L^2_\sigma(\Omega)) \ , \quad u \in L^s_{loc}([0,T); W^{2,2}(\Omega)^n), \tag{2.7.3}$$

and

$$u'(t) + Au(t) = Pf(t) \tag{2.7.4}$$

for almost all $t \in [0,T)$.

There exists an associated pressure

$$p \in L^s_{loc}([0,T); L^2_{loc}(\overline{\Omega})) \ , \quad \nabla p \in L^s_{loc}([0,T); L^2(\Omega)^n), \tag{2.7.5}$$

satisfying

$$u_t - \nu\Delta u + \nabla p = f \tag{2.7.6}$$

in the sense of distributions, and

$$\begin{aligned}\|u_t\|_{2,s;T'} + \|u\|_{2,s;T'} + \|\nabla^2 u\|_{2,s;T'} + \|\nabla p\|_{2,s;T'} \\ \le C\left(\|u_0\|_2 + \|A^{\frac{1}{2}}u_0\|_2 + \|A^{1-\frac{1}{s}}u_0\|_2 + \|f\|_{2,s;T'}\right)\end{aligned} \tag{2.7.7}$$

with $0 < T' < T$, $C = C(\nu, \Omega, s, T') > 0$.

Proof. First we apply Theorem 2.5.4 for each interval $[0, T')$, $0 < T' < T$. This yields (2.7.3), and (2.7.7) follows without the term $\|\nabla p\|_{2,s;T'}$. This term is treated in Theorem 2.6.3 and together with (2.6.17) we obtain (2.7.7). This proves the result. □

Moreover, from (2.5.30), (2.5.37) we obtain

$$u \in L^s_{loc}([0,T); W^{2,2}(\Omega)^n \cap W^{1,q}(\Omega)^n), \tag{2.7.8}$$

with $2 \le q < \infty$, $\frac{1}{q} + \frac{1}{n} \ge \frac{1}{2}$, and

$$\|u\|_{q,s;T'} + \|\nabla u\|_{q,s;T'} \;\le\; C\left(\|u_0\|_2 + \|A^{\frac{1}{2}}u_0\|_2 + \|A^{1-\frac{1}{s}}u_0\|_2 + \|f\|_{2,s;T'}\right) \tag{2.7.9}$$

with $0 < T' < T$, $C = C(\nu, \Omega, s, T') > 0$.

In the next step we prove regularity properties of higher order first only in the time direction. Here we use the method of differentiating the equation

$$u_t + Au = Pf \tag{2.7.10}$$

in the time direction. Even if f and u_0 are arbitrarily smooth, we cannot expect that the time derivatives of u are continuous up to $t = 0$. This is possible only if the data f and u_0 satisfy additional **compatibility conditions**. See [Sol68, page 97], [Rau83] concerning this problem. To explain this fact we consider the following example:

Let Ω be a bounded C^2- domain and assume that u, Au, Pf are continuous on $[0,T) \times \overline{\Omega}$, that u_t exists as a continuous function in the classical sense, and that (2.7.10) is satisfied on $[0,T) \times \overline{\Omega}$ together with $u|_{\partial\Omega} = 0$, $u(0) = u_0$. Since $u(t,x) = 0$ for $x \in \partial\Omega$, $t \in [0,T)$, we see that $u_t(0)|_{\partial\Omega} = 0$ and therefore that

$$Au_0|_{\partial\Omega} = Pf(0)|_{\partial\Omega}. \tag{2.7.11}$$

This is an additional condition on the given data f, u_0 which must be satisfied under the above smoothness assumptions. Further compatibility conditions are obtained considering $u_{tt}(0)|_{\partial\Omega} = 0$, and so on.

To avoid these complicated compatibility conditions on the data at $t = 0$, we consider here only regularity properties for $t > 0$. The formulation of these properties requires us to define the following loc-spaces.

Let $\Omega \subseteq \mathbb{R}^n$, $n \geq 2$, be any domain, and let $0 < T \leq \infty$, $1 \leq s < \infty$. We consider "cut-off" functions $\varphi \in C_0^\infty((0,T))$, and define the spaces

$$L^s_{loc}((0,T); L^2_\sigma(\Omega)) \ , \ L^s_{loc}((0,T); W^{2,2}(\Omega)^n) \ , \ L^s_{loc}((0,T); L^2(\Omega)^n) , \quad (2.7.12)$$

by setting

$$\begin{aligned} u \in L^s_{loc}((0,T); L^2_\sigma(\Omega)) \quad &\text{iff} \quad \varphi u \in L^s(0,T; L^2_\sigma(\Omega)), \\ u \in L^s_{loc}((0,T); W^{2,2}(\Omega)^n) \quad &\text{iff} \quad \varphi u \in L^s(0,T; W^{2,2}(\Omega)^n), \\ u \in L^s_{loc}((0,T); L^2(\Omega)^n) \quad &\text{iff} \quad \varphi u \in L^s(0,T; L^2(\Omega)^n) \end{aligned}$$

for all $\varphi \in C_0^\infty((0,T))$.

In particular the conditions

$$u, \ u', \ \ldots, \ (\frac{d}{dt})^k u \ \in \ L^s_{loc}((0,T); L^2_\sigma(\Omega))$$

are well defined for all $k \in \mathbb{N}$. We set

$$u^{(k)} := (\frac{d}{dt})^k u \ , \quad k \in \mathbb{N} \ , \quad u^{(0)} = u.$$

The next theorem yields a regularity result of higher order only in the time direction.

2.7.2 Theorem *Let $k \in \mathbb{N}$, let $\Omega \subseteq \mathbb{R}^n$, $n \geq 2$, be a uniform C^2-domain or let $\Omega = \mathbb{R}^n$, let $0 < T \leq \infty$, $1 < s < \infty$, $u_0 \in D(A^{1-\frac{1}{s}}) \cap D(A^{\frac{1}{2}})$, $f \in L^s_{loc}([0,T); L^2(\Omega)^n)$, and suppose u is a solution of the Stokes system (2.7.1) with data f, u_0 together with an associated pressure p as in Theorem 2.7.1.*

Then

$$f', \ f'', \ldots, f^{(k)} \in L^s_{loc}((0,T); L^2(\Omega)^n) \quad (2.7.13)$$

implies

$$u', \ u'' \ \ldots, u^{(k+1)}, Au', \ Au'', \ldots, Au^{(k)} \in L^s_{loc}((0,T); L^2_\sigma(\Omega)), \quad (2.7.14)$$

and

$$u', \ u'', \ \ldots, u^{(k)} \in L^s_{loc}((0,T); W^{2,2}(\Omega)^n). \quad (2.7.15)$$

Proof. From Theorem 2.7.1 we get the properties (2.7.3)–(2.7.7) for u and p. Let $\varphi \in C_0^\infty((0,T))$. Then from $u_t + Au = Pf$, (2.7.4), we get

$$(\varphi u)_t + A(\varphi u) = P(\varphi f) + \varphi' u,$$

and since φ only depends on t, it is obvious that φu is a weak solution with data $P(\varphi f) + \varphi' u$ and $u_0 = 0$.

Set $v := \varphi u,\ g := P(\varphi f) + \varphi' u$. Then the representation formula, Theorem 2.4.1, shows that

$$v = \mathcal{J} g \ , \quad v(t) = \int_0^t S(t-\tau)\, g(\tau)\, d\tau \ , \quad t \in [0,T).$$

Let $k = 1$ and assume $f' \in L^s_{loc}((0,T); L^2(\Omega)^n)$. Then we get $v' + Av = g,\ g' = P\varphi f' + P\varphi' f + \varphi'' u + \varphi' u'$, and we will show that

$$v' = \mathcal{J} g' \ , \quad v'(t) = \int_0^t S(t-\tau)\, g'(\tau)\, d\tau \ , \quad t \in [0,T). \tag{2.7.16}$$

Since $v, g \in W^{1,s}(0,T; L^2_\sigma(\Omega))$ and $D(A) \subseteq L^2_\sigma(\Omega)$ is dense, it is sufficient for (2.7.16) to show that

$$\begin{aligned} < v'(t), w >_\Omega &= < \int_0^t S(t-\tau)\, g'(\tau)\, d\tau, w >_\Omega \\ &= \int_0^t < g'(\tau), S(t-\tau) w >_\Omega \, d\tau \end{aligned} \tag{2.7.17}$$

for almost all $t \in [0,T)$, and all $w \in D(A)$. Since $w \in D(A)$, the derivative

$$\frac{\partial}{\partial \tau} S(t-\tau) w \ = \ -\frac{\partial}{\partial t} S(t-\tau) w \ = \ A S(t-\tau) w = S(t-\tau) A w$$

exists for $0 \le \tau \le t < T$ and is strongly continuous, see Lemma 1.5.1.

This enables us to carry out the following calculation. To prove (2.7.17) it suffices to show that

$$< v(\rho), w >_\Omega = \int_0^\rho \int_0^t < g'(\tau), S(t-\tau) w >_\Omega \, d\tau\, dt \tag{2.7.18}$$

for $0 \le t \le \rho < T,\ w \in D(A)$. This implies (2.7.17), see Lemma 1.3.1.

Using the rule of integration by parts, see Lemma 1.3.2, and Fubini's theorem, we get

$$\begin{aligned} &\int_0^\rho \int_0^t < g'(\tau), S(t-\tau) w >_\Omega \, d\tau\, dt \\ &= \int_0^\rho (< g(t), w >_\Omega - \int_0^t < g(\tau), \frac{\partial}{\partial \tau} S(t-\tau) w >_\Omega \, d\tau)\, dt \end{aligned}$$

$$
\begin{aligned}
&= \int_0^\rho < g(t), w >_\Omega \, dt + \int_0^\rho \int_0^t < g(\tau), \frac{\partial}{\partial t} S(t-\tau) w >_\Omega \, d\tau \, dt \\
&= \int_0^\rho < g(t), w >_\Omega \, dt + \int_0^\rho \int_\tau^\rho \frac{\partial}{\partial t} < g(\tau), S(t-\tau) w >_\Omega \, dt \, d\tau \\
&= \int_0^\rho < g(t), w >_\Omega \, dt + \int_0^\rho (< g(\tau), S(\rho-\tau) w >_\Omega \, d\tau - < g(\tau), w >_\Omega) \, d\tau \\
&= \int_0^\rho < S(\rho-\tau) g(\tau), w >_\Omega \, d\tau \ = \ < \int_0^\rho S(\rho-\tau) g(\tau) \, d\tau, w >_\Omega \\
&= \ < v(\rho), w >_\Omega .
\end{aligned}
$$

This proves (2.7.18), and therefore (2.7.17), (2.7.16).

Thus $v' = \mathcal{J} g'$ has the same properties as $v = \mathcal{J} g$. In particular, v' has the properties of u in Theorem 2.7.1, and therefore

$$
\begin{aligned}
v'' &= (\varphi u)'' \in L^s(0,T; L^2_\sigma(\Omega)) , \\
A(\varphi u)' &\in L^s(0,T; L^2_\sigma(\Omega)) , \\
(\varphi u)' &\in L^s(0,T; W^{2,2}(\Omega)^n) .
\end{aligned}
$$

This proves the lemma for $k = 1$.

In the next step $k = 2$ we consider the equation $v'' = \mathcal{J} g''$ and repeat the above argument. This yields the result for $k = 2$, and so on. The general result of the theorem follows by induction on k. □

Finally we investigate regularity properties in the spatial direction. For this purpose we have to combine the above result with regularity properties of the stationary Stokes system, see Theorem 1.5.1, III.

For simplicity we consider only the C^∞- regularity. From the proof it will be clear how to get results on $W^{k,2}_{loc}$-regularity in the spatial direction.

2.7.3 Theorem *Let $\Omega \subseteq \mathbb{R}^n$, $n \geq 2$, be a uniform C^2-domain or let $\Omega = \mathbb{R}^n$, let $0 < T \leq \infty$, $u_0 \in L^2_\sigma(\Omega)$, and let*

$$
f \in C_0^\infty\big(\overline{(0,T) \times \Omega}\big)^n . \tag{2.7.19}
$$

Suppose additionally that Ω is a C^∞-domain, and suppose

$$
u \in L^1_{loc}([0,T); W^{1,2}_{0,\sigma}(\Omega))
$$

is a weak solution of the Stokes system (2.7.1) with data f, u_0.

Then, after a redefinition on a null set of $[0,T) \times \Omega$,

$$
u \in C^\infty_{loc}\big(\overline{(\varepsilon,T') \times \Omega}\big)^n \tag{2.7.20}
$$

for all ε, T' with $0 < \varepsilon < T' < T$. In particular it follows that

$$u \in C^\infty((0,T) \times \Omega)^n \,. \tag{2.7.21}$$

Moreover, there exists an associated pressure p of u satisfying

$$p \in C^\infty_{loc}\big(\overline{(\varepsilon, T') \times \Omega}\big) \tag{2.7.22}$$

for all ε, T' with $0 < \varepsilon < T' < T$. It follows that

$$p \in C^\infty((0,T) \times \Omega) \,. \tag{2.7.23}$$

Proof. First we conclude that $u \in L^2_{loc}([0,T); W^{1,2}_{0,\sigma}(\Omega))$, see (2.3.2). Consider some $\varphi \in C_0^\infty((0,T))$ and the Stokes system

$$v_t - \nu\Delta v + \nabla \widehat{p} = g \;, \quad \operatorname{div} v = 0 \;, \quad v|_{\partial\Omega} = 0 \;, \quad v(0) = 0 \tag{2.7.24}$$

with $v = \varphi u$, $g := \varphi f + \varphi' u$, $\widehat{p} = \varphi p$. Then $v \in L^2(0,T; W^{1,2}_{0,\sigma}(\Omega))$ is a weak solution of (2.7.24) with data g and $v_0 = 0$. It holds that

$$g \in L^2(0,T; L^2(\Omega)^n)$$

and we may apply Theorem 2.7.1. This yields v', $Av \in L^2(0,T; L^2_\sigma(\Omega))$ and $v \in L^2(0,T; W^{2,2}(\Omega)^n)$. Thus we get $(\varphi u)'$, $A\varphi u \in L^2(0,T; L^2_\sigma(\Omega))$ and $\varphi u \in L^2(0,T; W^{2,2}(\Omega)^n)$ for all $\varphi \in C_0^\infty((0,T))$. This yields

$$u', \; Au \in L^2_{loc}((0,T); L^2_\sigma(\Omega)) \;, \quad u \in L^2_{loc}((0,T); W^{2,2}(\Omega)^n) \,.$$

In the next step we apply Theorem 2.7.2 with $k = 1$ to (2.7.24) for all φ and conclude that

$$u'', \; Au' \in L^2_{loc}((0,T); L^2_\sigma(\Omega)) \;, \quad u' \in L^2_{loc}((0,T); W^{2,2}(\Omega)^n) \,.$$

Now we may use this argument for $k = 2$, and so on. In this way we conclude that the properties (2.7.14) and (2.7.15) with $s = 2$ hold for all $k \in \mathbb{N}$.

This enables us to differentiate k times the equations (2.7.24) in the time direction as in the preceding proof. This yields

$$v^{(k+1)} - \nu\Delta v^{(k)} + \nabla \widehat{p}^{(k)} = g^{(k)} \;, \quad k \in \mathbb{N}.$$

We write these equations in the form

$$-\nu\Delta v^{(k)} + \nabla \widehat{p}^{(k)} = g^{(k)} - v^{(k+1)}, \tag{2.7.25}$$

and apply the stationary regularity results of Theorem 1.5.1, III. Using (2.7.14), (2.7.15) we conclude that

$$g^{(k)} - v^{(k+1)} \in L^2(0,T; W^{2,2}(\Omega)^n)$$

for all $k \in \mathbb{N}$, and this theorem implies

$$v^{(k)} \in L^2(0,T; W^{4,2}_{loc}(\overline{\Omega})^n) \tag{2.7.26}$$

for all $k \in \mathbb{N}$. To see this we consider bounded subdomains as in (1.5.4), III, write (1.5.4) with $k = 2$ for almost all $t \in [0,T)$ and take the L^2- norm over $[0,T)$.

Using (2.7.26) for all $\varphi \in C_0^\infty((0,T))$ we conclude

$$g^{(k)} - v^{(k+1)} \in L^2(0,T; W^{4,2}_{loc}(\overline{\Omega})^n)$$

for all $k \in \mathbb{N}$. Applying again Theorem 1.5.1, III, to (2.7.25) now yields

$$v^{(k)} \in L^2(0,T; W^{6,2}_{loc}(\overline{\Omega})^n)$$

for all $k \in \mathbb{N}$. Proceeding in this way we see that

$$v^{(k)} \in L^2(0,T; W^{j,2}_{loc}(\overline{\Omega})^n)$$

for all $k, j \in \mathbb{N}$. This shows that

$$v \in W^{k,2}_{loc}\big(\overline{(0,T) \times \Omega}\big)^n$$

for all $k \in \mathbb{N}$. Using this property for all $\varphi \in C_0^\infty((0,T))$, and applying the embedding estimate (1.3.10), II, we conclude that (2.7.20) and (2.7.21) are valid.

To investigate the pressure, we start with p constructed as in Theorem 2.6.1; here we admit the case $T = \infty$, see Corollary 2.6.2. From (2.6.7) we get

$$\nabla p = f - u_t + \nu \Delta u$$

which leads to $\nabla p \in C^\infty_{loc}(\overline{(\varepsilon,T') \times \Omega})^n$ for all $0 < \varepsilon < T' < T$. To get a representation of p itself, we use the elementary line integral. Let $x_0, x \in \Omega$ and let $\Gamma(x)$ be a smooth curve connecting x_0 with $x \in \Omega$; let L be the unit tangential vector of the curve and let $\int_{\Gamma(x)} \cdots dS$ be the usual line integral over $\Gamma(x)$.

Consider the integral

$$p_{x_0}(t,x) := \int_{\Gamma(x)} (\nabla p) \cdot L \, dS \ , \quad x \in \Omega, \ t \in [0,T).$$

Then an elementary calculation yields the following result: Each function p_{x_0} is well defined on $(0,T) \times B(x_0)$ where $B(x_0) \subseteq \Omega$ is any open ball with center x_0. It holds that $\nabla p_{x_0} = \nabla p$ and

$$p_{x_0} \in C_0^\infty\big(\overline{(\varepsilon,T') \times B(x_0)}\big)$$

for all $0 < \varepsilon < T' < T$. Since $\nabla p_{x_0} = \nabla p$, we conclude that we can redefine each p_{x_0}, adding a smooth function depending only on $t \in (0, T)$ in such a way that two functions coincide on nonempty intersections of different balls.

This yields a well defined (single valued) function $\tilde{p}$ satisfying (2.7.22) and (2.7.23). It holds that $\nabla \tilde{p} = \nabla p$, and therefore we can redefine p and obtain the properties (2.7.22), (2.7.23). This proves the lemma. □

Chapter V

The Full Nonlinear Navier-Stokes Equations

1 Weak solutions

1.1 Definition of weak solutions

In this chapter, $\Omega \subseteq \mathbb{R}^n$ means a domain with $n = 2$ or $n = 3$, and $[0, T)$ is a fixed time interval with $0 < T \leq \infty$. The full nonlinear Navier-Stokes system in $[0, T) \times \Omega$ has the form

$$\begin{aligned} u_t - \nu \Delta u + u \cdot \nabla u + \nabla p &= f \ , \quad \text{div } u = 0 \ , \\ u|_{\partial \Omega} &= 0 \ , \quad u(0) = u_0 \end{aligned} \tag{1.1.1}$$

where the boundary condition $u|_{\partial\Omega} = 0$ is omitted if $\Omega = \mathbb{R}^n$. We refer to [Lad69, Chap. 6], [Tem77, Chap. III, 3], [Ama00] concerning weak solutions of these equations.

As before, $u(t, x) = (u_1(t, x), \ldots, u_n(t, x))$ means the unknown velocity, $p(t, x)$ the unknown pressure, $f(t, x) = (f_1(t, x), \ldots, f_n(t, x))$ the given exterior force and $u_0(x)$ the given initial velocity with $t \in [0, T)$, $x = (x_1, \ldots, x_n) \in \Omega$, $\nu > 0$ means the viscosity constant.

The exterior force f will be a distribution of the form $f = f_0 +$ div F with

$$f_0 \in L^1_{loc}([0, T); L^2(\Omega)^n) \ , \quad F \in L^1_{loc}([0, T); L^2(\Omega)^{n^2}).$$

Thus it holds that $f_0(t) \in L^2(\Omega)^n$ for almost all $t \in [0, T)$, and the Helmholtz projection $Pf_0(t) \in L^2_\sigma(\Omega)$ is well defined. $F(t) = (F_{jl}(t))_{j,l=1}^n \in L^2(\Omega)^{n^2}$ is a matrix function for almost all $t \in [0, T)$ and div $F(t) = (D_1 F_{1l}(t) + \cdots + D_n F_{nl}(t))_{l=1}^n$ is a distribution.

We consider test functions $v = (v_1, \ldots, v_n)$ from $C_0^\infty((0, T) \times \Omega)^n$, from $C_0^\infty((0, T); C_{0,\sigma}^\infty(\Omega))$, or from $C_0^\infty([0, T); C_{0,\sigma}^\infty(\Omega))$, see (1.4.1), IV, and (1.4.2), IV, for these spaces. See Section 2.1, III, concerning $L^2_\sigma(\Omega)$ and $W^{1,2}_{0,\sigma}(\Omega)$. As before we consider f as a functional defined by

$$\begin{aligned}
[f,v]_{\Omega,T} &:= \ < f_0, v >_{\Omega,T} \ + \ [\mathrm{div}\ F, v]_{\Omega,T} \\
&:= \ < f_0, v >_{\Omega,T} \ - \ < F, \nabla v >_{\Omega,T} \\
&:= \ \int_0^T < f_0, v >_\Omega \ dt \ - \int_0^T < F, \nabla v >_\Omega \ dt \\
&:= \ \int_0^T \int_\Omega f_0 \cdot v \, dx \, dt \ - \int_0^T \int_\Omega F \cdot (\nabla v) \, dx \, dt \,,
\end{aligned}$$

$v \in C_0^\infty([0,T); C_{0,\sigma}^\infty(\Omega))$.

Here we consider only solutions u of (1.1.1) with

$$u \in L^\infty_{loc}([0,T); L^2_\sigma(\Omega)) \cap L^2_{loc}([0,T); W^{1,2}_{0,\sigma}(\Omega)). \tag{1.1.2}$$

Recall, this means that

$$u \in L^\infty(0,T'; L^2_\sigma(\Omega)) \cap L^2(0,T'; W^{1,2}_{0,\sigma}(\Omega))$$

for all T' with $0 < T' < T$. In particular we conclude that the (energy) quantity

$$E_{T'}(u) := \frac{1}{2} \, \|u\|^2_{2,\infty;T'} + \nu \, \|\nabla u\|^2_{2,2;T'} \tag{1.1.3}$$

is finite for $0 < T' < T$. An important role plays the condition

$$E_\infty(u) := \frac{1}{2} \, \|u\|^2_{2,\infty;\infty} + \nu \, \|\nabla u\|^2_{2,2;\infty} \ < \infty. \tag{1.1.4}$$

We obtain the notion of a weak solution u, see below, when we treat each term of (1.1.1) as a functional defined on $C_0^\infty([0,T); C_{0,\sigma}^\infty(\Omega))$. These functionals are given by

$$[u_t, v]_{\Omega,T} \ := \ - < u_0, v(0) >_\Omega - < u, v_t >_{\Omega,T}$$

with $v(0) = v(0,\cdot)$, $v' = v_t = \frac{d}{dt}\, v$, by

$$\begin{aligned}
[-\nu\Delta u, v]_{\Omega,T} &:= \ \nu < \nabla u, \nabla v >_{\Omega,T}, \\
[\nabla p, v]_{\Omega,T} &:= \ - < p, \ \mathrm{div}\ v >_{\Omega,T} = 0 \,,
\end{aligned}$$

and by

$$\begin{aligned}
[u \cdot \nabla u, v]_{\Omega,T} &:= \ < u \cdot \nabla u, v >_{\Omega,T} = < (u_1 D_1 + \ldots + u_n D_n) u, v >_{\Omega,T} \\
&= \ < \mathrm{div}\ (u\,u), v >_{\Omega,T} = \ - < u\,u, \nabla v >_{\Omega,T} \\
&= \ - \int_0^T \int_\Omega (u\,u) \cdot \nabla v \, dx \, dt
\end{aligned}$$

with $u\,u := (u_j u_l)_{j,l=1}^n$, $v \in C_0^\infty([0,T); C_{0,\sigma}^\infty(\Omega))$. See Section 3.1, III, concerning this calculation.

Using (1.1.2) and Hölder's inequality we see that

$$\|u \cdot \nabla u\|_{1,1;T'} = \int_0^{T'} \int_\Omega |u \cdot \nabla u| \, dx \, dt \le C \, \|u\|_{2,\infty;T'} \|\nabla u\|_{2,2;T'} < \infty \tag{1.1.5}$$

for all T' with $0 < T' < T$; $C > 0$ is a constant. This enables us to justify the following calculation, see Lemma 1.2.1 below:

$$\begin{aligned}
\operatorname{div}\,(u\,u) &= (D_1(u_1 u_l) + \cdots + D_n(u_n u_l))_{l=1}^n \\
&= D_1(u_1 u) + \cdots + D_n(u_n u) \\
&= u_1 D_1 u + \cdots + u_n D_n u \\
&= (u_1\, D_1 u_l + \cdots + u_n\, D_n u_l)_{l=1}^n \\
&= (u_1 D_1 + \cdots + u_n D_n)\, u = u \cdot \nabla u\,.
\end{aligned}$$

In particular we see that

$$< u \cdot \nabla u, v >_{\Omega,T} \; = \int_0^T \int_\Omega (u \cdot \nabla u) \cdot v \, dx \, dt$$

is well defined and the definition below is meaningful.

If u possesses some smoothness properties and satisfies (1.1.1) in the classical sense, then the condition (1.1.7) below easily follows using partial integration. This motivates the notion of a weak solution.

1.1.1 Definition *Let $\Omega \subseteq \mathbb{R}^n$, $n = 2, 3$, be any domain , let $0 < T \le \infty$, $u_0 \in L^2_\sigma(\Omega)$, and let $f = f_0 + \operatorname{div} F$ with*

$$f_0 \in L^1_{loc}([0,T); L^2(\Omega)^n) \;, \quad F \in L^1_{loc}([0,T); L^2(\Omega)^{n^2}).$$

Then

$$u \in L^\infty_{loc}([0,T); L^2_\sigma(\Omega)) \cap L^2_{loc}([0,T); W^{1,2}_{0,\sigma}(\Omega))$$

is called a **weak solution** *of the Navier-Stokes system* (1.1.1) *with data f, u_0, iff*

$$\begin{aligned}
- < u, v_t >_{\Omega,T} \; + \; \nu < \nabla u, \nabla v >_{\Omega,T} \; + < u \cdot \nabla u, v >_{\Omega,T} \\
= \; < u_0, v(0) >_\Omega + [f, v]_{\Omega,T}
\end{aligned} \tag{1.1.6}$$

holds for all $v \in C_0^\infty([0,T); C_{0,\sigma}^\infty(\Omega))$.

If u is such a weak solution, and p a distribution such that

$$u_t - \nu\Delta u + u \cdot \nabla u + \nabla p = f \tag{1.1.7}$$

holds in the sense of distributions in $(0,T) \times \Omega$, then p is called an **associated pressure** *of u.*

Our aim is to develop the theory of weak solutions using the Stokes operator A, the fractional powers A^α, $-1 \le \alpha \le 1$, and the semigroup $\{S(t); t \ge 0\}$, $S(t) = e^{-tA}$, see Section 2, III, and (1.5.7), IV, for these operators.

In particular, we need the properties

$$\nu < \nabla u, \nabla v >_\Omega \; = \; \nu \int_\Omega (\nabla u) \cdot (\nabla v)\, dx \; = < A^{\frac{1}{2}} u, A^{\frac{1}{2}} v >_\Omega \tag{1.1.8}$$

and $\nu^{\frac{1}{2}} \|\nabla u\|_2 = \|A^{\frac{1}{2}} u\|_2$ for $u, v \in D(A^{\frac{1}{2}}) = W^{1,2}_{0,\sigma}(\Omega)$, see Lemma 2.2.1, III.

Recall that the operators P and $A^{-\frac{1}{2}}$ have a natural extended meaning, see Section 2.5, III, and Lemma 2.6.1, III. $A^{-\frac{1}{2}}$ is extended by closure from the domain $D(A^{-\frac{1}{2}}) = R(A^{\frac{1}{2}})$ to the completion $\widehat{D}(A^{-\frac{1}{2}})$ with respect to the norm $\|A^{-\frac{1}{2}} u\|_2$. P div F means simply the restriction of the distribution div F to the test space $C^\infty_{0,\sigma}(\Omega)$. With F and u as above we get the important estimates

$$\|A^{-\frac{1}{2}} P \text{ div } F(t)\|_2 \; \le \; \nu^{-\frac{1}{2}} \|F(t)\|_2 \tag{1.1.9}$$

and similarly

$$\|A^{-\frac{1}{2}} P \text{ div } (u(t)\, u(t))\|_2 \; \le \; \nu^{-\frac{1}{2}} \|u(t)\, u(t)\|_2 \tag{1.1.10}$$

for almost all $t \in [0,T)$, see Lemma 2.6.1, III.

First we investigate several properties of weak solutions, an existence proof will be given in Section 3.

Regularity and uniqueness of weak solutions u are unsolved problems in the three-dimensional case $n = 3$. To prove these properties we need Serrin's additional condition

$$\|u\|_{q,s;T} = \Big(\int_0^T \|u(t)\|_q^s \, dt\Big)^{\frac{1}{s}} < \infty \tag{1.1.11}$$

with $n < q < \infty$, $2 < s < \infty$, $\frac{n}{q} + \frac{2}{s} \le 1$, see Theorem 1.5.1. The (special) class of weak solutions u defined by (1.1.11) is called Serrin's class, see [Ser63].

We will see that (1.1.11) is always satisfied if $n = 2$, see Theorem 1.5.3.

If $n = 3$, we have no general existence result for weak solutions within Serrin's class. Up to now the existence in this class can be shown for $n = 3$ only under an additional smallness condition on the data f, u_0, see Section 4.

In the two-dimensional case we have a general uniqueness and regularity result, see Theorem 4.2.1. Thus we have the following important open problem in the three-dimensional case:

Prove uniqueness and regularity (under smoothness assumptions on f, u_0) of a given weak solution u for each interval $[0,T)$, or prove the existence of at least one weak solution u in Serrin's class.

1.2 Properties of the nonlinear term $u \cdot \nabla u$

In this subsection we prepare some technical properties of the nonlinear term $u \cdot \nabla u$. They are essentially based on embedding estimates of the fractional powers A^α of the Stokes operator A, see Lemma 2.4.2, III. We consider the more general expression $u \cdot \nabla v$.

For $u = (u_1, \dots, u_n)$ and $v = (v_1, \dots v_n)$ contained in the class defined by (1.1.2), we define the matrix function

$$u\,v := (u_j v_l)_{j,l=1}^n \,,$$

and we apply the operator div to the columns of this matrix. Thus we set

$$\begin{aligned} \operatorname{div}(u\,v) = \operatorname{div} u\,v &:= (D_1(u_1 v_l) + \cdots + D_n(u_n v_l))_{l=1}^n \\ &= D_1(u_1 v) + \cdots + D_n(u_n v)\,. \end{aligned}$$

Since div $u = 0$ we get

$$\begin{aligned} \operatorname{div}(u\,v) &= u_1 D_1 v + \cdots + u_n D_n v \\ &= (u_1 D_1 + \cdots + u_n D_n) v \\ &= u \cdot \nabla v. \end{aligned} \tag{1.2.1}$$

Recall the notation

$$\begin{aligned} E_T(u) &:= \frac{1}{2}\,\|u\|^2_{2,\infty;T} + \nu\,\|\nabla u\|^2_{2,2;T} \\ &= \frac{1}{2}\Big(\operatorname*{ess-sup}_{t\in[0,T)} \|u(t)\|_2\Big)^2 + \nu \int_0^T \|\nabla u\|_2^2\,d\tau \end{aligned} \tag{1.2.2}$$

for $0 < T \le \infty$.

1.2.1 Lemma *Let $\Omega \subseteq \mathbb{R}^n$, $n = 2, 3$, be any domain, let $0 < T \le \infty$, and let*

$$u\,,\ v \in L^\infty_{loc}([0,T); L^2_\sigma(\Omega)) \cap L^2_{loc}([0,T); W^{1,2}_{0,\sigma}(\Omega)). \tag{1.2.3}$$

Then we have:

a) $|v(t)|\,|\nabla u(t)|\,|u(t)|\ ,\quad |v(t)|\,|\nabla|u(t)|^2|\ ,\quad |\nabla v(t)|\,|u(t)|^2\ \in L^1(\Omega)$

and

$$\begin{aligned} < v(t)\cdot\nabla u(t), u(t) >_\Omega &= < \text{div}\,(v(t)\,u(t)), u(t) >_\Omega \\ &= - < v(t)\,u(t), \nabla u(t) >_\Omega \\ &= -\frac{1}{2} < v(t), \nabla|u(t)|^2 >_\Omega \\ &= \frac{1}{2} < \text{div}\; v(t), |u(t)|^2 >_\Omega = 0 \end{aligned}$$

for almost all $t \in [0,T)$.

b)
$$\begin{aligned} \|u\|_{q,s;T'} &\le C\,\nu^{-\frac{1}{s}}\,\|A^{\frac{1}{s}}u\|_{2,s;T'} \\ &\le C\,\nu^{-\frac{1}{s}}\,\|A^{\frac{1}{2}}u\|_{2,2;T'}^{2/s}\,\|u\|_{2,\infty;T'}^{1-2/s} \\ &\le C\,\nu^{-\frac{1}{s}}\,(\|A^{\frac{1}{2}}u\|_{2,2;T'} + \|u\|_{2,\infty;T'}) \\ &\le C'\,\nu^{-\frac{1}{s}}E_{T'}(u)^{\frac{1}{2}}\ < \infty \end{aligned}$$

with $2 \le q < \infty$, $2 \le s \le \infty$ *satisfying*

$$\frac{n}{q} + \frac{2}{s} = \frac{n}{2}\,,$$

and with $0 < T' < T$; $C = C(s,n) > 0$, $C' = C'(s,n) > 0$.

c)
$$\begin{aligned} \|u\,v\|_{q,s;T'} &\le C\nu^{-\frac{1}{s}}\|A^{\frac{1}{2s}}u\|_{2,2s;T'}\,\|A^{\frac{1}{2s}}v\|_{2,2s;T'} \\ &\le C\,\nu^{-\frac{1}{s}}(\|A^{\frac{1}{2}}u\|_{2,2;T'} + \|u\|_{2,\infty;T'})\,(\|A^{\frac{1}{2}}v\|_{2,2;T'} + \|v\|_{2,\infty;T'}) \\ &\le C'\nu^{-\frac{1}{s}}E_{T'}(u)^{\frac{1}{2}}\,E_{T'}(v)^{\frac{1}{2}}\ < \infty \end{aligned}$$

with $1 \le q < \infty$, $1 \le s \le \infty$ *satisfying*

$$\frac{n}{q} + \frac{2}{s} = n\,,$$

and with $0 < T' < T$, $C = C(s,n) > 0$, $C' = C'(s,n) > 0$.

d)
$$\begin{aligned} \|u\,u\|_{2,s;T'} &\le C\,\nu^{-\frac{n}{4}}\,\|A^{\frac{n}{8}}u\|_{2,2s;T'}^2 \\ &\le C\,\nu^{-\frac{n}{4}}\,\|A^{\frac{1}{2}}u\|_{2,sn/2;T'}^{n/2}\,\|u\|_{2,\infty;T'}^{2-n/2} \\ &\le C\,\nu^{-\frac{n}{4}}\,(T')^{\frac{1}{s}-\frac{n}{4}}\,\|A^{\frac{1}{2}}u\|_{2,2;T'}^{n/2}\,\|u\|_{2,\infty;T'}^{2-n/2} \\ &\le C'\,\nu^{-\frac{n}{4}}\,(T')^{\frac{1}{s}-\frac{n}{4}}\,E_{T'}(u)\ < \infty \end{aligned}$$

with $1 \le s \le \frac{4}{n}$, $0 < T' < T$, $C = C(n) > 0$, $C' = C'(n) > 0$.

e) $$\|u \cdot \nabla v\|_{q,s;T'} \leq C\,\nu^{-\frac{1}{s}}\,\|A^{\frac{1}{2}}u\|_{2,2;T'}^{\frac{2}{s}-1}\,\|u\|_{2,\infty;T'}^{2-\frac{2}{s}}\,\|A^{\frac{1}{2}}v\|_{2,2;T'}$$
$$\leq C'\,\nu^{-\frac{1}{s}} E_{T'}(u)^{\frac{1}{2}} E_{T'}(v)^{\frac{1}{2}} < \infty$$

with $1 \leq s < 2,\ 1 \leq q < 2$ *satisfying*

$$\frac{n}{q} + \frac{2}{s} = n + 1,$$

and with $0 < T' < T,\ C = C(s,n) > 0, C' = C'(s,n) > 0$.

Proof. To prove a), we use the embedding estimate (2.4.6), III, with $\alpha = \frac{n}{8}$, $q = 4$, $2\alpha + \frac{n}{4} = \frac{n}{2}$, the interpolation inequality (2.2.8), III, and Hölder's inequality. This yields

$$\begin{aligned} \|u(t)\|_4 &\leq C\,\nu^{-\alpha}\,\|A^{\alpha}u(t)\|_2 \\ &\leq C\,\nu^{-\alpha}\,\|A^{\frac{1}{2}}u(t)\|_2^{2\alpha}\,\|u(t)\|_2^{1-2\alpha} \\ &= C\,\|\nabla u(t)\|_2^{2\alpha}\,\|u(t)\|_2^{1-2\alpha} < \infty \end{aligned} \tag{1.2.4}$$

for almost all $t \in [0,T)$, $C = C(n,\nu) > 0$, and correspondingly with $u(t)$ replaced by $v(t)$. Thus we obtain

$$\begin{aligned} \|\,|v(t)|\,|\nabla u(t)|\,|u(t)|\,\|_1 &\leq C_1\,\|v(t)\|_4\,\|\nabla u(t)\|_2\,\|u(t)\|_4 \\ &\leq C_2\,\|\nabla v(t)\|_2^{2\alpha}\,\|v(t)\|_2^{1-2\alpha}\,\|\nabla u(t)\|_2\,\|\nabla u(t)\|_2^{2\alpha}\,\|u(t)\|_2^{1-2\alpha} \\ &< \infty, \end{aligned}$$

$$\begin{aligned} \|\,|v(t)|\,|\nabla |u(t)|^2|\,\|_1 &\leq C_1\,\|v(t)\|_4\,\|\nabla|u(t)|^2\|_{\frac{4}{3}} \\ &\leq C_2\,\|\nabla v(t)\|_2^{2\alpha}\,\|v(t)\|_2^{1-2\alpha}\,\|u(t)\nabla u(t)\|_{\frac{4}{3}} \\ &\leq C_3\,\|\nabla v(t)\|_2^{2\alpha}\,\|v(t)\|_2^{1-2\alpha}\,\|u(t)\|_4\|\nabla u(t)\|_2 \\ &\leq C_4\,\|\nabla v(t)\|_2^{2\alpha}\,\|v(t)\|_2^{1-2\alpha}\,\|\nabla u(t)\|_2^{2\alpha}\|u(t)\|_2^{1-2\alpha}\,\|\nabla u(t)\|_2 \\ &< \infty, \end{aligned}$$

and

$$\begin{aligned} \|\,|\nabla v(t)|\,|u(t)|^2\,\|_1 &\leq C_1\,\|\nabla v(t)\|_2\,\|u(t)\|_4^2 \\ &\leq C_2\,\|\nabla v(t)\|_2\,\|\nabla u(t)\|_2^{4\alpha}\,\|u(t)\|_2^{2-4\alpha} < \infty, \end{aligned}$$

for almost all $t \in [0,T)$, with constants $C_1, C_2, C_3, C_4 > 0$ depending on n and ν. This proves the first assertion under a).

To prove the second assertion we use the definition

$$W^{1,2}_{0,\sigma}(\Omega) = \overline{C^\infty_{0,\sigma}(\Omega)}^{\|\cdot\|_{W^{1,2}}},$$

and for almost all $t \in [0,T)$ we can choose sequences $(u_j(t))_{j=1}^\infty$, $(v_j(t))_{j=1}^\infty$ in $C^\infty_{0,\sigma}(\Omega)$ such that

$$\lim_{j\to\infty} \|u(t) - u_j(t)\|_{W^{1,2}(\Omega)^n} = 0 \ , \quad \lim_{j\to\infty} \|v(t) - v_j(t)\|_{W^{1,2}(\Omega)^n} = 0 \, .$$

Further, the above estimates hold with $u(t), v(t)$ replaced by $u(t) - u_j(t)$, $v(t) - v_j(t)$. This yields

$$< v(t) \cdot \nabla u(t), u(t) >_\Omega = \lim_{j\to\infty} < v_j(t) \cdot \nabla u_j(t), u_j(t) >_\Omega \, ,$$

and correspondingly for $<\operatorname{div}\,(v(t)\,u(t)), u(t)>_\Omega$, $< v(t)\,u(t), \nabla u(t) >_\Omega$, and $< v(t), \nabla |u(t)|^2 >_\Omega$. The desired equations under a) are clear by an elementary calculation with $u(t), v(t)$ replaced by $u_j(t), v_j(t)$. The approximation property above shows that the limit as $j \to \infty$ exists. This proves a).

To prove b) we use (2.4.6), III, with $\alpha = \frac{1}{s}$, $2\alpha + \frac{n}{q} = \frac{n}{2}$, and together with the interpolation inequality (2.2.8), III, we see that

$$\|u(t)\|_q \leq C\,\nu^{-\alpha}\,\|A^\alpha u(t)\|_2 \leq C\,\nu^{-\alpha}\,\|A^{\frac{1}{2}} u(t)\|_2^{2\alpha}\,\|u(t)\|_2^{1-2\alpha} \tag{1.2.5}$$

for almost all $t \in [0,T)$ and $C = C(s,n) > 0$. This leads to

$$\begin{aligned}
\|u\|_{q,s;T'} &= \Big(\int_0^{T'} \|u\|_q^s \, dt \Big)^{\frac{1}{s}} \\
&\leq C\,\nu^{-\alpha} \Big(\int_0^{T'} \|A^{\frac{1}{2}} u\|_2^{2\alpha s}\, \|u\|_2^{s(1-2\alpha)}\, dt \Big)^{\frac{1}{s}} \\
&\leq C\,\nu^{-\alpha}\, \|A^{\frac{1}{2}} u\|_{2,2;T'}^{2/s}\, \|u\|_{2,\infty;T'}^{1-\frac{2}{s}} \\
&\leq C\,\nu^{-\alpha}\,(\|A^{\frac{1}{2}} u\|_{2,2;T'} + \|u\|_{2,\infty;T'}) \\
&\leq C\,\nu^{-\alpha}\sqrt{2}\,(\|A^{\frac{1}{2}} u\|_{2,2;T}^2 + \|u\|_{2,\infty;T}^2)^{\frac{1}{2}} \\
&\leq C'\,\nu^{-\alpha}\,(\nu\,\|\nabla u\|_{2,2;T'}^2 + \frac{1}{2}\,\|u\|_{2,\infty;T'}^2)^{\frac{1}{2}}
\end{aligned}$$

with C, C' depending on s, n.

To prove c) we apply Hölder's inequality, and get

$$\|u\,v\|_{q,s;T'} \leq C\,\|u\|_{2q,2s;T'}\,\|v\|_{2q,2s;T'}$$

with $C = C(n)$. Next we observe that $\frac{n}{2q} + \frac{2}{2s} = \frac{n}{2}$, and we apply the estimate above for u and for v with q, s replaced by $2q, 2s$. This proves c).

To prove d) we apply Hölder's inequality, inequality (1.2.4), and Young's inequality (3.3.8), I. This yields

$$\begin{aligned}
\|u\,u\|_{2,s;T'} &\le C_1\,\|u\|^2_{4,2s;T'} \le C_2\,\nu^{-\frac{n}{4}}\,\|A^{\frac{n}{8}}u\|^2_{2,2s;T'} \\
&\le C_3\,\nu^{-\frac{n}{4}} \left(\int_0^{T'} \|A^{\frac{1}{2}}u\|_2^{2s\cdot\frac{n}{4}}\,\|u\|_2^{2s(1-\frac{n}{4})}\,dt \right)^{\frac{1}{s}} \\
&\le C_3\,\nu^{-\frac{n}{4}}\,\|A^{\frac{1}{2}}u\|^{n/2}_{2,\frac{sn}{2};T'}\,\|u\|^{2-\frac{n}{2}}_{2,\infty;T'} \\
&\le C_3\,\nu^{-\frac{n}{4}} \left((T')^{1-\frac{sn}{4}} \left(\int_0^{T'} \|A^{\frac{1}{2}}u\|_2^2\,dt \right)^{\frac{sn}{4}} \right)^{\frac{2}{sn}\cdot\frac{n}{2}} \|u\|^{2-\frac{n}{2}}_{2,\infty;T'} \\
&\le C_3\,\nu^{-\frac{n}{4}}\,(T')^{\frac{1}{s}-\frac{n}{4}}\,\|A^{\frac{1}{2}}u\|^{\frac{n}{2}}_{2,2;T'}\,\|u\|^{2-\frac{n}{2}}_{2,\infty;T'} \\
&\le C_4\,\nu^{-\frac{n}{4}}\,(T')^{\frac{1}{s}-\frac{n}{4}}\,E_{T'}(u)
\end{aligned}$$

with constants C_1, C_2, C_3, C_4 depending on s, n. This proves d).

To prove e) we use Hölder's inequality and get

$$\|u \cdot \nabla v\|_{q,s;T'} \le C_1\,\|u\|_{(\frac{1}{q}-\frac{1}{2})^{-1},(\frac{1}{s}-\frac{1}{2})^{-1};T'}\,\|\nabla v\|_{2,2;T'}.$$

Next we observe that $n(\frac{1}{q} - \frac{1}{2}) + 2(\frac{1}{s} - \frac{1}{2}) = n + 1 - \frac{n}{2} - 1 = \frac{n}{2}$, and we apply b) with q, s replaced by $(\frac{1}{q} - \frac{1}{2})^{-1}$, $(\frac{1}{s} - \frac{1}{2})^{-1}$. This yields

$$\begin{aligned}
\|u\|_{(\frac{1}{q}-\frac{1}{2})^{-1},(\frac{1}{s}-\frac{1}{2})^{-1};T'} &\le C_2\,\nu^{-(\frac{1}{s}-\frac{1}{2})}\,\|A^{\frac{1}{2}}u\|^{2(\frac{1}{s}-\frac{1}{2})}_{2,2;T'}\,\|u\|^{1-2(\frac{1}{s}-\frac{1}{2})}_{2,\infty;T'} \\
&\le C_3\,\nu^{-(\frac{1}{s}-\frac{1}{2})}\,E_{T'}(u)^{\frac{1}{2}} < \infty\,,
\end{aligned}$$

and with $\|\nabla v\|_{2,2:T'} = \nu^{-\frac{1}{2}}\,\|A^{\frac{1}{2}}v\|_{2,2;T'}$ we get

$$\begin{aligned}
\|u \cdot \nabla v\|_{q,s;T'} &\le C_4\,\nu^{-\frac{1}{s}}\,\|A^{\frac{1}{2}}u\|^{\frac{2}{s}-1}_{2,2;T'}\,\|u\|^{2-\frac{2}{s}}_{2,\infty;T'}\,\|A^{\frac{1}{2}}v\|_{2,2;T'} \\
&\le 2C_4\,\nu^{-\frac{1}{s}}\,E_{T'}(u)^{\frac{1}{s}-\frac{1}{2}}\,E_{T'}(u)^{1-\frac{1}{s}}\,E_{T'}(v)^{\frac{1}{2}} \\
&\le 2C_4\,\nu^{-\frac{1}{s}}\,E_{T'}(u)^{\frac{1}{2}}\,E_{T'}(v)^{\frac{1}{2}} < \infty
\end{aligned}$$

with constants C_1, C_2, C_3, C_4 depending on s, n. The proof of the lemma is complete. □

1.2.2 Remark Suppose $T = \infty$ in Lemma 1.2.1, and suppose that

$$E_\infty(u) < \infty \ , \quad E_\infty(v) < \infty. \tag{1.2.6}$$

Then the estimates in b), c), and e) of this lemma remain valid for $T' = \infty$. To prove this we observe that the constants C, C' in these estimates do not depend on T'. This enables us to let $T' \to \infty$.

The estimate in d) does not depend on T' only if $s = \frac{4}{n}$. Thus in this case we get

$$\|u\,u\|_{2,\frac{4}{n};\infty} \;\leq\; C\,\nu^{-\frac{n}{4}}E_\infty(u) \;<\infty \tag{1.2.7}$$

with $C = C(n) > 0$.

1.3 Integral equation for weak solutions and weak continuity

In the following subsections we investigate important properties of weak solutions. A basic property is the integral equation (1.3.5) in the theorem below. To prove it, we write the Navier-Stokes system (1.1.1) in the form

$$u_t - \nu\Delta u + \nabla p = \tilde{f} \ , \quad \text{div } u = 0 \ , \quad u|_{\partial\Omega} = 0 \ , \quad u(0) = u_0 \tag{1.3.1}$$

with

$$\tilde{f} = f_0 + \ \text{div } \tilde{F} \ , \ \ \tilde{F} := F - u\,u,$$

and apply the representation formula (2.4.4), IV, from the linear theory. Using the properties of the nonlinear term $u \cdot \nabla u = \text{div } (u\,u)$ in the preceding subsection, we will show that a weak solution u of the nonlinear system (1.1.1) with data f, u_0 is also a weak solution of the linear system (1.3.1) with data $\tilde{f}, u_0$.

This fact enables us to apply the linear theory of Chapter IV to get basic properties of weak solutions of the nonlinear system.

The first property we develop in this way is the weak continuity of a weak solution u. This means by definition that

$$t \mapsto < u(t), w >_\Omega \ , \quad t \in [0, T)$$

is continuous for each fixed $w \in L^2_\sigma(\Omega)$.

1.3.1 Theorem *Let $\Omega \subseteq \mathbb{R}^n$, $n = 2, 3$, be any domain, let $0 < T \leq \infty$, $1 < s \leq \frac{4}{n}$, let $u_0 \in L^2_\sigma(\Omega)$, $f = f_0 +$ div F with*

$$f_0 \in L^1_{loc}([0, T); L^2(\Omega)^n) \ , \quad F \in L^s_{loc}([0, T); L^2(\Omega)^{n^2}) \,, \tag{1.3.2}$$

and let

$$u \in L^\infty_{loc}([0, T); L^2_\sigma(\Omega)) \,\cap\, L^2_{loc}([0, T); W^{1,2}_{0,\sigma}(\Omega)) \,. \tag{1.3.3}$$

Suppose u is a weak solution of the Navier-Stokes system (1.1.1) *with data f, u_0.*

Then, after redefinition on a null set of $[0,T)$, $u : [0,T) \to L^2_\sigma(\Omega)$ is weakly continuous with $u(0) = u_0$, and

$$\int_0^t S(t-\tau)A^{-\frac{1}{2}}P\,\mathrm{div}\,(F(\tau) - u(\tau)\,u(\tau))\,d\tau \in D(A^{\frac{1}{2}}), \tag{1.3.4}$$

$$\begin{aligned} u(t) &= S(t)u_0 + \int_0^t S(t-\tau)Pf_0(\tau)\,d\tau \\ &\quad + A^{\frac{1}{2}} \int_0^t S(t-\tau)A^{-\frac{1}{2}}P\,\mathrm{div}\,(F(\tau) - u(\tau)\,u(\tau))\,d\tau \end{aligned} \tag{1.3.5}$$

for all $t \in [0,T)$.

Conversely, let u satisfy the conditions (1.3.4) *and* (1.3.5) *at least for almost all $t \in [0,T)$, then u is a weak solution of the Navier-Stokes system* (1.1.1) *with data f, u_0.*

Proof. Note that (1.3.5) can be written in the form

$$u = S(\cdot)u_0 + \mathcal{J}Pf_0 + A^{\frac{1}{2}}\mathcal{J}A^{-\frac{1}{2}}P\ \mathrm{div}\ (F - u\,u) \tag{1.3.6}$$

where $\mathcal{J}$ is the integral operator defined in (1.6.3), IV.

Let u be a weak solution of (1.1.1). Using (1.3.3) and Lemma 1.2.1, d), we see that $u\,u \in L^s_{loc}([0,T); L^2(\Omega)^{n^2})$, and with $\tilde{F} := F - u\,u$, $\tilde{f} := f_0 +\ \mathrm{div}\ \tilde{F}$ we obtain

$$[\tilde{f}, v]_{\Omega,T} = [f, v]_{\Omega,T} - [\mathrm{div}\ (u\,u), v]_{\Omega,T} \tag{1.3.7}$$

for all $v \in C_0^\infty([0,T); C^\infty_{0,\sigma}(\Omega))$. Inserting this in (1.1.6) we see that u is a weak solution of the linear Stokes system (1.3.1) with data $\tilde{f}, u_0$.

Let $0<T'<T$. Then we get $f_0 \in L^1(0,T'; L^2(\Omega)^n)$, $F \in L^s(0,T';\ L^2(\Omega)^{n^2})$, $u\,u \in L^s(0,T'; L^2(\Omega)^{n^2})$ and therefore $\tilde{F} \in L^s(0,T'; L^2(\Omega)^{n^2})$. Using the representation formula in Theorem 2.4.1, IV, we conclude that (1.3.4), (1.3.5) are valid for almost all $t \in [0,T')$. Since $u \in L^\infty(0,T'; L^2_\sigma(\Omega))$, we obtain from Lemma 2.4.2, b), IV, that $u : [0,T') \to L^2_\sigma(\Omega)$ is weakly continuous after a corresponding redefinition, and that (1.3.5) holds for all $t \in [0,T')$. Since T', $0 < T' < T$, is arbitrary, we see that u is weakly continuous on the whole interval $[0,T)$, and that (1.3.5) holds for all $t \in [0,T)$.

Suppose u satisfies (1.3.4), (1.3.5) for almost all $t \in [0,T)$. Then from Theorem 2.4.1, IV, we obtain that u is a weak solution of the linear system (1.3.1) with data $\tilde{f}, u_0$. Using (1.3.7) we see that u is a weak solution of the Navier-Stokes system (1.1.1) with data f, u_0. This proves the theorem. □

1.4 Energy equality and strong continuity

From the physical point of view we would expect that a weak solution u satisfies the energy equality

$$\frac{1}{2}\|u(t)\|_2^2 + \nu \int_0^t \|\nabla u\|_2^2 \, d\tau \;=\; \frac{1}{2}\|u_0\|_2^2 + \int_0^t [f, u]_\Omega \, d\tau$$

for $0 \leq t < T$. Note that the last term

$$\int_0^t [f, u]_\Omega \, d\tau \;=\; \int_0^t < f_0, u >_\Omega \; d\tau - \int_0^t < F, \nabla u >_\Omega \; d\tau \;, \quad t \in [0, T)$$

is well defined under the assumption of the theorem below. To see this we use Hölder's inequality and get

$$\begin{aligned} \int_0^t | < f_0, u >_\Omega | \, d\tau \;&\leq\; \|f_0\|_{2,1;t} \, \|u\|_{2,\infty;t} \; < \infty , \\ \int_0^t | < F, \nabla u >_\Omega | \, d\tau \;&\leq\; \|F\|_{2,2;t} \, \|\nabla u\|_{2,2;t} \; < \infty \end{aligned}$$

for $0 \leq t < T$.

However, if $n = 3$ we do not know in general whether the energy equality is satisfied for each weak solution u. This is an open problem up to now. To prove this equality we need the additional condition (1.4.2), see the next theorem. This condition is always satisfied if $n = 2$, see Theorem 1.4.2. See [Shi74] for this result.

1.4.1 Theorem *Let* $\Omega \subseteq \mathbb{R}^n$, $n = 2, 3$, *be any domain, let* $0 < T \leq \infty$, $u_0 \in L^2_\sigma(\Omega)$, $f = f_0 + \operatorname{div} F$ *with*

$$f_0 \in L^1_{loc}([0, T); L^2(\Omega)^n) \;, \quad F \in L^2_{loc}([0, T); L^2(\Omega)^{n^2}) , \tag{1.4.1}$$

and let

$$u \in L^\infty_{loc}([0, T); L^2_\sigma(\Omega)) \cap L^2_{loc}([0, T); W^{1,2}_{0,\sigma}(\Omega))$$

be a weak solution of the Navier-Stokes system (1.1.1) *with data* f, u_0. *Suppose additionally that*

$$u\,u \in L^2_{loc}([0, T); L^2(\Omega)^{n^2}). \tag{1.4.2}$$

Then, after a redefinition on a null set of $[0, T)$, $u : [0, T) \to L^2_\sigma(\Omega)$ *is strongly continuous, we obtain the energy equality*

$$\frac{1}{2}\|u(t)\|_2^2 + \nu \int_0^t \|\nabla u\|_2^2 \, d\tau \;=\; \frac{1}{2}\|u_0\|_2^2 + \int_0^t [f, u]_\Omega \, d\tau \tag{1.4.3}$$

for all $t \in [0,T)$, *and the inequality*

$$\frac{1}{2}\,\|u\|^2_{2,\infty;T'} + \nu\,\|\nabla u\|^2_{2,2;T'} \;\le\; 2\,\|u_0\|^2_2 + 8\,\|f_0\|^2_{2,1;T'} + 4\nu^{-1}\,\|F\|^2_{2,2;T'} \quad (1.4.4)$$

for all T' *with* $0 < T' < T$.

Proof. Setting $\tilde F := F - u\,u$ as in (1.3.1), and using (1.4.2), (1.4.1), we see that $\tilde F \in L^2_{loc}([0,T); L^2(\Omega)^{n^2})$. As in the proof of Theorem 1.3.1 we conclude that u is a weak solution of the linear system (1.3.1) with data $\tilde f, u_0$ where $\tilde f = f_0 +$ div $\tilde F$.

For each interval $[0,T')$, $0 < T' < T$, we obtain from Theorem 2.3.1, IV, that $u : [0,T') \to L^2_\sigma(\Omega)$ is strongly continuous after a corresponding redefinition, and that

$$\frac{1}{2}\,\|u\|^2_2 + \nu\int_0^t \|\nabla u\|^2_2 \;=\; \frac{1}{2}\,\|u_0\|^2_2 + \int_0^t [\tilde f, u]_\Omega\, d\tau \qquad (1.4.5)$$

for all $t \in [0,T')$. Since T', $0 < T' < T$, is arbitrary, we see that (1.4.5) is valid for all $t \in [0,T)$.

Using Lemma 1.2.1, a), we get

$$\begin{aligned}[\text{div}\; u(t)\,u(t), u(t)]_\Omega &= \;< \text{div}\; u(t)\,u(t), u(t) >_\Omega \\ &= \;- < u(t)\,u(t), \nabla u(t) >_\Omega \\ &= \;-\frac{1}{2} < u(t), \nabla |u(t)|^2 >_\Omega = 0\end{aligned}$$

for almost all $t \in [0,T)$. Therefore we obtain

$$[\tilde f(t), u(t)]_\Omega \;= < f_0, u(t) >_\Omega + [\text{div}\; F(t), u(t)]_\Omega \;=\; [f(t), u(t)]_\Omega$$

for almost all $t \in [0,T)$, and this yields equality (1.4.3).

Inequality (1.4.4) is a consequence of (1.4.3). To prove this we use the same calculation as in the proof of (2.3.4), IV. We get as in (2.3.8)–(2.3.9), IV, that

$$\begin{aligned}\frac{1}{2}\,\|u\|^2_{2,\infty;T'} + \nu\,\|\nabla u\|^2_{2,2;T'} \;&\le\; \|u_0\|^2_2 + 4\,\|f_0\|^2_{2,1;T'} + \frac{1}{4}\,\|u\|^2_{2,\infty;T'} \\ &\quad + 2\nu^{-1}\,\|F\|^2_{2,2;T'} + \frac{\nu}{2}\,\|\nabla u\|^2_{2,2;T'},\end{aligned}$$

and this leads to the inequality (1.4.4). This proves the theorem. □

Using Hölder's inequality we get

$$\|u\,u\|_{2,2;T'} \;\le\; C\,\|u\|_{4,4;T'}\,\|u\|_{4,4;T'}$$

with $0 < T' < T$, $C = C(n) > 0$. Therefore,

$$u \in L^4_{loc}([0,T); L^4(\Omega)^n) \tag{1.4.6}$$

is a sufficient condition for (1.4.2) in Theorem 1.4.1.

The following theorem yields the validity of the energy equality in the case $n = 2$.

1.4.2 Theorem ($n = 2$) *Let $\Omega \subseteq \mathbb{R}^2$ be any two-dimensional domain, let $0 < T \le \infty$, $u_0 \in L^2_\sigma(\Omega)$, $f = f_0 + \operatorname{div} F$ with*

$$f_0 \in L^1_{loc}([0,T); L^2(\Omega)^2) \;, \quad F \in L^2_{loc}([0,T); L^2(\Omega)^4),$$

and let

$$u \in L^\infty_{loc}([0,T); L^2_\sigma(\Omega)) \cap L^2_{loc}([0,T); W^{1,2}_{0,\sigma}(\Omega))$$

be a weak solution of the Navier-Stokes system (1.1.1) with data f, u_0.

Then, after a redefinition on a null set of $[0,T)$, $u : [0,T) \to L^2_\sigma(\Omega)$ is strongly continuous, we obtain the energy equality

$$\frac{1}{2}\,\|u(t)\|_2^2 + \nu \int_0^t \|\nabla u\|_2^2\, d\tau \;=\; \frac{1}{2}\,\|u_0\|_2^2 + \int_0^t [f, u]_\Omega\, d\tau \tag{1.4.7}$$

for all $t \in [0,T)$, and the inequality

$$\frac{1}{2}\,\|u\|^2_{2,\infty;T'} + \nu\,\|\nabla u\|^2_{2,2;T'} \;\le\; 2\,\|u_0\|_2^2 + 8\,\|f_0\|^2_{2,1;T'} + 4\nu^{-1}\,\|F\|^2_{2,2;T'} \tag{1.4.8}$$

for all T' with $0 < T' < T$.

Proof. Using Lemma 1.2.1, d), with $s = \frac{4}{n} = 2$ we get the inequality

$$\|u\,u\|_{2,2;T'} \;\le\; C\,\nu^{-\frac{1}{2}} E_{T'}(u) \;<\; \infty \tag{1.4.9}$$

with $E_{T'}(u)$ in (1.2.2) and with $C = C(n) > 0$. Thus (1.4.2) is satisfied and the result follows from Theorem 1.4.1. This proves the result. $\square$

Consider a weak solution u as in Theorem 1.4.1, and suppose the additional condition

$$u \in L^s_{loc}([0,T); L^q(\Omega)^n) \tag{1.4.10}$$

with $n \le q < \infty$, $2 < s \le \infty$, $q > 2$ if $n = 2$, and with

$$\frac{n}{q} + \frac{2}{s} \;\le\; 1. \tag{1.4.11}$$

Then we will show that

$$u\,u \in L^2_{loc}([0,T); L^2(\Omega)^{n^2}), \tag{1.4.12}$$

and we obtain the result of Theorem 1.4.1.

Without loss of generality we may assume that (1.4.10) holds with $\frac{n}{q} + \frac{2}{s} = 1$. Indeed, if $\frac{n}{q} + \frac{2}{s} < 1$, we choose $2 < s_1 < s$ so that $\frac{n}{q} + \frac{2}{s_1} = 1$. Then we choose $\gamma > 1$ with $\frac{1}{s_1} = \frac{1}{s} + \frac{1}{\gamma}$, and with Hölder's inequality we get

$$\|u\|_{q,s_1;T'} \;\le\; (T')^{\frac{1}{\gamma}}\, \|u\|_{q,s;T'} \;<\; \infty$$

for $0 < T' < T$, and (1.4.10) holds with s replaced by s_1. Thus we may assume that (1.4.10) holds with

$$\frac{n}{q} + \frac{2}{s} = 1. \tag{1.4.13}$$

Now we choose $2 < q_1 < \infty$, $2 \le s_1 \le \infty$ in such a way that $\frac{1}{2} = \frac{1}{q_1} + \frac{1}{q}$, $\frac{1}{2} = \frac{1}{s_1} + \frac{1}{s}$. Then

$$\frac{n}{q_1} + \frac{2}{s_1} \;=\; n\,(\frac{1}{2} - \frac{1}{q}) + 2\,(\frac{1}{2} - \frac{1}{s}) \;=\; \frac{n}{2} + 1 - 1 \;=\; \frac{n}{2}\,,$$

and from Lemma 1.2.1, b), we obtain

$$\|u\|_{q_1,s_1;T'} \;\le\; C\,\nu^{-\frac{1}{s_1}}\, E_{T'}(u)^{\frac{1}{2}} \;<\; \infty \tag{1.4.14}$$

with $0 < T' < T$, $C = C(s_1, n) > 0$. Using Hölder's inequality we get

$$\|u\,u\|_{2,2;T'} \;\le\; C\, \|u\|_{q_1,s_1;T'}\, \|u\|_{q,s;T'} \;<\; \infty \tag{1.4.15}$$

with $C = C(n) > 0$.

Thus (1.4.10) with (1.4.11) is a sufficient condition for (1.4.2) in Theorem 1.4.1.

Under the assumptions of Theorem 1.4.1 we can prove the more general energy equality

$$\frac{1}{2}\, \|u(t)\|_2^2 + \nu \int_r^t \|\nabla u\|_2^2\, d\tau \;=\; \frac{1}{2}\, \|u(r)\|_2^2 + \int_r^t [f, u]_\Omega\, d\tau \tag{1.4.16}$$

for all r, t with $0 \le r \le t < T$. To prove this we let $r \ge 0$ be fixed, set $\tilde{u}(t) := u(t + r)$, $t \ge 0$, $\tilde{u}_0 := \tilde{u}(r)$, and apply Theorem 1.4.1 with u replaced by $\tilde{u}$.

The equality (1.4.16) is called the **generalized energy equality**. The corresponding inequality plays an important role in the theory, see Section 3.6.

1.5 Serrin's uniqueness condition

The uniqueness of weak solutions of the Navier-Stokes system (1.1.1) is an open problem if $n = 3$. The following theorem shows the uniqueness under the additional conditions (1.5.3) and (1.5.4) below. If $n = 2$, these conditions are always satisfied. If $n = 3$, they determine a restricted class of weak solutions.

The inequality (1.5.3) is called the **energy inequality**. We know that there exists at least one weak solution u satisfying the energy inequality, see Section 3. The property (1.5.4) is called **Serrin's condition**.

If $n = 3$, the existence of a weak solution satisfying Serrin's condition has been shown up to now only under an additional smallness assumption on the data, see Section 3.

The following uniqueness result is due to Serrin [Ser63] and Masuda [Mas84], see also [KoS96].

1.5.1 Theorem (Serrin, Masuda) *Let $\Omega \subseteq \mathbb{R}^n$, $n = 2, 3$, be any domain, let $0 < T \leq \infty$, $u_0 \in L^2_\sigma(\Omega)$, $f = f_0 + \operatorname{div} F$ with*

$$f_0 \in L^1_{loc}([0,T); L^2(\Omega)^n) \ , \quad F \in L^2_{loc}([0,T); L^2(\Omega)^{n^2}), \tag{1.5.1}$$

and let

$$u\, ,\ w\ \in\ L^\infty_{loc}([0,T); L^2_\sigma(\Omega)) \cap L^2_{loc}([0,T); W^{1,2}_{0,\sigma}(\Omega)) \tag{1.5.2}$$

be two weak solutions of the Navier-Stokes system (1.1.1) *with the same data f, u_0. Suppose additionally that*

$$\frac{1}{2}\, \|u(t)\|_2^2 + \nu \int_0^t \|\nabla u\|_2^2\, d\tau\ \leq\ \frac{1}{2}\, \|u_0\|_2^2 + \int_0^t [f, u]_\Omega\, d\tau \tag{1.5.3}$$

for almost all $t \in [0, T)$, and that

$$w \in L^s_{loc}([0,T); L^q(\Omega)^n) \tag{1.5.4}$$

with $n < q < \infty$, $2 < s < \infty$ such that

$$\frac{n}{q} + \frac{2}{s}\ \leq\ 1\, .$$

Then $u = w$ in $[0, T)$.

1.5.2 Remark Consider the case $n = 3$ and assume in this theorem that (1.5.4) holds with $q = 3,\ s = \infty$. Then from the proof below we obtain the following result:

Suppose the assumptions of Theorem 1.5.1 for $n = 3$ with the modification that (1.5.4) holds with $q = 3,\ s = \infty$. Then there is a constant $C = C(\nu) > 0$ such that

$$\|w\|_{3,\infty;T}\ \leq\ C \tag{1.5.5}$$

implies $u = w$ in $[0, T)$.

Proof of Theorem 1.5.1. If $u = w$ holds in each interval $[0, T')$ with $0 < T' < T$, then we get $u = w$ in the whole interval $[0, T)$. Therefore we may assume in the following that $0 < T < \infty$, that $f_0 \in L^1(0,T; L^2(\Omega)^n)$, $F \in L^2(0,T; L^2(\Omega)^{n^2})$, and that $u, w \in L^\infty(0,T; L^2_\sigma(\Omega)) \cap L^2(0,T; W^{1,2}_{0,\sigma}(\Omega))$, $w \in L^s(0,T; L^q(\Omega)^n)$.

Theorem 1.3.1 shows that $u : [0,T) \to L^2_\sigma(\Omega)$ is weakly continuous, after a corresponding redefinition, and that $u(0) = u_0$. We suppose this property. Then we can show that the inequality (1.5.3) holds for all $t \in [0,T)$. To prove this we consider an arbitrary $t \in [0,T)$ and a sequence $(t_j)_{j=1}^\infty$ with $t = \lim_{j\to\infty} t_j$ such that (1.5.3) holds for each t_j. Then we let $j \to \infty$ and use the weak continuity of u and the property (3.1.3), II.

We know that $w \in L^s(0,T; L^q(\Omega)^n)$ implies $w\,w \in L^2(0,T; L^2(\Omega)^{n^2})$, see (1.4.10), (1.4.12). Therefore we can apply Theorem 1.4.1 and obtain, after a corresponding redefinition, that $w : [0,T) \to L^2_\sigma(\Omega)$ is strongly continuous, that $w(0) = u_0$, and that

$$\frac{1}{2}\,\|w(t)\|_2^2 + \nu \int_0^t \|\nabla w\|_2^2\, d\tau \;=\; \frac{1}{2}\,\|u_0\|_2^2 + \int_0^t [f, w]_\Omega\, d\tau \tag{1.5.6}$$

holds for all $t \in [0,T)$.

The main step of the proof is to show that the equality

$$\begin{aligned} &< u(t), w(t) >_\Omega + 2\nu \int_0^t < \nabla u, \nabla w >_\Omega \, d\tau \\ &= \|u_0\|_2^2 + \int_0^t [f, w]_\Omega\, d\tau + \int_0^t [f, u]_\Omega\, d\tau \\ &\quad + \int_0^t < u\,u, \nabla w >_\Omega \, d\tau + \int_0^t < w\,w, \nabla u >_\Omega \, d\tau \end{aligned} \tag{1.5.7}$$

is satisfied for $0 \le t < T$.

First we show that (1.5.7) is well defined. As in (1.4.13) we may assume without loss of generality that $\frac{n}{q} + \frac{2}{s} = 1$. Then we choose exponents q_1, s_1 satisfying $2 < q_1 < \infty$, $2 < s_1 < \infty$, $\frac{1}{2} = \frac{1}{q_1} + \frac{1}{q}$, $\frac{1}{2} = \frac{1}{s_1} + \frac{1}{s}$, $\frac{n}{q_1} + \frac{2}{s_1} = \frac{n}{2}$. Then we use Lemma 1.2.1, b), and get, see (1.4.14), that

$$\|u\|_{q_1,s_1;T} < \infty\;, \quad \|w\|_{q_1,s_1;T} < \infty\,. \tag{1.5.8}$$

Hölder's inequality yields

$$|< w\,w, \nabla u >_{\Omega,T}| \;\le\; C\,\|w\|_{q_1,s_1;T}\, \|w\|_{q,s;T}\, \|\nabla u\|_{2,2;T} \;<\; \infty \tag{1.5.9}$$

with $C = C(n) > 0$.

Next we use the relation

$$< u\,u, \nabla w >_\Omega = \; - < \text{div}\,(u\,u), w >_\Omega = \; - < u\cdot\nabla u, w >_\Omega\,, \qquad (1.5.10)$$

for almost all $t \in [0,T)$, see Lemma 1.2.1, a), and define the exponents $q_2, s_2 < 2$ by $\frac{1}{q_2} = \frac{1}{2} + \frac{1}{q_1}$ and $\frac{1}{s_2} = \frac{1}{2} + \frac{1}{s_1}$. This yields $1 = \frac{1}{q_2} + \frac{1}{q}$, $1 = \frac{1}{s_2} + \frac{1}{s}$, and with Hölder' s inequality we get

$$\begin{aligned} |< u\,u, \nabla w >_{\Omega,T}| &= |< \text{div}\,(u\,u), w >_{\Omega,T}| \qquad (1.5.11)\\ &\le C_1\,\|\text{div}\,(u\,u)\|_{q_2,s_2;T}\,\|w\|_{q,s;T}\\ &= C_1\,\|u\cdot\nabla u\|_{q_2,s_2;T}\,\|w\|_{q,s:T}\\ &\le C_2\,\|u\|_{q_1,s_1;T}\,\|\nabla u\|_{2,2;T}\,\|w\|_{q,s;T} < \infty \end{aligned}$$

with $C_1, C_2 > 0$ depending only on n.

This shows that

$$\int_0^t |< w\,w, \nabla u >_\Omega|\,d\tau < \infty\;,\quad \int_0^t |< u\,u, \nabla w >_\Omega|\,d\tau < \infty\,,$$

$t \in [0,T)$, and (1.5.7) is well defined.

To prove (1.5.7) we use the mollification method in Section 1.7, II, with $n = 1$, and consider a function $\mathcal{F} \in C_0^\infty(\mathbb{R})$ with the properties (1.7.2), II. Then $\mathcal{F}_\varepsilon \in C_0^\infty(\mathbb{R})$, $\varepsilon > 0$, is defined by $\mathcal{F}_\varepsilon(t) := \varepsilon^{-1}\mathcal{F}(\varepsilon^{-1}t)$, $t \in \mathbb{R}$. The mollified function $u^\varepsilon : [0,T) \to L^2_\sigma(\Omega)$ is defined by

$$u^\varepsilon(t) := (\mathcal{F}_\varepsilon \star u)(t) := \int_0^T \mathcal{F}_\varepsilon(t-\tau)\,u(\tau)\,d\tau\;,\quad t \in [0,T).$$

In the same way, we define $w^\varepsilon, f_0^\varepsilon, F^\varepsilon, (u\,u)^\varepsilon, (w\,w)^\varepsilon$, and so on.

Consider $0 < t_0 < T' < T$, some fixed $\varepsilon_0 > 0$ with $\varepsilon_0 < t_0$, $\varepsilon_0 < T - T'$, and let $0 < \varepsilon < \varepsilon_0$. Then the elementary properties in Section 1.7, II, show that $u^\varepsilon, w^\varepsilon$ have strongly continuous derivatives in $[t_0, T']$. We insert the special test functions

$$v = \varphi^\varepsilon h \;\text{ with }\; \varphi \in C_0^\infty((t_0, T'))\;,\quad h \in C_{0,\sigma}^\infty(\Omega)$$

in the condition (1.1.6) of a weak solution. Then a calculation shows that

$$\begin{aligned} < \frac{d}{dt}u^\varepsilon(t), h >_\Omega \;+\; \nu < (\nabla u)^\varepsilon(t), \nabla h >_\Omega \;+\; < (\text{div}\;u\,u)^\varepsilon(t), h >_\Omega\\ = \; < f_0^\varepsilon(t), h >_\Omega \;-\; < F^\varepsilon(t), \nabla h >_\Omega \end{aligned}$$

holds for all $t \in [t_0, T']$. A closure argument shows that this is valid for all $h \in W^{1,2}_{0,\sigma}(\Omega)$. Inserting $h = w^\varepsilon(t)$ for each $t \in [t_0, T']$, we get

$$\begin{aligned} < \frac{d}{dt} u^\varepsilon, w^\varepsilon >_\Omega &+ \nu < (\nabla u)^\varepsilon, (\nabla w)^\varepsilon >_\Omega + < (\text{div}\, u\, u)^\varepsilon, w^\varepsilon >_\Omega \\ &= < f_0^\varepsilon, w^\varepsilon >_\Omega - < F^\varepsilon, (\nabla w)^\varepsilon >_\Omega, \end{aligned}$$

and correspondingly

$$\begin{aligned} < \frac{d}{dt} w^\varepsilon, u^\varepsilon >_\Omega &+ \nu < (\nabla w)^\varepsilon, (\nabla u)^\varepsilon >_\Omega + < (\text{div}\, w\, w)^\varepsilon, u^\varepsilon >_\Omega \\ &= < f_0^\varepsilon, u^\varepsilon >_\Omega - < F^\varepsilon, (\nabla u)^\varepsilon >_\Omega \end{aligned}$$

for all $t \in [t_0, T']$. Taking the sum, integrating over $[t_0, T']$ and using the rule

$$\begin{aligned} &\int_{t_0}^{T'} (< \frac{d}{dt} u^\varepsilon, w^\varepsilon >_\Omega + < u^\varepsilon, \frac{d}{dt} w^\varepsilon >_\Omega)\, dt \\ &\quad = < u^\varepsilon(T'), w^\varepsilon(T') >_\Omega - < u^\varepsilon(t_0), w^\varepsilon(t_0) >_\Omega \end{aligned}$$

of integration by parts, see Lemma 1.3.2, IV, we obtain

$$\begin{aligned} < u^\varepsilon(T'), w^\varepsilon(T') >_\Omega &- < u^\varepsilon(t_0), w^\varepsilon(t_0) >_\Omega + 2\nu \int_{t_0}^{T'} < (\nabla u)^\varepsilon, (\nabla w)^\varepsilon >_\Omega dt \\ &+ \int_{t_0}^{T'} (< (\text{div}\, u\, u)^\varepsilon, w^\varepsilon >_\Omega - < (w\, w)^\varepsilon, (\nabla u)^\varepsilon >_\Omega)\, dt \\ = \int_{t_0}^{T'} (< f_0^\varepsilon, w^\varepsilon >_\Omega &+ < f_0^\varepsilon, u^\varepsilon >_\Omega - < F^\varepsilon, (\nabla w)^\varepsilon >_\Omega - < F^\varepsilon, (\nabla u)^\varepsilon >_\Omega)\, dt. \end{aligned}$$

In the next step we consider the limit as $\varepsilon \to 0$. Using the convergence property (1.7.13), II, together with the above estimates (1.5.8), (1.5.9), (1.5.11), we see that

$$\begin{aligned} &\|\nabla u - (\nabla u)^\varepsilon\|_{2,2;T}\ , \quad \|\nabla w - (\nabla w)^\varepsilon\|_{2,2;T}\ , \quad \|w\, w - (w\, w)^\varepsilon\|_{2,2;T}\ , \\ &\quad \|\text{div}\, u\, u - (\text{div}\, u\, u)^\varepsilon\|_{q_2,s_2;T}\ , \quad \|w - w^\varepsilon\|_{q,s;T} \end{aligned}$$

tend to zero as $\varepsilon \to 0$. The convergence property (1.7.8), II, and the strong continuity of $w : [0, T) \to L^2_\sigma(\Omega)$ yields that

$$\lim_{\varepsilon \to 0} \|w(T') - w^\varepsilon(T')\|_2 = 0.$$

The weak continuity of $u : [0, T) \to L^2_\sigma(\Omega)$ leads to

$$\lim_{\varepsilon \to 0} |< u(T') - u^\varepsilon(T'), h >_\Omega| = 0\ , \quad h \in L^2_\sigma(\Omega).$$

This yields

$$\lim_{\varepsilon\to 0} < u^\varepsilon(T'), w^\varepsilon(T') >_\Omega = < u(T'), w(T') >_\Omega ,$$

and correspondingly with T' replaced by t_0. The same argument shows that

$$\lim_{t_0\to 0} < u(t_0), w(t_0) >_\Omega = < u(0), w(0) >_\Omega = \|u_0\|_2^2 .$$

Now we may let $\varepsilon \to 0$ in each term of the equation above and then we let $t_0 \to 0$. This proves the equation (1.5.7) with t replaced by T'.

In the next step we prove the inequality

$$\begin{aligned} \frac{1}{2}\|u(t) - w(t)\|_2^2 + \nu \int_0^t \|\nabla(u-w)\|_2^2\, d\tau \\ \le \int_0^t < w, (u-w)\cdot\nabla(u-w) >_\Omega\, d\tau \end{aligned} \tag{1.5.12}$$

for all $t \in [0,T)$.

First we show that this is well defined. Using q_1, s_1, see (1.5.8), we get with Lemma 1.2.1, b), that

$$\begin{aligned} \|u-w\|_{q_1,s_1;T} &\le C\nu^{-\frac{1}{s_1}} \|A^{\frac{1}{2}}(u-w)\|_{2,2;T}^{\frac{2}{s_1}} \|u-w\|_{2,\infty;T}^{1-\frac{2}{s_1}} \\ &= C\|\nabla(u-w)\|_{2,2;T}^{\frac{2}{s_1}} \|u-w\|_{2,\infty;T}^{1-\frac{2}{s_1}} , \end{aligned}$$

$C = C(s_1, n) > 0$, and as in (1.5.9) we conclude that

$$\begin{aligned} &\int_0^T |< w, (u-w)\cdot\nabla(u-w) >_\Omega|\, d\tau \\ &\le C_1 \|w\|_{q,s;T} \|u-w\|_{q_1,s_1;T} \|\nabla(u-w)\|_{2,2;T} \\ &\le C_2 \|w\|_{q,s;T} \|\nabla(u-w)\|_{2,2;T}^{1+\frac{2}{s_1}} \|u-w\|_{2,\infty;T}^{1-\frac{2}{s_1}} < \infty \end{aligned} \tag{1.5.13}$$

with $C_1 = C_1(n) > 0$, $C_2 = C_2(s_1, n) > 0$. This shows that (1.5.12) is well defined.

To prove (1.5.12) we use (1.5.10) and the relation

$$\begin{aligned} < u\,u, \nabla w >_\Omega + < w\,w, \nabla u >_\Omega &= < w, w\cdot\nabla u >_\Omega - < w, u\cdot\nabla u >_\Omega \\ &= < w, (w-u)\cdot\nabla u >_\Omega . \end{aligned}$$

Using Lemma 1.2.1, a), we get

$$\begin{aligned} < w, (w-u)\cdot\nabla w >_\Omega &= < (w-u)w, \nabla w >_\Omega \\ = \frac{1}{2} < w-u, \nabla|w|^2 >_\Omega &= -\frac{1}{2} < \mathrm{div}\,(w-u), |w|^2 > = 0. \end{aligned}$$

This shows that

$$< u\,u, \nabla w >_\Omega + < w\,w, \nabla u >_\Omega = < w, (w-u)\cdot\nabla(u-w) >_\Omega . \quad (1.5.14)$$

Next we take the sum of (1.5.3) and (1.5.6), substract (1.5.7), and use (1.5.14). This yields (1.5.12).

Consider any T' with $0 < T' \le T$, set $W := u - w$, and set

$$\|W\|_{T'}^2 := \frac{1}{2}\left(\sup_{0\le t\le T'} \|W(t)\|_2^2\right) + \nu \int_0^{T'} \|\nabla W\|_2^2 \, d\tau .$$

Taking $\sup_{0\le t\le T'}$ on both sides of (1.5.12) we obtain the inequality

$$\|W\|_{T'}^2 \le 2\int_0^{T'} | < w, W\cdot\nabla W >_\Omega | \, d\tau.$$

Using (1.5.13) with T replaced by T' and Young's inequality (3.3.8), I, we get the estimate

$$\begin{aligned} \|W\|_{T'}^2 &\le C_1 \|w\|_{q,s;T'} \|\nabla W\|_{2,2;T'} \|\nabla W\|_{2,2;T'}^{2/s_1} \|W\|_{2,\infty;T'}^{1-2/s_1} \\ &\le C_1 \|w\|_{q,s;T'} \|\nabla W\|_{2,2;T'} (\|\nabla W\|_{2,2;T'} + \|W\|_{2,\infty;T'}) \\ &\le C_2 \|w\|_{q,s;T'} \|W\|_{T'}^2 \end{aligned}$$

with constants $C_1 = C_1(n) > 0$ and $C_2 = C_2(s, n, \nu) > 0$.

Since $w \in L^s(0,T; L^q(\Omega)^n)$ and since C_2 does not depend on T', we can choose T' in such a way that

$$C_2 \|w\|_{q,s;T'} < 1. \quad (1.5.15)$$

This yields

$$(1 - C_2 \|w\|_{q,s;T'}) \|W\|_{T'}^2 \le 0, \quad (1.5.16)$$

therefore $\|W\|_{T'} = 0$, and $u = w$ in $[0, T')$.

Since C_2 does not depend on T', we can repeat this procedure if $T' < T$. We define $\tilde u, \tilde w$ by setting $\tilde u(t) := u(T'+t)$, $\tilde w(t) := w(T'+t)$, $0 \le t \le T - T'$, get $\tilde u(0) = \tilde w(0)$, and the above proof shows that $u = w$ in some interval $[T', T'')$ with $T' < T'' \le T$, and so on. In a finite number of steps we get $u = w$ in $[0, T)$. This proves the theorem. □

The case $s = \infty$, $q = n$ is excluded in Theorem 1.5.1. In the interesting case $n = 3$, $q = 3$, $s = \infty$, the calculations above remain valid up to (1.5.15). In this case we cannot always find some T', $0 < T' \leq T$, such that (1.5.15) is satisfied. Thus we have to require (1.5.15) as an additional assumption. This proves Remark 1.5.2.

The next theorem yields the uniqueness of weak solutions in the two-dimensional case.

1.5.3 Theorem $(n = 2)$ *Let* $\Omega \subseteq \mathbb{R}^2$ *be any two-dimensional domain, let* $0 < T \leq \infty$, $u_0 \in L^2_\sigma(\Omega)$, $f = f_0 + \operatorname{div} F$ *with*

$$f_0 \in L^1_{loc}([0,T); L^2(\Omega)^2) \ , \quad F \in L^2_{loc}([0,T); L^2(\Omega)^4),$$

and let

$$u, \, w \in L^\infty_{loc}([0,T); L^2_\sigma(\Omega)) \cap L^2_{loc}([0,T); W^{1,2}_{0,\sigma}(\Omega))$$

be two weak solutions of the Navier-Stokes system (1.1.1) *with the same data* f, u_0.

Then $u = w$ *in* $[0,T)$, *and Serrin's condition* (1.5.4) *is satisfied with* $q = s = 4$.

Proof. We use Lemma 1.2.1, b) with $q = s = 4$. This yields

$$\|u\|_{4,4;T'} \ \leq \ C\nu^{-\frac{1}{4}}\, E_{T'}(u)^{\frac{1}{2}} \ < \ \infty \tag{1.5.17}$$

with $0 < T' < T$. Thus we get

$$u \in L^4_{loc}([0,T); L^4(\Omega)^2) \tag{1.5.18}$$

with $\frac{n}{4} + \frac{2}{4} = 1$, $n = 2$. Thus Serrin's condition (1.5.4) is satisfied and the result is a consequence of Theorem 1.5.1. □

1.6 Integrability properties of weak solutions in space and time, the scale of Serrin's quantity

In this subsection we prove integrability properties of weak solutions u of the form

$$\|u\|_{q,\rho;T} \ = \left(\int_0^T \|u\|_q^\rho \, dt \right)^{\frac{1}{\rho}} \ < \ \infty \tag{1.6.1}$$

and

$$\|A^\alpha u\|_{2,\rho;T} \ = \left(\int_0^T \|A^\alpha u\|_2^\rho \, dt \right)^{\frac{1}{\rho}} \ < \ \infty \tag{1.6.2}$$

with certain values $q,\ \rho > 1,\ -1 \le \alpha \le 1$. In particular we are interested in the case $T = \infty$ and in the case of unbounded domains Ω, see [Soh99].

In these cases the properties above give us some important information on the asymptotic behavior of u as $t \to \infty,\ |x| \to \infty$. In particular we can prove certain algebraic decay estimates of u in the time direction, see Section 3.4.

An elementary consideration shows: The smaller the above exponents q and ρ, the stronger the decay of u in space and time. We can treat q and ρ as a measure for the quality of the asymptotic decay of u in space and time.

We will see in this context, that Serrin's quantity

$$S(q,\rho) := \frac{n}{q} + \frac{2}{\rho} \tag{1.6.3}$$

plays again an important role. We already know this quantity from the uniqueness result, see Theorem 1.5.1. In Subsection 1.8 we will prove local regularity properties of weak solutions u if (1.6.1) is satisfied with

$$S(q,\rho) \ \le \ 1.$$

From Lemma 1.2.1 we know that (1.6.1) always holds with

$$S(q,\rho) = \frac{n}{2}.$$

Uniqueness and regularity of u follow if (1.6.1) holds with

$$S(q,\rho) \ \le \ 1.$$

Thus we see, if $n = 3$, we have a gap of $\frac{1}{2}$ between that which we know, namely (1.6.1) with $S(q,\rho) = \frac{3}{2}$ and that which we need in order to get regularity and uniqueness, namely (1.6.1) with $S(q,\rho) \le 1$. There is no such gap if $n = 2$.

To get local regularity results on u, we are interested to know (1.6.1) with "small" values $S(q,\rho)$. To get global asymptotic decay properties on u we need (1.6.1) with values $S(q,\rho)$ as large as possible. The best possible result we can prove in the latter case is that (1.6.1) holds with

$$\frac{n}{2} \le S(q,\rho) < \frac{n}{2} + 1\,,$$

see Theorem 1.6.2 and Theorem 1.6.3.

Further we recall that if (1.6.1) is satisfied with $q = \rho = 4$, and therefore with

$$S(q,\rho) = \frac{n}{4} + \frac{1}{2},$$

then u is strongly continuous and satisfies the energy equality, see Theorem 1.4.1 and the conditions (1.4.6), (1.4.12).

Thus we get again $S(q,\rho) = 1$ if $n = 2$, but

$$S(q,\rho) = 1 + \frac{1}{4}$$

if $n = 3$. We have to consider the last properties as certain regularity information which is already available if $S(q,\rho)$ lies in the middle of the gap between $S(q,\rho) = 1$ and $S(q,\rho) = 1 + \frac{1}{2}$ for $n = 3$.

We see that the scale of Serrin's number $S(q,\rho)$ for which (1.6.1) is satisfied gives us some important information on properties of a weak solution u.

The following theorem is the first step to get integrability results (1.6.1), (1.6.2). Recall that

$$E_{T'}(u) = \frac{1}{2}\,\|u\|^2_{2,\infty;T'} + \nu\,\|\nabla u\|^2_{2,2;T'} < \infty$$

with $0 < T' < T$ if

$$u \in L^\infty_{loc}([0,T); L^2_\sigma(\Omega)) \cap L^2_{loc}([0,T); W^{1,2}_{0,\sigma}(\Omega)),$$

see Lemma 1.2.1. We suppose f has the special form $f = \operatorname{div} F$. See Section 1.6, II, for sufficient conditions for this representation.

1.6.1 Theorem *Let $\Omega \subseteq \mathbb{R}^n$, $n = 2, 3$, be an arbitrary domain, let $0 < T \leq \infty$, $s = \frac{4}{n}$, $s \leq \rho < \infty$, and let $u_0 \in D(A^{-\frac{1}{2}})$, $f = \operatorname{div} F$ with*

$$F \in L^s_{loc}([0,T); L^2(\Omega)^{n^2})\,.$$

Suppose

$$u \in L^\infty_{loc}([0,T); L^2_\sigma(\Omega)) \cap L^2_{loc}([0,T); W^{1,2}_{0,\sigma}(\Omega)) \tag{1.6.4}$$

is a weak solution of the Navier-Stokes system (1.1.1) with data f, u_0.

Then, after a redefinition on a null set of $[0,T)$, $u : [0,T) \to L^2_\sigma(\Omega)$ is weakly continuous, $u(0) = u_0$, and we obtain the following properties:

a) $\quad u(t) \in D(A^{-\frac{1}{2}})$ *for all* $t \in [0,T)$, $A^{-\frac{1}{2}}u : [0,T) \to L^2_\sigma(\Omega)$

is strongly continuous, and

$$\|A^{-\frac{1}{2}}u\|_{2,\infty;T'} \leq \|A^{-\frac{1}{2}}u_0\|_2 + (T')^{\frac{4-n}{4}}\,C\,(\|F\|_{2,s;T'} + E_{T'}(u)) \tag{1.6.5}$$

with $0 < T' < T$, $C = C(n,\nu) > 0$.

b) $$\|(A^{-\frac{1}{2}}u)_t\|_{2,s;T'} + \|A^{\frac{1}{2}}u\|_{2,s;T'} \le C\,(\|u_0\|_2 + \|A^{-\frac{1}{2}}u_0\|_2 + \|F\|_{2,s;T'} + E_{T'}(u)) \tag{1.6.6}$$

with $0 < T' < T$, $C = C(n,\nu) > 0$.

c) $$\|A^{\alpha}u\|_{2,\rho;T'} \;\le\; C\,(\|u_0\|_2 + \|A^{-\frac{1}{2}}u_0\|_2 + \|F\|_{2,s;T'} + E_{T'}(u)) \tag{1.6.7}$$

with $0 < T' < T$, $\alpha := \frac{1}{2} - \frac{n}{4} + \frac{1}{\rho}$, $\alpha \ge 0$, $C = C(n,\nu) > 0$.

d) $$\|u\|_{q,\rho;T'} \;\le\; C\,(\|u_0\|_2 + \|A^{-\frac{1}{2}}u_0\|_2 + \|F\|_{2,s;T'} + E_{T'}(u)) \tag{1.6.8}$$

with $0 < T' < T$, $C = C(n,\nu,\rho) > 0$ *and with* $2 \le q < \infty$ *determined by the condition*

$$\frac{n}{q} + \frac{2}{\rho} = n - 1.$$

Proof. We use Theorem 1.3.1 and conclude that u is weakly continuous after a corresponding redefinition, that $u(0) = u_0$, and that the integral equation

$$u(t) = S(t)u_0 + A^{\frac{1}{2}} \int_0^t S(t-\tau)A^{-\frac{1}{2}}P\,\mathrm{div}\,\tilde{F}(\tau)\,d\tau \;\;,\quad t \in [0,T) \tag{1.6.9}$$

is satisfied with $\tilde{F} := F - u\,u$. Using (1.1.9) we see that

$$\|A^{-\frac{1}{2}}P\,\mathrm{div}\,\tilde{F}(\tau)\|_2 \;\le\; \nu^{-\frac{1}{2}}\,\|\tilde{F}(\tau)\|_2 \tag{1.6.10}$$

for almost all $\tau \in [0,T)$.

From Lemma 1.2.1, c), we get with $q = 2$, $s = \frac{4}{n}$, $\frac{n}{q} + \frac{2}{s} = n$ that

$$\|u\,u\|_{2,s;T'} \;\le\; C\,E_{T'}(u) \;<\; \infty \tag{1.6.11}$$

with $0 < T' < T$, $C = C(n,\nu) > 0$.

This yields

$$\|\tilde{F}\|_{2,s;T'} \;\le\; \|F\|_{2,s;T'} + C\,E_{T'}(u) \;<\; \infty. \tag{1.6.12}$$

Since $u_0 \in D(A^{-\frac{1}{2}}) = R(A^{\frac{1}{2}})$, see Lemma 2.2.1, III, we deduce from (1.6.9) that $u(t) \in D(A^{-\frac{1}{2}})$ for all $t \in [0,T)$. Thus we get

$$\begin{aligned} A^{-\frac{1}{2}}u(t) &= S(t)A^{-\frac{1}{2}}u_0 + \int_0^t S(t-\tau)\tilde{f}(\tau)\,d\tau \\ &= S(t)A^{-\frac{1}{2}}u_0 + (\mathcal{J}\tilde{f})(t) \end{aligned} \tag{1.6.13}$$

for all $t \in [0,T')$ with $\tilde{f} := A^{-\frac{1}{2}} P \operatorname{div} \tilde{F}$. See (1.6.3), IV, concerning the integral operator $\mathcal{J}$. Using (1.6.10) and (1.6.12) we obtain

$$\|\tilde{f}\|_{2,s;T'} \leq \nu^{-\frac{1}{2}} \|\tilde{F}\|_{2,s;T'} \leq C\,(\|F\|_{2,s;T'} + E_{T'}(u)) \tag{1.6.14}$$

with $C = C(n,\nu) > 0$.

To the equation (1.6.13) we may apply the linear theory of Theorem 2.5.1, IV, with u_0 replaced by $A^{-\frac{1}{2}} u_0$, and Theorem 2.5.3, IV. It follows that $A^{-\frac{1}{2}} u : [0,T) \to L^2_\sigma(\Omega)$ is strongly continuous. Theorem 2.5.1, IV, may be applied to $S(\cdot)A^{-\frac{1}{2}} u_0$. Here we need that $\|A^{1-\frac{1}{s}}(A^{-\frac{1}{2}} u_0)\|_2 = \|u_0\|_2$ if $s = 2$, $n = 2$, and that $\|A^{\frac{1}{2}}(A^{-\frac{1}{2}} u_0)\|_2 = \|u_0\|_2$ if $s = \frac{4}{3}$, $n = 3$. This shows that

$$(A^{-\frac{1}{2}} u)_t \;,\quad A^{\frac{1}{2}} u \in L^s(0,T'; L^2_\sigma(\Omega)) \;,\quad (A^{-\frac{1}{2}} u)(0) = A^{-\frac{1}{2}} u_0 \;,$$

and that the evolution equation

$$(A^{-\frac{1}{2}} u)_t + A^{\frac{1}{2}} u = \tilde{f} = A^{-\frac{1}{2}} P \operatorname{div}(F - u\,u) \tag{1.6.15}$$

is satisfied in $L^s(0,T'; L^2_\sigma(\Omega))$, $0 < T' < T$.

To prove (1.6.5) we apply (1.5.8), IV, to $S(\cdot)A^{-\frac{1}{2}} u_0$, (2.5.17), IV, with u, f replaced by $\mathcal{J}\tilde{f}$, $\tilde{f}$, and get with Hölder's inequality and (1.6.12) that

$$\begin{aligned} \|A^{-\frac{1}{2}} u\|_{2,\infty;T'} &\leq \|A^{-\frac{1}{2}} u_0\|_2 + C_1 \|\tilde{f}\|_{2,1;T'} \\ &\leq \|A^{-\frac{1}{2}} u_0\|_2 + C_2 (T')^{\frac{4-n}{4}} (\|F\|_{2,s;T'} + E_{T'}(u)) \end{aligned} \tag{1.6.16}$$

with $C_2 = C_2(n,\nu) > 0$, $C_1 > 0$.

The inequality (1.6.6) follows when we apply (2.5.5), IV, to $S(\cdot)A^{-\frac{1}{2}} u_0$, and (2.5.13), IV to $\mathcal{J}\tilde{f}$; further we use (1.6.11).

To prove (1.6.7) we apply (2.5.7), IV, with $s = \frac{4}{n}$, u_0 replaced by $A^{-\frac{1}{2}} u_0$, and obtain

$$\begin{aligned} \|A^\alpha S(\cdot) u_0\|_{2,\rho;T'} &= \|A^{1+\frac{1}{\rho}-\frac{n}{4}} S(\cdot) A^{-\frac{1}{2}} u_0\|_{2,\rho;T'} \\ &\leq C\,(\|A^{-\frac{1}{2}} u_0\|_2 + \|u_0\|_2) \end{aligned}$$

since $1 < s \leq 2$. The condition $\alpha \geq 0$ in (1.6.7) leads to $\frac{1}{2} + \frac{1}{\rho} \geq \frac{1}{s}$ which is needed in (2.5.7), IV. Next we apply (2.5.15), IV, and get

$$\begin{aligned} \|A^\alpha A^{\frac{1}{2}} \mathcal{J}\tilde{f}\|_{2,\rho;T'} &= \|A^{1+\frac{1}{\rho}-\frac{n}{4}} \mathcal{J}\tilde{f}\|_{2,\rho;T'} \\ &\leq C\,\|\tilde{f}\|_{2,s;T'} \,, \end{aligned}$$

$C = C(s) > 0$. Using (1.6.9) and (1.6.14) we get the desired result (1.6.7).

To prove d) we use (2.5.9), IV, with $s = \frac{4}{n}$, and get

$$\begin{aligned} \|S(\cdot)u_0\|_{q,\rho;T'} &= \|A^{\frac{1}{2}}S(\cdot)A^{-\frac{1}{2}}u_0\|_{q,\rho;T'} \\ &\leq C\,(\|A^{-\frac{1}{2}}u_0\|_2 + \|u_0\|_2) \end{aligned}$$

with $C = C(n,\nu,q) > 0$, $1 + \frac{n}{q} + \frac{2}{\rho} = n = \frac{n}{2} + \frac{2}{s}$. Note that $\frac{1}{2} + \frac{1}{\rho} \geq \frac{1}{s}$ in this case; this is needed in (2.5.9), IV.

Next we use the inequality (2.5.24), IV, with u replaced by $A^{\frac{1}{2}}\mathcal{J}\tilde{f}$, and obtain

$$\|A^{\frac{1}{2}}\mathcal{J}\tilde{f}\|_{q,\rho;T'} \leq C\,\|\tilde{F}\|_{2,s;T'}$$

with $C = C(n,\nu,\rho) > 0$. Using (1.6.12) we obtain the desired inequality (1.6.8). This proves the theorem. □

In the next step we will improve the integrability properties of the theorem above. We will prove the properties (1.6.1), (1.6.2) with $T = \infty$, and with q,ρ such that Serrin's number $S(q,\rho)$ is larger as above. Thus we can improve the asymptotic decay of u as $t \to \infty$, $|x| \to \infty$. Our method is again to write the Navier-Stokes system in the form

$$u_t - \nu\Delta u + \nabla p = \tilde{f}\ , \quad \text{div}\ u = 0\ , \quad u|_{\partial\Omega} = 0\ , \quad u(0) = u_0 \tag{1.6.17}$$

with $\tilde{f} = \text{div}\ \tilde{F}$, $\tilde{F} = F - u\,u$. The information of the last theorem can now be used on the right side of (1.6.17). This yields

$$\|\tilde{F}\|_{2,s;\infty} \leq \|F\|_{2,s;\infty} + \|u\,u\|_{2,s;\infty} < \infty \tag{1.6.18}$$

even for all s with $1 \leq s \leq \frac{4}{n}$. Then we only have to apply the linear theory to the equation (1.6.17). This leads to the next results for $n = 3$ and $n = 2$ separately. Here we are not interested in explicit bounds on the right sides of the estimates below.

1.6.2 Theorem $(n = 3)$ *Let $\Omega \subseteq \mathbb{R}^3$ be any three-dimensional domain, let $T = \infty$, $1 < s \leq \frac{4}{3}$, $1 < \rho < \infty$, $2 \leq q \leq 6$, $0 \leq \alpha \leq \frac{1}{2}$, and let $u_0 \in D(A^{-\frac{1}{2}})$, $f = \text{div}\ F$ with*

$$F \in L^1(0,\infty;L^2(\Omega)^9) \cap L^2(0,\infty;L^2(\Omega)^9). \tag{1.6.19}$$

Suppose

$$u \in L^\infty_{loc}([0,\infty);L^2_\sigma(\Omega)) \cap L^2_{loc}([0,\infty);W^{1,2}_{0,\sigma}(\Omega))$$

is a weak solution of the Navier-Stokes system (1.1.1) *with data f, u_0. Suppose additionally that*

$$E_\infty(u) = \frac{1}{2}\,\|u\|^2_{2,\infty;\infty} + \nu\,\|\nabla u\|^2_{2,2;\infty} < \infty. \tag{1.6.20}$$

Then there is a constant $C > 0$, depending on $F, u_0, E_\infty(u), \nu, s, \rho, q, \alpha$ such that, after redefining u on a null set of $[0,\infty)$, the following properties are satisfied:

a) $\qquad u(t) \in D(A^{-\frac{1}{2}})$ *for all* $t \in [0,\infty)$, $A^{-\frac{1}{2}}u : [0,\infty) \to L^2_\sigma(\Omega)$

is strongly continuous in $[0,\infty)$, $(A^{-\frac{1}{2}}u)(0) = A^{-\frac{1}{2}}u_0$, *and*

$$\|A^{-\frac{1}{2}}u\|_{2,\infty;\infty} + \|u\|_{2,2;\infty} \;\le\; C. \tag{1.6.21}$$

b)
$$\|(A^{-\frac{1}{2}}u)_t\|_{2,s;\infty} + \|A^{\frac{1}{2}}u\|_{2,s;\infty} \;\le\; C. \tag{1.6.22}$$

c)
$$\|A^\alpha u\|_{2,\rho;\infty} \;\le\; C \tag{1.6.23}$$

with $-\frac{1}{2} + \frac{1}{\rho} < \alpha \le \frac{1}{\rho}$, $0 \le \alpha \le \frac{1}{2}$.

d)
$$\|u\|_{q,\rho;\infty} \;\le\; C \tag{1.6.24}$$

with

$$\frac{3}{2} \le \frac{3}{q} + \frac{2}{\rho} < \frac{3}{2} + 1\,. \tag{1.6.25}$$

Proof. First we recall the interpolation inequality

$$\|F\|_{2,s;\infty} \le \|F\|^{\beta}_{2,1;\infty}\,\|F\|^{1-\beta}_{2,2;\infty} \;\le\; \|F\|_{2,1;\infty} + \|F\|_{2,2;\infty}\,, \tag{1.6.26}$$

valid for all $1 \le s \le 2$, where $0 \le \beta \le 1$ is chosen so that $\frac{1}{s} = \beta + (1-\beta)\frac{1}{2}$, see (3.3.7), I. In the same way we get the inequality

$$\begin{aligned}\|A^{\frac{1}{2}}u\|_{2,\gamma;\infty} &\le \|A^{\frac{1}{2}}u\|^{\beta}_{2,4/3;\infty}\,\|A^{\frac{1}{2}}u\|^{1-\beta}_{2,2;\infty} \\ &\le \|A^{\frac{1}{2}}u\|_{2,4/3;\infty} + \|A^{\frac{1}{2}}u\|_{2,2;\infty}\end{aligned} \tag{1.6.27}$$

with $\frac{4}{3} \le \gamma \le 2$, $0 \le \beta \le 1$ such that $\frac{1}{\gamma} = \beta\frac{3}{4} + (1-\beta)\frac{1}{2}$.

From (1.6.20) we obtain

$$\|A^{\frac{1}{2}}u\|_{2,2;\infty} = \nu^{\frac{1}{2}}\|\nabla u\|_{2,2;\infty} \;\le\; E_\infty(u)^{\frac{1}{2}} \;<\infty \tag{1.6.28}$$

and (1.6.6) leads to

$$\begin{aligned}&\|A^{\frac{1}{2}}u\|_{2,4/3;\infty} \\ &\quad\le C\,\Big(\|u_0\|_2 + \|A^{-\frac{1}{2}}u_0\|_2 + \|F\|_{2,\frac{4}{3};\infty} + E_\infty(u)\Big) \;<\; \infty\end{aligned} \tag{1.6.29}$$

with $C = C(\nu) > 0$. Since the constant C in (1.6.6) does not depend on T', we may set $T' = T = \infty$ and get (1.6.29).

Using (1.6.27), (1.6.28), (1.6.29) we obtain

$$\|A^{\frac{1}{2}}u\|_{2,\gamma;\infty} \leq \|A^{\frac{1}{2}}\|_{2,4/3;\infty} + \|A^{\frac{1}{2}}u\|_{2,2;\infty} < \infty \tag{1.6.30}$$

with $\frac{4}{3} \leq \gamma \leq 2$.

Using Lemma 1.2.1, d), with $T' = \infty$, we obtain the inequality

$$\begin{aligned} \|u\,u\|_{2,s;\infty} &\leq C\,\|A^{\frac{1}{2}}u\|^{3/2}_{2,s3/2;\infty}\,\|u\|^{2-3/2}_{2,\infty;\infty} \\ &\leq C\,(\|A^{\frac{1}{2}}u\|_{2,4/3;\infty} + \|A^{\frac{1}{2}}u\|_{2,2;\infty})^{3/2}\,\|u\|^{2-3/2}_{2,\infty;\infty} < \infty \end{aligned} \tag{1.6.31}$$

with $1 \leq s \leq \frac{4}{3}$, $C = C(\nu) > 0$. Using (1.6.26) and (1.6.31) we conclude that

$$\|\tilde{F}\|_{2,s;\infty} \leq \|F\|_{2,s;\infty} + \|u\,u\|_{2,s;\infty} < \infty \tag{1.6.32}$$

with $1 \leq s \leq \frac{4}{3}$ and $\tilde{F} = F - u\,u$.

This enables us to apply the results of Section 2.5, IV, from the linear theory to the system (1.6.17). We use the same arguments as in the proof of Theorem 1.6.1. Note that the constants C in this theorem are independent of T'. Thus we may let $T' \to \infty$.

We apply Theorem 2.5.1, IV, with u_0 replaced by $A^{-\frac{1}{2}}u_0$, and Theorem 2.5.3, IV, to the system (1.6.17). Thus we get the representation (1.6.13) with $T = \infty$. $A^{-\frac{1}{2}}u$ is strongly continuous in $[0,\infty)$, $(A^{-\frac{1}{2}}u)(0) = A^{-\frac{1}{2}}u_0$, $(A^{-\frac{1}{2}}u)_t$, $A^{\frac{1}{2}}u \in L^s(0,\infty;L^2_\sigma(\Omega))$, and

$$(A^{-\frac{1}{2}}u)_t + A^{\frac{1}{2}}u = \tilde{f} = A^{-\frac{1}{2}}P\,\mathrm{div}\,(F - u\,u) \tag{1.6.33}$$

as in (1.6.15). Using (2.5.5), IV, and (2.5.21), IV, we obtain with (1.6.32) the validity of (1.6.22) with some $C > 0$.

To prove (1.6.21) we use (2.5.10), IV, with u_0 replaced by $A^{-\frac{1}{2}}u_0$, and (2.5.25), IV.

To prove c) we consider first the case $1 < \rho \leq \frac{4}{3}$. Then for each $\alpha \in (-\frac{1}{2} + \frac{1}{\rho}, \frac{1}{2}]$ we find some s, $1 < s \leq \rho$, such that $\alpha = \frac{1}{2} + \frac{1}{\rho} - \frac{1}{s}$. We apply (2.5.7), IV, with u_0 replaced by $A^{-\frac{1}{2}}u_0$, and we use (2.5.23), IV, with F replaced by $\tilde{F}$. This yields

$$\|A^\alpha u\|_{2,\rho;\infty} \leq C \tag{1.6.34}$$

with $C > 0$.

Next we consider the case $\frac{4}{3} < \rho \leq 2$. Then for each $\alpha \in (-\frac{1}{2} + \frac{1}{\rho}, -\frac{1}{4} + \frac{1}{\rho}]$ we find some s, $1 < s \leq \frac{4}{3}$, such that $\alpha = \frac{1}{2} + \frac{1}{\rho} - \frac{1}{s}$. This yields (1.6.34) in the same way as before. Let now $2 < \rho < \infty$. Using the interpolation inequality (2.2.8), III, we obtain

$$\|A^{\frac{1}{\rho}}u\|_2 = \|A^{\frac{1}{2}(\frac{2}{\rho})}u\|_2 \leq \|A^{\frac{1}{2}}u\|_2^{\frac{2}{\rho}}\,\|u\|_2^{1-\frac{2}{\rho}}$$

for almost all $t \in [0,\infty)$. Integrating over $[0,\infty)$ leads to

$$\|A^{\frac{1}{\rho}}u\|_{2,\rho;\infty} \leq \|A^{\frac{1}{2}}u\|_{2,2;\infty}^{\frac{2}{\rho}} \|u\|_{2,\infty;\infty}^{1-\frac{2}{\rho}} < \infty. \tag{1.6.35}$$

For each given $\alpha \in [0, \frac{1}{\rho}]$ we can choose some $\rho_2 \geq \rho$ with $\alpha = \frac{1}{\rho_2}$, and we choose ρ_1, s with $1 < s \leq \rho_1 \leq \rho$, $s \leq \frac{4}{3}$, satisfying $\alpha = \frac{1}{2} + \frac{1}{\rho_1} - \frac{1}{s}$. In this case, $\rho_2 = \infty$, $\alpha = 0$ is admitted. Using (1.6.35) and the argument in (1.6.34), we obtain

$$\|A^\alpha u\|_{2,\rho_1;\infty} \leq C \ , \ \ \|A^\alpha u\|_{2,\rho_2;\infty} \leq C.$$

Since $\rho_1 \leq \rho \leq \rho_2$, the same interpolation argument as in (1.6.27) yields

$$\|A^\alpha u\|_{2,\rho;\infty} \leq \|A^\alpha u\|_{2,\rho_1;\infty} + \|A^\alpha u\|_{2,\rho_2;\infty} \leq C.$$

If $\rho = 2$, we can use this argument and obtain (1.6.23) for all $\alpha \in (0, \frac{1}{2}]$.

The result (1.6.23) is now clear for $1 < \rho \leq \frac{4}{3}$ and for $2 \leq \rho < \infty$. In the case $\frac{4}{3} < \rho < 2$ we know (1.6.23) up to now only for $-\frac{1}{2} + \frac{1}{\rho} < \alpha \leq -\frac{1}{4} + \frac{1}{\rho}$. It remains to prove (1.6.23) for $\alpha \in (-\frac{1}{4} + \frac{1}{\rho}, \frac{1}{2}]$. In this case we choose $\rho_1 = \frac{4}{3}$ and $\rho_2 = 2$. Then we see that

$$\alpha \in (-\frac{1}{2} + \frac{1}{\rho_1}, \frac{1}{2}] \ , \ \ \alpha \in (0, \frac{1}{2}] \, ,$$

and from above we obtain

$$\|A^\alpha u\|_{2,\rho_1;\infty} \leq C \ , \ \ \|A^\alpha u\|_{2,\rho_2;\infty} \leq C \, .$$

The above interpolation yields

$$\|A^\alpha u\|_{2,\rho;\infty} \leq C$$

and this completes the proof of c). Here $C > 0$ is always a constant depending on F, u_0, $E_\infty(u)$, s, ρ, α.

To prove (1.6.24) we consider some fixed $1 < \rho < \infty$ and all α with $0 \leq \alpha \leq \frac{1}{2}$, $-\frac{1}{2} + \frac{1}{\rho} < \alpha \leq \frac{1}{\rho}$. For each such α we find a unique q with $2 \leq q \leq 6$, so that $\alpha = \frac{3}{2}(\frac{1}{2} - \frac{1}{q})$, $2\alpha + \frac{3}{q} = \frac{3}{2}$. The embedding inequality (2.4.6), III, now yields

$$\|u\|_q \leq C' \, \|A^\alpha u\|_2$$

for almost all $t \in [0,\infty)$ and with (1.6.23) we get

$$\|u\|_{q,\rho;\infty} \leq C' \, \|A^\alpha u\|_{2,\rho;\infty} \leq C \tag{1.6.36}$$

with $C, C' > 0$. Since $\alpha \le \frac{1}{\rho}$ we get $\frac{3}{2} \le \frac{3}{q} + \frac{2}{\rho}$, and since $\frac{1}{\rho} - \frac{1}{2} < \alpha$,

$$\frac{3}{q} + \frac{2}{\rho} < \frac{3}{2} + 1 .$$

Conversely, consider any $2 \le q \le 6$, $1 < \rho < \infty$ satisfying

$$\frac{3}{2} \le \frac{3}{q} + \frac{2}{\rho} < \frac{3}{2} + 1 .$$

Then we set $\alpha := \frac{1}{2}(\frac{3}{2} - \frac{3}{q})$, get

$$\frac{1}{\rho} - \frac{1}{2} < \alpha \le \frac{1}{\rho} , \quad 0 \le \alpha \le \frac{1}{2} ,$$

and (1.6.36) holds. This proves (1.6.24) and the proof of the theorem is complete. □

A similar result holds in the two-dimensional case. In this case we need an additional argument (**absorption principle**, see (1.6.54)).

1.6.3 Theorem ($n = 2$) *Let $\Omega \subseteq \mathbb{R}^2$ be any two-dimensional domain, let $T = \infty$, $1 < s \le 2$, $1 < \rho < \infty$, $2 \le q < \infty$, $0 \le \alpha \le \frac{1}{2}$, and let $u_0 \in D(A^{-\frac{1}{2}})$, $f = \operatorname{div} F$ with*

$$F \in L^1(0, \infty; L^2(\Omega)^4) \cap L^2(0, \infty; L^2(\Omega)^4). \tag{1.6.37}$$

Suppose

$$u \in L^\infty_{loc}([0, \infty); L^2_\sigma(\Omega)) \cap L^2_{loc}([0, \infty); W^{1,2}_{0,\sigma}(\Omega))$$

is a weak solution of the Navier-Stokes system (1.1.1) with data f, u_0. Suppose additionally that

$$E_\infty(u) = \frac{1}{2} \|u\|^2_{2,\infty;\infty} + \nu \|\nabla u\|^2_{2,2;\infty} < \infty. \tag{1.6.38}$$

Then there is a constant $C > 0$, depending on $F, u_0, \nu, s, \rho, \alpha$ and q, such that, after redefining u on a null set of $[0, \infty)$, the following properties are satisfied:

a) $\quad u(t) \in D(A^{-\frac{1}{2}})$ *for all* $t \in [0, \infty)$, $A^{-\frac{1}{2}} u : [0, \infty) \to L^2_\sigma(\Omega)$

is strongly continuous, $(A^{-\frac{1}{2}} u)(0) = A^{-\frac{1}{2}} u_0$, *and*

$$\|A^{-\frac{1}{2}} u\|_{2,\infty;\infty} + \|u\|_{2,2;\infty} \le C. \tag{1.6.39}$$

b)
$$\|(A^{-\frac{1}{2}}u)_t\|_{2,s;\infty} + \|A^{\frac{1}{2}}u\|_{2,s;\infty} \le C. \tag{1.6.40}$$

c)
$$\|A^\alpha u\|_{2,\rho;\infty} \le C \tag{1.6.41}$$

with $-\frac{1}{2} + \frac{1}{\rho} < \alpha \le \frac{1}{\rho}$, $0 \le \alpha \le \frac{1}{2}$.

d)
$$\|u\|_{q,\rho;\infty} \le C \tag{1.6.42}$$

with
$$1 \le \frac{2}{q} + \frac{2}{\rho} < 2. \tag{1.6.43}$$

Proof. First we admit that the following constant $C > 0$ also depends on the given weak solution u. Since u is uniquely determined in the case $n = 2$ by the given data f, u_0, see Theorem 1.5.3, we see that C depends in fact only on $F, u_0, s, \rho, \alpha, q$.

Using Theorem 1.6.1 with $s = 2$ we see that $A^{-\frac{1}{2}}u$ is strongly continuous, after a corresponding redefinition, and that $(A^{-\frac{1}{2}}u)(0) = A^{-\frac{1}{2}}u_0$. Using (1.6.6) we get the inequality (1.6.40) with $s = 2$ and some $C > 0$. From (1.6.13) we obtain the integral equation

$$A^{-\frac{1}{2}}u(t) = S(t)A^{-\frac{1}{2}}u_0 + \int_0^t S(t-\tau)\tilde{f}(\tau)\,d\tau \tag{1.6.44}$$

for all $t \in [0,\infty)$ with $\tilde{f} = A^{-\frac{1}{2}}P \operatorname{div} \tilde{F}$, $\tilde{F} = F - u\,u$. Further we get $(A^{-\frac{1}{2}}u)_t$, $A^{\frac{1}{2}}u \in L^2(0,\infty; L^2_\sigma(\Omega))$, and the evolution equation

$$(A^{-\frac{1}{2}}u)_t + A^{\frac{1}{2}}u = \tilde{f} \tag{1.6.45}$$

holds, see (1.6.15). From Lemma 1.2.1, b), we see with $q = s = 4$ that

$$\|u\|_{4,4;\infty} \le C_1 E_\infty(u)^{\frac{1}{2}} \tag{1.6.46}$$

with some $C_1 = C_1(\nu) > 0$, and with Hölder's inequality we get

$$\|u\,u\|_{2,2;\infty} \le C_2\,\|u\|^2_{4,4;\infty} \le C_3\,E_\infty(u) \tag{1.6.47}$$

with $C_2 > 0$, $C_3 = C_3(\nu) > 0$.

Our next purpose is to prove that

$$\|u\,u\|_{2,s;\infty} \le C \tag{1.6.48}$$

with $1 \le s \le 2$ and some $C > 0$. Here we cannot use the same argument as for $n = 3$.

Let $1 < s \le 2$, choose $s_1 > s$ so that $\frac{1}{s} = \frac{1}{4} + \frac{1}{s_1}$, and set $\alpha = \frac{1}{2} + \frac{1}{4}$.

Using the embedding inequality (2.4.6), III, with $2 \cdot \frac{1}{4} + \frac{2}{4} = \frac{2}{2}$, we get

$$\|u\|_{4,s_1;T'} \le C_1\,\|A^{\frac{1}{4}}u\|_{2,s_1;T'} = C_1\,\|A^\alpha\,A^{-\frac{1}{2}}u\|_{2,s_1;T'} \tag{1.6.49}$$

with $0 < T' < \infty$, $C_1 = C_1(\nu) > 0$.

Applying the estimates (2.5.5), IV, and (2.5.13), IV, from the linear theory to the equation (1.6.44), we obtain the inequality

$$\begin{aligned} &\|(A^{-\frac{1}{2}}u)_t\|_{2,s;T'} + \|A^{\frac{1}{2}}u\|_{2,s;T'} \\ &\quad \le \; C_2\,(\|u_0\|_2 + \|A^{-\frac{1}{2}}u_0\|_2 + \|F\|_{2,s;T'} + \|u\,u\|_{2,s;T'}). \end{aligned} \tag{1.6.50}$$

Inserting (1.6.45) in (1.6.44) we see that

$$A^{-\frac{1}{2}}u(t) \;=\; S(t)A^{-\frac{1}{2}}u_0 + \int_0^t S(t-\tau)((A^{-\frac{1}{2}}u)_\tau(\tau) + A^{\frac{1}{2}}u(\tau))\,d\tau \tag{1.6.51}$$

for all $t \in [0,\infty)$. Now we apply the estimates (2.5.7), IV, and (2.5.15), IV, with $\alpha = \frac{1}{2} + \frac{1}{4} = 1 + \frac{1}{s_1} - \frac{1}{s}$ to this equation. This yields

$$\begin{aligned} \|A^\alpha A^{-\frac{1}{2}}u\|_{2,s_1;T'} \le C_2\,(\|u_0\|_2 + \|A^{-\frac{1}{2}}u_0\|_2 + \|(A^{-\frac{1}{2}}u)_t\|_{2,s;T'} \\ + \|A^{\frac{1}{2}}u\|_{2,s;T'}) \end{aligned} \tag{1.6.52}$$

with $C_2 = C_2(s) > 0$.

Using Hölder's inequality, (1.6.49), and (1.6.52), we obtain with constants C_3, C_4 depending only on s that

$$\begin{aligned} &C_2\|u\,u\|_{2,s;T'} \le C_3\,\|u\|_{4,4;T'}\,\|u\|_{4,s_1;T'} \\ &\quad \le \; C_4\|u\|_{4,4;T'}\,(\|u_0\|_2 + \|A^{-\frac{1}{2}}u_0\|_2 + \|(A^{-\frac{1}{2}}u)_t\|_{2,s;T'} + \|A^{\frac{1}{2}}u\|_{2,s;T'}). \end{aligned}$$

Inserting this in (1.6.50), we get with some C depending on u_0, F and u, but not on $T' < \infty$, that

$$\begin{aligned} &\|(A^{-\frac{1}{2}}u)_t\|_{2,s;T'} + \|A^{\frac{1}{2}}u\|_{2,s;T'} \\ &\quad \le \; C + C_4\,\|u\|_{4,4;T'}\,(\|(A^{-\frac{1}{2}}u)_t\|_{2,s;T'} + \|A^{\frac{1}{2}}u\|_{2,s;T'}). \end{aligned}$$

We know from (1.6.46) that $\|u\|_{4,4;\infty} < \infty$. Therefore we may assume without loss of generality that

$$C_4\,\|u\|_{4,4;\infty} < 1. \tag{1.6.53}$$

Otherwise we can choose some $T_0 > 0$ so that $C_4\,(\int_{T_0}^\infty \|u\|_4^4\,dt)^{\frac{1}{4}} < 1$, and setting $\tilde{u}(t) := u(T_0 + t)$, $t \ge 0$, we can carry out the same procedure with u replaced by $\tilde{u}$. Using (1.6.53) we see (**absorption argument**) that

$$(\|(A^{-\frac{1}{2}}u)_t\|_{2,s;T'} + \|A^{\frac{1}{2}}u\|_{2,s;T'})(1 - C_4\|u\|_{4,4;\infty}) \;\le\; C \tag{1.6.54}$$

for $0 < T' < \infty$. Letting $T' \to \infty$ we obtain

$$\|(A^{-\frac{1}{2}}u)_t\|_{2,s;\infty} + \|A^{\frac{1}{2}}u\|_{2,s;\infty} \;\le\; C \tag{1.6.55}$$

with $1 < s \leq 2$ and $C > 0$ depending on u_0, F, u, s. This yields the property (1.6.40). Now we conclude from the inequality above that

$$\|u\,u\|_{2,s;\infty} \leq C \tag{1.6.56}$$

for $1 < s \leq 2$ with some $C > 0$.

To prove (1.6.41) we choose ρ, s with $1 < s \leq \rho < \infty$, $s \leq 2$, and set $\alpha := \frac{1}{2} + \frac{1}{\rho} - \frac{1}{s}$. Let $\alpha \geq 0$. Then we apply (2.5.7), IV, and (2.5.15), IV, to the equation (1.6.51) with α replaced by $\alpha + \frac{1}{2}$. This yields with $0 < T' < \infty$ the estimate

$$\begin{aligned} \|A^\alpha u\|_{2,\rho;T'} &= \|A^{\alpha+\frac{1}{2}} A^{-\frac{1}{2}} u\|_{2,s;T'} \\ &\leq C' \left(\|u_0\|_2 + \|A^{-\frac{1}{2}} u_0\|_2 + \|(A^{-\frac{1}{2}} u)_t\|_{2,s;T'} + \|A^{\frac{1}{2}} u\|_{2,s;T'}\right) \end{aligned}$$

where $C' = C'(\rho, s) > 0$.

If $1 < \rho \leq 2$, then for each $\alpha \in (-\frac{1}{2} + \frac{1}{\rho}, \frac{1}{2}]$, there is some s, $1 < s \leq \rho$, such that $\alpha = \frac{1}{2} + \frac{1}{\rho} - \frac{1}{s}$. The last inequality yields (1.6.41) in this case. If $2 < \rho < \infty$, then for each $\alpha \in [0, \frac{1}{\rho}]$, there is some s, $1 < s \leq 2$, such that $\alpha = \frac{1}{2} + \frac{1}{\rho} - \frac{1}{s}$. The last inequality yields (1.6.41) in this case. Note that negative values α are excluded in (1.6.41) since we use (2.5.7), IV, where $\frac{1}{2} + \frac{1}{\rho} - \frac{1}{s} \geq 0$ is needed. The proof of (1.6.41) is now complete.

To prove (1.6.42) we consider $2 \leq q < \infty$, $1 < \rho < \infty$ satisfying (1.6.43), and set $\alpha := \frac{1}{2}(1 - \frac{2}{q})$. Then $0 \leq \alpha \leq \frac{1}{2}$, $2\alpha + \frac{2}{q} = \frac{2}{2}$, and $\frac{1}{\rho} - \frac{1}{2} < \alpha \leq \frac{1}{\rho}$. The embedding inequality (2.4.6), III, now yields

$$\|u\|_q \leq C' \, \|A^\alpha u\|_2$$

for almost all $t \in [0, \infty)$ with $C' = C'(q) > 0$. Together with (1.6.41) we obtain

$$\|u\|_{q,\rho;T'} \leq C' \, \|A^\alpha u\|_{2,\rho;T'} \leq C \tag{1.6.57}$$

with $C > 0$ depending on u_0, F, q, ρ.

It remains to prove (1.6.39). For this purpose we use (1.6.42) with $q = 4$, $\rho = 2$. Then (1.6.43) is satisfied and therefore,

$$\|u\|_{4,2;\infty} \leq C.$$

Together with Hölder's inequality we see that

$$\|u\,u\|_{2,1;\infty} \leq C_1 \, \|u\|^2_{4,2;\infty} \leq C_1 \, C^2 \tag{1.6.58}$$

with $C_1 > 0$. To prove (1.6.39) we apply (2.5.10), IV, (2.5.17), IV, with u replaced by $A^{-\frac{1}{2}} u$, f replaced by $\tilde{f}$, and we use (1.6.58). Further we use that $\|\nabla A^{-\frac{1}{2}} u\|_2 = \nu^{-\frac{1}{2}} \|u\|_2$. The proof of the theorem is complete. □

1.7 Associated pressure of weak solutions

To construct an associated pressure p of a weak solution u we can go back to the linear theory and write the Navier-Stokes system (1.1.1) in the form

$$u_t - \nu\Delta u + \nabla p = \tilde{f} \ , \quad \text{div } u = 0 \ , \quad u|_{\partial\Omega} = 0 \ , \quad u(0) = u_0 \tag{1.7.1}$$

with $\tilde{f} = f_0 +$ div $\tilde{F}$, $\tilde{F} = F - u\,u$. From Lemma 1.2.1 we obtain the information

$$u\,u \in L^s_{loc}([0,T); L^2(\Omega)^{n^2}) \tag{1.7.2}$$

with $s = \frac{4}{n}$, and the linear theory then leads to the following result.

1.7.1 Theorem *Let $\Omega \subseteq \mathbb{R}^n$, $n = 2,3$, be any domain, let $0 < T \leq \infty$, $s = \frac{4}{n}$, $u_0 \in L^2_\sigma(\Omega)$, $f = f_0 + \text{div } F$ with*

$$f_0 \in L^1_{loc}([0,T); L^2(\Omega)^n) \ , \quad F \in L^s_{loc}([0,T); L^2(\Omega)^{n^2}),$$

and let

$$u \in L^\infty_{loc}([0,T); L^2_\sigma(\Omega)) \cap L^2_{loc}([0,T); W^{1,2}_{0,\sigma}(\Omega))$$

be a weak solution of the Navier-Stokes system (1.1.1) with data f, u_0.

Then there exists a function $\widehat{p} \in L^s_{loc}([0,T); L^2_{loc}(\Omega))$ such that the time derivative

$$p = \frac{\partial}{\partial t}\widehat{p} = \widehat{p}_t$$

is an associated pressure of u. This means, p satisfies the equation

$$u_t - \nu\Delta u + u \cdot \nabla u + \nabla p = f \tag{1.7.3}$$

in the sense of distributions in $[0,T) \times \Omega$.

Proof. Lemma 1.2.1, d), yields

$$\|u\,u\|_{2,s;T'} \leq C\,E_{T'}(u) = C\,(\frac{1}{2}\,\|u\|^2_{2,\infty;T'} + \nu\|\nabla u\|^2_{2,2;T'}) \ < \infty$$

with $0 < T' < T$, $s = \frac{4}{n}$, $C = C(n,\nu) > 0$. This proves (1.7.2). It follows that

$$\tilde{F} = F - u\,u \in L^s_{loc}([0,T); L^2(\Omega)^{n^2}),$$

and the result is a consequence of Theorem 2.6.1, IV, if $0 < T < \infty$ and of Corollary 2.6.2, IV, if $T = \infty$. This proves the theorem. See [Tem77, Chap. III, 3 (3.128)] concerning the pressure of weak solutions. □

1.8 Regularity properties of weak solutions

Our aim is to prove local regularity properties of a weak solution u, if Serrin's condition (1.8.1) and some smoothness properties are satisfied. See [Ser62], [Ser63], [Tem77, Chap. III, 3], [Hey80], [vWa85], [GaM88], [Neu99], [BdV95], [BdV97] concerning several regularity results. Our first regularity step, see the next theorem, requires a rather complicated proof. We use again the absorption principle, see (1.6.54). After this first step, the regularity properties of higher order can be shown by the linear theory.

1.8.1 Theorem *Let $\Omega = \mathbb{R}^n$ or let $\Omega \subseteq \mathbb{R}^n$, $n = 2, 3$, be a uniform C^2-domain, let $0 < T \le \infty$, $u_0 \in W^{1,2}_{0,\sigma}(\Omega)$, $f \in L^2_{loc}([0,T); L^2(\Omega)^n)$, and let*

$$u \in L^\infty_{loc}([0,T); L^2_\sigma(\Omega)) \cap L^2_{loc}([0,T); W^{1,2}_{0,\sigma}(\Omega))$$

be a weak solution of the Navier-Stokes system (1.1.1) *with data f, u_0. Suppose additionally that*

$$u \in L^s_{loc}([0,T); L^q(\Omega)^n) \tag{1.8.1}$$

with $n < q < \infty$, $2 < s < \infty$, $\frac{n}{q} + \frac{2}{s} \le 1$.
Then we get

$$u \in L^\infty_{loc}([0,T); W^{1,2}_{0,\sigma}(\Omega)) \cap L^2_{loc}([0,T); W^{2,2}(\Omega)^n), \tag{1.8.2}$$

$$u_t \in L^2_{loc}([0,T); L^2_\sigma(\Omega)), \tag{1.8.3}$$

and

$$u \cdot \nabla u \in L^2_{loc}([0,T); L^2(\Omega)^n). \tag{1.8.4}$$

Further there exists an associated pressure p satisfying

$$p \in L^2_{loc}([0,T); L^2_{loc}(\overline{\Omega}))\ ,\quad \nabla p \in L^2_{loc}([0,T); L^2(\Omega)^n). \tag{1.8.5}$$

Proof. Since T can be replaced by T', $0 < T' < T$, we may assume without loss of generality that $0 < T < \infty$ and that $f \in L^2(0,T; L^2(\Omega)^n)$, $u \in L^\infty(0,T; L^2_\sigma(\Omega)) \cap L^2(0,T; W^{1,2}_{0,\sigma}(\Omega))$, $u \in L^s(0,T; L^q(\Omega)^n)$. Further we may assume that q and s satisfy $\frac{n}{q} + \frac{2}{s} = 1$. Otherwise we can choose some s_1 with $\frac{n}{q} + \frac{2}{s_1} = 1$, $2 < s_1 \le s$, and we replace s by s_1.

Theorem 1.3.1 shows that u is weakly continuous after a corresponding redefinition, and that the integral equation

$$u(t) = S(t)u_0 + (\mathcal{J}Pf)(t) - A^{\frac{1}{2}}(\mathcal{J}A^{-\frac{1}{2}}P \operatorname{div} u\,u)(t) \tag{1.8.6}$$

is satisfied for all $t \in [0,T)$ with $\mathcal{J}$ defined in (1.6.3), IV.

First we have to prepare several estimates.

We use Yosida's operators $J_k = (I + \frac{1}{k}A^{\frac{1}{2}})^{-1}$, $k \in \mathbb{N}$, see Section 3.4, II, and set $u_{0k} := J_k u_0$, $u_k := J_k u$, $f_k := J_k P f$, $\hat{f}_k := J_k P$ div $(u\, u)$. Then $\|J_k\| \le 1$, and therefore

$$\|A^{\frac{1}{2}} J_k\| \;=\; k\|A^{\frac{1}{2}}(kI + A^{\frac{1}{2}})^{-1}\| \;=\; k\|I - J_k\| \;\le\; 2k. \tag{1.8.7}$$

With $\|A^{-\frac{1}{2}} P \text{ div } (u\, u)\|_2 \;\le\; \nu^{-\frac{1}{2}}\|u\, u\|_2$, see (1.1.10), we obtain

$$\|J_k P \text{ div } (u\, u)\|_2 = \|A^{\frac{1}{2}} J_k A^{-\frac{1}{2}} P \text{ div } (u\, u)\|_2 \;\le 2k\, \nu^{-\frac{1}{2}}\|u\, u\|_2\,, \tag{1.8.8}$$

and therefore

$$\|f_k - \hat{f}_k\|_{2,2;T} \;\le\; \|f\|_{2,2;T} + 2k\, \nu^{-\frac{1}{2}}\|u\, u\|_{2,2;T}. \tag{1.8.9}$$

Let $q_1,\, s_1 > 2$ be defined by $\frac{1}{2} = \frac{1}{q} + \frac{1}{q_1}$, $\frac{1}{2} = \frac{1}{s} + \frac{1}{s_1}$ so that $\frac{n}{q_1} + \frac{2}{s_1} = \frac{n}{2} + 1 - 1 = \frac{n}{2}$. Then from Lemma 1.2.1, b), we get

$$\|u\|_{q_1,s_1;T} \;\le\; C\, E_T(u)^{\frac{1}{2}} \;<\; \infty \tag{1.8.10}$$

with $C = C(n, \nu, s_1) > 0$. Using Hölder's inequality, we see that

$$\|u\, u\|_{2,2;T} \;\le\; C\, \|u\|_{q,s;T}\, \|u\|_{q_1,s_1;T} \;<\; \infty \tag{1.8.11}$$

with $C = C(n) > 0$. This shows that

$$f_k - \hat{f}_k \in L^2(0,T; L^2_\sigma(\Omega)) \;\;,\quad k \in \mathbb{N}. \tag{1.8.12}$$

Next we set $\alpha := \frac{1}{2} + \frac{1}{s_1}$ and use the embedding estimate (2.4.18), III, with $2\alpha + \frac{n}{q_1} = 1 + \frac{n}{2}$. This yields

$$\|\nabla u_k\|_{q_1} \;\le\; C\,(\|A^\alpha u_k\|_2 + \|u_k\|_2) \tag{1.8.13}$$

with $C = C(\Omega, \nu, q_1, s_1) > 0$. Using inequality (2.4.6), III, we obtain with $\frac{2}{s_1} + \frac{n}{q_1} = \frac{n}{2}$ that

$$\|A^{\frac{1}{2}} u_k\|_{q_1} \;\le\; C\, \|A^{\frac{1}{s_1}}(A^{\frac{1}{2}} u_k)\|_2 = C\, \|A^{\frac{1}{2}+\frac{1}{s_1}} u_k\|_2 = C\, \|A^\alpha u_k\|_2,$$

$C = C(q_1, s_1) > 0$. Together with (1.8.13) and (2.2.8), III, we get

$$\begin{aligned}
\|\nabla u_k\|_{q_1} + \|A^{\frac{1}{2}} u_k\|_{q_1} \;&\le\; C\,(\|A^\alpha u_k\|_2 + \|u_k\|_2) \\
&=\; C\,(\|(A^{\frac{1}{2}})^{\frac{2}{s_1}}(A^{\frac{1}{2}} u_k)\|_2 + \|u_k\|_2) \\
&\le\; C\,(\|A u_k\|_2^{2/s_1}\, \|A^{\frac{1}{2}} u_k\|_2^{1-2/s_1} + \|u_k\|_2)
\end{aligned}$$

with $C = C(\Omega, q_1, s_1) > 0$. This leads to the inequality

$$\begin{aligned} \|\nabla u_k\|_{q_1,s_1;T} &+ \|A^{\frac{1}{2}} u_k\|_{q_1,s_1;T} \\ &\le C\,(\|Au_k\|_{2,2;T}^{2/s_1}\, \|A^{\frac{1}{2}} u_k\|_{2,\infty;T}^{1-2/s_1} + \|u_k\|_{2,s_1;T}). \end{aligned} \tag{1.8.14}$$

Applying J_k to the representation (1.8.6), we obtain

$$\begin{aligned} u_k &= S(\cdot)u_{0k} + \mathcal{J} f_k - J_k A^{\frac{1}{2}} \mathcal{J} A^{-\frac{1}{2}} P \text{ div } (u\,u) \\ &= S(\cdot)u_{0k} + \mathcal{J} f_k - \mathcal{J} J_k P \text{ div } (u\,u) \\ &= S(\cdot)u_{0k} + \mathcal{J}(f_k - \hat{f}_k). \end{aligned}$$

Here we use that $J_k = J_k A^{\frac{1}{2}}\, A^{-\frac{1}{2}} = A^{\frac{1}{2}} J_k A^{-\frac{1}{2}}$, see Section 3.4, II.

To this equation we apply the basic estimates of the linear theory, see Section 2.5, IV, with $s = 2$. Combining (2.5.5), (2.5.13), (2.5.18), IV, setting

$$|||u_k|||_T := \|u_k'\|_{2,2;T} + \|A^{\frac{1}{2}} u_k\|_{2,\infty;T} + \|Au_k\|_{2,2;T} + \|u_k\|_{2,\infty;T}\ ,$$

and using

$$\begin{aligned} \|u_k(t)\|_2 &\le \|u_{0k}\|_2 + \int_0^t \|u_k'(\tau)\|_2\, d\tau, \quad t \in [0,T), \\ \|u_k\|_{2,\infty;T} &\le \|u_{0k}\|_2 + T^{\frac{1}{2}} \|u_k'\|_{2,2;T}\ , \end{aligned}$$

we thus obtain the inequality

$$|||u_k|||_T \le C\,(1+T)\,(\|u_{0k}\|_2 + \|A^{\frac{1}{2}} u_{0k}\|_2 + \|f_k - \hat{f}_k\|_{2,2;T}) \tag{1.8.15}$$

with $C > 0$ not depending on T.

Next we use the calculations

$$\begin{aligned} u = J_k^{-1} u_k &= (I + \frac{1}{k} A^{\frac{1}{2}}) u_k = u_k + \frac{1}{k} A^{\frac{1}{2}} u_k, \\ J_k P \text{ div } (u\,u) &= J_k P \text{ div } (u\,u_k) + k^{-1} J_k P \text{ div } (u(A^{\frac{1}{2}} u_k)) \\ &= J_k P(u \cdot \nabla u_k) + \frac{1}{k} J_k P \text{ div } (u(A^{\frac{1}{2}} u_k)) \\ &= J_k P(u \cdot \nabla u_k) + (kI + A^{\frac{1}{2}})^{-1} A^{\frac{1}{2}} A^{-\frac{1}{2}} P \text{ div } (u(A^{\frac{1}{2}} u_k)), \end{aligned}$$

and the estimates (1.8.8), (1.8.7), (1.8.11), (1.8.14).

With $\|J_k\| \le 1$, $\|P\| \le 1$, $\|A^{-\frac{1}{2}} P \text{ div }\| \le \nu^{-\frac{1}{2}}$, we thus obtain

$$\begin{aligned} \|J_k P \text{ div } (u\,u)\|_{2,2;T} &\le \|u \cdot \nabla u_k\|_{2,2;T} + 2\,\nu^{-\frac{1}{2}} \|u(A^{\frac{1}{2}} u_k)\|_{2,2;T} \\ &\le C\, \|u\|_{q,s;T}\, (\|\nabla u_k\|_{q_1,s_1;T} + \|A^{\frac{1}{2}} u_k\|_{q_1,s_1;T}) \end{aligned}$$

$$\begin{aligned}
&\leq \; C\,\|u\|_{q,s;T}\,(\|Au_k\|_{2,2;T}^{2/s_1}\,\|A^{\frac{1}{2}}u_k\|_{2,\infty;T}^{1-2/s_1} \;+\; \|u_k\|_{2,s_1;T}) \\
&\leq \; C\,\|u\|_{q,s;T}\,(\|Au_k\|_{2,2;T} + \|A^{\frac{1}{2}}u_k\|_{2,\infty;T} \;+\; (1+T)\,\|u_k\|_{2,\infty;T}) \\
&\leq \; C\,(1+T)\,\|u\|_{q,s;T}\,|||u_k|||_T
\end{aligned}$$

where $C = C(\Omega, \nu, q_1, s_1) > 0$.

Consider now any T' with $0 < T' \leq T$. The above estimates also hold with T replaced by T'. Combining (1.8.15), T replaced by T', with the last estimate, we obtain

$$\begin{aligned}
|||u_k|||_{T'} \;\; &\leq \;\; C\,(1+T)\big(\|u_0\|_2 + \|A^{\frac{1}{2}}u_0\|_2 + \|f\|_{2,2;T} \\
&\qquad\qquad + (1+T)\|u\|_{q,s;T'}\,|||u|||_{T'}\big)
\end{aligned}$$

with $C = C(\Omega, \nu, q_1, s_1) > 0$.

Next we choose T' in such a way that

$$C\,(1+T)^2\,\|u\|_{q,s;T'} < 1. \tag{1.8.16}$$

Then we get (absorption argument)

$$\begin{aligned}
&|||u_k|||_{T'}(1 - C\,(1+T)^2\|u\|_{q,s;T'}) \\
&\quad \leq C\,(1+T)\,(\|u_0\|_2 + \|A^{\frac{1}{2}}u_0\|_2 + \|f\|_{2,2;T}).
\end{aligned} \tag{1.8.17}$$

Letting $k \to \infty$ and using the argument in (3.1.8), II, we see that

$$|||u|||_{T'} = \|u'\|_{2,2;T'} + \|A^{\frac{1}{2}}u\|_{2,\infty;T'} + \|Au\|_{2,2;T'} + \|u\|_{2,\infty;T'} \;<\infty. \tag{1.8.18}$$

Using the inequality (2.1.9), III, for uniform C^2-domains, and the interpolation

$$\|\nabla u\|_2 \;=\nu^{-\frac{1}{2}}\,\|A^{\frac{1}{2}}u\|_2 \;\leq\; \nu^{-\frac{1}{2}}\|Au\|_2^{\frac{1}{2}}\,\|u\|_2^{\frac{1}{2}} \;\leq\; \nu^{-\frac{1}{2}}\,(\|Au\|_2 + \|u\|_2),$$

see (2.2.8), III, we obtain for the second order derivatives the inequality

$$\|\nabla^2 u\|_2 \;\leq\; C\,(\|Au\|_2 + \|u\|_2) \tag{1.8.19}$$

with $C = C(\Omega, \nu) > 0$. Together with (1.8.18), this shows that

$$\|u'\|_{2,2;T'} + \|\nabla u\|_{2,\infty;T'} + \|\nabla^2 u\|_{2,2;T'} + \|u\|_{2,\infty;T'} \;<\infty. \tag{1.8.20}$$

This proves the assertions (1.8.2) and (1.8.3) with T replaced by T'. Since C in (1.8.16) does not depend on T', we can repeat this procedure, if $T' < T$, with u replaced by $\tilde{u}$ defined by $\tilde{u}(t) = u(T'+t), t \geq 0$. After finitely many steps, we get (1.8.2) and (1.8.3) for $0 < T < \infty$.

To prove (1.8.4), we use (1.8.11), (1.8.14) and obtain

$$\begin{aligned}\|u\cdot\nabla u\|_{2,2;T} &\le C\,\|u\|_{q,s;T}\,\|\nabla u\|_{q_1,s_1;T} \qquad (1.8.21)\\ &\le C\,\|u\|_{q,s;T}(\|Au\|_{2,2;T}^{2/s_1}\,\|A^{\frac{1}{2}}u\|_{2,\infty;T}^{1-2/s_1}+\|u\|_{2,s_1;T})<\infty.\end{aligned}$$

To prove (1.8.5), we use (1.8.4) and apply Theorem 2.6.3, IV, with $s=2$. Another possibility is to use p from Theorem 1.7.1, and to write the equation (1.7.3) in the form

$$\nabla p = f - u_t + \nu\Delta u - u\cdot\nabla u. \qquad (1.8.22)$$

Using (1.8.2), (1.8.3) and (1.8.4), we get

$$\nabla p \in L^2_{loc}(0,T;L^2(\Omega)^n),$$

and Lemma 1.4.2, IV, yields

$$p \in L^2_{loc}(0,T;L^2_{loc}(\overline{\Omega})).$$

The proof of the theorem is complete. □

In the next step we will improve the regularity of u by applying the linear theory to the system

$$u_t - \nu\Delta u + \nabla p = f - u\cdot\nabla u\ ,\quad \operatorname{div} u = 0\ ,\quad u|_{\partial\Omega}=0\ ,\quad u(0)=u_0.$$

For simplicity we consider only the case of smooth exterior forces f.

1.8.2 Theorem *Let $\Omega=\mathbb{R}^n$, or let $\Omega\subseteq\mathbb{R}^n$, $n=2,3$, be a uniform C^2-domain. Suppose Ω is also a C^∞-domain if $\Omega\neq\mathbb{R}^n$. Let $0<T\le\infty$, $u_0\in W^{1,2}_{0,\sigma}(\Omega)$, $f\in C_0^\infty(\,\overline{(0,T)\times\Omega}\,)^n$, and let*

$$u \in L^\infty_{loc}([0,T);L^2_\sigma(\Omega)) \cap L^2_{loc}([0,T);W^{1,2}_{0,\sigma}(\Omega))$$

be a weak solution of the Navier-Stokes system (1.1.1) with data f,u_0. Assume additionally that

$$u \in L^s_{loc}([0,T);L^q(\Omega)^n) \qquad (1.8.23)$$

with $n<q<\infty$, $2<s<\infty$, $\frac{n}{q}+\frac{2}{s}\le 1$.

Then, after a redefinition on a null set of $[0,T)\times\Omega$, we obtain

$$u\in C^\infty_{loc}\big(\,\overline{(\varepsilon,T')\times\Omega}\,\big)^n \qquad (1.8.24)$$

for all ε, T' with $0<\varepsilon<T'<T$. In particular,

$$u\in C^\infty((0,T)\times\Omega)^n.$$

Moreover, there exists an associated pressure p of u satisfying

$$p \in C^\infty_{loc}\big(\overline{(\varepsilon, T') \times \Omega}\big) \tag{1.8.25}$$

for all ε, T' with $0 < \varepsilon < T' < T$. In particular,

$$p \in C^\infty((0,T) \times \Omega). \tag{1.8.26}$$

Proof. We will improve the regularity of u in several steps. As in the previous proof we may assume that $0 < T < \infty$ and that $u \in L^s(0,T;L^q(\Omega)^n)$. Then we get from Theorem 1.8.1 that

$$u \in L^\infty(0,T;W^{1,2}_{0,\sigma}(\Omega)) \cap L^2(0,T;W^{2,2}(\Omega)^n)\ , \quad u_t \in L^2(0,T;L^2_\sigma(\Omega)) \tag{1.8.27}$$

and

$$u \cdot \nabla u \in L^2(0,T;L^2(\Omega)^n). \tag{1.8.28}$$

Using the property (1.8.28) we may write (1.8.6) in the form

$$u = S(\cdot)u_0 + \mathcal{J}Pf - \mathcal{J}P\ \mathrm{div}\ (u\,u) = S(\cdot)u_0 + \mathcal{J}Pf - \mathcal{J}Pu \cdot \nabla u. \tag{1.8.29}$$

It follows, see Theorem 2.5.1, IV, and Theorem 2.5.2, IV, that the evolution equation

$$u_t + Au = Pf - Pu \cdot \nabla u \tag{1.8.30}$$

is satisfied in $L^2(0,T;L^2_\sigma(\Omega))$. Using a (cut-off) function $\varphi \in C_0^\infty((0,T))$ in the same way as in the proof of Theorem 2.7.2, IV, we conclude that

$$(\varphi u)_t + A(\varphi u) = P(\varphi f) - P(\varphi u) \cdot \nabla u + \varphi_t u, \tag{1.8.31}$$

and because of $\varphi(0) = 0$, we get the representation

$$\varphi u = \mathcal{J}P(\varphi f - \varphi u \cdot \nabla u + \varphi_t u)\,, \tag{1.8.32}$$

see (2.4.4), IV.

Using (1.8.27), Sobolev's embedding property (1.3.10), II, and the embedding inequality (2.4.18), III, we obtain for almost all $t \in [0,T)$ that

$$\|\varphi u \cdot \nabla u\|_2 \ \le\ C\, \|u\|_\infty\, \|\nabla u\|_2\,,$$

$C = C(\Omega) > 0$, and

$$\begin{aligned} \|u\|_\infty \ &\le\ C_1\, \|\nabla u\|_q \ \le\ C_2\, (\|A^{\frac{1}{2}+\alpha} u\|_2 + \|u\|_2) \\ &\le\ C_2\, (\|Au\|_2^{2\alpha}\, \|A^{\frac{1}{2}} u\|_2^{1-2\alpha} + \|u\|_2) \end{aligned}$$

with $0 < \alpha \le \frac{1}{2}$, $2\alpha + \frac{n}{q} = \frac{n}{2}$, $q > n$, and C_1, C_2 depending on Ω, q, n.

This yields

$$\|\varphi u \cdot \nabla u\|_{2,s;T} \le C\,(\|Au\|_{2,2;T}^{2\alpha}\|A^{\frac{1}{2}}u\|_{2,\infty;T}^{1-2\alpha} + \|u\|_{2,s;T})\,\|\nabla u\|_{2,\infty;T} < \infty$$

with $\alpha s = 1,\ s = \frac{1}{\alpha}$. We see that

$$\|\varphi u \cdot \nabla u\|_{2,s;T} < \infty$$

with $2 \le s < 4$ if $n = 3$, and with $2 \le s < \infty$ if $n = 2$. From Theorem 2.7.1, IV, we get

$$\|(\varphi u)_t\|_{2,s;T} + \|\nabla^2(\varphi u)\|_{2,s;T} < \infty$$

with $2 \le s < \infty$ if $n = 2$, and with $2 \le s < 4$ if $n = 3$. Using this property for all $\varphi \in C_0^\infty((0,T))$ we conclude that

$$u_t \in L^s_{loc}((0,T); L^2_\sigma(\Omega))\ ,\quad u \in L^s_{loc}((0,T); W^{2,2}(\Omega)^n) \tag{1.8.33}$$

for these values s. See (2.7.12), IV, for this notation.

If $n = 3$, we can repeat this procedure and get (1.8.33) for $4 \le s < 2\cdot 4$, then for $8 \le s < 2\cdot 8$, and so on. Thus we get (1.8.33) for all s with $2 \le s < \infty$ if $n = 2, 3$.

In the next step we improve the regularity of u in the time direction. For this purpose we use the method of differentiating (1.8.30) in the time direction in the same way as in the proof of Theorem 2.7.2, IV. Differentiating (1.8.30) in the sense of distributions and setting $v = \varphi u_t$, $\varphi \in C_0^\infty((0,T))$, we obtain the equation

$$v_t + Av = P(\varphi f_t) - P\ \mathrm{div}\ \varphi(u_t\, u + u\, u_t) + \varphi_t u_t. \tag{1.8.34}$$

Using the above properties of u we see that v is a weak solution of this linear equation with data $\tilde f, u_0 = 0$ where

$$\tilde f := P(\varphi f_t) + \varphi_t u_t - P\ \mathrm{div}\ \varphi(u_t\, u + u\, u_t)\,.$$

From (1.8.33) we get with similar estimates as above that

$$\|\varphi(u_t\, u + u\, u_t)\|_{2,s;T} < \infty$$

with $2 \le s < \infty$. Theorem 2.5.3, IV, from the linear theory now shows that

$$\|(A^{-\frac{1}{2}}v)_t\|_{2,s;T} + \|A^{\frac{1}{2}}v\|_{2,s;T} < \infty$$

with $2 \le s < \infty$. Since this holds for all $\varphi \in C_0^\infty((0,T))$ we get

$$A^{-\frac{1}{2}}u_{tt}\ ,\ \ A^{\frac{1}{2}}u_t \in L^s_{loc}((0,T); L^2_\sigma(\Omega))$$

with $2 \le s < \infty$. Using similar estimates as above we now conclude that

$$\|\varphi(u_t \nabla u + u \nabla u_t)\|_{2,s;T} < \infty,$$

$2 \le s < \infty$. We write div $\varphi(u_t\, u + u\, u_t) = \varphi(u_t \cdot \nabla u + u \cdot \nabla u_t)$, see Section 3.2, III, and apply Theorem 2.5.2, IV, to (1.8.34). This yields

$$\|v_t\|_{2,s;T} + \|Av\|_{2,s;T} < \infty$$

for all $\varphi \in C_0^\infty((0,T))$, and therefore

$$u_{tt}\,,\ \ Au_t \ \in\ L^s_{loc}((0,T); L^2_\sigma(\Omega)),$$

$2 \le s < \infty$. We may repeat this procedure and obtain

$$(d/dt)^{k+1} u\,,\ A(d/dt)^k u \ \in\ L^s_{loc}((0,T); L^2_\sigma(\Omega)), \tag{1.8.35}$$

$2 \le s < \infty$, $k \in \mathbb{N}$.

In the last step of the proof we improve the regularity of u in the spatial direction. For this purpose we use the same argument as in the proof of Theorem 2.7.3, IV. We write (1.8.30) with some associated pressure p in the form

$$-\nu\Delta v + \nabla(\varphi p) = \tilde{f} \tag{1.8.36}$$

with $v = \varphi u$, $\varphi \in C_0^\infty((0,T))$, and $\tilde{f} := -\varphi u_t + \varphi f - \varphi u \cdot \nabla u$. From (1.8.35) we conclude with $k = 1$ that

$$u_t \ \in\ L^s_{loc}((0,T); W^{2,2}(\Omega)^n), \tag{1.8.37}$$

$2 \le s < \infty$, see (2.1.9), III. Using similar embedding estimates as above we conclude that

$$\tilde{f} \in L^s_{loc}((0,T); W^{1,2}(\Omega)^n),$$

$2 \le s < \infty$. To (1.8.36) we can apply the linear stationary theory of Theorem 1.5.1, III. Applying the estimate (1.5.4), III, to bounded subdomains of Ω for almost all $t \in [0,T)$, and taking the L^s-norm over $[0,T)$, we see that

$$u \in L^s_{loc}((0,T); W^{3,2}_{loc}(\overline{\Omega})^n),$$

$2 \le s < \infty$. In the same way as above we now conclude that

$$\tilde{f} \in L^s_{loc}((0,T); W^{2,2}_{loc}(\overline{\Omega})^n).$$

Using again (1.5.4), III, we see that

$$u \in L^s_{loc}((0,T); W^{4,2}_{loc}(\overline{\Omega})^n). \tag{1.8.38}$$

Next we set $v = \varphi u_t$ and use instead of (1.8.36) the equation

$$-\nu\Delta v + \nabla(\varphi p_t) = \tilde{f} \tag{1.8.39}$$

with $\tilde{f} := -\varphi u_{tt} + \varphi f_t - \varphi u_t \cdot \nabla u - \varphi u \cdot \nabla u_t$. Using (1.8.35) with $k = 2$, and (1.8.37), (1.8.38), we see that

$$\tilde{f} \in L^s_{loc}((0,T); W^{1,2}_{loc}(\overline{\Omega})^n),$$

$2 \leq s < \infty$. Applying Theorem 1.5.1, III, to (1.8.39) we conclude that

$$u_t \in L^s_{loc}((0,T); W^{3,2}_{loc}(\overline{\Omega})^n).$$

Next we apply Theorem 1.5.1, III, to (1.8.36), and see that

$$u \in L^s_{loc}((0,T); W^{5,2}_{loc}(\overline{\Omega})^n).$$

Repeating the procedure in this way we obtain that

$$u \in W^{k,2}_{loc}\left(\overline{(\varepsilon, T') \times \Omega}\right)^n$$

for all ε, T' with $0 < \varepsilon < T' < T$, and all $k \in \mathbb{N}$. Using the embedding property (1.3.10), II, we obtain the desired property (1.8.24), after a corresponding redefinition.

The property concerning p follows in the same way as in the proof of Theorem 2.7.3, IV, from the linear theory. This completes the proof of Theorem 1.8.2. □

The assumption (1.8.23) is always satisfied in the two-dimensional case, see (1.5.18). This leads to the following result:

1.8.3 Theorem ($n = 2$) *Let $\Omega = \mathbb{R}^2$, or let $\Omega \subseteq \mathbb{R}^2$ be any uniform C^2-domain, let $0 < T \leq \infty$, $u_0 \in W^{1,2}_{0,\sigma}(\Omega)$, $f \in L^2_{loc}([0,T); L^2(\Omega)^2)$, and let*

$$u \in L^\infty_{loc}([0,T); L^2_\sigma(\Omega)) \cap L^2_{loc}([0,T); W^{1,2}_{0,\sigma}(\Omega))$$

be a weak solution of the Navier-Stokes system (1.1.1) with data f, u_0.

Then

$$\begin{aligned} u &\in L^\infty_{loc}([0,T); W^{1,2}_{0,\sigma}(\Omega)) \cap L^2_{loc}([0,T); W^{2,2}(\Omega)^2), \\ u_t &\in L^2_{loc}([0,T); L^2_\sigma(\Omega)) \quad , \quad u \cdot \nabla u \in L^2_{loc}([0,T); L^2(\Omega)^2), \end{aligned} \tag{1.8.40}$$

and there exists an associated pressure p satisfying

$$p \in L^2_{loc}([0,T); L^2_{loc}(\overline{\Omega})) \quad , \quad \nabla p \in L^2_{loc}([0,T); L^2(\Omega)^2). \tag{1.8.41}$$

If Ω is a C^∞-domain and if $f \in C_0^\infty(\overline{(0,T)\times\Omega})^2$, then, after a redefinition of u on a null set of $[0,T)\times\Omega$,

$$u \in C^\infty_{loc}\big(\overline{(\varepsilon,T')\times\Omega}\big)^2 \tag{1.8.42}$$

for all ε, T' with $0<\varepsilon<T'<T$, and

$$u \in C^\infty((0,T)\times\Omega)^2.$$

Further, there exists an associated pressure p satisfying

$$p \in C^\infty_{loc}\big(\overline{(\varepsilon,T')\times\Omega}\big)$$

for all ε, T' with $0<\varepsilon<T'<T$, and

$$p \in C^\infty((0,T)\times\Omega).$$

Proof. Using (1.5.18), the result follows from Theorem 1.8.1 and Theorem 1.8.2. □

2 Approximation of the Navier-Stokes equations

2.1 Approximate Navier-Stokes system

Our aim is to approximate the Navier-Stokes system

$$u_t - \nu\Delta u + u\cdot\nabla u + \nabla p = f\ ,\quad \operatorname{div} u = 0\ ,\quad u|_{\partial\Omega} = 0\ ,\quad u(0) = u_0 \tag{2.1.1}$$

in a certain sense by a sequence of systems which have unique global solutions. To construct such systems we use again Yosida's approximation procedure, see Section 3.4, II. The idea is to replace the nonlinear term $u\cdot\nabla u$ by the "regularized term"

$$(J_k u)\cdot\nabla u\ ,\quad J_k := (I + \frac{1}{k}A^{\frac{1}{2}})^{-1}\ ,\quad k = 1,2,\dots$$

where A means the Stokes operator. We will see that the approximate systems, obtained in this way, are uniquely solvable and that the solutions $u = u_k$ have certain important convergence properties.

In this context this method was originally used in [Soh83], [Soh84]. The approximate systems can be solved by Banach's fixed point principle.

As an application we use the approximate solutions u_k in order to construct a weak solution u of the original system (2.1.1). This yields special important properties, see Section 2.2. In particular, this u satisfies the energy inequality and has a special asymptotic behavior, see Section 3.4.

In the literature there are several other approximation procedures in order to construct weak solutions of (2.1.1). The first existence proofs go back to Leray [Ler33], [Ler34], and Hopf [Hop41], [Hop50]. The Galerkin-procedure is mainly used to approximate the Navier-Stokes system, see [Lad69, Chap. 6, Theorem 13], [Tem77, Chap. III, Theorem 3.1], [Mas84, Chap. 3]. Other constructions are used by Caffarelli-Kohn-Nirenberg [CKN82, Appendix], by Temam [Tem77, Chap. III, 5 and 8] and by Borchers-Miyakawa [BMi92].

The notion of weak solutions of the approximate systems is, replacing $u \cdot \nabla u$ by $(J_k u) \cdot \nabla u$, the same as for the original system (2.1.1). We admit exterior forces $f = f_0 +$ div F with

$$f_0 \in L^1_{loc}([0,T); L^2(\Omega)^n) \ , \quad F \in L^2_{loc}([0,T); L^2(\Omega)^{n^2})$$

where $F = (F_{jl})^n_{j,l=1}$ is a matrix field. The matrix field $(J_k u)u$ is defined as before by

$$(J_k u)u := ((J_k u)_j u_l)^n_{j,l=1}$$

where $u = (u_1, \dots, u_n)$ and $J_k u = ((J_k u)_1, \dots, (J_k u)_n)$.

Using the calculations in Lemma 1.2.1, a), we obtain the relations

$$(J_k u) \cdot \nabla u = ((J_k u)_1 D_1 + \dots + (J_k u)_n D_n) u = \text{ div } ((J_k u)u)$$

with $\text{div}\,((J_k u)u) = \text{div}\,(J_k u)u = (D_1((J_k u)_1 u_l) + \dots + D_n((J_k u)_n u_l))^n_{l=1}$, and

$$\begin{aligned} < (J_k u) \cdot \nabla u, v >_{\Omega,T} \quad &= \quad < \text{div } (J_k u)u, v >_{\Omega,T} \qquad (2.1.2) \\ = - < (J_k u)u, \nabla v >_{\Omega,T} \quad &= \quad - < u, (J_k u) \cdot \nabla v >_{\Omega,T} \end{aligned}$$

for all test functions $v \in C_0^\infty([0,T); C^\infty_{0,\sigma}(\Omega))$, see (1.4.2), IV, for this test space.

We will show, see (2.2.14), that

$$\begin{aligned} \|(J_k u)u\|_{2,s;T'} \quad &\leq \quad C\, k^{\frac{2}{s}} \|A^{\frac{1}{2}} u\|^{2/s}_{2,2;T'} \, \|u\|^{2-2/s}_{2,\infty;T'} \qquad (2.1.3) \\ &\leq \quad C'\, k^{\frac{2}{s}} E_{T'}(u) \ < \infty \end{aligned}$$

where $s = \frac{8}{n}$, $0 < T' < T$,

$$E_{T'}(u) = \frac{1}{2} \, \|u\|^2_{2,\infty;T'} + \nu \, \|\nabla u\|^2_{2,2;T'} \, ,$$

$C = C(\nu, n) > 0$, $C' = C'(\nu, n) > 0$. Thus (2.1.2) is well defined. Since $s \geq 2$, we conclude that

$$(J_k u)u \, \in \, L^2_{loc}([0,T); L^2(\Omega)^{n^2}) \, , \qquad (2.1.4)$$

and from Lemma 1.2.1 we get

$$\begin{aligned} < (J_k u) \cdot \nabla u, u >_{\Omega,T'} &= - < (J_k u)u, \nabla u >_{\Omega,T'} \\ &= -\frac{1}{2} < J_k u, \nabla |u|^2 >_{\Omega,T'} \\ &= \frac{1}{2} < \text{div } (J_k u), |u|^2 >_{\Omega,T'} = 0. \end{aligned} \tag{2.1.5}$$

2.1.1 Definition *Let $\Omega \subseteq \mathbb{R}^n$, $n = 2,3$, be any domain, let $k \in \mathbb{N}$, $0 < T \leq \infty$, $u_0 \in L^2_\sigma(\Omega)$, and $f = f_0 + \text{div } F$ with*

$$f_0 \in L^1_{loc}([0,T); L^2(\Omega)^n) \ , \quad F \in L^2_{loc}([0,T); L^2(\Omega)^{n^2}). \tag{2.1.6}$$

Then

$$u \in L^\infty_{loc}([0,T); L^2_\sigma(\Omega)) \cap L^2_{loc}([0,T); W^{1,2}_{0,\sigma}(\Omega))$$

is called a **weak solution** *of the* **approximate Navier-Stokes system**

$$u_t - \nu\Delta u + (J_k u) \cdot \nabla u + \nabla p = f \ , \quad \text{div } u = 0 \ , \quad u|_{\partial\Omega} = 0 \ , \quad u(0) = u_0 \tag{2.1.7}$$

with data f, u_0, iff

$$\begin{aligned} - < u, v_t >_{\Omega,T} + \nu < \nabla u, \nabla v >_{\Omega,T} + < (J_k u) \cdot \nabla u, v >_{\Omega,T} \\ = < u_0, v(0) >_\Omega + [f, v]_{\Omega,T} \end{aligned} \tag{2.1.8}$$

for all $v \in C_0^\infty([0,T); C^\infty_{0,\sigma}(\Omega))$. If a distribution p together with a weak solution u satisfy the equation

$$u_t - \nu\Delta u + (J_k u) \cdot \nabla u + \nabla p = f \tag{2.1.9}$$

in the sense of distributions in $(0,T) \times \Omega$, p is called an **associated pressure** *of u.*

Our purpose is to investigate existence, uniqueness and regularity properties of these weak solutions. Then we investigate the limit as $k \to \infty$, and prove the existence of a weak solution of the original Navier-Stokes system (2.1.1).

2.2 Properties of approximate weak solutions

The integral equation in the next lemma, written in the form

$$u = S(\cdot)u_0 + \mathcal{J} P f_0 + A^{\frac{1}{2}} \mathcal{J} A^{-\frac{1}{2}} P \text{ div } (F - (J_k u)u), \tag{2.2.1}$$

is basic for the approximate systems (2.1.7). Here $\mathcal{J}$ means the integral operator defined in (1.6.3), IV, A means the Stokes operator, $S(t) = e^{-tA}$, $t \geq 0$, and Pf_0 means the Helmholtz projection. The operator $A^{-\frac{1}{2}}P$ div with the extended meaning of $A^{-\frac{1}{2}}$ and P has been treated in Section 2.6, III. In particular we get $\|A^{-\frac{1}{2}}P \text{ div }\| \le \nu^{-\frac{1}{2}}$, see (2.6.4), III, and therefore

$$\|A^{-\frac{1}{2}}P \text{ div } (F-(J_k u)u)\|_2 \le \nu^{-\frac{1}{2}}(\|F\|_2 + \|(J_k u)u\|_2) \tag{2.2.2}$$

for almost all $t \in [0,T)$.

2.2.1 Lemma *Let $\Omega \subseteq \mathbb{R}^n$, $n = 2,3$, be any domain, let $k \in \mathbb{N}$, $0 < T \le \infty$, $u_0 \in L^2_\sigma(\Omega)$, $f = f_0 + \text{div } F$ with*

$$f_0 \in L^1_{\text{loc}}(0,T;L^2(\Omega)^n)\ , \quad F \in L^2_{loc}([0,T);L^2(\Omega)^{n^2})\,,$$

and let

$$u = u_k \in L^\infty_{loc}([0,T);L^2_\sigma(\Omega)) \cap L^2_{loc}([0,T);W^{1,2}_{0,\sigma}(\Omega))$$

be a weak solution of the approximate Navier-Stokes system (2.1.7) with data f, u_0.

Then, after a redefinition on a null set of $[0,T)$, $u : [0,T) \to L^2_\sigma(\Omega)$ has the following properties:

a) *$u : [0,T) \to L^2_\sigma(\Omega)$ is strongly continuous with $u(0) = u_0$.*

b) *$(\mathcal{J}A^{-\frac{1}{2}}P\text{div } (F-(J_k u)u))(t) \in D(A^{\frac{1}{2}})$ and*

$$u(t) = S(t)u_0 + (\mathcal{J}Pf_0)(t) + A^{\frac{1}{2}}(\mathcal{J}A^{-\frac{1}{2}}P \text{ div } (F-(J_k u)u))(t) \tag{2.2.3}$$

for all $t \in [0,T)$.

c)
$$\frac{1}{2}\|u(t)\|_2^2 + \nu\int_0^t \|\nabla u\|_2^2\, d\tau = \frac{1}{2}\|u_0\|_2^2 + \int_0^t < f_0, u >_\Omega d\tau - \int_0^t < F, \nabla u >_\Omega d\tau \tag{2.2.4}$$

for all $t \in [0,T)$.

d)
$$\frac{1}{2}\|u\|^2_{2,\infty;T'} + \nu\|\nabla u\|^2_{2,2;T'} \le 2\|u_0\|_2^2 + 4\nu^{-1}\|F\|^2_{2,2;T'} + 8\|f_0\|^2_{2,1;T'} \tag{2.2.5}$$

for all T' with $0 < T' < T$.

Proof. First we have to prepare several inequalities.

Using Lemma 2.4.2, III, we choose $0 \le \alpha \le \frac{1}{2}$, $2 \le q < \infty$ with $2\alpha + \frac{n}{q} = \frac{n}{2}$, and get

$$\|u\|_q \le C\|A^\alpha u\|_2 \tag{2.2.6}$$

for almost all $t \in [0,T)$ with $C = C(\nu, q, n) > 0$. The interpolation inequality (2.2.8), III, yields

$$\|A^\alpha u\|_2 = \|(A^{\frac{1}{2}})^{2\alpha} u\|_2 \le \|A^{\frac{1}{2}} u\|_2^{2\alpha} \, \|u\|_2^{1-2\alpha} \le \|A^{\frac{1}{2}} u\|_2 + \|u\|_2 . \qquad (2.2.7)$$

Using (2.2.6) with $n = 3$, $2 \le q \le 6$, or with $n = 2$, $2 \le q < \infty$, we get

$$\|u\|_q \le C \, \|A^{\frac{1}{2}} u\|_2^{2\alpha} \, \|u\|_2^{1-2\alpha} \le C \, (\|A^{\frac{1}{2}} u\|_2 + \|u\|_2) \qquad (2.2.8)$$

for $2 \le q \le 6$ if $n = 3$, and for $2 \le q < \infty$ if $n = 2$.

Using the properties of J_k, see Section 3.4, II, we get the relation

$$J_k u - u = (J_k - J_k^{-1} J_k) u = -k^{-1} A^{\frac{1}{2}} J_k u \, , \qquad (2.2.9)$$

and the estimates

$$\begin{aligned} \|J_k u\|_2 &\le \|u\|_2 \, , \quad \|(kI + A^{\frac{1}{2}})^{-1} u\|_2 \le k^{-1} \|u\|_2 \, , \qquad (2.2.10) \\ \|A^{\frac{1}{2}} J_k u\|_2 &\le k \, \|u\|_2 . \end{aligned}$$

Using (2.2.7) we see that

$$\|A^\alpha J_k u\|_2 \le \|A^{\frac{1}{2}} J_k u\|_2^{2\alpha} \, \|J_k u\|_2^{1-2\alpha} \le k^{2\alpha} \|u\|_2 \qquad (2.2.11)$$

for $k \in \mathbb{N}$, $0 \le \alpha \le \frac{1}{2}$.

From (2.2.11) we get

$$\begin{aligned} \|A^\alpha J_k u\|_{2,s;T'} &\le \|A^{\frac{1}{2}} J_k u\|_{2,2;T'}^{2\alpha} \, \|J_k u\|_{2,\infty;T'}^{1-2\alpha} \\ &\le \|A^{\frac{1}{2}} J_k u\|_{2,2;T'} + \|J_k u\|_{2,\infty;T'} \end{aligned}$$

with $s = \frac{1}{\alpha}$, $0 \le \alpha \le \frac{1}{2}$, $0 < T' < T$.

Using (2.2.8), (2.2.10) we obtain

$$\begin{aligned} \|J_k u\|_{q,\infty;T'} &\le C \, (\|A^{\frac{1}{2}} J_k u\|_{2,\infty;T'} + \|J_k u\|_{2,\infty;T'}) \qquad (2.2.12) \\ &\le C \, (k \, \|u\|_{2,\infty;T'} + \|u\|_{2,\infty;T'}) < \infty \end{aligned}$$

with q as in (2.2.8), and it follows that

$$\|J_k u\|_{q,s;T'} < \infty$$

with $s \ge 1$, $0 < T' < T$.

This means in particular that Serrin's condition

$$J_k u \in L^s_{loc}([0,T); L^q(\Omega)^n) \qquad (2.2.13)$$

is satisfied with certain values q, s such that $n < q < \infty$, $2 < s < \infty$, $\frac{n}{q} + \frac{2}{s} \le 1$.

Next we use $\nu^{\frac{1}{2}} \|\nabla u\|_2 = \|A^{\frac{1}{2}} u\|_2$, see (2.2.2), III, and we use (2.2.6) with $\alpha = \frac{n}{8}$, $q = 4$, $2\alpha + \frac{n}{4} = \frac{n}{2}$. This leads to

$$\begin{aligned}
\|(J_k u)u\|_2 &\leq C \, \|J_k u\|_4 \, \|u\|_4 \\
&\leq C' \, \|A^\alpha J_k u\|_2 \, \|A^\alpha u\|_2 \\
&= C' \, \|(A^{\frac{1}{2}})^{2\alpha} \, J_k u\|_2 \, \|(A^{\frac{1}{2}})^{2\alpha} u\|_2 \\
&\leq C' \, \|A^{\frac{1}{2}} J_k u\|_2^{2\alpha} \, \|J_k u\|_2^{1-2\alpha} \, \|A^{\frac{1}{2}} u\|_2^{2\alpha} \, \|u\|_2^{1-2\alpha} \\
&\leq C' \, k^{2\alpha} \, \|A^{\frac{1}{2}} u\|_2^{2\alpha} \, \|u\|_2^{2-2\alpha},
\end{aligned}$$

and therefore we get

$$\begin{aligned}
\|(J_k u)u\|_{2,8/n;T'} &\leq C' \, k^{\frac{n}{4}} \, \|A^{\frac{1}{2}} u\|_{2,2;T'}^{n/4} \, \|u\|_{2,\infty;T'}^{2-n/4} \\
&\leq 2C' \, k^{\frac{n}{4}} \, E_{T'}(u)
\end{aligned} \tag{2.2.14}$$

with $0 < T' < T$, $C = C(\nu, n) > 0$, $C' = C'(\nu, n) > 0$. This yields (2.1.3) and therefore (2.1.4), (2.1.5). Using

$$\|A^\alpha J_k u\|_2 = \|J_k A^\alpha u\|_2 \leq \|A^\alpha u\|_2$$

we get similarly that

$$\|(J_k u)u\|_2 \leq C' \|A^\alpha u\|_2 \, \|A^\alpha u\|_2 \leq C' \, \|A^{\frac{1}{2}} u\|_2^{\frac{n}{2}} \, \|u\|_2^{2-\frac{n}{2}},$$

and this leads to

$$\|(J_k u)u\|_{2,4/n;T'} \leq C \, \|A^{\frac{1}{2}} u\|_{2,2;T'}^{n/2} \, \|u\|_{2,\infty;T'}^{2-\frac{n}{2}} \leq C \, E_{T'}(u), \tag{2.2.15}$$

with $0 < T' < T$ and $C = C(\nu, n) > 0$ not depending on k.

To prove the property a) we use (2.1.4), observe that u is also a weak solution of the linear system

$$u_t - \nu \Delta u + \nabla p = \tilde{f} \ , \quad \text{div } u = 0 \ , \quad u|_{\partial\Omega} = 0 \ , \quad u(0) = u_0$$

with $\tilde{f} = f -$ div $((J_k u)u)$, and apply Theorem 2.3.1, IV. This shows that u is strongly continuous, after a corresponding redefinition, and that

$$\begin{aligned}
&\frac{1}{2} \|u(t)\|_2^2 + \nu \int_0^t \|\nabla u\|_2^2 \, d\tau \\
&= \frac{1}{2} \|u_0\|_2^2 + \int_0^t [f, u]_\Omega \, d\tau + \int_0^t < (J_k u)u, \nabla u >_\Omega \, d\tau
\end{aligned}$$

with $0 < t < T$. Using (2.1.5) we get the energy equality (2.2.4). This proves a) and c).

The property d) is a consequence of c), see the proof of (1.4.4), Theorem 1.4.1.

To prove the integral equation (2.2.3), we argue as in the proof of (1.3.5), Theorem 1.3.1; we only have to replace $u\,u$ by $(J_k u)u$. This proves the lemma. $\square$

Consider the case $T = \infty$ in Lemma 2.2.1, and assume additionally that

$$f_0 \in L^1(0,\infty; L^2(\Omega)^n) \ , \quad F \in L^2(0,\infty; L^2(\Omega)^{n^2}).$$

Then we may let $T' \to \infty$ in (2.2.5) and get the inequality

$$\begin{aligned} E_\infty(u) &= \frac{1}{2}\|u\|^2_{2,\infty;\infty} + \nu\|\nabla u\|^2_{2,2;\infty} \\ &\le 2\|u_0\|^2_2 + 4\nu^{-1}\|F\|^2_{2,2;\infty} + 8\|f_0\|^2_{2,1;\infty} \,. \end{aligned} \tag{2.2.16}$$

2.3 Regularity properties of approximate weak solutions

The nonlinear term in the approximate system (2.1.7) has the form $(J_k u)\cdot\nabla u$, and we know, see (2.2.13), that for each weak solution $u = u_k$, $J_k u$ satisfies Serrin's uniqueness and regularity condition. Therefore, we can apply the complete uniqueness and regularity theory of the Navier-Stokes system, see Section 1.5 and Section 1.8. In particular we get the following result.

2.3.1 Theorem *Let $\Omega = \mathbb{R}^n$, or let $\Omega \subseteq \mathbb{R}^n$, $n = 2,3$, be a uniform C^2-domain, let $0 < T \le \infty$, $u_0 \in W^{1,2}_{0,\sigma}(\Omega)$, $f \in L^2_{loc}([0,T); L^2(\Omega)^n)$, $k \in \mathbb{N}$, and let*

$$u = u_k \in L^\infty_{loc}([0,T); L^2_\sigma(\Omega)) \cap L^2_{loc}([0,T); W^{1,2}_{0,\sigma}(\Omega))$$

be a weak solution of the approximate Navier-Stokes system (2.1.7) with data f, u_0.

Then u has the properties

$$u \in L^\infty_{loc}([0,T); W^{1,2}_{0,\sigma}(\Omega)) \cap L^2_{loc}([0,T); W^{2,2}(\Omega)^n), \tag{2.3.1}$$

$$u_t \in L^2_{loc}([0,T); L^2_\sigma(\Omega)), \tag{2.3.2}$$

and

$$u\cdot\nabla u \in L^2_{loc}([0,T); L^2(\Omega)^n). \tag{2.3.3}$$

Further, there exists an associated pressure $p = p_k$ satisfying

$$p \in L^2_{loc}([0,T); L^2_{loc}(\overline{\Omega})) \ , \quad \nabla p \in L^2_{loc}([0,T); L^2(\Omega)^n). \tag{2.3.4}$$

Proof. The proof is the same as the proof of Theorem 1.8.1, if we replace $u \cdot \nabla u$ by $(J_k u) \cdot \nabla u$, and use for $J_k u$ Serrin's regularity condition (2.2.13). □

In the same way we can also prove for (2.1.7) the regularity result of Theorem 1.8.2. In particular, if additionally Ω is a C^∞- domain and $f \in C_0^\infty(\,\overline{(0,T) \times \Omega}\,)^n$, we obtain the regularity properties

$$u \in C^\infty((0,T) \times \Omega)^n \;\; , \quad p \in C^\infty((0,T) \times \Omega). \tag{2.3.5}$$

2.4 Smooth solutions of the Navier-Stokes equations with "slightly" modified forces

Approximate weak solutions u have some further important properties. Using the relation (2.2.9) we can write

$$\begin{aligned} (J_k u) \cdot \nabla u &= (J_k u - u) \cdot \nabla u + u \cdot \nabla u \\ &= -\frac{1}{k}(A^{\frac{1}{2}} J_k u) \cdot \nabla u + u \cdot \nabla u, \end{aligned} \tag{2.4.1}$$

and the approximate Navier-Stokes system (2.1.7) can be written in the form

$$u_t - \nu \Delta u + u \cdot \nabla u + \nabla p = f_k \;\; , \quad \text{div } u = 0 \;\; , \quad u|_{\partial\Omega} = 0 \;\; , \quad u(0) = u_0 \tag{2.4.2}$$

which is the original Navier-Stokes system with the modified force

$$f_k = f + r_k \;\; , \quad r_k = \frac{1}{k}(A^{\frac{1}{2}} J_k u) \cdot \nabla u \, , \tag{2.4.3}$$

$k \in \mathbb{N}$. We will see in (2.4.8) that

$$\lim_{k \to \infty} \|r_k\|_{q,s;T} = 0 \tag{2.4.4}$$

if $1 < q < 2$, $1 < s < 2$, $n + 1 < \frac{n}{q} + \frac{2}{s} < n + 2$.

If the data f, u_0, and the domain Ω satisfy the assumptions of Theorem 2.3.1, then we obtain the following result:

Each weak solution of the approximate system (2.1.7) is a smooth solution of the original Navier-Stokes system (2.1.1) with the "slightly" perturbed force $f + r_k$, where the "error" r_k tends to zero as $k \to \infty$ in the norm $\|\cdot\|_{q,s;T}$.

In other words: After a "small" modification of the given force f, the Navier-Stokes system has a smooth solution. Theorem 2.4.1 yields the precise formulation of this fact. We are only interested in the case $n = 3$.

Fursikov [Fur80] investigated similar problems using a different approach. See also [Tem83, Part I, 3.4] and [SvW87] for similar results.

Let $u_0 \in W^{1,2}_{0,\sigma}(\Omega)$, $0 < T < \infty$. Then we consider the nonlinear operator

$$u \mapsto u_t - \nu P\Delta u + Pu \cdot \nabla u \ , \quad u \in D_{u_0} \ ,$$

with domain

$$D_{u_0} := \{u \in L^2(0,T; W^{1,2}_{0,\sigma}(\Omega) \cap W^{2,2}(\Omega)^n); \\ u_t \in L^2(0,T; L^2_\sigma(\Omega)),\ u(0) = u_0\}$$

and range

$$R_{u_0} := \{u_t - \nu P\Delta u + Pu \cdot \nabla u;\ u \in D_{u_0}\}.$$

Up to now we do not know whether the Navier-Stokes system (2.1.1) has a solution $u \in D_{u_0}$ for each given $f \in L^2(0,T; L^2_\sigma(\Omega))$, $u_0 \in W^{1,2}_{0,\sigma}(\Omega)$. Thus we do not know whether

$$R_{u_0} = L^2(0,T; L^2_\sigma(\Omega))$$

which means that the surjectivity of this operator is an open problem. However, we can give a partial answer. The following theorem shows a certain density property of R_{u_0} within the space $L^2(0,T; L^2_\sigma(\Omega))$.

2.4.1 Theorem $(n = 3)$ *Let $\Omega = \mathbb{R}^3$, or let $\Omega \subseteq \mathbb{R}^3$ be any three-dimensional uniform C^2- domain, let $0 < T < \infty$,*

$$1 < q < 2 \ , \quad 1 < s < 2 \quad \text{with } 4 < \frac{3}{q} + \frac{2}{s} < 5, \tag{2.4.5}$$

and let $u_0 \in W^{1,2}_{0,\sigma}(\Omega)$, $f \in L^2(0,T; L^2(\Omega)^3)$.

Then for each $k \in \mathbb{N}$, there exists some

$$r_k \in L^s(0,T; L^q(\Omega)^3) \cap L^2(0,T; L^2(\Omega)^3)$$

such that the Navier-Stokes system

$$u_t - \nu\Delta u + u \cdot \nabla u + \nabla p = f + r_k \ , \quad u(0) = u_0 \tag{2.4.6}$$

is uniquely solvable with

$$u = u_k \in D_{u_0} \ , \quad \nabla p = \nabla p_k \in L^2(0,T; L^2(\Omega)^3) .$$

We get

$$\|r_k\|_{q,s;T} \le C\, k^{4-(\frac{3}{q}+\frac{2}{s})} \left(\|u_0\|_2^2 + \|f\|^2_{2,1;T}\right) \tag{2.4.7}$$

with $C = C(\nu, q) > 0$. Therefore, the "error" r_k tends to zero as $k \to \infty$ in the sense that

$$\lim_{k\to\infty} \|r_k\|_{q,s;T} = 0 . \tag{2.4.8}$$

2.4.2 Remark Consider a bounded C^2-domain $\Omega \subseteq \mathbb{R}^3$ in Theorem 2.4.1. Then we know, see [Gal94, III.1, Th. 1.2], that $\|Pr_k\|_{q,s;T} \le C\|r_k\|_{q,s;T}$ with some constant $C = C(\Omega, q) > 0$, and we obtain the following result:

The range R_{u_0} is dense in $L^2(0,T; L^2_\sigma(\Omega))$ with respect to the norm $\|\cdot\|_{q,s;T}$.

Proof. Theorem 2.5.1 in the next subsection yields the existence of a weak solution $u = u_k$ of the approximate system (2.1.7) with data f, u_0. Using (2.4.1) we see that u is a weak solution of the Navier-Stokes system (2.1.1) with data $f + r_k$ and u_0, where $r_k = \frac{1}{k}(A^{\frac{1}{2}} J_k u) \cdot \nabla u$. Theorem 2.3.1 shows that

$$u \in D_{u_0} \ , \quad u \cdot \nabla u \in L^2(0,T; L^2(\Omega)^3) \, ,$$

and therefore that

$$\begin{aligned} r_k \ &:= \ \frac{1}{k}(A^{\frac{1}{2}} J_k u) \cdot \nabla u \ \in \ L^2(0,T; L^2(\Omega)^3) \, , \\ f + r_k \ &\in \ L^2(0,T; L^2(\Omega)^3) \, ; \end{aligned}$$

observe that $0 < T < \infty$.

Since $u_t \in L^2(0,T; L^2_\sigma(\Omega))$, we see that $u : [0,T) \to L^2_\sigma(\Omega)$ is strongly continuous, after a redefinition on a null set, and $u(0) = u_0$. Theorem 2.3.1 also yields the associated pressure p.

To prove (2.4.7) we set $\alpha := \frac{3}{q} + \frac{2}{s} - 4$ so that $0 < \alpha < 1$, choose $\gamma, \rho > 2$ with $\frac{1}{q} = \frac{1}{2} + \frac{1}{\gamma}$, $\frac{1}{s} = \frac{1}{2} + \frac{1}{\rho}$, and set $\beta := \frac{3}{2}(\frac{1}{2} - \frac{1}{\gamma})$. Then we get $2\beta + \frac{3}{\gamma} = \frac{3}{2}$, $\alpha = \frac{3}{\gamma} + \frac{2}{\rho} - \frac{3}{2} = \frac{2}{\rho} - 2\beta$, $0 < \frac{2}{\rho} - 2\beta < 1$, $\beta < \frac{1}{\rho}$, $0 < \rho\beta < 1$, and $0 < \beta < \frac{1}{\rho} < \frac{1}{2}$.

Now we use Hölder's inequality and the inequalities (2.2.6), (2.2.11) and (2.2.5) with $F = 0$, $f_0 = f$. This yields

$$\begin{aligned} \|r_k\|_{q,s;T} \ &= \ \|(A^{\frac{1}{2}}(kI + A^{\frac{1}{2}})^{-1}u) \cdot \nabla u\|_{q,s;T} \\ &\le \ C_1 \, \|A^{\frac{1}{2}}(kI + A^{\frac{1}{2}})^{-1}u\|_{\gamma,\rho;T} \, \|\nabla u\|_{2,2;T} \\ &\le \ C_2 \, \|A^{\beta}(kI + A^{\frac{1}{2}})^{-1}A^{\frac{1}{2}}u\|_{2,\rho;T} \, \|\nabla u\|_{2,2;T} \\ &= \ C_2 \, \|((A^{\beta\rho})^{\frac{1}{2}})^{\frac{2}{\rho}}(kI + A^{\frac{1}{2}})^{-1}A^{\frac{1}{2}}u\|_{2,\rho;T} \, \|\nabla u\|_{2,2;T} \\ &\le \ C_2 \, \|A^{\frac{\beta\rho}{2}}(kI + A^{\frac{1}{2}})^{-1}A^{\frac{1}{2}}u\|^{\frac{2}{\rho}}_{2,2;T} \\ &\qquad \cdot \|(kI + A^{\frac{1}{2}})^{-1}A^{\frac{1}{2}}u\|^{1-\frac{2}{\rho}}_{2,\infty;T} \, \|\nabla u\|_{2,2;T} \\ &= \ C_2 \, k^{-1} \, \|A^{\frac{\beta\rho}{2}} J_k A^{\frac{1}{2}}u\|^{\frac{2}{\rho}}_{2,2;T} \, \|A^{\frac{1}{2}} J_k u\|^{1-\frac{2}{\rho}}_{2,\infty;T} \, \|\nabla u\|_{2,2;T} \\ &\le \ C_2 \, k^{-(1-\rho\beta)\frac{2}{\rho}} \, \|A^{\frac{1}{2}}u\|^{\frac{2}{\rho}}_{2,2;T} \|u\|^{1-\frac{2}{\rho}}_{2,\infty;T} \, \|\nabla u\|_{2,2;T} \end{aligned}$$

$$\leq \; C_2\, k^{-\alpha}\,(\|A^{\frac{1}{2}}u\|_{2,2;T} + \|u\|_{2,\infty;T})\,\|\nabla u\|_{2,2;T}$$
$$\leq \; C_3\, k^{-\alpha}(\|u_0\|_2^2 + \|f\|_{2,1;T}^2)$$

with constants $C_1 = C_1(n) > 0,\; C_2 = C_2(q,\nu) > 0,\; C_3 = C_3(q,\nu) > 0$. This proves (2.4.7). The proof of the theorem is complete. □

2.5 Existence of approximate weak solutions

The existence result below rests on Banach's fixed point principle.

2.5.1 Theorem *Let $\Omega \subseteq \mathbb{R}^n$, $n = 2,3$, be any domain, let $k \in \mathbb{N}$, $0 < T \leq \infty$, $u_0 \in L^2_\sigma(\Omega)$, and let $f = f_0 + \operatorname{div} F$ with*

$$f_0 \in L^1_{loc}([0,T); L^2(\Omega)^n) \;, \quad F \in L^2_{loc}([0,T); L^2(\Omega)^{n^2}).$$

Then there exists a uniquely determined weak solution

$$u = u_k \in L^\infty_{loc}([0,T); L^2_\sigma(\Omega)) \cap L^2_{loc}([0,T); W^{1,2}_{0,\sigma}(\Omega))$$

of the approximate Navier-Stokes system (2.1.7) *with data f, u_0.*

Proof. Without loss of generality we may assume that $0 < T < \infty$ and that $f_0 \in L^1(0,T; L^2(\Omega)^n)$, $F \in L^2(0,T; L^2(\Omega)^{n^2})$. In the general case, we prove the result for arbitrary T' with $0 < T' < T$, and using the uniqueness we get the desired solution on the whole interval $[0,T)$.

The space

$$X_T := L^\infty(0,T; L^2_\sigma(\Omega)) \cap L^2(0,T; W^{1,2}_{0,\sigma}(\Omega))$$

is a Banach space with norm

$$|||u|||_T \;:=\; \|u\|_{2,\infty;T} + \|A^{\frac{1}{2}}u\|_{2,2;T} \;=\; \|u\|_{2,\infty;T} + \nu^{\frac{1}{2}}\|\nabla u\|_{2,2;T}. \qquad (2.5.1)$$

We will prove the existence of a unique weak solution $u \in X_T$ of (2.1.7).

Let $u \in X_T$. Set $\hat{F} := F - (J_k u)u$. Then from Theorem 2.4.1, IV, we conclude that

$$\hat{u} \;:=\; \mathcal{F}_T(u) \;:=\; S(\cdot)u_0 + \mathcal{J}Pf_0 + A^{\frac{1}{2}}\mathcal{J}A^{-\frac{1}{2}}P \operatorname{div} \hat{F}$$

is a weak solution of the linear system

$$\hat{u}_t - \nu\Delta\hat{u} + \nabla p = f_0 + \operatorname{div} \hat{F}, \quad \operatorname{div} \hat{u} = 0, \quad \hat{u}|_{\partial\Omega} = 0, \quad \hat{u}(0) = u_0 \qquad (2.5.2)$$

with data $f_0 + \operatorname{div} \hat{F}$, u_0. Our aim is to find some $u \in X_T$ in such a way that

$$\hat{u} = \mathcal{F}_T(u) = u.$$

In the first step we show that there exists some T', $0 < T' \le T$, such that

$$u = \mathcal{F}_{T'}(u) \tag{2.5.3}$$

has a unique solution $u \in X_{T'}$. Then u is a weak solution of (2.5.2) in $[0, T')$, and therefore also a weak solution of the approximate system (2.1.7) in $[0, T')$ with data f, u_0. Then we will repeat this procedure with $\tilde{u}$ defined by

$$\tilde{u}(t) = u(T' + t) \ , \quad t \ge 0 \ , \quad \tilde{u}(0) = u(T'). \tag{2.5.4}$$

This yields the desired solution in the next interval if $T' < T$, and so on. In this way we get the desired solution on the whole interval $[0, T)$.

To solve (2.5.3), we have to prepare several inequalities. Let $0 < T' \le T$, and let $u \in X_{T'}$ be given. Since $\hat{u} = \mathcal{F}_{T'}(u)$ is a weak solution of (2.5.2), we get from Theorem 2.3.1, IV, c), the estimates

$$\begin{aligned} |||\hat{u}|||_{T'} &= \|\hat{u}\|_{2,\infty;T'} + \nu^{\frac{1}{2}} \|\nabla \hat{u}\|_{2,2;T'} \\ &\le 2 \big(\frac{1}{2} \|\hat{u}\|^2_{2,\infty;T'} + \nu \|\nabla \hat{u}\|^2_{2,2;T'}\big)^{\frac{1}{2}} \\ &\le 4 \big(\|u_0\|_2^2 + 2\nu^{-1} \|\hat{F}\|^2_{2,2;T'} + 4 \|f_0\|^2_{2,1;T'}\big)^{\frac{1}{2}} \\ &\le 4 \big(\|u_0\|_2 + \sqrt{2}\, \nu^{-\frac{1}{2}} \|\hat{F}\|_{2,2;T'} + 2 \|f_0\|_{2,1;T'}\big). \end{aligned}$$

Next we choose $s := \frac{8}{n}$, $\rho := (\frac{1}{2} - \frac{n}{8})^{-1}$. Then $\frac{1}{2} = \frac{1}{s} + \frac{1}{\rho}$, $\rho = 8$ if $n = 3$ and $\rho = 4$ if $n = 2$.

We use Hölder's inequality, inequality (2.2.6) with $q = 4$, $\alpha = \frac{n}{8}$, $2\alpha + \frac{n}{4} = \frac{n}{2}$, and the inequalities (2.2.11), (2.2.12) in a slightly modified way. This yields

$$\begin{aligned} \|(J_k u) u\|_{2,2;T'} &\le C_1 \|J_k u\|_{4,\rho;T'} \|u\|_{4,s;T'} \\ &\le C_2 \|A^\alpha J_k u\|_{2,\rho;T'} \|A^\alpha u\|_{2,s;T'} \\ &\le C_2\, k^{2\alpha} \|u\|_{2,\rho;T'} \|A^{\frac{1}{2}} u\|^{2\alpha}_{2,2;T'} \|u\|^{1-2\alpha}_{2,\infty;T'} \\ &\le C_2\, k^{2\alpha} (T')^{\frac{1}{\rho}} \|u\|_{2,\infty;T'} (\|A^{\frac{1}{2}} u\|_{2,2;T'} + \|u\|_{2,\infty;T'}) \\ &\le C_2\, k^{2\alpha} \, (T')^{\frac{1}{\rho}} \, |||u|||^2_{T'} \end{aligned}$$

with $C_1 = C_1(n) > 0$, $C_2 = C_2(\nu, n) > 0$.

Together with

$$\|\hat{F}\|_{2,2;T'} \le \|F\|_{2,2;T'} + \|(J_k u) u\|_{2,2;T'}$$

and the estimate above we obtain the inequality

$$|||\hat{u}|||_{T'} = |||\mathcal{F}_{T'}(u)|||_{T'} \le a \, |||u|||^2_{T'} + b \tag{2.5.5}$$

with $a := 4\sqrt{2}\ \nu^{-\frac{1}{2}}\, C_2\, k^{2\alpha}(T')^{\frac{1}{\rho}}$ and $b := 4(\|u_0\|_2 + \sqrt{2}\,\nu^{-\frac{1}{2}}\|F\|_{2,2;T'} + 2\|f_0\|_{2,1;T'})$.

Let now $u, v \in X_{T'}$. Then we see in the same way as above that $\hat{u} - \hat{v} = \mathcal{F}_{T'}(u) - \mathcal{F}_{T'}(v)$ is a weak solution of the linear system

$$\begin{aligned}(\hat{u}-\hat{v})_t - \nu\Delta(\hat{u}-\hat{v}) + \nabla p &= \text{div}\ (J_k v)v - \text{div}\ (J_k u)u, \quad \text{div}(\hat{u}-\hat{v}) = 0,\\ \hat{u}-\hat{v}|_{\partial\Omega} &= 0\ , \quad (\hat{u}-\hat{v})(0) = 0.\end{aligned}$$

Using

$$\text{div}\ (J_k v)v - \text{div}\ (J_k u)u = \text{div}\ ((J_k v)(v-u) + (J_k(v-u))u),$$

and the same estimates as above, we obtain instead of (2.5.5) that

$$\begin{aligned}|||\hat{u}-\hat{v}|||_{T'} &= |||\mathcal{F}_{T'}(u) - \mathcal{F}_{T'}(v)|||_{T'}\\ &\le a\,|||u-v|||_{T'}\,(|||u|||_{T'} + |||v|||_{T'}).\end{aligned} \tag{2.5.6}$$

We choose T' with $0 < T' \le T$ in such a way that

$$4ab < 1, \tag{2.5.7}$$

and we consider the equations

$$ay^2 + b = y\ ,\quad y^2 - \frac{1}{a}y + \frac{b}{a} = 0\ ,\quad y \in \mathbb{R}. \tag{2.5.8}$$

An elementary calculation shows that

$$y_1 = \frac{1}{2a}\,(1 - \sqrt{1-4ab}) = 2b\,(1 + \sqrt{1-4ab})^{-1} > 0 \tag{2.5.9}$$

is the minimal root of (2.5.8). This argument is well known, see [Sol77, Lemma 10.2, p. 522]. We see, $y_1 < 2b$.

Consider the closed set

$$D_{T'} := \{u \in X_{T'};\ |||u|||_{T'} \le y_1\}. \tag{2.5.10}$$

If $u \in D_{T'}$ we conclude with (2.5.5) that

$$|||\mathcal{F}_{T'}(u)|||_{T'} \le a\,|||u|||_{T'}^2 + b \le ay_1^2 + b = y_1,$$

and therefore that $\mathcal{F}_{T'}(u) \in D_{T'}$. From (2.5.6) we get for $u, v \in D_{T'}$ that

$$|||\mathcal{F}_{T'}(u) - \mathcal{F}_{T'}(v)|||_{T'} \le 2y_1\,a\,|||u-v|||_{T'} \le 4ab\,|||u-v|||_{T'}.$$

Since $4ab < 1$, we may apply Banach's fixed point principle and get a unique solution $u \in D_{T'}$ with $u = \mathcal{F}_{T'}(u)$, u being a weak solution of the approximate system (2.1.7) for $[0,T')$.

We can repeat this procedure if $T' < T$, with u replaced by $\tilde{u}$ defined in (2.5.4). This yields the existence of a weak solution $\tilde{u}$ of (2.1.7) with $\tilde{u}(0) = u(T')$ in some interval $[0,T'')$, $T'' > 0$, which is determined by (2.5.7). Now $\|u_0\|_2$ is replaced by $\|u(T')\|_2$. Note that $u(T')$ is well defined since u is strongly continuous after a corresponding redefinition, see Lemma 2.2.1, a). From (2.2.5) we see that

$$\frac{1}{2}\,\|u(T')\|_2^2 \;\leq\; 2\|u_0\|_2^2 + 4\nu^{-1}\,\|F\|_{2,2;T}^2 + 8\,\|f_0\|_{2,1;T}^2,$$

and this shows that $T'' > 0$ can be chosen independently of T'. A calculation yields that $u : [0,T'+T'') \to L^2_\sigma(\Omega)$, defined by $u(t)$ if $t \in [0,T')$, and by $u(T'+t) := \tilde{u}(t)$ if $t \in [0,T'']$, is a weak solution of (2.1.7) in the interval $[0,T'+T'')$. For this purpose we can use the characterization of weak solutions in Lemma 2.2.1, b), IV, for the linear case. If $T'+T'' < T$, we can repeat this procedure, and so on. After finitely many steps this yields a weak solution u of (2.1.7) in the whole interval $[0,T)$.

To prove the uniqueness of u, we suppose there is another weak solution $v \in X_T$ of (2.1.7) in $[0,T)$. Using (2.2.3) we conclude that $v = \mathcal{F}_T(v)$, and from (2.2.5) we obtain with T' as above the estimate

$$\begin{aligned} |||v|||_{T'} &= \|v\|_{2,\infty;T'} + \nu^{\frac{1}{2}}\,\|\nabla v\|_{2,2;T'} \\ &\leq 4\,(\|u_0\|_2 + \sqrt{2}\nu^{-\frac{1}{2}}\,\|F\|_{2,2;T'} + 2\,\|f_0\|_{2,1;T'}) \\ &= b \,<\, a y_1^2 + b \,=\, y_1 . \end{aligned}$$

Then $|||u|||_{T'} \leq y_1$ and $|||v|||_{T'} \leq y_1$. The uniqueness of the fixed point in $D_{T'}$ shows that $u = v$ in $[0,T')$. Repeating this conclusion as above, we see that $u = v$ in $[0,T)$. This completes the proof. □

2.6 Uniform norm bounds of approximate weak solutions

In the next section we will use the approximate solutions $u = u_h$, $k \in \mathbb{N}$, in order to construct a weak solution of the Navier-Stokes system. This construction leads to some further important properties, for example decay estimates in the time direction. For this purpose we need certain uniform norm bounds, see the next lemma. This means that the constant C below does not depend on k. Here we are interested only in the cases $n = 3$, $T = \infty$.

2.6.1 Lemma $(n = 3)$ *Let $\Omega \subseteq \mathbb{R}^3$ be any three-dimensional domain, let $k \in \mathbb{N}$, $T = \infty$, $1 < s \leq \frac{4}{3}$, $2 \leq q \leq 6$, $1 < \rho < \infty$, $0 \leq \alpha \leq \frac{1}{2}$, and let*

$u_0 \in D(A^{-\frac{1}{2}})$, $f = \operatorname{div} F$ *with*

$$F \in L^1(0,\infty; L^2(\Omega)^9) \cap L^2(0,\infty; L^2(\Omega)^9).$$

Suppose

$$u_k \in L^\infty_{loc}([0,\infty); L^2_\sigma(\Omega)) \cap L^2_{loc}([0,\infty); W^{1,2}_{0,\sigma}(\Omega))$$

is a weak solution of the approximate Navier-Stokes system (2.1.7) *with data* f, u_0.

Then there exists a constant $C = C(\nu, \rho, \alpha, q, s, u_0, F) > 0$ *not depending on* k, *such that, after redefinition on a null set of* $[0,T)$, u_k *has the following properties:*

a) $u_k(t) \in D(A^{-\frac{1}{2}})$ *for all* $t \in [0,\infty)$, $A^{-\frac{1}{2}} u_k$ *is strongly continuous,* $(A^{-\frac{1}{2}} u_k)(0) = A^{-\frac{1}{2}} u_0$, *and*

$$\|A^{-\frac{1}{2}} u_k\|_{2,\infty;\infty} + \|u_k\|_{2,2;\infty} \;\le\; C. \tag{2.6.1}$$

b)
$$\|(A^{-\frac{1}{2}} u_k)_t\|_{2,s;\infty} + \|A^{\frac{1}{2}} u_k\|_{2,s;\infty} \;\le\; C. \tag{2.6.2}$$

c)
$$\|A^\alpha u_k\|_{2,\rho;\infty} \;\le\; C \tag{2.6.3}$$

with $\frac{1}{\rho} - \frac{1}{2} < \alpha \le \frac{1}{\rho}$, $0 \le \alpha \le \frac{1}{2}$.

d)
$$\|u_k\|_{q,\rho;\infty} \;\le\; C \tag{2.6.4}$$

with

$$\frac{3}{2} \le \frac{3}{q} + \frac{2}{\rho} < \frac{3}{2} + 1.$$

Proof. Theorem 1.6.2 contains the corresponding inequalities for a weak solution u of the Navier-Stokes system (1.1.1) under the assumption $E_\infty(u) < \infty$. In order to prove the above result, we only have to use, see (2.2.5), that

$$\begin{aligned} E_\infty(u_k) &= \frac{1}{2}\|u_k\|^2_{2,\infty;\infty} + \nu\,\|\nabla u_k\|^2_{2,2;\infty} \\ &\le 2\|u_0\|_2^2 + 4\nu^{-1}\|F\|^2_{2,2;\infty} \;<\; \infty \end{aligned} \tag{2.6.5}$$

holds with a bound on the right side which does not depend on k.

Then we investigate the proof of Theorem 1.6.2, replacing the term $u\cdot\nabla u$ by $(J_k u)\cdot\nabla u$, and see that the constant C in this proof can be chosen independently of k. This proves the lemma. □

3 Existence of weak solutions of the Navier-Stokes system

3.1 Main result

To prove the existence of (at least one) weak solution u of the Navier-Stokes system, we consider the approximate weak solutions u_k, $k \in \mathbb{N}$, and carry out the limit as $k \to \infty$ in a certain weak sense. This leads to the theorem below; the proof will be given later on. First we need some preliminary compactness results which are prepared in the next subsection.

This special construction of the weak solution u given here enables us to prove some additional properties of u. In particular we prove the validity of the energy inequality, see below, and we prove some properties concerning the asymptotic behavior as $t \to \infty$, see Section 3.4.

In the literature there are several approaches to the existence of weak solutions, see [Ler33], [Ler34], [Hop41], [Hop50], [Lad69, Chap. 6], [Hey80], [Tem77, Chap. III], [CKN82, Appendix], [Mas84], [BMi92]. These sources have used mainly the Galerkin approximation, see e.g. [Hey80].

3.1.1 Theorem *Let $\Omega \subseteq \mathbb{R}^n$, $n = 2, 3$, be any domain, let $0 < T \le \infty$, $u_0 \in L^2_\sigma(\Omega)$, and $f = f_0 + \operatorname{div} F$ with*

$$f_0 \in L^1_{loc}([0,T); L^2(\Omega)^n) \;,\quad F \in L^2_{loc}([0,T); L^2(\Omega)^{n^2}).$$

Then there exists a weak solution

$$u \in L^\infty_{loc}([0,T); L^2_\sigma(\Omega)) \cap L^2_{loc}([0,T); W^{1,2}_{0,\sigma}(\Omega))$$

of the Navier-Stokes system

$$u_t - \nu \Delta u + u \cdot \nabla u + \nabla p = f \;,\quad \operatorname{div} u = 0 \;,\quad u|_{\partial\Omega} = 0 \;,\quad u(0) = u_0 \qquad (3.1.1)$$

satisfying the following properties:

a) *$u : [0,T) \mapsto L^2_\sigma(\Omega)$ is weakly continuous and $u(0) = u_0$.*

b)
$$\frac{1}{2}\|u(t)\|_2^2 + \nu \int_0^t \|\nabla u\|_2^2\, d\tau \qquad (3.1.2)$$
$$\le \frac{1}{2}\|u_0\|_2^2 + \int_0^t < f_0, u >_\Omega d\tau - \int_0^t < F, \nabla u >_\Omega d\tau$$

for all t with $0 \le t < T$.

c)
$$\frac{1}{2}\|u\|^2_{2,\infty;T'} + \nu \|\nabla u\|^2_{2,2;T'} \qquad (3.1.3)$$
$$\le 2\|u_0\|_2^2 + 4\nu^{-1}\|F\|^2_{2,2;T'} + 8\|f_0\|^2_{2,1;T'}$$

for all T' with $0 \le T' < T$.

Proof. See Section 3.3. □

Let u be as in Theorem 3.1.1. We mention some further properties of this weak solution u.

Theorem 1.3.1 yields the validity of the integral representation

$$\begin{aligned} u(t) &= S(t)u_0 + \int_0^t S(t-\tau)Pf_0(\tau)\,d\tau \\ &\quad + A^{\frac{1}{2}}\int_0^t S(t-\tau)A^{-\frac{1}{2}}P \text{ div } F(\tau)\,d\tau \\ &\quad - A^{\frac{1}{2}}\int_0^t S(t-\tau)A^{-\frac{1}{2}}P \text{ div } (u(\tau)u(\tau))\,d\tau \end{aligned} \tag{3.1.4}$$

for all $t \in [0,T)$. Using the integral operator $\mathcal{J}$, see (1.6.3), IV, we write this equation in the form

$$u = S(\cdot)u_0 + \mathcal{J}Pf_0 + A^{\frac{1}{2}}\mathcal{J}A^{-\frac{1}{2}} \text{ div } F - A^{\frac{1}{2}}\mathcal{J}A^{-\frac{1}{2}}P \text{ div } (u\,u). \tag{3.1.5}$$

Each of these terms has special properties, the critical term is

$$U := A^{\frac{1}{2}}\mathcal{J}A^{-\frac{1}{2}}P \text{ div } (u\,u). \tag{3.1.6}$$

Using $\|A^{-\frac{1}{2}}P \text{ div }\| \le \nu^{-\frac{1}{2}}$, see (2.6.2), III, we get

$$\|A^{-\frac{1}{2}}P \text{ div } (u\,u)\|_2 \le \nu^{-\frac{1}{2}}\,\|u\,u\|_2 \tag{3.1.7}$$

for almost all $t \in [0,T)$, and using Lemma 1.2.1, d), we see that

$$\|u\,u\|_{2,s;T'} \le C\,\nu^{-\frac{n}{4}}\,(T')^{\frac{1}{s}-\frac{n}{4}}\,E_{T'}(u) < \infty \tag{3.1.8}$$

with $1 \le s \le \frac{4}{n}$, $0 < T' < T$,

$$E_{T'}(u) := \frac{1}{2}\,\|u\|^2_{2,\infty;T'} + \nu\,\|\nabla u\|^2_{2,2;T'},$$

$C = C(n) > 0$. From Theorem 2.5.3, IV, we now obtain (with F replaced by $u\,u$) the following properties:

$U(t) \in D(A^{-\frac{1}{2}})$ for all $t \in [0,T)$, $A^{-\frac{1}{2}}U : [0,T) \to L^2_\sigma(\Omega)$ is strongly continuous, $(A^{-\frac{1}{2}}U)(0) = 0$,

$$(A^{-\frac{1}{2}}U)_t\,,\;\; A^{\frac{1}{2}}U \in L^s(0,T';L^2_\sigma(\Omega)), \tag{3.1.9}$$

$$(A^{-\frac{1}{2}}U)_t + A^{\frac{1}{2}}U = A^{-\frac{1}{2}}P \text{ div } u\,u, \tag{3.1.10}$$

and

$$\|(A^{-\frac{1}{2}}U)_t\|_{2,s;T'} + \|A^{\frac{1}{2}}U\|_{2,s;T'} \le C\,\nu^{-\frac{1}{2}-\frac{n}{4}}\,(T')^{\frac{1}{s}-\frac{n}{4}}\,E_{T'}(u) < \infty \tag{3.1.11}$$

with $1 < s \le \frac{4}{n}$, $0 < T' < T$, $C = C(n,s) > 0$.

Further we get from (2.5.25), IV, combined with (3.1.8) for $s = 1$, that

$$\frac{1}{2}\,\|A^{-\frac{1}{2}}U\|^2_{2,\infty;T'} + \|U\|^2_{2,2;T'} \;\leq\; C\,\nu^{-1-\frac{n}{2}}\,(T')^{2-\frac{n}{2}}\,E_{T'}(u)^2 \tag{3.1.12}$$

with $0 < T' < T$, $C = C(n) > 0$.

Consider now the case $T = \infty$ in Theorem 3.1.1, and assume that

$$f_0 \in L^1(0,\infty;L^2(\Omega)^n)\ ,\quad F \in L^2(0,\infty;L^2(\Omega)^{n^2}).$$

Then we may let $T' \to \infty$ in (3.1.3), and get the inequality

$$\begin{aligned} E_\infty(u) &= \frac{1}{2}\,\|u\|^2_{2,\infty;\infty} + \nu\|\nabla u\|^2_{2,2;\infty} \\ &\leq 2\,\|u_0\|_2^2 \,+ 4\nu^{-1}\,\|F\|^2_{2,2;\infty} + 8\,\|f_0\|^2_{2,1;\infty} \;< \infty\,. \end{aligned} \tag{3.1.13}$$

In the case $s = \frac{4}{n}$, we conclude from (3.1.10), (3.1.11) that

$$\|(A^{-\frac{1}{2}}U)_t\|_{2,4/n;\infty} + \|A^{\frac{1}{2}}U\|_{2,4/n;\infty} \;\leq\; C\,E_\infty(u) < \infty \tag{3.1.14}$$

with $C = C(\nu,n) > 0$.

Finally we consider Theorem 3.1.1 with $T = \infty$, $n = 3$, and assume that $u_0 \in D(A^{-\frac{1}{2}})$, $f_0 = 0$, and

$$F \in L^2(0,\infty;L^2(\Omega)^{n^2}) \cap L^1(0,\infty;L^2(\Omega)^{n^2}).$$

Then we apply Theorem 1.6.2, b), and get

$$\|A^{\frac{1}{2}}u\|_{2,4/3;\infty} \;\leq\; C \tag{3.1.15}$$

with some constant $C > 0$ depending on F, u_0, ν. From (1.6.31) it follows that

$$\|uu\|_{2,1;\infty} \;\leq\; C \tag{3.1.16}$$

with $C = C(F,u_0,\nu) > 0$. Applying Theorem 2.5.3, d), IV, (1.5.24), IV, and (3.1.16), we now obtain

$$\frac{1}{2}\,\|A^{-\frac{1}{2}}u\|^2_{2,\infty;\infty} + \|u\|^2_{2,2;\infty} \;\leq\; C \tag{3.1.17}$$

with $C = C(F,u_0,\nu) > 0$.

3.2 Preliminary compactness results

In this subsection we use some arguments from Temam's book [Tem77]. The first lemma below will be taken from [Tem77, Chap. III, 2] without proof. In the second lemma we apply the compactness result of Lemma 3.2.1 to the sequence of approximate weak solutions given by Theorem 2.5.1. In the proof of this lemma we use an important argument from [Tem77, Chap. III, (3.38)–(3.39)] concerning the function (3.2.11) below.

First we introduce some notations on the Fourier transform, see [Tem77, Chap. III, (2.25)] or [Miz73, Chap. 2, 5].

The use of the Fourier transform requires us to leave the real vector spaces, which we considered up to now, and to work in the corresponding complexifications of these spaces. We will do it below keeping the same notations as in the real case.

Let $0 < T < \infty$, and let X be a (complex) Hilbert space with scalar product $< \cdot, \cdot >_X$ and norm $\| \cdot \|_X$. Then we consider the (complex) Hilbert space $L^2(0,T;X)$ with scalar product $< u,v >_{X,T} := \int_0^T < u,v >_X \, dt$, norm $\|u\|_{X,T} := \left(\int_0^T \|u\|_X^2 \, dt \right)^{\frac{1}{2}}$, and correspondingly the Hilbert space $L^2(\mathbb{R}, X)$ with scalar product $< u,v >_{X,\mathbb{R}} := \int_{-\infty}^{\infty} < u,v >_X \, dt$ and norm $\|x\|_{X,\mathbb{R}} := \left(\int_{-\infty}^{\infty} \|u\|_X^2 \, dt \right)^{\frac{1}{2}}$, see Section 1.2, IV.

Let $v \in L^2(0,T;X)$. Then it is convenient to extend v by zero to get a function from $\mathbb{R}$ to X. Thus we define the Fourier transform $v^\sim$ of v by

$$v^\sim(\tau) := \int_{-\infty}^{\infty} v(t) e^{-2\pi i \tau t} \, dt = \int_0^T v(t) e^{-2\pi i \tau t} \, dt, \tag{3.2.1}$$

$\tau \in \mathbb{R}$. This definition can be extended to a class of distributions in $\mathbb{R}$ in the same way as for scalar functions, see [Yos80, VI, 1], [Miz73, Chap. 2, 5]. In particular we get the important Parseval equality

$$\int_{-\infty}^{\infty} \|v^\sim(\tau)\|_X^2 \, d\tau = \int_0^T \|v(t)\|_X^2 \, dt. \tag{3.2.2}$$

The following compactness lemma is a special case of [Tem77, Chap. III, Theorem 2.2].

3.2.1 Lemma (Temam) *Let X_0, X be Hilbert spaces with norms $\| \cdot \|_{X_0}$ and $\| \cdot \|_X$, respectively, and suppose that there is a compact embedding*

$$X_0 \subseteq X.$$

Let $0 < T < \infty$, $0 < \gamma \leq 1$, and let $(v_j)_{j=1}^{\infty}$ be a sequence in $L^2(0,T;X_0)$ satisfying

$$\sup_j \left(\int_0^T \|v_j\|_{X_0}^2 \, dt \right) < \infty \; , \quad \sup_j \left(\int_{-\infty}^{\infty} |\tau|^{2\gamma} \, \|v_j^{\sim}(\tau)\|_X^2 \, d\tau \right) < \infty. \quad (3.2.3)$$

Then there exists a subsequence of $(v_j)_{j=1}^{\infty}$ which converges strongly in $L^2(0,T;X)$ to some $v \in L^2(0,T;X)$.

Proof. See Temam [Tem77, Chap. III, Theorem 2.2]. □

This lemma is needed in the following special situation.

3.2.2 Lemma *Let $\Omega \subseteq \mathbb{R}^n$, $n = 2,3$, be any domain, let $0 < T < \infty$, $u_0 \in L^2_\sigma(\Omega)$, and let $f = f_0 + \operatorname{div} F$ with*

$$f_0 \in L^1(0,T;L^2(\Omega)^n) \; , \quad F \in L^2(0,T;L^2(\Omega)^{n^2}).$$

Let u_k, $k \in \mathbb{N}$, be the (uniquely determined) weak solution of the approximate Navier-Stokes system (2.1.7) *with data f, u_0, see Theorem* 2.5.1.

Then the sequence $(u_k)_{k=1}^{\infty}$ has the following properties:

a) *If Ω is bounded, then there exists a subsequence which converges strongly in $L^2(0,T;L^2_\sigma(\Omega))$.*

b) *If Ω is unbounded, then for each bounded Lipschitz subdomain $\Omega_0 \subseteq \Omega$ with $\overline{\Omega}_0 \subseteq \Omega$, there exists a subsequence which converges strongly in $L^2(0,T;L^2(\Omega_0)^n)$.*

Proof. We know, see Lemma 2.2.1, that each $u_k : [0,T) \to L^2_\sigma(\Omega)$ is strongly continuous, and that the integral equation

$$u_k = S(\cdot)u_0 + \mathcal{J}Pf_0 + A^{\frac{1}{2}}\mathcal{J}A^{-\frac{1}{2}}P\operatorname{div}F - A^{\frac{1}{2}}\mathcal{J}A^{-\frac{1}{2}}P \operatorname{div}\,(J_k u_k)u_k \quad (3.2.4)$$

is satisfied. We set $V_1 := S(\cdot)u_0$, $V_2 := \mathcal{J}Pf_0$, $V_3 := A^{\frac{1}{2}}\mathcal{J}A^{-\frac{1}{2}}P \operatorname{div} F$, $V := V_1 + V_2 + V_3$, and

$$U_k := -A^{\frac{1}{2}}\mathcal{J}A^{-\frac{1}{2}}P \operatorname{div}\,(J_k u_k)u_k. \quad (3.2.5)$$

Thus we get the representation

$$u_k \; = \; V + U_k \; , \quad k \in \mathbb{N}.$$

□

In the following, C, C', $C_1, \ldots$ are always positive constants depending on u_0, f_0, $F, T, \ldots$ but not on $k \in \mathbb{N}$.

Using the energy inequalities (2.5.10), (2.5.17), (2.5.26) in IV, we see that

$$E_T(V) = \frac{1}{2}\|V\|^2_{2,\infty;T} + \nu\|\nabla V\|^2_{2,2;T} \le C < \infty. \tag{3.2.6}$$

The energy inequality (2.2.5) yields, letting $T' \to T$, that

$$\begin{aligned} E_T(u_k) &= \frac{1}{2}\|u_k\|^2_{2,\infty;T} + \nu\|\nabla u_k\|^2_{2,2;T} \\ &\le 2\|u_0\|_2^2 + 4\nu^{-1}\|F\|^2_{2,2;T} + 8\|f_0\|^2_{2,1;T} < \infty \end{aligned} \tag{3.2.7}$$

and inequality (2.2.14) yields

$$\|(J_k u_k)u_k\|_{2,s;T} \le C\, k^{\frac{n}{4}}\, E_T(u_k) < \infty$$

with $1 \le s \le \frac{8}{n}$, $k \in \mathbb{N}$, $C = C(\nu, n, T) > 0$. In particular we may set $s = 2$.

Thus we can apply Theorem 2.5.3, IV, (with F replaced by $(J_k u_k)u_k$) and see that $A^{-\frac{1}{2}}U_k : [0,T) \to L^2_\sigma(\Omega)$ is strongly continuous, that

$$(A^{-\frac{1}{2}}U_k)_t\ ,\quad A^{\frac{1}{2}}U_k \in L^2(0,T;L^2_\sigma(\Omega))\ ,\quad (A^{-\frac{1}{2}}U_k)(0) = 0,$$

and that the evolution equation

$$(A^{-\frac{1}{2}}U_k)_t + A^{\frac{1}{2}}U_k = -A^{-\frac{1}{2}}P \operatorname{div}\,(J_k u_k)u_k \tag{3.2.8}$$

is satisfied. We get

$$\begin{aligned} \|A^{-\frac{1}{2}}P \operatorname{div}\,(J_k u_k)u_k\|_{2,2;T} &\le \nu^{-\frac{1}{2}}\|(J_k u_k)u_k\|_{2,2;T} \\ &\le \nu^{-\frac{1}{2}}\, C\, k^{\frac{n}{4}}\, E_T(u_k) < \infty. \end{aligned} \tag{3.2.9}$$

Applying the Fourier transform to (3.2.8), using integration by parts and that $A^{-\frac{1}{2}}U_k(0) = 0$, we obtain

$$\begin{aligned} U_k^{\sim}(\tau) &= \int_0^T U_k(t)e^{-2\pi i\tau t}\,dt\,, \\ (A^{-\frac{1}{2}}U_k)_t^{\sim}(\tau) &= \int_0^T (A^{-\frac{1}{2}}U_k)_t(t)e^{-2\pi i\tau t}\,dt \\ &= A^{-\frac{1}{2}}U_k(T)e^{-2\pi i\tau T} + 2\,\pi i\tau A^{-\frac{1}{2}}U_k^{\sim}(\tau)\,, \end{aligned}$$

$\tau \in \mathbb{R}$, and therefore

$$\begin{aligned} 2\pi i\tau A^{-\frac{1}{2}}\,U_k^{\sim}(\tau) &+ A^{\frac{1}{2}}\,U_k^{\sim}(\tau) \\ &= -\big(A^{-\frac{1}{2}}P \operatorname{div}\,(J_k u_k)u_k\big)^{\sim}(\tau) - A^{-\frac{1}{2}}U_k(T)e^{-2\pi i\tau T}. \end{aligned}$$

Taking the scalar product with $A^{\frac{1}{2}}U_k^{\sim}(\tau)$ yields

$$\begin{aligned}
2\pi|\tau|\,\|U_k^{\sim}(\tau)\|_2^2 \;&\leq\; \|A^{\frac{1}{2}}U_k^{\sim}(\tau)\|_2^2 \\
&\quad + |<(A^{-\frac{1}{2}}P \text{ div } (J_k u_k)u_k)^{\sim}(\tau)\,,\, A^{\frac{1}{2}}U_k^{\sim}(\tau)>_\Omega| \\
&\quad\; + |<A^{-\frac{1}{2}}U_k(T),\, A^{\frac{1}{2}}U_k^{\sim}(\tau)>_\Omega| \\
&\leq\; \|A^{\frac{1}{2}}U_k^{\sim}(\tau)\|_2^2 \\
&\quad + \left(\int_0^T \|A^{-\frac{1}{2}}P \text{ div } (J_k u_k)u_k\|_2\, dt\right)\|A^{\frac{1}{2}}U_k^{\sim}(\tau)\|_2 \\
&\quad\; + \|A^{-\frac{1}{2}}U_k(T)\|_2\, \|A^{\frac{1}{2}}U_k^{\sim}(\tau)\|_2 \\
&\leq\; \|A^{\frac{1}{2}}U_k^{\sim}(\tau)\|_2^2 + \nu^{-\frac{1}{2}}\left(\int_0^T \|(J_k u_k)u_k\|_2\, dt\right)\|A^{\frac{1}{2}}U_k^{\sim}(\tau)\|_2 \\
&\quad + \|A^{-\frac{1}{2}}U_k(T)\|_2\|A^{\frac{1}{2}}U_k^{\sim}(\tau)\|_2 .
\end{aligned}$$

Using (3.2.5) leads to

$$A^{-\frac{1}{2}}\,U_k(T) \;=\; -\int_0^T S(T-\tau)A^{-\frac{1}{2}}P \text{ div } ((J_k u_k)u_k)\,d\tau\,,$$

and therefore, with (3.2.9) we get

$$\begin{aligned}
\|A^{-\frac{1}{2}}U_k(T)\|_2 \;&\leq\; \int_0^T \|A^{-\frac{1}{2}}P \text{ div } ((J_k u_k)u_k)\|_2\, d\tau \\
&\leq\; \nu^{-\frac{1}{2}}\int_0^T \|(J_k u_k)u_k\|_2\, d\tau.
\end{aligned}$$

From (2.2.15) we obtain

$$\|(J_k u_k)u_k\|_{2,1;T} \;\leq\; C\,E_T(u_k)$$

with $C = C(\nu, n, T) > 0$. Using (3.2.7) we see that

$$\|(J_k u_k)u_k\|_{2,1;T} \;\leq\; C$$

with $C > 0$ not depending on k. This leads to

$$\|A^{-\frac{1}{2}}U_k(T)\|_2 \;\leq\; C\,.$$

From above we now get

$$2\pi|\tau|\,\|U_k^{\sim}(\tau)\|_2^2 \;\leq\; \|A^{\frac{1}{2}}U_k^{\sim}(\tau)\|_2^2 + C\,\|A^{\frac{1}{2}}U_k^{\sim}(\tau)\|_2 \qquad (3.2.10)$$

with $C > 0$ not depending on k.

Let $0 < \gamma < \frac{1}{4}$. Then an elementary calculation shows that

$$|\tau|^{2\gamma} \leq C' \left(\frac{1 + |\tau|}{1 + |\tau|^{1-2\gamma}} \right) \tag{3.2.11}$$

holds for all $\tau \in \mathbb{R}$ with $C' > 0$ not depending on τ. We use this estimate in a similar way as in [Tem77, Chap. III, (3.38)–(3.39)]. Using (3.2.10) and (3.2.2) we obtain

$$\begin{aligned}
\int_{-\infty}^{\infty} |\tau|^{2\gamma} \, \|U_k^{\sim}(\tau)\|_2^2 \, d\tau &\leq C' \int_{-\infty}^{\infty} \left(\frac{1 + |\tau|}{1 + |\tau|^{1-2\gamma}} \right) \|U_k^{\sim}(\tau)\|_2^2 \, d\tau \\
&\leq C' \left(\int_{-\infty}^{\infty} \|U_k^{\sim}(\tau)\|_2^2 \, d\tau \right. \\
&\quad \left. + \int_{-\infty}^{\infty} \left(\frac{|\tau|}{1 + |\tau|^{1-2\gamma}} \right) \|U_k^{\sim}(\tau)\|_2^2 \, d\tau \right) \\
&\leq C' \int_0^T \|U_k(t)\|_2^2 \, dt + C' \int_0^T \|A^{\frac{1}{2}} U_k(t)\|_2^2 \, dt \\
&\quad + C' \int_{-\infty}^{\infty} \left(\frac{1}{1 + |\tau|^{1-2\gamma}} \right) \|A^{\frac{1}{2}} U_k^{\sim}(\tau)\|_2 \, d\tau.
\end{aligned}$$

Since $U_k = u_k - V$, we get from (3.2.6), (3.2.7) that

$$\|U_k\|_{2,2;T}^2 \leq C \quad , \quad \nu \, \|\nabla U_k\|_{2,2;T}^2 = \|A^{\frac{1}{2}} U_k\|_{2,2;T}^2 \leq C.$$

Since $0 < \gamma < \frac{1}{4}$, $2\,(1 - 2\gamma) > 1$, we see that

$$\int_{-\infty}^{\infty} \left(\frac{1}{1 + |\tau|^{1-2\gamma}} \right)^2 d\tau < \infty.$$

This yields

$$\begin{aligned}
&\int_{-\infty}^{\infty} |\tau|^{2\gamma} \|U_k^{\sim}(\tau)\|_2^2 \, d\tau \\
&\leq C_1 + C_2 + C_3 \left(\int_{-\infty}^{\infty} \left(\frac{1}{1 + |\tau|^{1-2\gamma}} \right)^2 d\tau \right)^{\frac{1}{2}} \left(\int_{-\infty}^{\infty} \|A^{\frac{1}{2}} U_k^{\sim}(\tau)\|_2^2 d\tau \right)^{\frac{1}{2}} \\
&= C_4 + C_5 \left(\int_0^T \|A^{\frac{1}{2}} U_k(t)\|_2^2 \, dt \right)^{\frac{1}{2}} \leq C_6
\end{aligned}$$

with constants $C, C', C_1, \ldots, C_6$ not depending on $k \in \mathbb{N}$.

Thus we obtain

$$\sup_k \left(\int_0^T \|\nabla U_k\|_2^2 \, dt \right) < \infty \ , \quad \sup_k \left(\int_{-\infty}^{+\infty} |\tau|^{2\gamma} \, \|U_k^{\sim}(\tau)\|_2^2 \, d\tau \right) < \infty. \quad (3.2.12)$$

Consider any bounded Lipschitz subdomain $\Omega_0 \subseteq \Omega$ with $\overline{\Omega}_0 \subseteq \Omega$. Then

$$\begin{aligned} U_k^{\sim}(\tau, x) &= \int_0^T U_k(t,x) \, e^{-2\pi i \tau t} \, dt \ , \quad \tau \in \mathbb{R} \ , \quad x \in \Omega, \\ \|U_k^{\sim}(\tau)\|_{L^2(\Omega_0)}^2 &= \int_{\Omega_0} |U_k^{\sim}(\tau, x)|^2 \, dx \ \le \ \int_{\Omega} |U_k^{\sim}(\tau, x)|^2 \, dx, \end{aligned}$$

and therefore,

$$\begin{aligned} &\sup_k \left(\int_{-\infty}^{+\infty} |\tau|^{2\gamma} \, \|U_k^{\sim}(\tau)\|_{L^2(\Omega_0)}^2 \, d\tau \right) \\ &\le \ \sup_k \left(\int_0^T |\tau|^{2\gamma} \, \|U_k^{\sim}(\tau)\|_2^2 \, d\tau \right) \ < \ \infty. \end{aligned} \quad (3.2.13)$$

Further we get

$$\|\nabla U_k(t)\|_{L^2(\Omega_0)}^2 = \int_{\Omega_0} |\nabla U_k(t,x)|^2 \, dx \ \le \ \int_{\Omega} |\nabla U_k(t,x)|^2 \, dx,$$

and therefore

$$\sup_k \left(\int_0^T \|\nabla U_k\|_{L^2(\Omega_0)}^2 \, dt \right) \ \le \ \sup_k \left(\int_0^T \|\nabla U_k\|_2^2 \, dt \right) \ < \ \infty. \quad (3.2.14)$$

If Ω is bounded, we know by Lemma 1.5.1, II, that the embedding $X_0 \subseteq X$, $X_0 := W_0^{1,2}(\Omega)^n$, $X := L^2(\Omega)^n$, is compact. From (3.2.12) we see that $(U_k)_{k=1}^{\infty}$ satisfies the condition (3.2.3). Then Lemma 3.2.1 yields the existence of a subsequence which converges strongly in $L^2(0,T; L^2_\sigma(\Omega))$. Since $u_k = U_k + V$, $V \in L^2(0,T; W_{0,\sigma}^{1,2}(\Omega))$, the sequence $(u_k)_{k=1}^{\infty}$ has the property in Lemma 3.2.2, a).

If Ω is unbounded, we choose $X_0 := W^{1,2}(\Omega_0)^n$, $X := L^2(\Omega_0)^n$, and get by Lemma 1.5.3, II, the compact embedding $W^{1,2}(\Omega_0)^n \subseteq L^2(\Omega_0)^n$. Consider $(U_k)_{k=1}^{\infty}$ as a sequence in $L^2(0,T; X_0)$. Then from (3.2.13), (3.2.14) we get the validity of (3.2.3), and Lemma 3.2.1 shows the existence of a subsequence satisfying the property in Lemma 3.2.2, b). This completes the proof.

3.3 Proof of Theorem 3.1.1

To construct a weak solution u as required in this theorem, let $(u_k)_{k=1}^\infty$ be the sequence of solutions of the approximate system as in Theorem 2.5.1. We investigate the convergence properties as $k \to \infty$.

First let $0 < T < \infty$ and

$$f_0 \in L^1(0,T;L^2(\Omega)^n) \ , \quad F \in L^2(0,T;L^2(\Omega)^{n^2}).$$

Then we let $T' \to T$ in the energy estimate (2.2.5) and obtain the inequality

$$\frac{1}{2}\|u_k\|_{2,\infty;T}^2 + \nu \, \|\nabla u_k\|_{2,2;T}^2 \ \leq \ 2\,\|u_0\|_2^2 + 4\nu^{-1}\|F\|_{2,2;T}^2 + 8\,\|f_0\|_{2,1;T}^2 \ . \tag{3.3.1}$$

In particular it follows that $(u_k)_{k=1}^\infty$ is a bounded sequence in the Hilbert space $L^2(0,T;W_{0,\sigma}^{1,2}(\Omega))$. Since this space is reflexive, we find a subsequence which converges weakly in this space to some element $u \in L^2(0,T;W_{0,\sigma}^{1,2}(\Omega))$. For simplicity we may assume that the sequence itself has this property.

Let Ω be bounded. Then Lemma 3.2.2 shows the existence of a subsequence of $(u_k)_{k=1}^\infty$ which converges to u strongly in $L^2(0,T;L_\sigma^2(\Omega))$. Again we may assume that the sequence itself has this property. The Fischer-Riez theorem, see [Apo74, Note at the end of Chapter 10.25], yields the existence of a subsequence which converges strongly to $u(t)$ for almost all $t \in [0,T)$. We may assume that the sequence itself has this property. Thus we get the existence of a null set $N \subseteq [0,T)$ such that

$$u(t) = s - \lim_{k\to\infty} u_k(t)$$

for all $t \in [0,T)\backslash N$. We thus obtain the following convergence properties:

$$\left.\begin{array}{l} (u_k)_{k=1}^\infty \text{ converges to } u \text{ weakly in } L^2(0,T;W_{0,\sigma}^{1,2}(\Omega)) \text{ and} \\ \text{strongly in } L^2(0,T;L_\sigma^2(\Omega)); \\ (\nabla u_k)_{k=1}^\infty \text{ converges to } \nabla u \text{ weakly in } L^2(0,T;L^2(\Omega)^{n^2}); \\ (u_k(t))_{k=1}^\infty \text{ converges to } u(t) \text{ strongly in } L_\sigma^2(\Omega) \\ \text{ for all } t \in [0,T)\backslash N. \end{array}\right\} \tag{3.3.2}$$

Consider any test function $v \in C_0^\infty([0,T);C_{0,\sigma}^\infty(\Omega))$. Since u_k is a weak solution, see Definition 2.1.1, we get from (2.1.8) that

$$\begin{aligned} -<u_k,v_t>_{\Omega,T} + \nu <\nabla u_k,\nabla v>_{\Omega,T} + <(J_k u_k)\cdot\nabla u_k, v>_{\Omega,T} \\ = <u_0,v(0)>_\Omega + <f_0,v>_{\Omega,T} - <F,\nabla v>_{\Omega,T} \ . \end{aligned} \tag{3.3.3}$$

Using (3.3.2) we see that

$$< u, v_t >_{\Omega,T} = \lim_{k\to\infty} < u_k, v_t >_{\Omega,T} \tag{3.3.4}$$

and

$$< \nabla u, \nabla v >_{\Omega,T} = \lim_{k\to\infty} < \nabla u_k, \nabla v >_{\Omega,T} . \tag{3.3.5}$$

To treat the limit of $< (J_k u_k) \cdot \nabla u_k, v >_{\Omega,T}$ we write

$$< (J_k u_k) \cdot \nabla u_k, v >_{\Omega,T} = - < (J_k u_k) u_k, \nabla v >_{\Omega,T} ,$$

and get the representation

$$\begin{aligned} &< (J_k u_k) u_k, \nabla v >_{\Omega,T} - < u\, u, \nabla v >_{\Omega,T} \\ &\quad = < (J_k u_k)(u_k - u), \nabla v >_{\Omega,T} + < (J_k(u_k - u))u, \nabla v >_{\Omega,T} \\ &\quad + < ((J_k - I)u)u, \nabla v >_{\Omega,T} . \end{aligned}$$

Using the estimates

$$\begin{aligned} |< (J_k u_k)(u_k - u), \nabla v >_{\Omega,T}| &\le \| < (J_k u_k)(u_k - u)\|_{1,1;T} \, \|\nabla v\|_{\infty,\infty;T} \\ &\le C \, \|J_k u_k\|_{2,2;T} \, \|u_k - u\|_{2,2;T} \, \|\nabla v\|_{\infty,\infty;T} \\ &\le C \, \|u_k\|_{2,2;T} \, \|u_k - u\|_{2,2;T} \, \|\nabla v\|_{\infty,\infty;T} , \end{aligned}$$

$$|< (J_k(u_k - u))u, \nabla v >_{\Omega,T}| \le C \, \|u_k - u\|_{2,2;T} \, \|u\|_{2,2;T} \, \|\nabla v\|_{\infty,\infty;T} ,$$

and

$$|< ((J_k - I)u)u, \nabla v >_{\Omega,T}| \le C \, \|(J_k - I)u\|_{2,2;T} \, \|u\|_{2,2;T} \, \|\nabla v\|_{\infty,\infty;T}$$

with $C = C(n) > 0$ not depending on T, we get from (3.3.2) that

$$\begin{aligned} &\lim_{k\to\infty} < (J_k u_k)(u_k - u), \nabla v >_{\Omega,T} = 0 , \\ &\lim_{k\to\infty} < (J_k(u_k - u))u, \nabla v >_{\Omega,T} = 0 . \end{aligned} \tag{3.3.6}$$

Using (3.4.8), II, we get

$$\lim_{k\to\infty} \|(J_k - I)u(t)\|_2 = 0 \quad \text{for all} \ \ t \in [0,T)\backslash N. \tag{3.3.7}$$

Further we obtain

$$\|(J_k - I)u(t)\|_2 \le \|J_k u(t)\|_2 + \|u(t)\|_2 \le 2 \, \|u(t)\|_2 \tag{3.3.8}$$

for almost all $t \in [0,T)$. Therefore, we may use Lebesgue's dominated convergence lemma, see [Apo74, Chapter 10.10], and get

$$\lim_{k\to\infty} \|(J_k - I)u\|_{2,2;T} = 0 .$$

It follows that

$$\lim_{k\to\infty} < ((J_k - I)u)u, \nabla v >_{\Omega,T} = 0 . \tag{3.3.9}$$

Thus we may let $k \to \infty$ in each term of (3.3.3), and obtain

$$\begin{aligned} - < u, v_t >_{\Omega,T} + \nu < \nabla u, \nabla v >_{\Omega,T} - < u\,u, \nabla v >_{\Omega,T} & \qquad (3.3.10)\\ = \; < u_0, v(0) >_{\Omega} + < f_0, v >_{\Omega,T} - < F, \nabla v >_{\Omega,T} . \end{aligned}$$

This shows that u is a weak solution of the Navier-Stokes system (3.1.1).

To prove (3.1.2) we use Lemma 2.2.1 and conclude that each u_k is strongly continuous, after a corresponding redefinition, and that

$$\begin{aligned} \frac{1}{2}\|u_k(t)\|_2^2 + \nu \int_0^t \|\nabla u_k\|_2^2 \, d\tau & \qquad (3.3.11)\\ = \frac{1}{2}\|u_0\|_2^2 + \int_0^t < f_0, u_k >_{\Omega} \, d\tau - \int_0^t < F, \nabla u_k >_{\Omega} \, d\tau \end{aligned}$$

for all $t \in [0,T)$. The weak convergence property in (3.3.2) concerning $(\nabla u_k)_{k=1}^{\infty}$ shows that

$$\|\nabla u\|_{2,2;t} \;\le\; \liminf_{k\to\infty} \|\nabla u_k\|_{2,2;t} \tag{3.3.12}$$

for all $t \in [0,T)$, see (3.1.3), II, and the property concerning $(u_k(t))_{k=1}^{\infty}$ shows that

$$\|u(t)\|_2^2 \;=\; \lim_{k\to\infty} \|u_k(t)\|_2^2 \tag{3.3.13}$$

for all $t \in [0,T)\backslash N$. The properties in (3.3.2) also show that

$$\begin{aligned} \int_0^t < f_0, u >_{\Omega} \, d\tau &= \lim_{k\to\infty} \int_0^t < f_0, u_k > \, d\tau, & \qquad (3.3.14)\\ \int_0^t < F, \nabla u >_{\Omega} \, d\tau &= \lim_{k\to\infty} \int_0^t < F, \nabla u_k > \, d\tau \end{aligned}$$

for all $t \in [0,T)$.

Taking $\liminf_{k\to\infty}$ in each term of (3.3.11), we get the energy inequality

$$\begin{aligned} \frac{1}{2}\|u(t)\|_2^2 + \nu \int_0^t \|\nabla u\|_2^2 \, d\tau & \qquad (3.3.15)\\ \le \frac{1}{2}\|u_0\|_2^2 + \int_0^t < f_0, u >_{\Omega} \, d\tau - \int_0^t < F, \nabla u >_{\Omega} \, d\tau \end{aligned}$$

for all $t \in [0,T)\backslash N$.

From Theorem 1.3.1 we conclude that $u : [0,T) \to L^2_\sigma(\Omega)$ is weakly continuous, after a corresponding redefinition. Therefore, for each $t \in [0,T)$ we find a sequence $(t_j)_{j=1}^\infty$ in $[0,T)\backslash N$ so that $u(t_j)$ tends to $u(t)$, weakly in $L^2_\sigma(\Omega)$ as $j \to \infty$. It follows that

$$\|u(t)\|_2 \leq \lim_{j\to\infty} \inf \|u(t_j)\|_2. \tag{3.3.16}$$

Inserting $t = t_j$ in (3.3.15) and taking $\liminf_{j\to\infty}$ in each term, we conclude that (3.3.15) holds for all $t \in [0,T)$.

The inequality (3.1.3) is a consequence of (3.3.15). This has been shown in the proof of Theorem 1.4.1, see (1.4.4).

This proves Theorem 3.1.1 for the case that Ω is bounded, that $0 < T < \infty$, and that $f_0 \in L^1(0,T;L^2(\Omega)^n)$, $F \in L^2(0,T;L^2(\Omega)^{n^2})$.

Consider now the general case. Then we use Lemma 1.4.1, II, and choose a sequence $(\Omega_j)_{j=1}^\infty$ of bounded Lipschitz domains $\Omega_j \subseteq \Omega$, with $\overline{\Omega}_j \subseteq \Omega_{j+1}$, $j \in \mathbb{N}$, and with

$$\Omega = \bigcup_{j=1}^\infty \Omega_j .$$

For each bounded Lipschitz subdomain Ω_0 with $\overline{\Omega}_0 \subseteq \Omega$, we find some $j \in \mathbb{N}$ so that $\Omega_0 \subseteq \Omega_j$, see Remark 1.4.2, II.

Further we choose a sequence $(T_j)_{j=1}^\infty$ of real numbers with $0 < T_j < T$, $T_j < T_{j+1}$, $j \in \mathbb{N}$, and with $T = \lim_{j\to\infty} T_j$. Then we get

$$f_0 \in L^1(0,T_j;L^2(\Omega_j)^n) \ , \quad F \in (0,T_j;L^2(\Omega)^{n^2})$$

for each $j \in \mathbb{N}$.

In the next step we will construct for each $j \in \mathbb{N}$ a subsequence $(u_k^{(j)})_{k=1}^\infty$ of $(u_k)_{k=1}^\infty$, a function $u^{(j)} \in L^2(0,T;W^{1,2}(\Omega_j)^n)$, and a null set $N_j \subseteq [0,T_j)$ with the following properties corresponding to (3.3.2):

$$\left.\begin{array}{l} (u_k^{(j)})_{k=1}^\infty \text{ converges to } u^{(j)} \text{ weakly in } L^2(0,T_j;W^{1,2}(\Omega_j)^n) \text{ and} \\ \text{strongly in } L^2(0,T_j;L^2(\Omega_j)^n); \\ (\nabla u_k^{(j)})_{k=1}^\infty \text{ converges to } \nabla u^{(j)} \text{weakly in } L^2(0,T_j;L^2(\Omega_j)^{n^2}); \\ (u_k^{(j)}(t))_{k=1}^\infty \text{ converges to } u^{(j)}(t) \text{ strongly in } L^2(\Omega_j) \\ \text{ for all } t \in [0,T_j)\backslash N_j. \end{array}\right\} \tag{3.3.17}$$

First let $j = 1$. Then we use Lemma 3.2.2, b), with Ω_0 replaced by Ω_1, and find in the same way as in (3.3.2) a subsequence $(u_k^{(1)})_{k=1}^\infty$ of $(u_k)_{k=1}^\infty$, a function $u^{(1)} \in L^2(0,T_1;L^2(\Omega_1)^n)$, and a null set $N_1 \subseteq [0,T_1)$ such that (3.3.17) holds with $j=1$. Then we choose $(u_k^{(2)})_{k=1}^\infty$ as a subsequence of $(u_k^{(1)})_{k=1}^\infty$, and we choose $u^{(2)} \in L^2(0,T_2;L^2(\Omega_2)^n)$, $N_2 \subseteq [0,T_2)$ such that (3.3.17) is satisfied with $j=2$. Next we choose $(u_k^{(3)})_{k=1}^\infty$ as a subsequence of $(u_k^{(2)})_{k=1}^\infty$ satisfying (3.3.17) together with some $u^{(3)}$, N_3, for $j=3$, and so on.

Thus we get a sequence of subsequences of $(u_k)_{k=1}^\infty$ which we can write as the lines of a matrix. Then we take the **diagonal sequence**, which is a subsequence of $(u_k)_{k=1}^\infty$ and satisfies (3.3.17) simultaneously for all $j \in \mathbb{N}$. For simplicity we may assume that $(u_k)_{k=1}^\infty$ itself has this property.

This construction shows that for all $j \in \mathbb{N}$, $u^{(j)}$ is the restriction of $u^{(j+1)}$ to $[0,T_j) \times \Omega_j$. Thus we get a well defined function $u \in L^2_{loc}([0,T);W^{1,2}_{loc}(\Omega)^n)$ so that $u^{(j)}$ coincides with the restriction of u to $[0,T_j) \times \Omega_j$, $j \in \mathbb{N}$.

Consider a fixed T' with $0 < T' < T$. Then (3.3.1) holds with T replaced by T'. Therefore, $(u_k)_{k=1}^\infty$ is also a bounded sequence in the Hilbert space $L^2(0,T';W^{1,2}_{0,\sigma}(\Omega))$ and we obtain a weakly convergent subsequence; we may assume that the sequence itself has this property. Now the above convergence properties in particular show that $(u_k)_{k=1}^\infty$ converges weakly in $L^2(0,T';W^{1,2}_{0,\sigma}(\Omega))$ to u, and that $u \in L^2(0,T';W^{1,2}_{0,\sigma}(\Omega))$. Further we conclude from (3.3.1) that

$$u \in L^\infty(0,T';L^2_\sigma(\Omega)) \cap L^2(0,T';W^{1,2}_{0,\sigma}(\Omega)).$$

Thus we get

$$u \in L^\infty_{loc}([0,T);L^2_\sigma(\Omega)) \cap L^2_{\mathrm{loc}}([0,T);W^{1,2}_{0,\sigma}(\Omega)).$$

Let the null set $N \subseteq [0,T)$ be defined as the union of all N_j, $j \in \mathbb{N}$, from (3.3.17). Then we conclude from (3.3.17) that

$$\lim_{k\to\infty} \|u(t) - u_k(t)\|_{L^2(\Omega_j)} = 0 \tag{3.3.18}$$

for all $t \in [0,T)\backslash N$ and for all $j \in \mathbb{N}$. In particular we conclude that $u_k(t)$ tends to $u(t)$ weakly in $L^2_\sigma(\Omega)$ for all $t \in [0,T)\backslash N$. Using (3.1.3), II, we get

$$\|u(t)\|_2 \le \liminf_{k\to\infty} \|u_k(t)\|_2 \tag{3.3.19}$$

for all $t \in [0,T)\backslash N$.

Consider now any test function $v \in C_0^\infty([0,T);C^\infty_{0,\sigma}(\Omega))$. Then we can choose some $j \in \mathbb{N}$ with

$$\mathrm{supp}\, v \subseteq [0,T_j) \times \Omega_j.$$

To prove (3.3.10) we may now use the same convergence properties as in the above case for bounded Ω, T. This yields (3.3.10) and shows that u is a weak solution of (3.1.1).

To prove the energy inequality (3.3.15) we use (3.3.11), (3.3.12) as above, but we replace the condition (3.3.13) now by (3.3.19). This proves (3.3.15) for all $t \in [0,T)\backslash N$.

Theorem 1.3.1 yields again that u is weakly continuous, and using (3.3.16) as above, we see that (3.3.15) holds for all $t \in [0,T)$. The inequality (3.1.3) is again a consequence of (3.3.15). The proof of Theorem 3.1.1 is complete.

3.4 Weighted energy inequalities and time decay

In this subsection we prove some important properties of the special weak solutions u from Theorem 3.1.1. These properties are not available up to now for general weak solutions if $n = 3$. First we consider the energy inequality below which contains a time dependent scalar weight function ϕ.

To explain the method, we derive from

$$u_t - \nu\Delta u + u \cdot \nabla u + \nabla p = f \ , \quad \text{div } u = 0 \ , \quad u|_{\partial\Omega} = 0 \ , \quad u(0) = u_0 \quad (3.4.1)$$

the **weighted system**

$$\begin{aligned} (\phi u)_t - \nu\Delta(\phi u) + (\phi u) \cdot \nabla u + \nabla(\phi p) &= \phi f + \phi_t u, \qquad (3.4.2) \\ \text{div } (\phi u) = 0 \ , \quad \phi u|_{\partial\Omega} &= 0 \ , \quad (\phi u)(0) = \phi(0)u_0. \end{aligned}$$

Formally this system follows when we multiply (3.4.1) by ϕ and use some elementary calculations.

Then we can apply the results of the preceding sections to this weighted system. This yields some further properties of u, in particular concerning the asymptotic behavior as $t \to \infty$. We need some assumptions on the derivative $\dot{\phi} = \frac{d}{dt}\phi = \phi_t$.

In the literature there are several other approaches to asymptotic results of weak solutions, mainly for special domains like exterior domains, see [Mas75], [Hey80], [Mar84], [GaM86], [Sch86], [MiSo88], [BMi91], [BMi92], [KOS92], [KoO93], [KoO94].

The following theorem is interesting mainly for unbounded domains.

A weak solution u of the Navier-Stokes system (3.4.1) as constructed in the proof of Theorem 3.1.1 is called a **suitable weak solution** of (3.4.1) with data f, u_0.

In the literature this notion is used for several special types of weak solutions, see for example [CKN82, p. 779].

3.4.1 Theorem *Let $\Omega \subseteq \mathbb{R}^n$, $n = 2,3$, be any domain, let $T = \infty$, $u_0 \in D(A^{-\frac{1}{2}})$, $f = \text{div } F$ with*

$$F \in L^1(0,\infty; L^2(\Omega)^{n^2}) \cap L^2(0,\infty; L^2(\Omega)^{n^2}),$$

and let $\phi : [0,\infty) \to \mathbb{R}$ be continuous with $\phi_t \in L^\infty_{loc}([0,\infty); \mathbb{R})$.

Suppose

$$u \in L^\infty_{loc}([0,\infty); L^2_\sigma(\Omega)) \cap L^2_{loc}([0,\infty); W^{1,2}_{0,\sigma}(\Omega))$$

is a weak solution of the Navier-Stokes system (3.4.1) *with data* f, u_0 *if* $n = 2$, *or a suitable weak solution of* (3.4.1) *if* $n = 3$.

Then, after redefinition on a null set of $[0,\infty)$, $u : [0,\infty) \to L^2_\sigma(\Omega)$ *is weakly continuous and satisfies the* **weighted energy inequality**

$$\phi^2(t)\,\|u(t)\|_2^2 + \nu\int_0^t \phi^2\|\nabla u\|_2^2\,d\tau \le \phi^2(0)\,\|u_0\|_2^2 + \nu^{-1}\int_0^t \phi^2\|F\|_2^2\,d\tau + C\sup_{0\le\tau\le t}|\dot{\phi}(\tau)\phi(\tau)| \tag{3.4.3}$$

for all $t \in [0,\infty)$ *with* $C = C(u_0, F) > 0$.

In particular it follows that

$$(1+t)\,\|u(t)\|_2^2 + \nu\int_0^t (1+\tau)\,\|\nabla u\|_2^2\,d\tau \le \|u_0\|_2^2 + \nu^{-1}\int_0^t (1+\tau)\,\|F\|_2^2\,d\tau + C \tag{3.4.4}$$

for all $t \in [0,\infty)$ *with* $C = C(u_0, F) > 0$.

If $\int_0^\infty (1+t)\,\|F\|_2^2\,dt < \infty$, *then we get*

$$\|u(t)\|_2 \le C'\,(1+t)^{-\frac{1}{2}} \tag{3.4.5}$$

for all $t \ge 0$ *with* $C' = C'(u_0, F) > 0$, *and*

$$\int_0^\infty (1+t)\,\|\nabla u\|_2^2\,dt < \infty.$$

Proof. Let $n = 2$. Using (2.1.4), IV, we conclude that ϕu is a weak solution of the linear system

$$(\phi u)_t - \nu\Delta(\phi u) + \nabla(\phi p) = \tilde{f}\ ,\quad \phi u|_{\partial\Omega} = 0\ ,\quad (\phi u)(0) = \phi(0)u_0 \tag{3.4.6}$$

with data $\tilde{f} = \dot{\phi}\, u + \operatorname{div}(\phi F - \phi u\, u)$ and $\phi(0)u_0$. Since $n = 2$, ϕu satisfies Serrin's condition and $u : [0,T) \to L^2_\sigma(\Omega)$ is strongly continuous. As in Theorem 1.4.2, (1.4.7), we obtain the energy equality

$$\frac{1}{2}\,\phi^2(t)\,\|u(t)\|_2^2 + \nu\int_0^t \phi^2\,\|\nabla u\|_2^2\,d\tau = \frac{1}{2}\phi^2(0)\,\|u_0\|_2^2 - \int_0^t \phi^2 < F, \nabla u >_\Omega\,d\tau + \int_0^t \dot{\phi}\,\phi\|u\|_2^2\,d\tau\,, \tag{3.4.7}$$

$t \in [0, \infty)$. From Theorem 1.6.3, (1.6.39), we conclude that

$$\|u\|_{2,2;\infty}^2 = \int_0^\infty \|u\|_2^2 \, dt < \infty. \tag{3.4.8}$$

Using the assumption on ϕ we see that

$$\left| \int_0^t \dot{\phi} \, \phi \, \|u\|_2^2 \, d\tau \right| \leq \|u\|_{2,2;\infty}^2 \sup_{0 \leq \tau \leq t} \left| \dot{\phi}(\tau) \phi(\tau) \right| < \infty. \tag{3.4.9}$$

Next we use Young's inequality and obtain

$$\left| \int_0^t \phi^2 < F, \nabla u >_\Omega \, d\tau \right| \leq \frac{\nu^{-1}}{2} \int_0^t \phi^2 \|F\|_2^2 \, d\tau + \frac{\nu}{2} \int_0^t \phi^2 \, \|\nabla u\|_2^2 \, d\tau. \tag{3.4.10}$$

This leads to (3.4.3).

If $n = 3$, u is a weak solution as constructed in the proof of Theorem 3.1.1. In this case, we use Lemma 2.2.1 and get the above equality (3.4.7) in the same way with u replaced by u_k, $k \in \mathbb{N}$, u_k as in Theorem 2.5.1. Using Lemma 2.6.1 we obtain

$$\|u_k\|_{2,2;\infty}^2 = \int_0^\infty \|u_k\|_2^2 \, dt \leq C \tag{3.4.11}$$

for all $k \in \mathbb{N}$ with $C = C(u_0, F) > 0$ not depending on k.

Replacing (3.4.8) by (3.4.11), we obtain the inequality (3.4.3) in the same way as above, now with u replaced by u_k. Letting $k \to \infty$ and using the convergence properties of $(u_k)_{k=1}^\infty$ in the proof of Theorem 3.1.1, we get (3.4.3) for almost all $t \in [0, \infty)$. After a corresponding redefinition as in this proof, we get (3.4.3) for all $t \in [0, \infty)$.

The inequality (3.4.4) follows if we choose

$$\phi(t) := (1 + t)^{\frac{1}{2}} \ , \quad t \geq 0. \tag{3.4.12}$$

Then $| \dot{\phi}(t) \phi(t)| = \frac{1}{2}$, and we get (3.4.4); (3.4.5) is a consequence of (3.4.4). This proves the theorem. □

3.5 Exponential decay for domains for which the Poincaré inequality holds

The weighted energy inequality in the preceding subsection yields a certain exponential decay if the Poincaré inequality is valid.

We say the Poincaré inequality holds for a domain $\Omega \subseteq \mathbb{R}^n$, iff there exists a constant $d = d(\Omega) > 0$ such that

$$\|u\|_2 \leq d \, \|\nabla u\|_2 \tag{3.5.1}$$

for all $u \in W_0^{1,2}(\Omega)^n$. In particular we know, see Lemma 1.1.1, II, that (3.5.1) holds if the domain Ω is bounded. Below we use the notation $e^x = \exp x$ for $x \in \mathbb{R}$.

3.5.1 Theorem *Let $\Omega \subseteq \mathbb{R}^n$, $n = 2, 3$, be a domain for which the Poincaré inequality (3.5.1) is valid with $d = d(\Omega) > 0$. Let*

$$\phi(t) := \exp\left(\frac{\nu}{8d^2} t\right),$$

$t \geq 0$, $u_0 \in L^2_\sigma(\Omega)$, and let $f = f_0 + \operatorname{div} F$ with

$$f_0 \in L^1(0,\infty; L^2(\Omega)^n) \ , \quad F \in L^2(0,\infty; L^2(\Omega)^{n^2}).$$

Suppose

$$u \in L^\infty_{loc}([0,\infty); L^2_\sigma(\Omega)) \cap L^2_{loc}([0,\infty); W^{1,2}_{0,\sigma}(\Omega))$$

is a weak solution of (3.4.1) with data f, u_0 if $n = 2$, or a suitable weak solution if $n = 3$.

Then, after redefining on a null set of $[0,T)$, $u : [0,\infty) \to L^2_\sigma(\Omega)$ is weakly continuous and satisfies the weighted energy inequality

$$\frac{1}{2}\phi^2(t)\,\|u(t)\|_2^2 + \nu \int_0^t \phi^2\,\|\nabla u\|_2^2\, d\tau \tag{3.5.2}$$

$$\leq 2\phi^2(0)\,\|u_0\|_2^2 + 8\nu^{-1}\int_0^t \phi^2\,\|F\|_2^2\, d\tau + 8\left(\int_0^t \phi\,\|f_0\|_2\, d\tau\right)^2$$

for all $t \in [0,\infty)$.

If $\int_0^\infty \phi^2\,\|F\|_2^2\, dt < \infty$, $\int_0^\infty \phi\,\|f_0\|_2\, dt < \infty$, then

$$\|u(t)\|_2 \leq C\exp\left(-\frac{\nu}{8d^2} t\right) \tag{3.5.3}$$

for all $t \geq 0$ with $C = C(u_0, f, \nu) > 0$, and

$$\int_0^\infty \phi^2\,\|\nabla u\|_2^2\, dt < \infty.$$

Proof. Let $n = 2$. Then we see as in the preceding proof, ϕu is a weak solution of the linear system (3.4.6), now with data $\tilde{f} = \dot{\phi}\, u + \phi f_0 + \operatorname{div}(\phi F - \phi u\, u)$, and $\phi(0)u_0$, u is strongly continuous, and

$$\frac{1}{2}\phi^2(t)\,\|u(t)\|_2^2 + \nu\int_0^t \phi^2 \|\nabla u\|_2^2 d\tau \tag{3.5.4}$$

$$= \frac{1}{2}\phi^2(0)\,\|u_0\|_2^2 + \int_0^t \phi^2 < f_0, u >_\Omega d\tau - \int_0^t \phi^2 < F, \nabla u >_\Omega d\tau + \int_0^t \dot{\phi}\phi \|u\|_2^2 d\tau$$

for all $t \in [0, \infty)$. We use the inequalities

$$\|u\|_2^2 \leq d^2 \|\nabla u\|_2^2 \ , \quad |\dot{\phi}(t)\phi(t)| \leq \frac{\nu}{8d^2} \phi^2(t) \ , \quad t \geq 0, \tag{3.5.5}$$

and get

$$\Big| \int_0^t \dot{\phi}\, \phi \, \|u\|_2^2 \, d\tau \Big| \leq \frac{\nu}{8} \int_0^t \phi^2 \, \|\nabla u\|_2^2 \, d\tau \ , \quad t \geq 0.$$

In the same way as in (2.3.7), IV, we see that

$$\begin{aligned} &\frac{1}{2}\Big(\sup_{0 \leq \tau \leq t} \phi^2 \, \|u\|_2^2\Big) + \nu \int_0^t \phi^2 \|\nabla u\|_2^2 \, d\tau \\ &\quad \leq \phi^2(0) \, \|u_0\|_2^2 + 2 \int_0^t \phi^2 \|f_0\|_2 \, \|u\|_2 d\tau \\ &\qquad + 2 \int_0^t \phi^2 \|F\|_2 \, \|\nabla u\|_2 \, d\tau + 2 \int_0^t \dot{\phi}\, \phi \, \|u\|_2^2 \, d\tau \end{aligned}$$

for all $t \in [0, \infty)$. Using Young's inequality (3.3.8), I, we get

$$\begin{aligned} 2 \int_0^t (2\phi \, \|f_0\|_2)(\frac{1}{2}\phi \|u\|_2) \, d\tau &\leq 2 \Big(\int_0^t 2\phi \, \|f_0\|_2 \, d\tau \Big) \cdot \Big(\frac{1}{4} \sup_{0 \leq \tau \leq t} \phi^2 \, \|u\|_2^2 \Big)^{\frac{1}{2}} \\ &\leq 4 \Big(\int_0^t \phi \, \|f_0\|_2 \, d\tau \Big)^2 + \frac{1}{4} \Big(\sup_{0 \leq \tau \leq t} \phi^2 \, \|u\|_2^2 \Big) \ , \end{aligned}$$

and

$$\begin{aligned} 2 \int_0^t \Big(\frac{2}{\sqrt{\nu}} \phi \, \|F\|_2 \Big) \Big(\frac{\sqrt{\nu}}{2} \phi \|\nabla u\|_2 \Big) \, d\tau &\leq 4\nu^{-1} \Big(\int_0^t \phi^2 \, \|F\|_2^2 \, d\tau \Big) \\ &\quad + \frac{\nu}{4} \Big(\int_0^t \phi^2 \, \|\nabla u\|_2^2 \, d\tau \Big) . \end{aligned}$$

Combining the last inequalities in a similar way as in the proof of (2.3.9), IV, we obtain in particular (3.5.2).

If $n = 3$, we obtain (3.5.4) first with u replaced by u_k, $k \in \mathbb{N}$, and with u_k as in the proof of Theorem 3.1.1. Then we get (3.5.5) and therefore (3.5.2) with u replaced by u_k. Letting $k \to \infty$ and using the convergence properties of u_k in the proof of Theorem 3.1.1, we obtain (3.5.2) for almost all $t \in [0, \infty)$. After a corresponding redefinition, $u : [0, \infty) \to L^2_\sigma(\Omega)$ is weakly continuous and (3.5.2) holds for all $t \in [0, \infty)$. Inequality (3.5.3) is a consequence. This proves the theorem. □

3.6 Generalized energy inequality

The **generalized energy inequality** (3.6.2) below has some important consequences in the regularity and decay theory of Navier-Stokes equations, see [ShK66], [Mas84], [GaM86], [SWvW86], [Hey88], [MiSo88], [BMi91], [BMi92], [KoO94], [Wie99].

In the problematic case $n = 3$ we are not able to prove this inequality for general unbounded domains. The reason is, we need that the sequence $(u_k)_{k=1}^{\infty}$, see the proof below, converges strongly in $L^2(0, T; L^2_\sigma(\Omega))$. Therefore, an easy proof of this inequality seems to be possible only for bounded domains. More complicated proofs are available up to now only for special unbounded domains like exterior domains, see [GaM86], [SWvW86], [MiSo88]. However, under the additional assumption of Shinbrot's condition (3.6.1), [Shi74], see also (1.4.2), we can prove the generalized energy equality (3.6.2) below. The proof is the same as for the usual energy equality, see Theorem 1.4.1. We know that the condition (3.6.1) is always satisfied if $n = 2$, see (1.4.15).

3.6.1 Theorem *Let $\Omega \subseteq \mathbb{R}^n$, $n = 2, 3$, be any domain, let $0 < T \le \infty$, $u_0 \in L^2_\sigma(\Omega)$, $f = f_0 + \operatorname{div} F$ with*

$$f_0 \in L^1_{loc}([0, T); L^2(\Omega)^n) \ , \quad F \in L^2_{loc}([0, T); L^2(\Omega)^{n^2}),$$

and let

$$u \in L^\infty_{loc}([0, T); L^2_\sigma(\Omega)) \cap L^2_{loc}([0, T); W^{1,2}_{0,\sigma}(\Omega))$$

be a weak solution of the Navier-Stokes system (3.4.1) with data f, u_0.

If $n = 3$, assume additionally that

$$u\,u \in L^2_{loc}([0, T); L^2(\Omega)^{n^2}) . \tag{3.6.1}$$

Then, after redefinition on a null set, $u : [0, T) \to L^2_\sigma(\Omega)$ is strongly continuous and satisfies the generalized energy equality

$$\begin{aligned} &\frac{1}{2}\, \|u(t)\|_2^2 \; + \; \nu \int_s^t \|\nabla u\|_2^2 \, d\tau \\ &\quad = \frac{1}{2}\, \|u(s)\|_2^2 \; + \int_s^t < f_0, u >_\Omega \, d\tau \; - \int_s^t < F, \nabla u >_\Omega \, d\tau \end{aligned} \tag{3.6.2}$$

for all s, t with $0 \le s \le t < T$.

Proof. See the proof of Theorem 1.4.1 now with $u(0) = u_0$ replaced by $u(s)$, $s \ge 0$. □

For bounded domains we know the following result.

3.6.2 Theorem *Let $\Omega \subseteq \mathbb{R}^3$ be a three-dimensional bounded domain, let* $0 < T \leq \infty$, $u_0 \in L^2_\sigma(\Omega)$, $f = f_0 + \operatorname{div} F$ *with*

$$f_0 \in L^1_{loc}([0,T); L^2(\Omega)^3) \ , \quad F \in L^2_{loc}([0,T); L^2(\Omega)^9),$$

and let

$$u \in L^\infty_{loc}([0,T); L^2_\sigma(\Omega)) \cap L^2_{loc}([0,T); W^{1,2}_{0,\sigma}(\Omega))$$

be a suitable weak solution of the Navier-Stokes system (3.4.1) *with data* f, u_0 *as in Theorem* 3.4.1.

Then $u : [0,T) \to L^2_\sigma(\Omega)$ *is weakly continuous after a redefinition on a null set of* $[0,T)$, *and*

$$\begin{aligned}\frac{1}{2}\|u(t)\|_2^2 \ + \ \nu \int_s^t \|\nabla u\|_2^2 \, d\tau \ &\leq \ \frac{1}{2}\|u(s)\|_2^2 \\ &+ \int_s^t < f_0, u >_\Omega \, d\tau - \int_s^t < F, \nabla u >_\Omega \, d\tau\end{aligned} \tag{3.6.3}$$

for almost all $s \geq 0$, *including* $s = 0$, *and all* t *with* $s \leq t < T$.

Proof. Let $(u_k)_{k=1}^\infty$ be the same sequence as used in the proof of Theorem 3.1.1 for the construction of u. It is sufficient to prove the assertion with T replaced by T', $0 < T' < T$. Then $T' < \infty$, and we get

$$f \in L^1([0,T'); L^2(\Omega)^3) \ , \quad F \in L^2([0,T'); L^2(\Omega)^9) \ .$$

Therefore, we may assume that $(u_k)_{k=1}^\infty$ converges to u strongly in $L^2(0,T'; L^2_\sigma(\Omega))$, see (3.3.2).

Consider any continuous function $\phi : [0,T') \to \mathbb{R}$ with $\phi_t \in L^\infty(0,T';\mathbb{R})$. As in the proof of Theorem 3.4.1, we obtain the weighted energy equality

$$\begin{aligned}\frac{1}{2}\phi^2(t)\|u_k(t)\|_2^2 + \nu \int_0^t \phi^2 \|\nabla u_k\|_2^2 \, d\tau \\ = \ \frac{1}{2}\phi^2(0)\|u_0\|_2^2 + \int_0^t \phi^2 < f_0, u_k >_\Omega \, d\tau \\ - \int_0^t \phi^2 < F, \nabla u_k >_\Omega \, d\tau + \int_0^t \dot{\phi}\, \phi \|u_k\|_2^2 \, d\tau\end{aligned}$$

for all $t \in [0,T')$ and $k \in \mathbb{N}$. The last term on the right side is problematic if we consider the limit as $k \to \infty$. Here we need the strong convergence in $L^2(0,T';L^2_\sigma(\Omega))$. Using (3.3.12), (3.3.13) and the convergence properties in (3.3.2), we obtain, letting $k\to\infty$, the inequality

$$\begin{aligned}\frac{1}{2}\,\phi^2(t)\|u(t)\|_2^2 \;+\; \nu\int_0^t \phi^2\|\nabla u\|_2^2\,d\tau \qquad\qquad (3.6.4)\\ \le\; \frac{1}{2}\,\phi^2(0)\|u_0\|_2^2 + \int_0^t \phi^2 < f_0, u >_\Omega\,d\tau\\ - \int_0^t \phi^2 < F,\nabla u >_\Omega\,d\tau \;+\; \int_0^t \dot{\phi}\,\phi\|u\|_2^2\,d\tau\end{aligned}$$

for almost all $t\in[0,T')$. After a corresponding redefinition, u is weakly continuous and (3.6.4) holds for all $t\in T'$.

Since $\tau\mapsto\|u(\tau)\|_2^2$, $\tau\in[0,T')$, is integrable, we get

$$\lim_{\varepsilon\to0}\frac{1}{\varepsilon}\int_s^{s+\varepsilon}\|u(\tau)\|_2^2\,d\tau = \|u(s)\|_2^2\;,\quad \varepsilon>0$$

for almost all $s\in[0,T')$. A value s satisfying this condition is called a **Lebesgue point**, see [HiPh57, Sec. 3.8, Theorem 3.8.5, Cor. 2], [Yos80, Chap. V, 5, Theorem 2], or [Miz73, Chap. 2, (2.17)].

For such a Lebesgue point $s\in[0,T')$ we can use the following elementary calculations. Set $v(\tau):=\|u(\tau)\|_2^2-\|u(s)\|_2^2$. Then we get

$$\begin{aligned}\lim_{\varepsilon\to0}\left(\frac{1}{\varepsilon^2}\int_s^{s+\varepsilon}(\tau-s)\,d\tau\right) &= \frac{1}{2}\,,\\ \lim_{\varepsilon\to0}\left(\frac{1}{\varepsilon}\int_s^{s+\varepsilon}v(\tau)\,d\tau\right) &= 0\,,\end{aligned}$$

$$\begin{aligned}&\left|\lim_{\varepsilon\to0}\left(\frac{1}{\varepsilon^2}\int_s^{s+\varepsilon}\left(\int_s^\tau v(\rho)\,d\rho\right)d\tau\right)\right|\\ &= \lim_{\varepsilon\to0}\left|\frac{1}{\varepsilon}\int_s^{s+\varepsilon}\frac{\tau-s}{\varepsilon}\left(\frac{1}{\tau-s}\int_s^\tau v(\rho)\,d\rho\right)d\tau\right|\\ &\le \lim_{\varepsilon\to0}\left(\frac{1}{\varepsilon}\int_s^{s+\varepsilon}\left|\frac{1}{\tau-s}\int_s^\tau v(\rho)\,d\rho\right|d\tau\right) = 0\,,\end{aligned}$$

and using intergration by parts we conclude that

$$\begin{aligned}&\lim_{\varepsilon\to0}\left(\frac{1}{\varepsilon^2}\int_s^{s+\varepsilon}(\tau-s)v(\tau)\,d\tau\right)\\ &= \lim_{\varepsilon\to0}\left(\frac{1}{\varepsilon}\int_s^{s+\varepsilon}v(\tau)\,d\tau - \frac{1}{\varepsilon^2}\int_s^{s+\varepsilon}\left(\int_s^\tau v(\rho)\,d\rho\right)d\tau\right) = 0.\end{aligned}$$

This yields the relation

$$\lim_{\varepsilon\to 0}\left(\frac{1}{\varepsilon^2}\int_s^{s+\varepsilon}(\tau - s)\,\|u(\tau)\|_2^2\,d\tau\right) = \frac{1}{2}\,\|u(s)\|_2^2. \tag{3.6.5}$$

Consider any points $s, t \in [0, T')$ with $0 < s < t < T'$, any $\varepsilon > 0$ with $s + \varepsilon < t - \varepsilon$, and define the function $\phi_\varepsilon : \tau \mapsto \phi_\varepsilon(\tau)$, $\tau \in [0, T')$, by

$$\begin{aligned}
\phi_\varepsilon(\tau) &= 0 \text{ for } 0 \le \tau \le s\ , \quad \phi_\varepsilon(\tau) = \frac{1}{\varepsilon}(\tau - s) \text{ for } s < \tau \le s+\varepsilon,\\
\phi_\varepsilon(\tau) &= 1 \text{ for } s+\varepsilon \le \tau \le t-\varepsilon\ , \quad \phi_\varepsilon(\tau) = \frac{1}{\varepsilon}(t - \tau) \text{ for } t-\varepsilon < \tau \le t,\\
\phi_\varepsilon(\tau) &= 0 \text{ for } t \le \tau \le T'.
\end{aligned}$$

Then $\dot\phi_\varepsilon \in L^\infty(0, T'; \mathbb{R})$ and $\dot\phi_\varepsilon(\tau) = 0$ if $0 < \tau < s$, $\dot\phi_\varepsilon(\tau) = \frac{1}{\varepsilon}$ if $s < \tau < s + \varepsilon$, $\dot\phi_\varepsilon(\tau) = 0$ if $s + \varepsilon < \tau < t - \varepsilon$, $\dot\phi_\varepsilon(\tau) = -\frac{1}{\varepsilon}$ if $t - \varepsilon < \tau < t$, and $\dot\phi_\varepsilon(\tau) = 0$ if $t < \tau < T'$.

Next we assume that s is a Lebesgue point. Then (3.6.5) is satisfied. Correspondingly we obtain that for almost all $t \in (s, T')$ the relation

$$\lim_{\varepsilon\to 0}\left(\frac{1}{\varepsilon^2}\int_{t-\varepsilon}^{t}(t - \tau)\,\|u(\tau)\|_2^2\,d\tau\right) = \frac{1}{2}\,\|u(t)\|_2^2.$$

is satisfied. We choose t in such a way.

Now we insert ϕ_ε in (3.6.4) and let $\varepsilon \to 0$ in each term, using Lebesgue's dominated convergence lemma. For the last term we obtain

$$\begin{aligned}
&\lim_{\varepsilon\to 0}\int_0^t \dot\phi_\varepsilon\,\phi_\varepsilon\|u(\tau)\|_2^2\,d\tau\\
&= \lim_{\varepsilon\to 0}\left(\frac{1}{\varepsilon^2}\int_s^{s+\varepsilon}(\tau - s)\|u(\tau)\|_2^2\,d\tau\right) - \lim_{\varepsilon\to 0}\left(\frac{1}{\varepsilon^2}\int_{t-\varepsilon}^{t}(t - \tau)\|u(\tau)\|_2^2\,d\tau\right)\\
&= \frac{1}{2}\|u(s)\|_2^2 - \frac{1}{2}\|u(t)\|_2^2
\end{aligned}$$

This leads to the inequality (3.6.3) for s and t.

Since u is weakly continuous, see Theorem 3.1.1, and the term $\frac{1}{2}\|u(t)\|_2^2$ appears on the left side of (3.6.3), we can use the argument in (3.3.16) and obtain the validity of (3.6.3) for all $t \in [s, T')$. The validity of (3.6.3) for $s = 0$ follows from (3.1.2). The proof is complete. □

3.6.3 Remark The proof shows that for bounded domains $\Omega \subseteq \mathbb{R}^3$ the generalized energy inequality (3.6.3) is valid for each Lebesgue point $s \in [0, T)$, see (3.6.5), and for all $t \in [s, T)$.

4 Strong solutions of the Navier-Stokes system

4.1 The notion of strong solutions

Up to now we have investigated weak solutions of the Navier-Stokes system. Our next purpose is to prove the existence of strong solutions under some restrictions on the data f, u_0 if $n = 3$. Strong solutions satisfy by definition Serrin's regularity condition, see below. They are classical solutions, at least for $t > 0$, if the data and the domain are sufficiently smooth, see Section 1.8.

4.1.1 Definition *Let* $\Omega \subseteq \mathbb{R}^n$, $n = 2, 3$, *be any domain, let* $0 < T \leq \infty$, $u_0 \in L^2_\sigma(\Omega)$, $f = f_0 + \text{div } F$ *with*

$$f_0 \in L^1_{loc}([0,T); L^2(\Omega)^n) \ , \quad F \in L^2_{loc}([0,T); L^2(\Omega)^{n^2}),$$

and let

$$u \in L^\infty_{loc}([0,T); L^2_\sigma(\Omega)) \cap L^2_{loc}([0,T); W^{1,2}_{0,\sigma}(\Omega)),$$

be a weak solution of the Navier-Stokes system

$$u_t - \nu\Delta u + u \cdot \nabla u + \nabla p = f \ , \quad \text{div } u = 0 \ , \quad u|_{\partial\Omega} = 0 \ , \quad u(0) = u_0 \tag{4.1.1}$$

with data f, u_0.

Then u *is a called a* **strong solution** *of this system with data* f, u_0, *iff Serrin's condition*

$$u \in L^s_{loc}([0,T); L^q(\Omega)^n) \tag{4.1.2}$$

is satisfied with $n < q < \infty$, $2 < s < \infty$, $\frac{n}{q} + \frac{2}{s} \leq 1$.

Theorem 1.5.1 shows that such a strong solution is uniquely determined. Theorem 1.8.1 and Theorem 1.8.2 yield regularity properties of strong solutions. We also know, see Theorem 1.4.1 and (1.4.10), that each strong solution u satisfies the energy equality (1.4.3).

Concerning strong solutions we refer to [KiL63], [FuK64], [Lad69, Chap. 6, 3], [Hey76], [Sol77], [Tem77, Chap. III, Theorem 3.7], [Hey80], [Miy82], [Kat84], [vWa85], [Gig86], [KoO93], [KoO94], [KoY95], [DvW95], [Can95], [Wie99], [Ama00]. In the literature, the notion of a strong solution includes sometimes additional regularity properties.

If $n = 2$, we know that each weak solution as above is a uniquely determined strong solution, see Theorem 1.5.3.

4.2 Existence results

If $n = 2$, the following existence result follows by a combination of theorems which we already know.

4.2.1 Theorem $(n = 2)$ *Let $\Omega \subseteq \mathbb{R}^2$ be any two-dimensional domain, let $0 < T \leq \infty$, $u_0 \in L^2_\sigma(\Omega)$, and let $f = f_0 + \operatorname{div} F$ with*

$$f_0 \in L^1_{loc}([0,T); L^2(\Omega)^2) \ , \quad F \in L^2_{loc}([0,T); L^2(\Omega)^4).$$

Then there exists a uniquely determined weak solution

$$u \in L^\infty_{loc}([0,T); L^2_\sigma(\Omega)) \cap L^2_{loc}([0,T); W^{1,2}_{0,\sigma}(\Omega)),$$

of the Navier-Stokes system (4.1.1) *with data f, u_0; u satisfies Serrin's condition*

$$u \in L^4_{loc}([0,T); L^4(\Omega)^2) \tag{4.2.1}$$

and is therefore a strong solution of (4.1.1).

Proof. Theorem 3.1.1 shows the existence and Theorem 1.5.3 yields the uniqueness of u. To prove (4.2.1), we use Lemma 1.2.1, b), with $q = 4$, $s = 4$, $n = 2$, see (1.5.18). This proves the result. □

If $n = 3$, we can prove only a local existence result, see below. Let A be the Stokes operator and let $S(t) = e^{-tA}$, $t \geq 0$, see (1.5.7), IV. Using Lemma 1.5.1, IV, we see that

$$v = s - \lim_{t \to 0} e^{-tA} v \, , \tag{4.2.2}$$

and that

$$0 = s - \lim_{t \to \infty} e^{-tA} v \tag{4.2.3}$$

for each $v \in L^2_\sigma(\Omega)$ in the strong sense.

Therefore, the existence condition (4.2.5) below can always be satisfied if T' with $0 < T' \leq T$ is chosen sufficiently small. Thus we always obtain a strong solution in a certain sufficiently small initial interval $[0, T')$.

We will admit, in the following smallness condition (4.2.5), the case $T = T' = \infty$. We set by definition

$$e^{-2T'A} \ := \ 0 \ \text{if} \ \ T' = \infty \, . \tag{4.2.4}$$

This means that in this case the strong solution u exists in the whole interval $[0, \infty)$. Thus there exists a global strong solution u on $[0, \infty)$ if the data f, u_0 satisfy the smallness condition

$$\|f_0\|_{2,\frac{4}{3};\infty} + \nu^{-\frac{1}{2}} \|F\|_{2,4;\infty} + \|A^{\frac{1}{4}} u_0\|_2 \ \leq \ K \nu^{1+\frac{1}{4}} .$$

For $n = 3$ it is an open problem, whether a strong solution u exists on the whole given interval without any smallness condition.

4.2.2 Theorem ($n = 3$) *Let $\Omega \subseteq \mathbb{R}^3$ be any three-dimensional domain, let $0 < T \leq \infty$, $u_0 \in D(A^{\frac{1}{4}})$, and let $f = f_0 + \operatorname{div} F$ with*

$$f_0 \in L^{\frac{4}{3}}(0,T;L^2(\Omega)^3) \ , \quad F \in L^4(0,T;L^2(\Omega)^9).$$

Then there exists a constant $K > 0$, not depending on Ω, u_0, f, T and ν, with the following property:

Choose any T', $0 < T' \leq T$, such that

$$\begin{aligned} \|f_0\|_{2,\frac{4}{3};T'} + \nu^{-\frac{1}{2}}\|F\|_{2,4;T'} + \|(I - e^{-2T'A})A^{\frac{1}{4}}u_0\|_2^{\frac{1}{8}}\,\|A^{\frac{1}{4}}u_0\|_2^{\frac{7}{8}} \qquad (4.2.5) \\ \leq K\,\nu^{1+\frac{1}{4}}. \end{aligned}$$

Then in the interval $[0,T')$ there exists a uniquely determined strong solution

$$u \in L^\infty_{loc}([0,T');L^2_\sigma(\Omega)) \cap L^2_{loc}([0,T');W^{1,2}_{0,\sigma}(\Omega)),$$

of the Navier-Stokes system (4.1.1) with data f, u_0. The solution u satisfies Serrin's condition

$$u \in L^8(0,T';L^4(\Omega)^3) \qquad (4.2.6)$$

with $\frac{3}{4} + \frac{2}{8} = 1$, and additionally

$$\nabla u \in L^4(0,T';L^2(\Omega)^9) \ , \quad u\,u \in L^4(0,T';L^2(\Omega)^9). \qquad (4.2.7)$$

4.2.3 Remarks In the literature there are several other approaches to the existence of strong solutions, see [KiL63], [FuK64], [Sol77], [Hey80], [Miy82], [vWa85], [KoO94], [KoY95], [Wie99], [Ama00].

In particular we mention the results of Fujita-Kato [FuK64] and Kozono-Ogawa [KoO94] for smooth bounded domains which requires the same condition $u_0 \in D(A^{\frac{1}{4}})$ on the initial value u_0 as the theorem above. Thus we may consider Theorem 4.2.2 as an extension of these results to completely general domains Ω and to more general forces f.

One aspect of Theorem 4.2.2 is that the conditions on the data f, u_0 are optimal concerning the regularity of the solution. If $0 < T' < T$, the solution u has, maybe, an extension to the whole interval $[0,T)$ – this is an open problem – but the regularity of u given by (4.2.6), (4.2.7) is optimal for the given class of data.

Another aspect is that K in (4.2.5) is an absolute constant which does not depend on anything. In particular, K does not depend on the domain Ω. Therefore, the existence interval $[0,T')$ can be chosen independently of Ω, if (4.2.5) holds uniformly for all Ω, see Corollary 4.2.4. This enables us, for example, to approximate a given nonsmooth domain by a sequence of smooth domains, and to investigate the convergence of the corresponding strong solutions on a fixed interval $[0,T')$.

Proof. Let $0 < T' \leq T$. From Theorem 1.3.1 we know that

$$u \in L^\infty_{loc}([0,T'); L^2_\sigma(\Omega)) \cap L^2_{loc}([0,T');\ W^{1,2}_{0,\sigma}(\Omega))$$

is a weak solution of the system (4.1.1) with data $f,\ u_0$, iff the integral equation

$$\begin{aligned} u(t) \;&=\; S(t)u_0 + \int_0^t S(t-\tau)Pf_0(\tau)\,d\tau \\ &\quad + A^{\frac{1}{2}}\int_0^t S(t-\tau)A^{-\frac{1}{2}}P \operatorname{div} F(\tau)\,d\tau \\ &\quad - A^{\frac{1}{2}}\int_0^t S(t-\tau)A^{-\frac{1}{2}}P \operatorname{div}(u(\tau)u(\tau))\,d\tau \end{aligned} \tag{4.2.8}$$

is satisfied for almost all $t \in [0,T')$. □

Using the integral operator $\mathcal{J}$, see (1.6.3), IV, we define the expressions V, V_1, V_2, V_3, and U by setting

$$\begin{aligned} V_1 &:= S(\cdot)u_0, \\ V_2 &:= \mathcal{J}Pf_0, \\ V_3 &:= A^{\frac{1}{2}}\mathcal{J}A^{-\frac{1}{2}}P \operatorname{div} F, \\ V &:= V_1+V_2+V_3, \\ U &:= A^{\frac{1}{2}}\mathcal{J}A^{-\frac{1}{2}}P \operatorname{div}(u\,u). \end{aligned}$$

Then we get $u = V - U$, and U satisfies the integral equation

$$U = A^{\frac{1}{2}}\mathcal{J}A^{-\frac{1}{2}}P \operatorname{div}(V-U)(V-U). \tag{4.2.9}$$

Our aim is to solve this equation with Banach's fixed point principle. The arguments are similar to those in the proof of Theorem 2.5.1.

For this purpose we define the Banach space $X_{T'}$ as the space of all $U : [0,T') \to L^2_\sigma(\Omega)$ with the following properties:

$U(t) \in D(A^{-\frac{1}{2}})$ for all $t \in [0,T')$, $A^{-\frac{1}{2}}U : [0,T') \to L^2_\sigma(\Omega)$ is strongly continuous, $(A^{-\frac{1}{2}}U)(0) = 0$, $(A^{-\frac{1}{2}}U)_t \in L^4(0,T';L^2_\sigma(\Omega))$, and $A^{\frac{1}{2}}U \in L^4(0,T';L^2_\sigma(\Omega))$.

The norm in $X_{T'}$ is defined by

$$|||U|||_{T'} := \|(A^{-\frac{1}{2}}U)_t\|_{2,4;T'} + \|A^{\frac{1}{2}}U\|_{2,4;T'}. \tag{4.2.10}$$

Set

$$\mathcal{F}_{T'}(U) := A^{\frac{1}{2}}\mathcal{J}A^{-\frac{1}{2}}P \text{ div } (V-U)(V-U) \tag{4.2.11}$$

for all $U \in X_{T'}$. We will show that Banach's fixed point principle is applicable to the equation

$$U = \mathcal{F}_{T'}(U).$$

To prove this we have to prepare several estimates. Let $U \in X_{T'}$. Using (2.6.4), III, we get the inequality

$$\|A^{-\frac{1}{2}}P \text{ div } (V-U)(V-U)\|_2 \leq \nu^{-\frac{1}{2}}\|(V-U)(V-U)\|_2 \tag{4.2.12}$$

for almost all $t \in [0,T)$. Applying Theorem 2.5.3, a), IV, we get with $s=4$ the estimate

$$|||\mathcal{F}_{T'}(U)|||_{T'} \leq C\nu^{-\frac{1}{2}}\|(V-U)(V-U)\|_{2,4;T'} \tag{4.2.13}$$

with some absolute constant $C>0$. Using Hölder's inequality leads to

$$\|(V-U)(V-U)\|_{2,4;T'} \leq C\,\|V-U\|_{4,8;T'}\|V-U\|_{4,8;T'} \tag{4.2.14}$$

with some absolute constant $C>0$.

Using the embedding inequality (2.4.6), III, with $\alpha = \frac{3}{8}$, $q=4$, $2\alpha+\frac{3}{4} = \frac{3}{2}$, we obtain

$$\|V-U\|_{4,8;T'} \leq C\nu^{-\frac{3}{8}}\|A^{\frac{3}{8}}(V-U)\|_{2,8;T'} \tag{4.2.15}$$

with some absolute constant $C>0$.

From (2.5.20), (2.5.22), IV, we obtain the representation

$$(A^{-\frac{1}{2}}U)(t) = \int_0^t S(t-\tau)((A^{-\frac{1}{2}}U)_\tau + A^{\frac{1}{2}}U)\,d\tau \ , \quad t \in [0,T').$$

We apply Theorem 2.5.2, b), IV, with $\rho=8$, $s=4$, $\alpha = \frac{3}{8}+\frac{1}{2}$. This yields

$$\begin{aligned} \|A^{\frac{3}{8}}U\|_{2,8;T'} &\leq C\,\|(A^{-\frac{1}{2}}U)_t + A^{\frac{1}{2}}U\|_{2,4;T'} \\ &\leq C\,|||U|||_{T'} \end{aligned} \tag{4.2.16}$$

with some $C>0$ as above.

Next we apply Theorem 2.5.2, b), IV, to $V_2 = \mathcal{J}Pf_0$, with $\rho=8$, $s=\frac{4}{3}$, $\alpha = \frac{3}{8} = 1+\frac{1}{8}-\frac{3}{4}$, and obtain

$$\|A^{\frac{3}{8}}V_2\|_{2,8;T'} \leq C\,\|f_0\|_{2,4/3;T'} \tag{4.2.17}$$

with $C>0$ as above.

To $V_3 = A^{\frac{1}{2}}\mathcal{J}A^{-\frac{1}{2}}P \operatorname{div} F$ we apply Theorem 2.5.3, b), IV, with $\rho = 8$, $s = 4$, $\alpha = \frac{3}{8} + \frac{1}{2} = 1 + \frac{1}{8} - \frac{1}{4}$, and get

$$\|A^{\frac{3}{8}}V_3\|_{2,8;T'} \le C\nu^{-\frac{1}{2}}\|F\|_{2,4;T'} \tag{4.2.18}$$

with $C > 0$ as above.

To estimate $V_1 = S(\cdot)u_0$ we use a similar calculation as in the proof of Lemma 1.5.3, IV. Using the interpolation inequality (2.2.8), III, we get

$$\begin{aligned}\|A^{\frac{3}{8}}S(t)u_0\|_2 = \|(A^{\frac{1}{2}})^{\frac{1}{4}}\, S(t)A^{\frac{1}{4}}u_0\|_2 &\le \|A^{\frac{1}{2}}S(t)A^{\frac{1}{4}}u_0\|_2^{\frac{1}{4}}\, \|S(t)A^{\frac{1}{4}}u_0\|_2^{\frac{3}{4}} \\ &\le \|A^{\frac{1}{2}}S(t)A^{\frac{1}{4}}u_0\|_2^{\frac{1}{4}}\, \|A^{\frac{1}{4}}u_0\|_2^{\frac{3}{4}}.\end{aligned}$$

This yields

$$\|A^{\frac{3}{8}}V_1\|_{2,8;T'} \le \|A^{\frac{1}{2}}S(\cdot)A^{\frac{1}{4}}u_0\|_{2,2;T'}^{\frac{1}{4}}\, \|A^{\frac{1}{4}}u_0\|_2^{\frac{3}{4}}.$$

Further we get

$$\begin{aligned}\|A^{\frac{1}{2}}S(\cdot)A^{\frac{1}{4}}u_0\|_{2,2;T'}^2 &= \int_0^{T'} \|A^{\frac{1}{2}}S(t)A^{\frac{1}{4}}u_0\|_2^2\, dt \\ &= \int_0^{T'} < AS^2(t)A^{\frac{1}{4}}u_0, A^{\frac{1}{4}}u_0 >_\Omega\, dt \\ &= -\frac{1}{2}\int_0^{T'} (\frac{d}{dt} < S^2(t)A^{\frac{1}{4}}u_0, A^{\frac{1}{4}}u_0 >_\Omega)\, dt \\ &= \frac{1}{2} < (I - S^2(T'))A^{\frac{1}{4}}u_0, A^{\frac{1}{4}}u_0 >_\Omega \\ &\le \|(I - S^2(T'))\, A^{\frac{1}{4}}u_0\|_2\, \|A^{\frac{1}{4}}u_0\|_2,\end{aligned}$$

and therefore we obtain

$$\begin{aligned}\|A^{\frac{3}{8}}V_1\|_{2,8;T'} &\le \|A^{\frac{1}{2}}S(\cdot)A^{\frac{1}{4}}u_0\|_{2,2;T'}^{2\cdot\frac{1}{8}}\, \|A^{\frac{1}{4}}u_0\|_2^{\frac{3}{4}} \\ &\le \|(I - S^2(T'))A^{\frac{1}{4}}u_0\|_2^{\frac{1}{8}}\, \|A^{\frac{1}{4}}u_0\|_2^{\frac{1}{8}}\, \|A^{\frac{1}{4}}u_0\|_2^{\frac{3}{4}} \\ &= \|(I - S^2(T'))A^{\frac{1}{4}}u_0\|_2^{\frac{1}{8}}\, \|A^{\frac{1}{4}}u_0\|_2^{\frac{7}{8}}.\end{aligned}$$

Combining the last inequality with (4.2.17), (4.2.18), we get with $V = V_1 + V_2 + V_3$ that

$$\begin{aligned}\|A^{\frac{3}{8}}V\|_{2,8;T'} &\le \|(I - S^2(T'))A^{\frac{1}{4}}u_0\|_2^{\frac{1}{8}}\, \|A^{\frac{1}{4}}u_0\|_2^{\frac{7}{8}} \\ &+ C(\|f_0\|_{2,4/3;T'} + \nu^{-\frac{1}{2}}\|F\|_{2,4:T'}).\end{aligned}$$

Combining (4.2.13), (4.2.14), (4.2.15), we obtain

$$|||\mathcal{F}_{T'}(U)|||_{T'} \leq C\nu^{-\frac{1}{2}-\frac{3}{4}} \|A^{\frac{3}{8}}(V-U)\|_{2,8;T'}^2 \tag{4.2.19}$$

with some absolute constant $C > 0$. Applying (4.2.16) and (4.2.19) leads to

$$\begin{aligned} |||\mathcal{F}_{T'}(U)|||_{T'} \leq \; & C\nu^{-\frac{1}{2}-\frac{3}{4}} \Big(\|(I - S^2(T'))A^{\frac{1}{4}}u_0\|_2^{\frac{1}{8}} \|A^{\frac{1}{4}}u_0\|_2^{\frac{7}{8}} \\ & + \|f_0\|_{2,4/3;T'} + \nu^{-\frac{1}{2}}\|F\|_{2,4;T'} + |||U|||_{T'} \Big)^2 \end{aligned} \tag{4.2.20}$$

with some absolute constant $C > 0$.

We set

$$\begin{aligned} a &:= C\nu^{-1-\frac{1}{4}}, \\ b &:= \|(I - S^2(T'))\,A^{\frac{1}{4}}u_0\|_2^{\frac{1}{8}} \|A^{\frac{1}{4}}u_0\|_2^{\frac{7}{8}} + \|f_0\|_{2,4/3;T'} + \nu^{-\frac{1}{2}}\|F\|_{2,4;T'} \end{aligned}$$

and get the inequality

$$|||\mathcal{F}_{T'}(U)|||_{T'} \leq a\,(|||U|||_{T'} + b)^2 \tag{4.2.21}$$

for all $U \in X_{T'}$.

Next we consider the equation

$$y = ay^2 + b\,, \quad y^2 - \frac{1}{a}y + \frac{b}{a} = 0\,, \quad y > 0.$$

Suppose that

$$4\,ab < 1. \tag{4.2.22}$$

Then the minimal root of this equation is given by

$$y_1 = \frac{1}{2a}(1 - \sqrt{1-4ab}) = 2b(1 + \sqrt{1-4ab})^{-1} > 0.$$

It follows that $y_1 < 2b$. See [Sol77, Lemma 10.2] for this argument, see also (2.5.9). We define the closed subset

$$D_{T'} := \{U \in X_{T'}\,; |||U|||_{T'} + b \leq y_1\}.$$

Since $y_1 = ay_1^2 + b > b$, we see that $D_{T'}$ is not empty.

Let $U \in D_{T'}$. Then from (4.2.21) we get

$$|||\mathcal{F}_{T'}(U)|||_{T'} + b \leq a\,(|||U|||_{T'} + b)^2 + b \leq ay_1^2 + b = y_1$$

and therefore that $\mathcal{F}_{T'}(U) \in D_{T'}$.

Consider any $U, W \in D_{T'}$. Then we obtain from (4.2.11) that

$$\begin{aligned} & \mathcal{F}_{T'}(U) - \mathcal{F}_{T'}(W) \\ = \; & A^{\frac{1}{2}} \mathcal{J} A^{-\frac{1}{2}} P \operatorname{div} \left[(V-U)(W-U) + (W-U)(V-W) \right]. \end{aligned}$$

The same estimates which lead above to (4.2.20) and (4.2.21) yield now the inequality

$$\begin{aligned} & |||\mathcal{F}_{T'}(U) - \mathcal{F}_{T'}(W)|||_{T'} \\ \leq \; & a \left[(|||U|||_{T'} + b) \, |||U - W|||_{T'} + |||U - W|||_{T'} \, (|||W|||_{T'} + b) \right] \\ \leq \; & a \, (y_1 + y_1) \, |||U - W|||_{T'} \; \leq \; 4 \, ab \, |||U - W|||_{T'}. \end{aligned}$$

Since $4ab < 1$, we are able to apply Banach's fixed point principle and obtain some $U \in D_{T'}$ satisfying $U = \mathcal{F}_{T'}(U)$. We define

$$u := V - U.$$

In the next step we prove some regularity properties of U and u. Using (4.2.13), (4.2.14), (4.2.15), (4.2.16), (4.2.17), (4.2.18), we obtain that

$$\|(V-U)(V-U)\|_{2,4;T'} < \infty \; , \quad \|V - U\|_{4,8;T'} < \infty,$$

and

$$\begin{aligned} \|V\|_{4,8;T'} &\leq C \nu^{-\frac{3}{8}} \, \|A^{\frac{3}{8}} V\|_{2,8;T'} < \infty \, , \\ \|U\|_{4,8;T'} &\leq C \nu^{-\frac{3}{8}} \, \|A^{\frac{3}{8}} U\|_{2,8;T'} < \infty \end{aligned}$$

with $C > 0$. Since $u = V - U$, we obtain the property (4.2.6).

Similarly we get

$$\begin{aligned} \|UU\|_{2,4;T'} < \infty \quad &, \quad \|VV\|_{2,4;T'} < \infty, \\ \|UV\|_{2,4;T'} < \infty \quad &, \quad \|VU\|_{2,4;T'} < \infty. \end{aligned}$$

This shows that $\|u \, u\|_{2,4;T'} < \infty$.

Since $|||U|||_{T'} < \infty$ we get

$$\|\nabla U\|_{2,4;T'} = \nu^{-\frac{1}{2}} \|A^{\frac{1}{2}} U\|_{2,4;T'} < \infty.$$

Applying (1.5.24), IV, with $s = 4$, u_0 replaced by $A^{\frac{1}{4}} u_0$, and using that $u_0 \in D(A^{\frac{1}{4}})$, we obtain

$$\|\nabla V_1\|_{2,4;T'} = \nu^{-\frac{1}{2}} \, \|A^{\frac{1}{2}} V_1\|_{2,4;T'} \leq \nu^{-\frac{1}{2}} \, \|A^{\frac{1}{4}} u_0\|_2 < \infty.$$

Applying (1.6.31), IV, with $\alpha = \frac{1}{2}$, $\rho = 4$, $s = \frac{4}{3}$, leads to

$$\|\nabla V_2\|_{2,4;T'} = \nu^{-\frac{1}{2}} \|A^{\frac{1}{2}} V_2\|_{2,4;T'} \le C\nu^{-\frac{1}{2}} \|f_0\|_{2,4/3;T'} < \infty.$$

Applying (2.5.21), IV, with $s = 4$ yields

$$\|\nabla V_3\|_{2,4;T'} = \nu^{-\frac{1}{2}} \|A^{\frac{1}{2}} V_3\|_{2,4;T'} \le C\nu^{-1} \|F\|_{2,4;T'} < \infty.$$

This leads to $\|\nabla u\|_{2,4;T'} < \infty$, and therefore we obtain (4.2.7).

Since $\|(V-U)(V-U)\|_{2,4;T'} < \infty$, we get $\|(V-U)(V-U)\|_{2,2;T'} < \infty$ if $T' < \infty$, and from (2.5.26), IV, we conclude with F replaced by $(V-U)(V-U)$ that

$$U \in L^\infty_{loc}([0,T'); L^2_\sigma(\Omega)) \cap L^2_{loc}([0,T'); W^{1,2}_{0,\sigma}(\Omega)).$$

Using (2.5.10), (2.5.17), IV, and again (2.5.26), IV, we see that

$$V \in L^\infty_{loc}([0,T'); L^2_\sigma(\Omega)) \cap L^2_{loc}([0,T'); W^{1,2}_{0,\sigma}(\Omega)).$$

This proves that

$$u \in L^\infty_{loc}([0,T'); L^2_\sigma(\Omega)) \cap L^2_{loc}([0,T'); W^{1,2}_{0,\sigma}(\Omega)).$$

Since $U = \mathcal{F}_{T'}(U)$, U satisfies (4.2.9), and $u = V - U$ satisfies the integral equation (4.2.8) for almost all $t \in [0, T')$.

Thus u is a weak solution of the system (4.1.1) with data f, u_0, see Theorem 1.3.1. Since Serrin's condition (4.2.6) holds, u is uniquely determined, see Theorem 1.5.1.

Set $K := (8C)^{-1}$ with C from (4.2.20). Then the condition (4.2.5) implies (4.2.22).

The proof is complete. □

The above proof yields several further properties of u. They can be derived from the above representation

$$u = V - U$$

where $V = S(\cdot)u_0 + \mathcal{J}Pf_0 + A^{\frac{1}{2}}\mathcal{J}A^{-\frac{1}{2}}P \operatorname{div} F$ is explicitly represented by the data f, u_0, and where U is contained in $X_{T'}$ so that

$$\|(A^{-\frac{1}{2}}U)_t\|_{2,4;T'} + \|A^{\frac{1}{2}}U\|_{2,4;T'} < \infty.$$

The smallness condition (4.2.5) for the existence of a strong condition in the interval $[0, T')$ is rather complicated. The following property (4.2.23) is an easy sufficient condition. Below, see the next proof, we will see that (4.2.5) in Theorem 4.2.2 can be replaced by the stronger condition

$$\|f_0\|_{2,\frac{4}{3};T'} + \nu^{-\frac{1}{2}}\|F\|_{2,4;T'} + 4(T')^{\frac{1}{32}}(\nu^{\frac{1}{2}}\|\nabla u_0\|_2 + \|u_0\|_2) \tag{4.2.23}$$
$$\leq K\,\nu^{1+\frac{1}{4}},$$

but here the case $T' = \infty$ is excluded.

Thus the following result holds:

Let $\Omega \subseteq \mathbb{R}^3$ be any domain, let $0 < T \leq \infty$, $u_0 \in W_{0,\sigma}^{1,2}(\Omega)$, and let $f = f_0 + \operatorname{div} F$ with

$$f_0 \in L^{\frac{4}{3}}(0,T;L^2(\Omega)^3)\ ,\quad F \in L^4(0,T;L^2(\Omega)^9).$$

Then there exists a constant $K > 0$, not depending on Ω, u_0, f, T, ν, with the following property:

Let T', $0 < T' < T$, satisfy the condition (4.2.23). Then there exists a uniquely determined strong solution

$$u \in L^\infty(0,T';L^2_\sigma(\Omega)) \cap L^2(0,T';W_{0,\sigma}^{1,2}(\Omega))$$

of the Navier-Stokes system (4.1.1) with data f, u_0, satisfying (4.2.6) and (4.2.7).

As in (4.2.5), K in (4.2.23) is an absolute constant which in particular does not depend on the domain Ω. Therefore we can extend slightly this formulation and get as a corollary of Theorem 2.2.2 the following result:

4.2.4 Corollary *Let $0 < T < \infty$, and for each domain $\Omega \subseteq \mathbb{R}^3$, choose some initial value $u_{0,\Omega} \in W_{0,\sigma}^{1,2}(\Omega)$ and some exterior force $f_\Omega = f_{0,\Omega} + \operatorname{div} F_\Omega$ with*

$$f_{0,\Omega} \in L^{\frac{4}{3}}(0,T;L^2(\Omega)^3)\ ,\quad F_\Omega \in L^4(0,T;L^2(\Omega)^9).$$

Then there exists a constant $K > 0$, not depending on Ω, $u_{0,\Omega}$, f_Ω, T, ν, with the following property:

If

$$\|f_{0,\Omega}\|_{2,\frac{4}{3};T} + \nu^{-\frac{1}{2}}\|F_\Omega\|_{2,4;T} + 4T^{\frac{1}{32}}(\nu^{\frac{1}{2}}\|\nabla u_{0,\Omega}\|_2 + \|u_{0,\Omega}\|_2) \tag{4.2.24}$$
$$\leq K\,\nu^{1+\frac{1}{4}}$$

is satisfied for all domains $\Omega \subseteq \mathbb{R}^3$, then for each such domain there exists a uniquely determined strong solution u_Ω in the (fixed) interval $[0,T)$ satisfying

$$\begin{aligned} u_\Omega &\in L^8(0,T;L^4(\Omega)^3),\\ \nabla u_\Omega &\in L^4(0,T;L^2(\Omega)^9),\\ u_\Omega u_\Omega &\in L^4(0,T;L^2(\Omega)^9). \end{aligned}$$

Proof. Let $u_0 = u_{0,\Omega}$. Using (1.5.15), IV, we obtain

$$
\begin{aligned}
\|(I - e^{-2TA})A^{\frac{1}{4}}u_0\|_2 &= \left\| \left(\int_0^{2T} \frac{d}{dt} e^{-tA}\, dt \right) A^{\frac{1}{4}} u_0 \right\|_2 \\
&= \left\| \int_0^{2T} A e^{-tA} A^{\frac{1}{4}} u_0 dt \right\|_2 \\
&= \left\| \int_0^{2T} A^{\frac{3}{4}} e^{-tA} A^{\frac{1}{2}} u_0 dt \right\|_2 \\
&\leq \left(\int_0^{2T} t^{-\frac{3}{4}}\, dt \right) \|A^{\frac{1}{2}} u_0\|_2 \\
&= 4(2T)^{\frac{1}{4}}\, \|A^{\frac{1}{2}} u_0\|_2 \,.
\end{aligned}
$$

The interpolation inequality (2.2.8), III, yields

$$
\|A^{\frac{1}{4}} u_0\|_2 = \|(A^{\frac{1}{2}})^{\frac{1}{2}} u_0\|_2 \leq \|A^{\frac{1}{2}} u_0\|_2^{\frac{1}{2}}\, \|u_0\|_2^{\frac{1}{2}} \leq \|A^{\frac{1}{2}} u_0\|_2 + \|u_0\|_2 \,.
$$

Thus we obtain with Young's inequality (3.3.8), I, that

$$
\begin{aligned}
&\|(I - e^{-2TA})A^{\frac{1}{4}}u_0\|_2^{\frac{1}{8}}\, \|\, A^{\frac{1}{4}} u_0\|_2^{\frac{7}{8}} \\
&\leq\; 2^{\frac{1}{4}+\frac{1}{32}}\, T^{\frac{1}{32}}\, \|A^{\frac{1}{2}} u_0\|_2^{\frac{1}{8}}\, \|A^{\frac{1}{4}} u_0\|_2^{\frac{7}{8}} \\
&\leq\; 2T^{\frac{1}{32}}\, (\|A^{\frac{1}{2}} u_0\|_2 + \|A^{\frac{1}{4}} u_0\|_2) \\
&\leq\; 2T^{\frac{1}{32}}\, (2\|A^{\frac{1}{2}} u_0\|_2 + \|u_0\|_2) \\
&\leq\; 4T^{\frac{1}{32}}\, (\|A^{\frac{1}{2}} u_0\|_2 + \|u_0\|_2) \\
&=\; 4T^{\frac{1}{32}}\, (\nu^{\frac{1}{2}}\, \|\nabla u_0\|_2 + \|u_0\|_2) \,.
\end{aligned}
$$

Therefore, $u_{0,\Omega}$, $f_{0,\Omega}$, F_Ω satisfy the condition (4.2.5) for each domain Ω, now with T' replaced by T. The result follows from Theorem 4.2.2. □

Bibliography

[Ada75] R. A. Adams. Sobolev Spaces. Academic Press, New York, 1975.

[Agm65] S. Agmon. Lectures on Elliptic Boundary Value Problems. Van Nostrand Comp., New York, 1965.

[Ama95] H. Amann. Linear and Quasilinear Parabolic Problems. Birkhäuser Verlag, Basel, 1995.

[Ama00] H. Amann. On the strong solvability of the Navier-Stokes equations. J. Math. Fluid Mech., 2 (2000), 16–98.

[Apo74] T. M. Apostol. Mathematical Analysis. Addison-Wesley, Amsterdam, 1974.

[AsSo94] A. Ashyralyev, P. E. Sobolevskii. Well-Posedness of Parabolic Difference Equations. Birkhäuser Verlag, Basel, 1994.

[BdV95] H. Beirão da Veiga. A new regularity class for the Navier-Stokes equations in $\mathbb{R}^n$. Chin. Ann. Math., 16B:4 (1995), 407–412.

[BdV97] H. Beirão da Veiga. Remarks on the smoothness of the $L^\infty(0,T;L^3)$-solutions of the $3-D$ Navier-Stokes equations. Portugaliae Mathematica, 54, (1997), 381–391.

[BMi91] W. Borchers, T. Miyakawa. Algebraic L^2 decay for Navier-Stokes flows in exterior domains. Hiroshima Math. J., 21 (1991), 621–640.

[BMi92] W. Borchers, T. Miyakawa. L^2-decay for Navier-Stokes flows in unbounded domains, with applications to exterior stationary flows. Arch. Rat. Mech. Anal., 118 (1992), 273–295.

[Bog79] M. E. Bogovski. Solution of the first boundary value problem for the equation of continuity of an incompressible medium. Soviet Math. Dokl., 20 (1979), 1094–1098.

[Bog80] M. E. Bogovski. Solution of some vector analysis problems connected with operators div and grad. Trudy Seminar N. L. Sobolev, 80, Academia Nauk SSSR, Sibirskoe Otdelenie Mathematiki, Nowosibirsk, 1980, 5–40.

[BSo87] W. Borchers, H. Sohr. On the semigroup of the Stokes operator for exterior domains in L^q-spaces. Math. Z., 196 (1987), 415–425.

[BuBe67] P. L. Butzer, H. Berens. Semi-groups of Operators and Approximation. Springer-Verlag, Heidelberg, 1967.

[Can95] M. Cannone. Ondelettes, paraproduits et Navier-Stokes. Diderot Editeur, Paris, 1995.

[Catt61] L. Cattabriga. Su un problema al contorno relativo al sistema di equazioni di Stokes. Rend. Sem. Mat. Univ. Padova, 31 (1961), 308–340.

[CaV86] P. Cannarsa, V. Vespri. On maximal L^p-regularity for the abstract Cauchy problem. Bollettino U. M. I., 5-B (1986), 165–175.

[CKN82] L. Caffarelli, R. Kohn, L. Nirenberg. Partial regularity of suitable weak solutions of the Navier-Stokes equations. Comm. Pure Appl. Math., 35 (1982), 771–831.

[dRh60] G. de Rham. Variétés Différentiables, Hermann, Paris, 1960.

[deS64] L. de Simon. Un' applicazione della teoria degli integrali singolari allo studio delle equazioni differenziali astratte del primo ordine. Rend. Sem, Mat. Univ. Padova, 34 (1964), 205–232.

[DLi55] J. Deny, J. L. Lions. Les espaces du type de Beppo Levi. Ann. Inst. Fourier Grenoble, 5 (1955), 305–370.

[DoVe87] G. Dore, A. Venni. On the closedness of the sum of two closed operators. Math. Z., 196 (1987), 189–201.

[DvW95] P. Deuring, W. v. Wahl. Strong solutions of the Navier-Stokes equations in Lipschitz bounded domains. Math. Nachr., 171 (1995), 111–148.

[FSS93] R. Farwig, C. G. Simader, H. Sohr. An L^q-theory for weak solutions of the Stokes system in exterior domains. Math. Meth. Appl. Sc., 16 (1993), 707–723.

[FaS94a] R. Farwig, H. Sohr. The stationary and non-stationary Stokes system in exterior domains with non-zero divergence and non-zero boundary values. Math. Meth. Appl. Sc., 17 (1994), 269–291.

[FaS94b] R. Farwig, H. Sohr. Generalized resolvent estimates for the Stokes system in bounded and unbounded domains. J. Math. Soc. Japan, 46 (1994), 607–643.

[Fin65] R. Finn. On the exterior stationary problem for the Navier-Stokes equations. Arch. Rat. Mech. Anal., 19 (1965), 363–406.

[Fri69] A. Friedman. Partial Differential Equations. Holt, Rinehart and Winston, New York, 1969.

[FuK64] H. Fujita, T. Kato. On the Navier-Stokes initial value problem. Arch. Rat. Mech. Anal., 16 (1964), 269–315.

[FuM77] D. Fujiwara, H. Morimoto. An L_r-theorem of the Helmholtz decomposition of vector fields. J. Fac. Sc. Univ. Tokyo, Sec. Math., 24 (1977), 685–700.

[Fur80] A. V. Fursikov. On some problems of control and results concerning the unique solvability of a mixed boundary value problem for the three-dimensional Navier-Stokes and Euler systems. Dokl. Akad. Nauk SSSR, 252 (1980), 1066–1070.

[Gal94a] G. P. Galdi. An Introduction to the Mathematical Theory of the Navier-Stokes equations, Vol. I, Linearized Steady Problems. Springer-Verlag, New York, 1994.

[Gal94b] G. P. Galdi. An Introduction to the Mathematical Theory of the Navier-Stokes equations, Vol. II, Nonlinear Steady Problems. Springer-Verlag, New York, 1994.

[Gal98] G. P. Galdi. Slow motion of a body in a viscous incompressible fluid with application to particle sedimentation. Recent developments in partial differential equations. Quaderni di mathematica, vol. 2 (1998), 3–35, Universita' di Napoli.

[GaM86] G. P. Galdi, P. Maremonti. Monotonic decreasing and asymptotic behaviour of the kinematic energy for weak solutions of the Navier-Stokes equations in exterior domains. Arch. Rat. Mech. Anal., 94 (1986), 253–266.

[GaM88] G. P. Galdi, P. Maremonti. Sulla regolarità delle soluzioni deboli al sistema di Navier-Stokes in domini arbitrari. Annali dell' Universita' di Ferrara, 34, Sez. VII (1988), 59–73.

[GSi90] G. P. Galdi, C. G. Simader. Existence, uniqueness and L^q-estimates for the Stokes problem in exterior domains. Arch. Rat. Mech. Anal., 112 (1990), 291–318.

[GSS94] G. P. Galdi, C. G. Simader, H. Sohr. On the Stokes problem in Lipschitz domains. Annali di Matematica pura ed applicata., CLXVII (1994), 147–163.

[Gig81] Y. Giga. Analyticity of the semigroup generated by the Stokes operator in L_r-spaces. Math. Z., 178 (1981), 297–329.

[Gig85] Y. Giga. Domains of fractional powers of the Stokes operator in L_r spaces. Arch. Rat. Mech. Anal., 89 (1985), 251–265.

[Gig86] Y. Giga. Solutions for semilinear parabolic equations in L^p and regularity of weak solutions of the Navier-Stokes system. J. Diff. Equ., 62 (1986), 186–212.

[GiSe91] V. Girault, A. Sequeira. A well-posed problem for the exterior Stokes equations in two and three dimensions. Arch. Rat. Mech. Anal., 114 (1991), 313–333.

[GiSo89] Y. Giga, H. Sohr. On the Stokes operator in exterior domains. J. Fac. Sci. Univ. Tokyo, Sec. IA, Vol. 36 (1989), 103–130.

[GiSo91] Y. Giga, H. Sohr. Abstract L^p estimates for the Cauchy problem with applications to the Navier-Stokes equations in exterior domains. J. Funct. Anal., 102 (1991), 72–94.

[GiRa86] V. Girault, P. A. Raviart. Finite Element Methods for the Navier-Stokes Equations. Springer-Verlag, Heidelberg, 1986.

[Hei51] E. Heinz. Beiträge zur Störungstheorie der Spektralzerlegung. Math. Ann., 123 (1951), 415–438.

[Heu75] H. Heuser. Funktionalanalysis. Teubner, Stuttgart, 1975.

[HiPh57] E. Hille, R. S. Phillips. Functional Analysis and Semi-groups. Amer. Math. Soc., Providence, 1957.

[Hey76] J. G. Heywood. On uniqueness questions in the theory of viscous flow. Acta Math., 136 (1976), 61–102.

[Hey80] J. G. Heywood. The Navier-Stokes Equations: On the Existence, Regularity and Decay of Solutions. Indiana Univ. Math. J., 29 (1980), 639–681.

[Hey88] J. G. Heywood. Epochs of regularity for weak solutions of the Navier-Stokes equations in unbounded domains. Tohoku Math. J., 40 (1988), 293–313.

[HeW94] J. G. Heywood, O. D. Walsh. A counter-example concerning the pressure in the Navier-Stokes equations, as $t \to 0$. Pacific J. Math., 164 (1994), 351–359.

[HLi56] L. Hörmander, J. L. Lions. Sur la complétion par rapport à une intégrale de Dirichlet. Math. Scand., 4 (1956), 259–270.

[Hop41] E. Hopf. Ein allgemeiner Endlichkeitssatz der Hydrodynamik. Math. Ann., 117 (1941), 764–775.

[Hop50] E. Hopf. Über die Anfangswertaufgabe für die hydrodynamischen Grundgleichungen. Math. Nachr., 4 (1950), 213–231.

[Kat66] T. Kato. Perturbation theory for linear operators. Springer-Verlag, 1966.

[Kat84] T. Kato. Strong L^p-solutions of the Navier-Stokes equations in $\mathbb{R}^m$, with application to weak solutions. Math. Z., 187 (1984), 471–480.

[KiL63] A. A. Kiselev, O. A. Ladyzhenskaya. On the existence and uniqueness of solutions of the non-stationary problems for flows of non-compressible fluids. Amer. Math. Soc. Transl. Ser. 2, Vol. 24 (1963), 79–106.

[Kom67] Y. Komura. Nonlinear semi-groups in Hilbert spaces. J. Math. Soc. Japan, 19 (1967), 493–507.

[KoO93] H. Kozono, T. Ogawa. Two-dimensional Navier-Stokes flow in unbounded domains. Math. Ann., 297 (1993), 1–31.

[KoO94] H. Kozono, T. Ogawa. Global strong solutions and its decay properties for the Navier-Stokes equations in three dimensional domains with non-compact boundaries. Math. Z., 216 (1994), 1–30.

[KOS92] H. Kozono, T. Ogawa, H. Sohr. Asymptotic behaviour in L^r for weak solutions of the Navier-Stokes equations in exterior domains. manuscripta math., 74 (1992), 253–275.

[KoS96] H. Kozono, H. Sohr. Remark on uniqueness of weak solutions to the Navier-Stokes equations. Analysis, 16 (1996), 255–271.

[KoY95] H. Kozono, M. Yamazaki. Local and global unique solvability of the Navier-Stokes exterior problem with Cauchy data in the space $L^{n,\infty}$. Houston J. Math., 21 (1995), 755–799.

[Kre71] S. G. Krein. Linear Differential Equations in Banach Space. Amer. Math. Soc., Rhode Island, 1971.

[Lad69] O. A. Ladyzhenskaya. The Mathematical Theory of Viscous Incompressible Flow. Gordon and Breach, New York, 1969.

[Ler33] J. Leray. Etude de diverses équations intégrales non linéaires et de quelques problémes que pose l' hydrodynamique. J. Math. Pures Appl., 12 (1933), 1–82.

[Ler34] J. Leray. Sur le mouvement d' un liquide visqueux emplissant l' espace. Acta Math., 63 (1934), 193–248.

[LeSch34] J. Leray, J. Schauder. Topologie et équations fonctionelles. Ann. Sci. École Norm. Sup., 13 (1934), 45–78.

[LiMa72] J. L. Lions, E. Magenes. Non-Homogeneous Boundary Value Problems and Applications, Vol. I. Springer-Verlag, Heidelberg, 1972.

[Mar84] P. Maremonti. Asymptotic stability theorems for viscous fluid motions in exterior domains. Rend. Sem. Mat. Univ. Padova, 71 (1984), 35–72.

[Mas75] K. Masuda. On the stability of incompressible viscous fluid motions past objects. J. Math. Soc. Japan, 27 (1975), 294–327.

[Mas84] K. Masuda. Weak solutions of Navier-Stokes equations. Tohoku Math. J., 36 (1984), 623–646.

[Miy82] T. Miyakawa. On nonstationary solutions of the Navier-Stokes equations in an exterior domain. Hiroshima Math. J., 12 (1982), 115–140.

[MiSo88] T. Miyakawa, H. Sohr. On energy inequality, smoothness and large time behaviour in L^2 for weak solutions of the Navier-Stokes equations in exterior domains. Math. Z., 199 (1988), 455–478.

[Miz73] S. Mizohata. The theory of partial differential equations. Cambridge, University Press, 1973.

[Monn99] S. Monniaux. Uniqueness of mild solutions of the Navier-Stokes equations and maximal L^p-regularity. C. R. Acad. Sc. Paris, 328 (1999), 663–668.

[MSol97] P. Maremonti, V. A. Solonnikov. On nonstationary Stokes problem in exterior domains. Ann. Sc. Norm. Sup. Pisa, Ser. IV, Vol. XXIV (1997), 395–449.

[Nav1827] C. L. Navier. Mémoire sur les lois du mouvement des fluides. Mem. Acad. Sci. de France., 6 (1827), 389–440.

[Nec67] J. Nečas. Les méthodes directes en théorie des équations elliptiques. Academia, Éditions de l' Acad. Tchécoslovaque des Sciences, Prague, 1967.

[Nec67b] J. Nečas. Sur les normes equivalentes dans $W^{k,p}(\Omega)$ et sur la coercivite des formes formellement positives. Les Presses de l'Universite de Montreal, 1967, 102–128.

[Neu99] J. Neustupa. Partial regularity of weak solutions to the Navier-Stokes equations in the class $L^\infty(0,T;L^3(\Omega)^3)$. J. Math. Fluid Mech., 1 (1999), 1–17.

[Nir59] L. Nirenberg. On elliptic partial differential equations. Ann. Scuo. Norm. Sup. Pisa, 13 (1959), 115–162.

[NST99] S. A. Nazarov, M. Specovius-Neugebauer, G. Thäter. Full steady Stokes system in domains with cylindrical outlets. Math. Ann., 314 (1999), 729–762.

[Pil80] K. Pileckas. On spaces of solenoidal vectors. Zap. Nauch. Sem. Leningr. Otd. Mat. Inst. Steklova 96 (1980), 237–239.

[Pil96] K. Pileckas. Strong solutions of the steady nonlinear Navier-Stokes system in domains with exits to infinity. Rend. Sem. Math. Univ. Padova, 97 (1996), 72–107.

[PrS90] J. Prüss, H. Sohr. On operators with bounded imaginary powers in Banach spaces. Math. Z., 203 (1990), 429–452.

[Rau83] R. Rautmann. On optimum regularity of Navier-Stokes solutions at time $t = 0$. Math. Z., 184 (1983), 141–149.

[ReSi75] M. Reed, B. Simon. Methods of Modern Mathematical Physics, Vol. II, Fourier Analysis, Selfadjointness. Academic Press, New York, 1975.

[Sch86] M. E. Schonbek. Large-time behavior of solutions of the Navier-Stokes equations. Comm. P. D. E., 11 (1986), 753–763.

[Ser62] J. Serrin. On the interior regularity of weak solutions of the Navier-Stokes equations. Arch. Rat. Mech. Anal., 9 (1962), 187–196.

[Ser63] J. Serrin. The initial value problem for the Navier-Stokes equations. Univ. Wisconsin Press, Nonlinear problems, Ed. R. E. Langer, 1963.

[Shi74] M. Shinbrot. The energy for the Navier-Stokes system. Siam J. Math. An., 5 (1974), 948–954.

[ShK66] M. Shinbrot, S. Kaniel. The initial value problem for the Navier-Stokes equations. Arch. Rat. Mech. Anal., 21 (1966), 270–285.

[SiSo92] C. G. Simader, H. Sohr. A new approach to the Helmholtz decomposition and the Neumann problem in L^q-spaces for bounded and exterior domains. Ser. Adv. Math. Appl. Sci., 11, Mathematical problems relating to the Navier-Stokes equation, World Scientific Publ., 1992.

[SiSo96] C. G. Simader, H. Sohr. The Dirichlet problem for the Laplacian in bounded and unbounded domains. Longman, Edinbourgh, 1996.

[SiZ98] C. G. Simader, T. Ziegler. The weak Dirichlet and Neumann problems in L^q for the Laplacian and the Helmholtz decomposition in infinite cylinders and layers. Recent developments in partial differential equations, quaderni di matematica, vol. 2 (1998), 37–161, Università di Napoli.

[Soh83] H. Sohr. Zur Regularitätstheorie der instationären Gleichungen von Navier-Stokes. Math. Z., 184 (1983), 359–375.

[Soh84] H. Sohr. Optimale lokale Existenzsätze für die Gleichungen von Navier-Stokes. Math. Ann., 267 (1984), 107–123.

[Soh99] H. Sohr. A special class of weak solutions of the Navier-Stokes equations in arbitrary three-dimensional domains. Progress in Nonlinear Differential Equations and their Applications, 35, Birkhäuser Verlag, Basel, 1999.

[SoS73] V. A. Solonnikov, V. E. Scadilov. On a boundary value problem for a stationary system of Navier-Stokes equations. Trudy Mat. Inst. Steklov, 125 (1973), 186–199.

[Sol68] V. A. Solonnikov. Estimates of the solutions of a nonstationary linearized system of Navier-Stokes equations. Am. Math. Soc. Transl., Ser. 2, 75 (1968), 1–116.

[Sol77] V. A. Solonnikov. Estimates for solutions of nonstationary Navier-Stokes equations. J. Soviet Math., 8 (1977), 467–529.

[Spe86] M. Specovius-Neugebauer. Exterior Stokes problem and decay at infinity. Math. Mech. Appl., 8 (1986), 351–367.

[SSp98] H. Sohr, M. Specovius-Neugebauer. The Stokes problem for exterior domains in homogeneous Sobolev spaces. Ser. Adv. Math. Appl. Sc., 47, World Scientific, 1998, 185–205.

[STh98] H. Sohr, G. Thäter. Imaginary powers of second order differential operators and L^q-Helmholtz decomposition in the infinite cylinder. Math. Ann., 311 (1998), 577–602.

[SoV88] H. Sohr, W. Varnhorn. On decay properties of the Stokes equations in exterior domains. Lecture Notes in Mathematics, 1431 (1988), 134–151.

[Ste70] E. M. Stein. Singular Integrals and Differentiability Properties of Functions. Princeton University Press, Princeton, 1970.

[Sto1845] G. G. Stokes. On the Theories of the Internal Friction of Fluids in Motion. Trans. Cambridge Phil. Soc., 8 (1845), 287–319.

[SvW87] H. Sohr, W. von Wahl. Generic solvability of the equation of Navier-Stokes. Hiroshima Math. J., 17 (1987), 613–625.

[SWvW86] H. Sohr, W. von Wahl, M. Wiegner. Zur Asymptotik der Gleichungen von Navier-Stokes. Nachr. Akad. Wiss. Göttingen, 1986, 45–69.

[Tan79] H. Tanabe. Equations of Evolution. Pitman, London, 1979.

[Tem77] R. Temam. Navier-Stokes Equations. North-Holland, Amsterdam, 1977.

[Tem83] R. Temam. Navier-Stokes Equations and Nonlinear Functional Analysis. CBMS-NSF, Regional Conference Series in Applied Mathematics, Philadelphia, Pennsylvania 19103, 1983.

[Tri78] H. Triebel. Interpolation Theory, Function Spaces, Differential Operators. North-Holland, 1978.

[Var94] W. Varnhorn. The Stokes Equations. Akademie Verlag, Berlin, 1994.

[vWa85] W. von Wahl. The Equations of Navier-Stokes and Abstract Parabolic Equations. Vieweg, Braunschweig, 1985.

[vWa88] W. von Wahl. On necessary and sufficient conditions for the solvability of the equations rot $u = \gamma$ and div $u = \varepsilon$ with u vanishing on the boundary. Lecture Notes in Mathematics, 1431 (1988), 152–167.

[Wei76] J. Weidmann. Lineare Operatoren in Hilberträumen. Teubner, Stuttgart, 1976.

[Wie87] M. Wiegner. Decay results for weak solutions of the Navier-Stokes equations in $\mathbb{R}^n$. J. London Math. Soc., 35 (1987), 303–313.

[Wie99] M. Wiegner. The Navier-Stokes equations – a neverending challenge? Jahresberichte DMV, 101 (1999), 1–25.

[Yos80] K. Yosida. Functional Analysis, Springer-Verlag, Heidelberg, 1980.

[Zei76] E. Zeidler. Vorlesungen über nichtlineare Funktionalanalysis I. Teubner, Leipzig, 1976.

Index